NANOFIBERS

EDITED BY
ASHOK KUMAR

Nanofibers
Edited by Ashok Kumar

Republished by InTech
Janeza Trdine 9, 51000 Rijeka, Croatia

Nanofibers, Edited by Ashok Kumar
p. cm.
ISBN 978-953-7619-86-2

For EU product safety concerns:
IN TECH d.o.o., Prolaz Marije Krucifikse Kozulić 3, 51000 Rijeka, Croatia,
info@intechopen.com or visit our website at intechopen.com.

Meet the editor

Prof. A. Kumar obtained his M. Tech and Ph.D. degree from IIT, Kanpur in Material Science. He had done masters in Physics with Electronics as specialization from Meerut University. His research interest involve Materials Science/Condensed Matter Physics: Polymer Electrolytes and Conducting Polymers. He has been a faculty member of the department of physics since 1999.

Preface

"There's Plenty of Room at the Bottom" — this was the title of the lecture, Prof. Richard Feynman delivered at California Institute of Technology on December 29, 1959 at an American Physical Society meeting. He considered the possibility to manipulate matter on an atomic scale. Indeed, the design and controllable synthesis of nanomaterials have attracted much attention because of their distinctive geometries, and novel physical and chemical properties. For the last two decades nano-scaled materials in the form of nanofibers, nanoparticles, nanotubes, nanoclays, nanorods, nanodisks, nanoribbons, nanowhiskers etc. have been investigated with increased interest due to their enormous advantages, such as large surface area and active surface sites. Among all nanostructures, nanofibers have attracted tremendous interest in nanotechnology and biomedical engineering owing to the ease of controllable production processes, low pore size and superior mechanical properties for a range of applications in diverse areas such as catalysis, sensors, medicine, pharmacy, drug delivery, tissue engineering, filtration, textile, adhesive, aerospace, capacitors, transistors, battery separators, energy storage, fuel cells, information technology, photonic structures and flat panel displays, just to mention a few.

Nanofibers are continuous filaments of generally less than about 1000 nm diameters. Nanofibers of a variety of cellulose and non-cellulose based materials can be produced by a variety of techniques such as phase separation, self assembly, drawing, melt fibrillation, template synthesis, electro-spinning, and solution spinning. They reduce the handling problems mostly associated with the nanoparticles. Nanoparticles can agglomerate and form clusters, whereas nanofibers form a mesh that stays intact even after regeneration.

The present book is a result of contributions of experts from international scientific community working in different areas and types of nanofibers. The book thoroughly covers latest topics on different varieties of nanofibers. It provides an up-to-date insightful coverage to the synthesis, characterization, functional properties and potential device applications of nanofibers in specialized areas. Based on the thematic topics, it contains the following chapters:

Chapter 1: Carbon Nanofibers as Macro-structured Catalytic Support gives an overview of the synthesis of carbon nanofibers and their efficient applications in different catalytic reactions.

Chapter 2: Nanofiber Reinforced Composite Polymer Electrolyte Membranes deals with the enhancement of ionic conductivity of poly(vinylidene fluoride hexafluoropropylene) based composite polymer electrolyte membranes due to addition of dedoped polyaniline nanofibers as insulating fillers.

Chapter 3: Nanoreinforced Adhesives reviews the polymer nanocomposites manufactured from an effective dispersion of nanofillers (nanoclays, carbon nanotubes or nanofibers, inorganic nanoparticles etc) into a polymeric matrix (thermoplastic or thermosetting) for generating new multifunctional materials with improved mechanical, physical and chemical properties. These nanoreinforced polymer composites have efficient application as adhesive.

Chapter 4: Fabrication of Bio-nanocomposite Nanofibers Mimicking the Mineralized Hard Tissues via Electrospinning Process describes electrospinning as a promising approach for fabrication of hydroxyapatite nanocrystals based bio nanocomposite fibers as hard tissue (bone and dentin) replacement and coating implants since they resemble to the nanostructure of living bone from the physiochemical point of view.

Chapter 5: Diversity of Nanofibers from Electrospinning: from Graphitic Carbons to Ternary Oxides discusses the electrospinning technique for synthesis of nanofibers and microfibers from graphitic carbon, binary, ternary and more complex oxides.

Chapter 6: Preparation of Functionalized Nanofibers and Their Applications provides an overview of the synthesis of functionalized nanofibers by electrospinning, their properties, theories and applications in energy storage devices, biomedical engineering and photocatalysis.

Chapter 7: Keratin-based Nanofibers deals with the extraction of keratin from wool and electro spinning of keratin-based blends with high molecular weight polymers like poly(ethylene-oxide), fibroin and polyamide 6 for biomedical applications such as tissue engineering and medical textiles and active filtration of air .

Chapter 8: In Situ Probing of Oxygen-containing Groups on Acid-treated Carbon Nanofibers using Aromatic Molecules reports in situ fluorescence measurements using oxygen-containing aromatic probe molecules for studying the physicochemical properties on the carbon nanofibers surface. The relationship between the carbon nanofiber dispersion throughout the solution of the probe molecules and their adsorption onto the carbon nanofibers has been discussed by analyzing the UV-visible and fluorescence spectra of the suspension containing the probe molecules and the untreated or acid-treated carbon nanofibers.

Chapter 9: Preparation of Cellulose-based Nanofibers Using Electrospinning describes the preparation of ethyl-cellulose and hydroxypropyl methylcellulose nanofibers using electro spinning technique. Effect of polymer concentration, tip-target distance, solution flow rate, and applied voltage on the morphology of nanofibers is also emphasized.

Chapter 10: Nanofibrous Scaffolds of Bio-polyesters: In vitro and In vivo Characterizations and Tissue Response aims to prepare bio-polyesters by electro spinning for fabricating tissue-engineering scaffold with enhanced mechanical properties, bioabsorption and biocompatibility. Its performance as a nanofibrous scaffold for tissue engineering was compared with electrospun homopolymers. All of these nanofibrous scaffolds were implanted subcutaneously in rats to evaluate their tissue response.

Chapter 11 Photocatalyst Nanofibers Obtained by Calcination of Organic-Inorganic Hybrids depicts the synthesis of TiO_2 - PVA organic-inorganic hybrids using electro spinning followed by calcinations. The hybrid nanofibers are used as photocatalyst in degradation of Methylene blue under white light irradiation.

Chapter 12: Electrochemical and Adsorption Properties of Catalytically Formed Carbon Nanofibers describes the synthesis of carbon nanofibers by Catalytic Chemical Vapour Deposition(CCVD) method. The electrochemical, adsorption and sensing properties of the fibers is also briefed.

Chapter 13: Synthesis of Carbon Nanofibers by a Glow-arc Discharge deals with the growth of carbon nanofibers by a glow-arc HF discharge. These nanofibers are used as toxic gas absorber.

Chapter 14: Morphology and Dispersion of Pristine and Modified Carbon Nanofibers in Water focuses on dispersion of untreated and acid-treated carbon nanofibers suspended in water. The morphology and dispersion of solubilized carbon nanofibers and dispersion of plasma-treated carbon nanofibers has been studied.

Chapter 15: Non-Catalytic, Low-Temperature Synthesis of Carbon Nanofibers by Plasma-Enhanced Chemical Vapor Deposition describes the non-catalytic, low temperature synthesis of carbon nanofibers by using DC plasma-enhanced chemical vapor deposition system and microwave plasma-enhanced chemical vapor deposition system, reaction mechanisms and growth models and comparison of properties of carbon nanofibers synthesized by the two systems.

Chapter 16: Carbon Nanotubes Reinforced Electrospun Polymer Nanofibers Synthesis of Alumina Nanofibers and Composites discusses the incorporation of carbon nanotubes into polymer nanofibers by electrospinning technique. Some significant property changes and applications are also briefed. A brief description on carbon nanotubes, its properties and composites are given. It also gives an overview of the electrospinning technique and electrospun nanofibers.

Chapter 17: On the Electron Transport in Conducting Polymer Nanofibers highlights the current understanding of the electron conduction mechanisms in conducting polymer nanofibers.

Chapter 18: Spectroscopy of Polyaniline Nanofibers examines the vibrational Raman spectroscopic studies of polyaniline nanofibers synthesized by interfacial polymerization and soft miceller templates.

Chapter 19: Fabrication of Ceramic Nanofibers Using Atrane Precursor provides sol-gel and electro spinning approach for synthesis of zirconia nanofibers from its inexpensive and moisture-stable zirconatrane precursor. The morphological and thermal properties of zirconia nanofibers have been studied.

Chapter 20: Organic Fluorescent Nanofibers and Sub-micrometer Rods: Interest of a Solvent-exchange Preparation Method describes the photo physical behavior and design of fluorescent nanofibers. An overview of the different types of fluorescent nanofibers bionanofibers, dendrimer nanofibers and nanofibers based on berberine ion pairs, coumarin series and low molecular weight fatty acids and their preparation methods are presented here.

Chapter 21: Synthesis of Alumina Nanofibers and Composites covers the different chemical and physical methods for synthesis of alumina nanofibers. The effect of alumina nanofibers on different polymers, inorganic and carbon substrates is also reviewed.

Chapter 22: Core-Shell Nanofibers: Nano Channel and Capsule by Coaxial Electrospinning deals with the synthesis of nanochannel, core-shell nano-capsule and tubular nanostructures of various organic or inorganic materials by coaxial electrospinning technique. Nanoencapsulation of stimulating responsive materials into core-shell nanofibers and their functional devices applications have also been discussed in detail.

We hope that this book will prove to be timely and thought provoking and will serve as a valuable reference for researchers working in different areas of nanofibers. Special thanks go to the authors for their valuable contributions.

Editor

Ashok Kumar
Department of Physics, Tezpur University,
Tezpur784028, Assam
India

Contents

1

Carbon Nanofibers as Macro-structured Catalytic Support

Ricardo Vieira
National Institute for Space Research
Brazil

1. Introduction

Since their discovery as a by-product of the arc-discharge process (Iijima, 1991), carbon nanotubes and their related materials, i.e. nanofibers and onion-like particles, have received an increasing academic and industrial interest due to their exceptional mechanical and electronic properties (Dresselhaus et al., 2001). Most commonly, carbon nanofilaments are produced by evaporating solid carbon in an arc discharge, by laser beams or by catalytic chemical vapour deposition (CCVD) of carbon-containing gases (Ebbesen, 1997). Depending on the nature of the metal catalyst, the composition of the carburizing mixture, and the synthesis temperature, carbon nanostructures with different shapes, i.e. nanotubes or nanofibers, can be prepared (de Jong & Geus, 2000). A carbon nanotube structure consists of cylindrical graphene layers with a hollow internal cavity, whilst a carbon nanofiber structure consists of a stacking of different graphite sheets oriented at an angle with respect to the fiber axis. The exposed surface of the carbon nanofibers mainly consists of prismatic planes with high surface reactivity when compared to the graphite basal planes of the carbon nanotubes.

Among the different potential applications of these materials, catalysis either within the gas or the liquid phase seems to be the most promising according to the results recently reported in literature (Salman et al., 1999). Metals supported on carbon nanofibers or nanotubes exhibit unusual catalytic activity and selectivity patterns when compared to those encountered with traditional catalyst supports such as alumina, silica or activated carbon. The extremely high external surface area displayed by these nanomaterials significantly reduce the mass transfer limitations, especially in liquid phase reactions (Pham-Huu et al., 2001), and the low interaction between the impregnated metallic phase and the exposed planes of the support which leads to the formation of active metallic faces (Rodriguez et al., 1994) were advanced to explain these catalytic behaviours.

However, these materials have only been synthesized in a powder form, making difficult their handling and large scale use, especially in a conventional fixed-bed catalytic reactor. The handling of the carbon nanostructures is hampered by the formation of dust and a severe pressure drop along the catalyst bed. It is of interest to find a method allowing *(i)* the synthesis of carbon nanostructures on a large scale and *(ii)* with a macroscopic shape in order for it to be used as a catalyst support. It is expected that the macroscopic shaping of such nano-structured materials will open-up a real opportunity for their use as a catalyst support in relation to the traditional catalysts carriers (Vieira et al., 2004). The macroscopic

support should also not change the physical properties of the carbon nanostructures deposited on it, i.e. high mechanical strength in order to avoid breaking and catalytic bed plugging, high specific volume in order to afford a high space velocity of the gaseous reactants, high thermal conductivity which is essential for catalysts operating in highly exothermic or endothermic reactions and finally, a high chemical resistance in order to be used in aggressive environments, i.e. highly acid or alkaline medium.
Macro-structured carbon nanofibers catalytic support was synthesized through a chemical vapour deposition technique described in next section, gowning carbon nanofibers on a macroscopic host, i.e. felt, cloth or foam, as well as the different applications of these new materials in the catalysis field.

2. Carbon nanofibers synthesis

The macroscopic host structures used in this work were mainly carbon felt or fabric, which was constituted of a dense entangled network of micrometer carbon fibers with a smooth surface. The starting graphite host had almost no porosity which was in good agreement with its extremely low specific surface area (< $1m^2.g^{-1}$). These materials were cut into a pre-defined shape, in cylindrical tablets or disks with appropriate dimensions to the desirable application. The tablets/disks were further impregnated with 1wt.% of nickel using an alcoholic solution (50vol.% ethanol) of nickel nitrate. The solid was dried at room temperature overnight, then at 100°C for 2h and finally calcinated at 350°C for 2h. The sample was placed inside a quartz reactor inserted in a tubular electric oven. After 1h of *in situ* reduction at 400°C, the hydrogen flow was replaced by a mixture of hydrogen and ethane with a molar ratio of 5:1 (H_2:C_2H_6). The carbon nanofibers were obtained by catalytic hydrocarbon decomposition over the supported nickel particles at 680°C for 2h, resulting in a yield of 100wt.%. Figure 1A presents the surface of the felt completely covered by a layer of carbon nanofibers after its synthesis. The formation of the carbon nanofibers on the macroscopic carbon felt led to a significant increase in the specific surface area, i.e. $85m^2.g^{-1}$, of the final composite, taking into account the low surface area ($1m^2.g^{-1}$) of the starting felt. The surface area was measured using a Coulter SA-3100 sorptometer with nitrogen as adsorbant at liquid nitrogen temperature. Before measurements were taken, the samples were degassed at 200°C for 2h in order to desorb moisture and other weakly adsorbed residues.

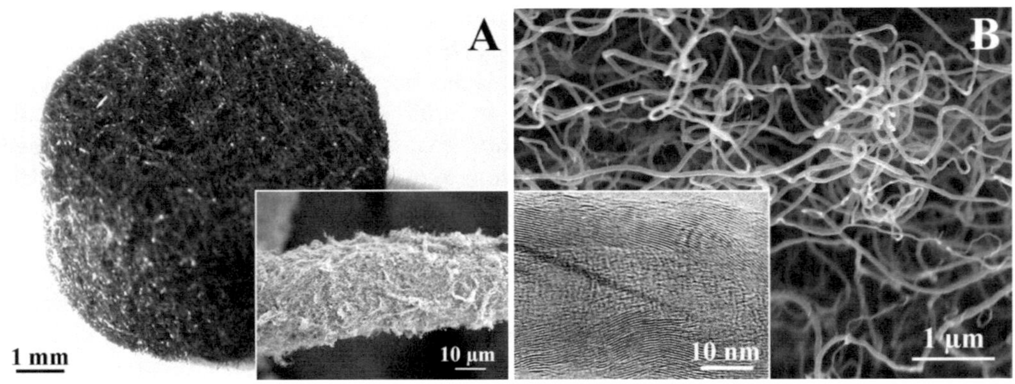

Fig. 1. (A) Image of a macro-structured carbon felt and detail of a fiber covered by carbon nanofibers; (B) Scanning Electron Microscopy image of carbon nanofibers with homogeneous diameters and detail of its fishbone structure.

The nanofibers produced by this technique showed homogeneous diameters (around 30nm) and fishbone structure with exposed prismatic planes (Figure 1B). The images were taken by Scanning Electron Microscopy LEO model 440.
The shape of this macro-structured material can be modified at will, depending on its required application (Figure 2). Thus, carbon nanofibers composites can be efficiently used as a catalytic support in reactions where diffusion phenomena of the reactants are essential, and in reactions with high mass and heat transfers. Moreover, the high thermal conductivity of this carbon based support allows a fast homogenization of the heat generated during the reaction throughout the catalyst body, preventing the formation of hot spots which are detrimental to the active phase dispersion and catalyst body conservation.

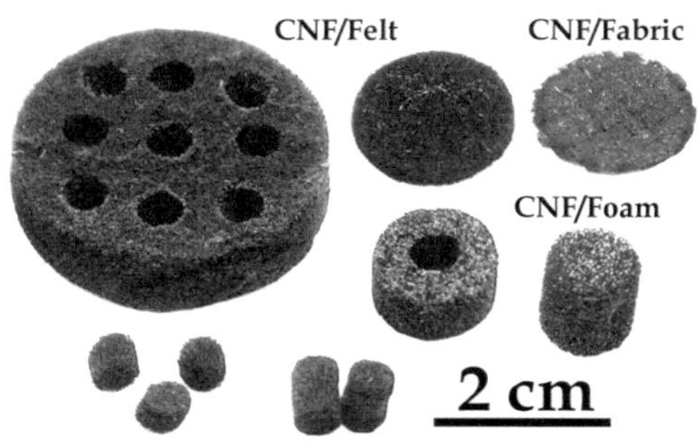

Fig. 2. Macro-structured carbon nanofibers with different shapes, sizes and micro-structures.

3. Catalytic reactions using carbon nanofibers

3.1 Catalytic hydrazine decomposition

Nowadays, most of the satellites in orbit use liquid monopropellant hydrazine in its propulsive subsystem for orbit correction and positioning operations. The thrust is obtained by catalytic decomposition of the monopropellant on a highly loaded catalyst containing about 30 to 40wt.% of iridium supported on alumina (traditional catalyst). This catalyst is mainly composed of macro and mesoporous network, in order to increase the active sites accessibility. Macro-structured carbon nanofibers were proposed as an alternative catalytic support for the iridium in the hydrazine decomposition (Vieira et al., 2005). The composite was cut in cylinder forms and impregnated with 30wt.% of iridium from a solution of hexachloroiridic acid. The tests were carried out with a 2N microthruster designed to operate with the traditional catalyst, placed in a vacuum chamber, which allowed simulating the pressure conditions in space. The dimensions of the microthruster reaction chamber are 6.7mm in diameter and 20mm in length. The microthruster was loaded with 1.14g of the traditional catalyst and with 0.15g of the carbon nanofibers based catalyst, due the different densities of these materials.
The parameters of performance evaluations of the different catalysts in the microthruster used were the steady-state pressure chamber and the ignition time delay, the latter one being defined as the time response from the valve response to 90% of steady-state pressure chamber signal.
The tests were carried out under two level of hydrazine injection pressures, i.e., 22 and 5.5bar, thus simulating the real conditions of the propellant tank pressure at the beginning

and the end of its lifespan. For each level of pressure 4 series of 100 short pulses and 1 duty cycle were performed. The catalyst bed was pre-heated at 120°C for all tests.
The carbon nanofibers based catalyst showed superior thrust performances than the traditional catalyst. The pressure chamber pulses delivered during the hydrazine decomposition over the Ir/CNF composite were invariably higher than those detected on the Ir/Al_2O_3 catalyst (Figure 3), regardless of the great difference between the iridium loading in the propulsion chamber with the traditional and carbon nanofiber based catalysts. The high performance of the Ir/CNF catalyst was attributed to the greater accessiblity of the reactant to the active sites, due to the open structure of the support, i.e. high external surface area and absence of ink-bottled pores. In this type of reaction, where the heat and mass transfers must be very fast, the reactant does not have time to penetrate into the pores, and the main reaction occurs on the external surface of the catalyst grains. In addiction, the high thermal conductivity of the carbon nanofibers also allowed a rapid homogenization of the temperature through the entire catalyst bed whereas on the alumina-based catalyst, due to its insulating character, especially during the continuous long duration tests, i.e., 5000ms.

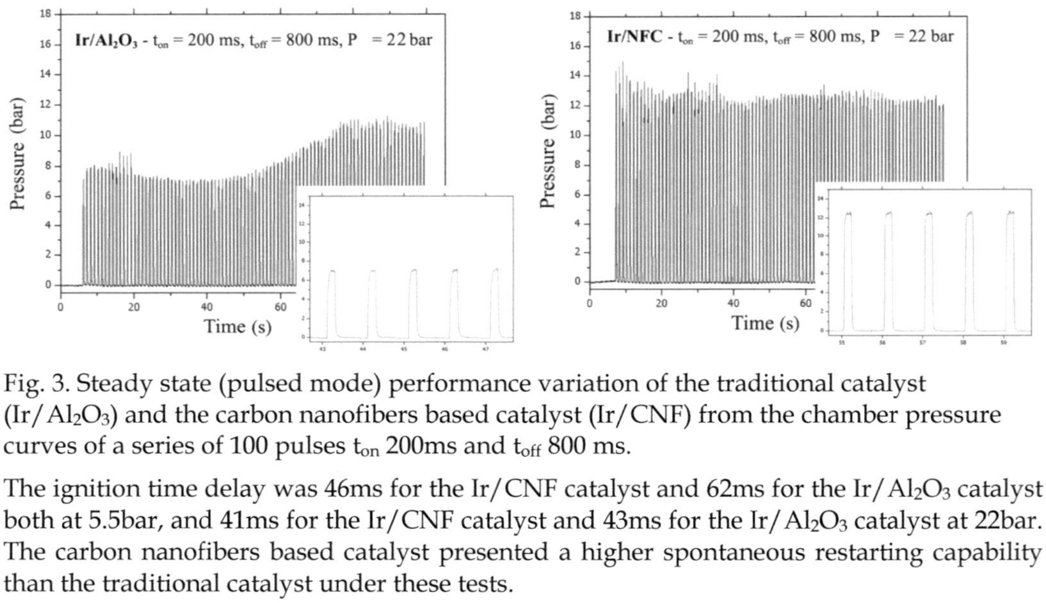

Fig. 3. Steady state (pulsed mode) performance variation of the traditional catalyst (Ir/Al_2O_3) and the carbon nanofibers based catalyst (Ir/CNF) from the chamber pressure curves of a series of 100 pulses t_{on} 200ms and t_{off} 800 ms.

The ignition time delay was 46ms for the Ir/CNF catalyst and 62ms for the Ir/Al_2O_3 catalyst both at 5.5bar, and 41ms for the Ir/CNF catalyst and 43ms for the Ir/Al_2O_3 catalyst at 22bar. The carbon nanofibers based catalyst presented a higher spontaneous restarting capability than the traditional catalyst under these tests.

3.2 Styrene synthesis

The styrene synthesis is one of the ten largest industrial processes used nowadays. This monomer is involved in several polymer synthesis, and is industrially produced by the direct dehydrogenation of ethylbenzene at 600 - 680°C over potassium promoted iron oxide catalyst. This strongly endothermic process suffers from drawbacks such as: thermodynamic limitations, an irreversible catalyst deactivation and a relatively low yield (> 50%). Carbon is reported to be a catalytically active material for this reaction (Guerrero-Ruiz & Rodriguez-Ramos, 1994). It has been shown that high performances could be obtained over onion-like carbons and carbon nanotubes (Pereira et al., 2004). Such non-planar sp^2-nanocarbons are very active and promising alternative catalysts to the industrial catalyst, with a higher styrene yield and a lower reaction temperature.

Delgado et al. (2006) have shown that just carbon felt with its surface covered by carbon nanofibers is catalytically active in the oxidative dehydrogenation of the ethylbenzene under low temperatures, combining the properties of the nanostructures and the advantages of the macro-structured felt support. Figure 4 shows the performances obtained at temperatures ranging from 375 to 530°C for total flows of 30ml.min^{-1} and 8ml.min^{-1}. Under both conditions, increasing the temperature led to a slight decrease in the styrene selectivity. The selectivity was not strongly influenced by the space velocity. In contrast, this parameter strongly influenced the ethylbenzene conversion, especially at lower temperatures. This implies that operating a low temperature (440°C) and a low space velocity results in upper styrene yields (38%) with a high styrene selectivity (81%), so a low flow favouring reactant adsorption and a low temperature minimizing the combustion, could make such supported carbon nanofibers attractive for industrial applications. Whatever the reaction conditions, and in particular the space velocity, no pressure drop occurred in the reactor and the catalyst bed.

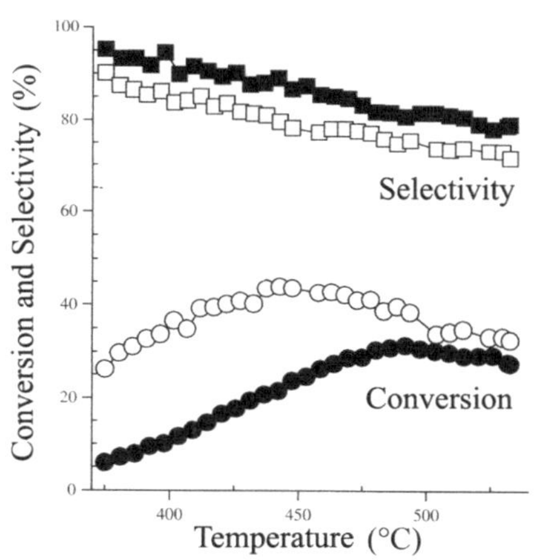

Fig. 4. Ethylbenzene conversion curve (○,●) and selectivity to styrene (□,■) on the macro-structured carbon nanofibers as a function of the reaction temperature in the range 375 – 530°C, using total flow rate of 30ml.min^{-1} (●,■) and 8ml.min^{-1} (○,□).

3.3 H_2S oxidation into elemental sulfur

Over the last few decades, sulfur recovery from the H_2S-containing acid gases have become more and more important due to the ever increasing standards of efficiency required by environmental protection measures. The general trend is to selectively transform H_2S into elemental sulfur which is a valuable product, by the modified Claus process. However, due to thermodynamic limitations of the Claus equilibrium reaction, typical sulfur recovery efficiencies are only 90–96% for a two stage reactor plant and 95–98% for a three stage reactor plant (Wieckowska, 1995).

Ledoux et al. (2000) have shown the high activity and selectivity of the NiS_2/β-SiC catalyst in the direct oxidation of the H_2S into elemental sulfur at a low reaction temperature (60°C). The catalytic desulfurization reaction conditions were: H_2S 2500ppm, O_2 5000ppm, H_2O

30vol.% and balance He, which is typically the industrial working concentrations in a discontinuous mode. The total selectivity for sulfur was attributed to the absence of micropores in the support. The heterogeneous nature of the support surface (hydrophilic and hydrophobic areas) could explain the important role played by water to maintain a high and stable H_2S conversion level. The catalyst was submitted at frequently regenerations cycles due to the sulfur fouling of the catalyst surface. Later, Nhut et al. (2004) have studied the use of NiS_2 nanoparticles encapsulated inside multi-walled carbon nanotubes for the trickle-bed oxidation of H_2S into sulfur at 60°C. The use of carbon nanotubes led to a significant increase in the overall catalytic performance, both in terms of desulfurization activity and resistance to the solid sulfur deposition onto the material, as compared to the reference catalyst supported on macroscopic SiC grains. The high capacity of solid sulfur storage on the carbon nanotubes compared to that of macroscopic SiC grains was thus attributed to the large void volume outside the nanotubes available for the sulfur storage.
Recently, we studied macro-structured carbon nanofibers as support for the catalytic selective oxidation of H_2S into elemental sulfur at low temperatures (Coelho, 2009). The catalyst was NiS_2 with a metal loading of 15wt.% and the reaction conditions were: H_2S 1000ppm, O_2 1000 to 4000ppm, H_2O 30vol.% and balance He. The results showed that the stoichiometric O_2/H_2S ratio did not significantly the reaction yield, suggesting that the partial order for O_2 is zero. The H_2S conversion at 120°C without water addition was complete during the first 390min (Figure 5A). The deactivation after this period of time was attributed to the encapsulation of the active sites by solid sulfur deposition, i.e. the absence of liquid water hindered the sulfur removal from the active sites; whereas at 180°C, the sulfur liquefaction allowed the maintenance of activity for an additional time (700min).
The water plays a special role of mechanical transport in this reaction. The reaction performed at 60°C with the addition of 30wt.% of water vapour, it was observed 100% of conversion and selectivity in sulfur, even after 288h. Despite these preliminary results, we notice that the sulfur is continuously removed outside of the composite by water added due the hydrophobicity of nanofibers (Figure 5B), letting the active sites exposed for a longer period of time.

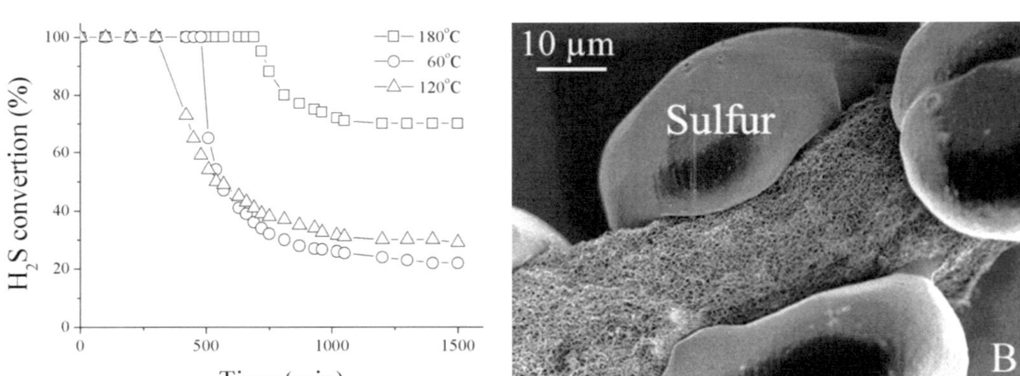

Fig. 5. (A) H_2S conversion as a function of time on 15% NiS_2/CNF catalyst at different temperatures. (B) SEM micrograph of the particular mode of sulfur deposition on carbon nanofibers.

3.4 Fuel cells electrodes

Fuel cells have received considerable interests because of its potential for power generation efficiency and a low pollutant emission. Among them, polymer electrolyte membrane (PEM) fuel cells are the most promising, particular for mobile applications.

The nature of the conductive macroscopic carbon support and a high surface area are fundamental, and conductivity is essential for good electrocatalytic performances. The peculiar electronic, adsorption, mechanical and thermal properties suggest that carbon nanotubes are a suitable material for electrodes and catalysts support in PEM fuel cells (Yu et al., 2002). It should also be noted that the good and reversible hydrogen storage capacity of carbon nanotube materials is very interesting for application in the PEM fuel cell. Recently, Gangeri et al. (2005) evaluated the electrocatalytic performance of platinum supported on felt covered with carbon nanofibers (Pt/CNF/felt) and fabric (Pt/CNF/fabric) as alternative electrodes for PEM fuel cells. The authors compared the performance of these new macro-structured platinum/carbon materials with those of a commercial Pt-carbon black on carbon cloth electrode (E-TEK Inc.) in a 1cm^2 fuel cell working at room temperature. The analysis of polarization curves (Figure 6) indicated that both carbon nanofibers composites (Pt/CNF/felt and Pt/CNF/fabric) are better electrocatalysts than commercial ones, and the Pt/CNF/fabric gives the lowest ohmic losses while Pt/CNF/felt materials give the lowest mass transfer losses. Further studies of a better membrane and electrodes assembly engineering are necessary to validate the results, but these preliminary analysis, indicate that new electrode material based both on carbon macro-structured composite could be very interesting for fuel cells applications.

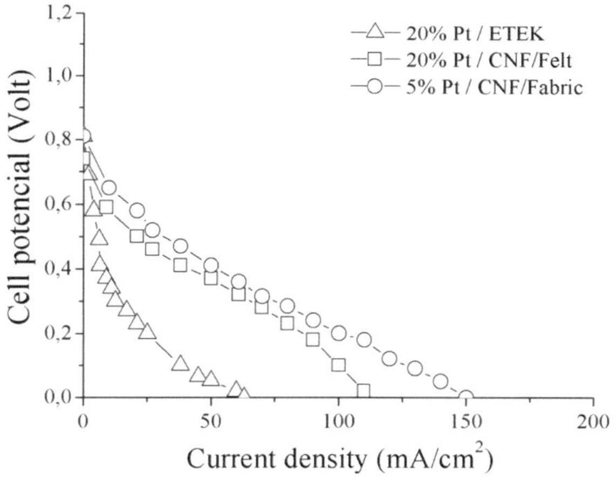

Fig. 6. Polarization curves for the different classes of electrodes.

3.5 Ammonia decomposition

The catalytic decomposition of ammonia appears to be an appealing process for clean hydrogen production. Ammonia decomposition for production of hydrogen in large-scale, would seem pointless, as ammonia itself is produced from hydrogen, but it appears to be promising from the point of view of specific fuel cell applications, because it can produce CO_x-free hydrogen. Recently, Yin et al. (2004) have reported that very small ruthenium nanoparticles on carbon nanotubes (CNT) present a high activity in ammonia catalytic decomposition. The good performances of Ru/CNT and K-Ru/CNT were related to the high dispersion of the metal and to the high graphitization and purity of carbon nanotubes. The authors have investigated also the influence of the support in catalyst activity. Moreover, the order of ammonia conversion was ranked as Ru/CNT > Ru/MgO > Ru/TiO_2 > Ru/Al_2O_3 > Ru/ZrO_2. Among the reported transition metals, ruthenium is the most active in ammonia catalytic decomposition. Various materials have been found to be effective supports for loading of ruthenium metal, but the carbon nanotubes was presented as the better performance as support for ammonia decomposition.

We used carbon nanofibers composite as a support for ruthenium for the catalytic decomposition of ammonia (Furtado et al., 2008). Macro-structured carbon nanofibers with a specific surface area of 85$m^2.g^{-1}$ were prepared using nickel catalyst impregnated onto macroscopic carbon felt. Carbon nanotubes in powder form were deployed as reference support. The above powder was prepared by vapour deposition using 20% Fe/Al_2O_3 as catalyst, under a reaction mixture of ethane and hydrogen with a molar ratio of 1:1 at 650°C for 2h. The carbon nanotube with a specific surface area of 220$m^2.g^{-1}$ was purified by refluxing in 5M HNO_3 solution for 2h, followed by exhaustive washing. Thereafter, the material was refluxed in 1M KOH solution for 2h, followed by washing, drying and calcination at 350°C. The ruthenium loaded catalyst (5wt.% metal) was prepared by incipient wetness impregnation of the carbon nanofibers composite and the carbon nanotubes powder with $RuCl_3$ using ethanol as solvent, followed by drying at 100°C for 12h and calcination at 350°C for 2h.

Catalytic testing was carried out in 10mm inner diameter continuous flow quartz reactor, loaded with 100mg of catalyst, under pure ammonia atmosphere at flow rate 50$ml.min^{-1}$. Prior to the reaction, the metal catalyst was reduced *in situ* at 500°C for 2h under hydrogen. The reaction temperature range was 300 to 500°C. Ammonia decomposition rate over carbon nanofibers based composite was about 30% higher than on carbon nanotubes powder support for reactions above 400°C. Figure 7 shows the better performance of the carbon nanofibers based support, due to its opened structure, thus reducing the diffusion problems.

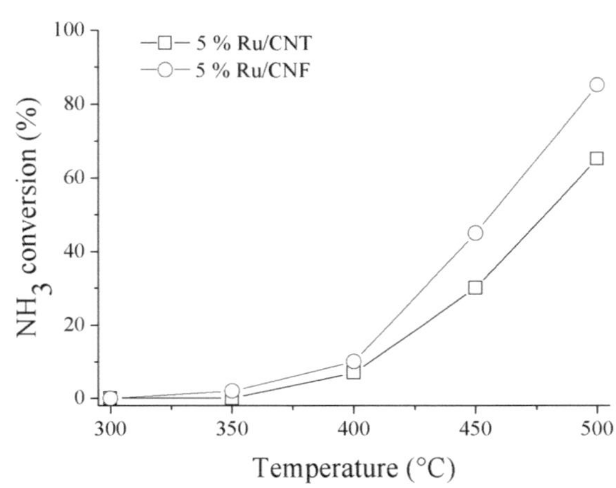

Fig. 7. Curve of ammonia conversion in function of the temperature on the 5% Ru/CNF (○) and 5% Ru/CNT catalyst (□).

3.6 Fischer-Tropsch synthesis

The Fischer-Tropsch synthesis is a process to convert syngas ($CO + H_2$) to a cleaner fuel. The nature of its products depends on the catalysts used (Fe, Co, Ni, Ru…), the temperature and the reaction pressure. Diffusion problems of the reactants and products within the catalytic bed are frequent, causing the deactivation of the catalysts by pore obstruction of oxide-based support, consequently affecting the reaction selectivity. As this reaction is highly exothermic, it is necessary to release the heat generated to avoid an eventual sintering of the active phase. Thus, we studied the activity and selectivity of the Co/CNF catalyst in the CO hydrogenation reaction (Furtado, 2009). A central composite design combined with response surface methodology was applied to determine the optimum operating condition on the production of diesel from Fischer-Tropsch synthesis. The variables used in this study were reaction pressure (8 to 22bar), temperature (194 to 236°C), contact time (1 to 7s) and concentration of the active phase (8 to 22%). The response chosen was the CO conversion and hydrocarbons yield. In all experiments the molar ratio CO/H_2 was kept equal to 2. It was verified that high temperatures and pressures, large contact time and metal loading favour the CO conversion. The subsequent figures show that the gasoline (C8-C13) yield is

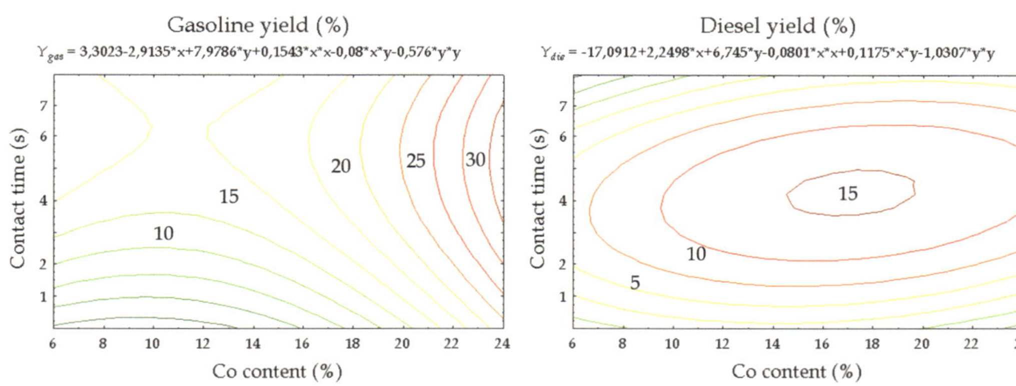

Fig. 8. Contour plot of the relationship: gasoline and diesel yield with contact time and Co content at fixed temperature of 215°C and pressure of 15bar.

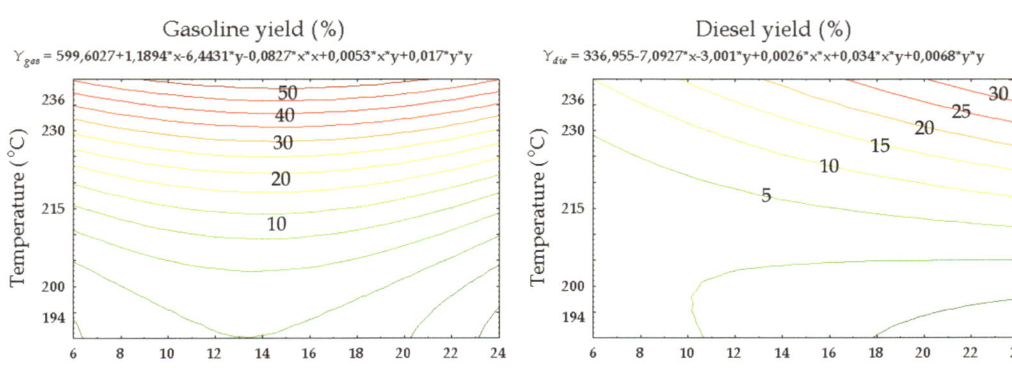

Fig. 9. Contour plot of the relationship: gasoline and diesel yield with temperature and reaction pressure at fixed contact time of 4s and Co content of 20wt.%.

improved mainly by high temperature and cobalt loading, and moderate contact times (around 4s). On the other hand, the pressure has little influence on the gasoline yield. However, diesel (C14-C19) is enhanced by high temperature and pressure, cobalt loading over 15% and moderate contact times.

No deactivation was observed with the Co/CNF catalyst over 75h of reaction. A previous comparative study was done with 15% Co/CNF and 15% Co/Al_2O_3 catalysts, maintaining the best reaction conditions for the first catalyst. It was observed that the Co/CNF catalyst presented a CO conversion (30%) of about tree times higher than that of the Co/Al_2O_3 catalyst. The good results obtained from this work are attributed to the high thermal conductivity and specific area, chemical inertia, absence of ink-bottle pores and carbon nanofibers hydrophobicity.

4. Conclusion

In summary, macro-structured carbon nanofibers can be efficiently employed as a new class of catalyst support which exhibits a high catalytic activity along with peculiar product selectivity when compared to those observed using traditional catalysts. The high catalytic performances of the carbon nanofibers based catalyst were mainly attributed to (i) the high external surface area and morphology of the support which allows rapid access of the reactants to the active sites and (ii) the high thermal conductibility of the support which allows the rapid evacuation of the heat formed during the reaction and favours the temperature homogenization throughout the catalyst bed. Finally, it should be noted that the complete absence of bottled pores allows the rapid 3D access of the entire volume of the catalyst to the reactants and also for the rapid evacuation of the products which significantly lowers secondary reactions.

5. References

Coelho, N. M. A. (2009). Carbon nanofibers based filter for sulfur sequestration in H_2S-rich gaseous effluents. *MSc Thesis*, National Institute for Space Research, São José dos Campos, Brazil.

De Jong, K. P. & Geus, J. W. (2000). Carbon Nanofibers: Catalytic Synthesis and Applications. *Catalysis Reviews Science and Engineering*, 42, 4, 481–510, 1520-5703.

Delgado, J. J.; Vieira, R.; Rebmann, G.; Su, D. S.; Keller, N.; Ledoux, M. J. & Schlögl, R. (2006). Supported carbon nanofibers for the fixed-bed synthesis of styrene. *Carbon*, 44, 4, (April 2006) 809-812, 0008-6223.

Dresselhaus, M. S.; Dresselhaus; G. & Avouris, Ph. (2001) Carbon Nanotubes: Synthesis, Structures, Properties and Applications. *Topics in Applied Physics*, vol. 80; Springer Berlin, 1437-0859, Heidelberg.

Ebbesen, T. W. (1997). *Carbon Nanotubes: Preparation and Properties*, CRC Press, 9780849396021, Boca Raton, USA.

Furtado, J. L. B.; Coelho, N. M. A. & Vieira, R. (2008). Ammonia decomposition using carbon nanofibers composites as catalytic support of ruthenium. *Proceedings of III International Symposium on Carbon for Catalysis*, pp. 95, Fritz Haber Institute, November 2008, Berlin.

Furtado, J. L. B. (2009). Development of a new monolith catalyst based carbon nanofibers for produce clean fuel by Fischer-Tropsch synthesis. *PhD Thesis*, National Institute for Space Research, São José dos Campos, Brazil.

Gangeri, M.; Centi, G.; La Malfa, A.; Perathoner, S.; Vieira, R.; Pham-Huu, C. & Ledoux M. J. (2005). Electrocatalytic performances of nanostructured platinum–carbon materials. *Catalysis Today*, 102-103, (May 2005) 50-57, 0920-5861.

Guerrero-Ruiz, A & Rodriguez-Ramos, I. (1994). Oxydehydrogenation of ethylbenzene to styrene catalyzed by graphites and activated carbons. *Carbon*, 32, 1994, 23-29, 0008-6223.

Iijima, S. (1991). Helical microtubules of graphitic carbon. *Nature*, 354, (November 1991), 56-58, 0028-0836.

Ledoux, M. J.; Pham-Huu, C.; Keller, N.; Nougayrède, J. B.; Savin-Poncet, S. & Bousquet, J. (2000). Selective oxidation of H_2S in Claus tail-gas over SiC supported NiS_2 catalyst. *Catalysis Today*, 61, 1-4, (August 2000) 157-163, 0920-5861.

Nhut, J. M.; Nguyen, P.; Pham-Huu, C.; Keller, N. & Ledoux, M. J. (2004). Carbon nanotubes as nanosized reactor for the selective oxidation of H_2S into elemental sulfur. *Catalysis Today*, 91-92, (July 2004) 91-97, 0920-5861.

Pham-Huu, C.; Keller, N.; Ehret, G.; Charbonnière, L. J.; Ziessel, R. & Ledoux, M. J. (2001). Carbon nanofiber supported palladium catalyst for liquid-phase reactions: An active and selective catalyst for hydrogenation of cinnamaldehyde into hydrocinnamaldehyde. *Journal of Molecular Catalysis A: Chemical*, 170, 1-2, (May 2001) 155-163, 1381-1169.

Pereira, M. F. R.; Figueiredo, J. L.; Órfão, J. J. M.; Serp, P.; Kalck, P. & Kihn, Y. (2004). Catalytic activity of carbon nanotubes in the oxidative dehydrogenation of ethylbenzene. *Carbon*, 42, 14, 2807-2813, 0008-6223.

Rodriguez, N. M.; Kim, M. S. & Baker R. T. K. (1994). Carbon Nanofibers: A Unique Catalyst Support Medium. *The Journal of Physical Chemistry*, 98, 50, (December 1994) 13108-13111, 0022-3654.

Salman, F.; Park, C. & Baker R. T. K. (1999). Hydrogenation of crotonaldehyde over graphite nanofiber supported nickel. *Catalysis Today*, 53, 3, (November 1999) 385-394 , 0920-5861.

Vieira, R.; Bastos-Netto, D.; Ledoux, M. J. & Pham-Huu, C. (2005). Hydrazine decomposition over iridium supported on carbon nanofibers composite for space applications: near actual flight conditions tests. *Applied Catalysis A: General*, 279, 1-2, (January 2005) 35-40, 0926-860X.

Vieira, R.; Ledoux, M. J. & Pham-Huu, C. (2004). Synthesis and characterisation of carbon nanofibres with macroscopic shaping formed by catalytic decomposition of C_2H_6/H_2 over nickel catalyst. *Applied Catalysis A: General*, 274, 1-2, (October 2004) 1-8, 0926-860X.

Wieckowska, J. (1995). Catalytic and adsorptive desulphurization of gases. *Catalysis Today*, 24, 4, (July 1995) 405-465, 0920-5861.

Yin, S. F.; Xu, B. Q.; Ng, C. F. & Au, C. T. (2004). Nano Ru/CNTs: a highly active and stable catalyst for the generation of CO_x-free hydrogen in ammonia decomposition. *Applied Catalysis B: Environmental*, 48, 4, (April 2004) 237-241, 0926-3373.

Yu, J. S.; Kang, S.; Yoon, S. B. & Chai, G. (2002). Fabrication of Ordered Uniform Porous Carbon Networks and Their Application to a Catalyst Supporter. *Journal of the American Chemical Society*, 124, 32, (August 2002) 9382–9383, 0002-7863.

2

Nanofiber Reinforced Composite Polymer Electrolyte Membranes

A. Kumar and M. Deka
Department of Physics, Tezpur University, Tezpur784028, Assam India

1. Introduction

The path breaking studies of Wright and Armand on ionically conducting polymers, called "polymer electrolytes" in the 1970s have opened an innovative area of materials research with potential applications in the power sources industry (Fenton et al., 1973). The main applications of the polymer electrolytes are in rechargeable lithium batteries as an alternative to liquid electrolytes (Chen et al., 2002; Lobitz et al., 1992). The advantages such as no leakage of electrolyte, higher energy density, flexible geometry and improved safety hazards have drawn the attention of many researchers on the development of lithium polymer batteries and other electrochemical devices such as supercapacitors, electrochromic windows, and sensors (Gray, 1991). In batteries being a separator membrane polymer electrolyte must meet the following requirements.

1. high ionic conductivity
2. high cationic transference number
3. good dimensional stability
4. high electrochemical stability and chemical compatibility with both Li anode and cathode material and
5. good mechanical stability.

The need for high ionic conductivity arises from the fact that at what rate or how fast energy from a Li-battery can be drained, which largely depends on the extent of ionic mobility in the electrolyte and hence on ionic conductivity. For battery applications, along with high ionic conductivity the electrolyte material must be dimensionally stable since the polymer electrolyte will also function as separator in the battery, which will provide electrical insulation between the cathode and the anode. This implies that it must be possible to process polymer electrolyte into freestanding film with adequate mechanical strength. Requirement of high cationic transport number rather than anionic is also important in view of the battery performance because concentration gradients caused by the mobility of both cations and anions in the electrolyte arise during discharging, which may result in premature battery failure.

Recent advances in nanotechnology have made materials and devices easier to be fabricated at the nanoscale. Nanofibres and nanowires with their huge surface area to volume ratio, about a thousand times higher than that of a human hair, have the potential to significantly improve current technology and find applications in new areas. Nanofibers in particular, have been used for a wide range of applications such as tissue engineering (Bhattarai et al., 2004), filter media (Suthar & Chase, 2001), reinforcement in composites (Chatterjee & Deopura, 2002) and micro/nano-electro-mechanical systems (MEMS/NEMS) (Sundararajan et al., 2002). Such fibers can be made from various materials such as polymers, carbon and semiconductors into the form of continuous nanofibers, nanofibrous networks or short nanowires and nanotubes. On the other hand polymer nanocomposite formation through the reinforcement of particles having dimensions less than 100 nm into polymer matrix finds important applications for the development of advanced materials. This reinforcement occurs for particles of spherical shape as well as for plate-like and fibrous nanofillers, the latter with their high aspect ratio exhibits a stronger effect. As a result, nanofibers because of their excellent mechanical properties coupled with very high aspect ratio theoretically are the ideal candidates for reinforcing a polymer matrix. The main challenge in this case lies in the fact of increased specific surface area due to the small size of the reinforcing particles, forces per unit mass resulting from interactions between this surface and surrounding area become more pronounced. As a result nanofibers cannot be easily dispersed in substances of different surface energy. However a strong interface between the reinforcing phase and the host matrix is always desirable in order to achieve desirable properties.

2. Types of polymer electrolytes

The development of polymer electrolytes has passed mainly three stages namely: (a) solid polymer electrolytes, (b) gel polymer electrolytes and (c) composite polymer electrolytes. In dry solid polymer electrolytes the polymer host itself is used as a solid solvent along with lithium salt and does not contain any organic liquids. The most commonly studied polymer electrolyte membranes are complexes of Li salts with a high molecular weight polyethylene oxide (PEO) (Ahn et al., 2003). PEO excels as a polymer host because of its high solvating power for lithium ions and its compatibility with the lithium electrode (Algamir & Abraham, 1994). However, it is also known that the high conductivity (10^{-3}-10^{-4} S/cm) of most PEObased polymer electrolytes requires operation in the temperature range of 80-100 °C (Ahn et al., 2003; Algamir & Abraham, 1994; Abraham, 1993; Kovac et al., 1998), below which these electrolytes suffer from low conductivity values in the range of 10^{-7}-10^{-8} S/cm because of the high crystallinity of PEO (Kovac et al., 1998). Gel electrolytes are formed by incorporating an electrolyte solution into polymer matrix (Algamir & Abraham, 1994; Tarascon et al., 1996). Since the electrolyte molecules such as ethylene carbonate (EC), propylene carbonate (PC), diethyl carbonate (DEC) can solvate ions, coordinating polymers like PEO are no longer necessary. The ion conduction in these electrolytes takes place through the liquid electrolytes where the host polymer mostly provides the structural support. The gel polymer electrolyte systems based on poly(methyl methacrylate) (PMMA) (Rajendran & Uma, 2000; Rajendran et al., 2001; Stephan et al., 2000; Feuillade & Perche, 1975) have been proposed for lithium battery application particularly because of their beneficial effects on stabilization of the lithium–electrolyte interface (Appetecch et al., 1995). However, reasonable conductivity achieved of such plasticized film is offset by poor mechanical properties at high plasticizer content. Rajendran et al. (Rajendran & Uma, 2000; Rajendran et al., 2001) were able to improve the mechanical property of PMMA by blending with poly(vinyl alcohol) (PVC). However, a decrease of ionic conductivity was observed due to higher viscosity and lower dissociability of lithium salt. Recently poly(vinylidenefluoride) (PVdF) as a host has drawn the attention of many researchers due to its high anodic stability and high dielectric constant ($\varepsilon = 8.4$) which helps in greater ionization of lithium salts (Choe et al., 1995). Unfortunately, PVdF-based polymer electrolytes suffer due to syneresis; a phenomenon by which the liquid component separates out from the host matrix in due course or upon application of pressure leading to battery leak and related safety problems. Very recently, poly(vinylidenefluoride-co-hexafluoropropylene) {P(VdF-HFP)} based systems have drawn the attention of many researchers because of its various appealing properties like high dielectric constant, low crystallinity and glass transition temperature (Song et al., 2000; Wu et al., 2006; Stephan et al., 2006; Stephan & Nahm, 2006; Stolarska et al., 2007; Michael & Prabaharan, 2004). P(VdF-HFP) has excellent chemical stability due to VdF unit and plasticity due to HFP unit (Aravindan & Vickraman, 2007). However, gelled or plasticized P(VdF-HFP) based electrolytes exhibit drawbacks, such as increased reactivity with lithium metal electrode, solvent volatility and poor mechanical properties at high degree of plasticization (Jacob et al., 2003). In order to retain the mechanical properties of polymer gel electrolytes, the gel films have to be hardened either by chemical or physical curing (high energy radiation), which results in high processing costs. Alternatively, the addition of inert oxides to the polymer electrolytes has recently become an ever increasing attractive approach, due to the improved mechanical stability, enhanced ionic conductivity and electrode-electrolyte interface stability (Croce et al., 1998; Quartarone et al., 1998). This type of electrolyte is known as composite polymer electrolyte. The increase in ionic conductivity in composite polymer electrolytes depends on the concentration and particle size of the inert solid phases. Generally, smaller the particle sizes of the oxides, the larger the conductivity enhancement. However, due to the absence of exact structure–property correlations in the polymer electrolyte systems, a complete understanding of the ion conduction phenomenon is still lacking. Nonetheless, to explain the mechanistic aspects of ion transport in micro/nanocomposite polymer electrolyte systems, a working hypothesis has been suggested accordingly to which, the dispersion of submicron or nano-size filler particles having large surface area, into the polymer host lowers the degree of crystallinity, which may also be thought to be due to Lewis acid–base interaction between ceramic surface states and polymer segments (Croce et al., 1998; Golodnitsky et al., 2002). Hence, in addition to the usual space charge effects of the dispersoid particles, the increased amorphosity would also support the conductivity enhancement in terms of increased ionic mobility through the amorphous phase. Essentially because of this idea, nanocomposite polymer electrolytes wherein nanosized inert solid particles are added to the polymer electrolytes are presently the focus of many studies, both practical as well as theoretical.

It has been reported that addition of nanoscale inorganic fillers, such as alumina (Al_2O_3), silica (SiO_2), titania (TiO_2) etc. to the polymer electrolytes resulted in the improvements of transport properties as well as mechanical and electrochemical properties (Kim et al., 2001; Kim et al., 2002; Scrosati et al., 2000). A novel composite micro-porous polymer electrolyte membrane based on optimized composition of P(VdF-HFP)-ZrO_2 was prepared by a preferential polymer dissolution process. The incorporation of ZrO_2 nanoparticles in the P(VdF-HFP) matrix, improved the ionic conductivity due to the availability of a large amount of oxygen vacancies on ZrO_2 surface which may act as the active Lewis acidic sites that interact with ions (Kalyana et al., 2007). A gel polymer electrolyte based on the blend of poly(methyl methacrylate-co-acrylonitrile-co-lithium methacrylate) (PMAML) and poly(vinylidene fluoride-co-hexafluoropropylene) P(VdF-HFP)) was prepared and characterized. The highest ionic conductivity achieved in the system was 2.6×10^{-3} S cm^{-1} at ambient temperature (Wang & Tang, 2004). Highly conducting porous polymer electrolytes comprised of P(VdF-HFP), metal oxide (TiO_2, MgO, ZnO)/or mesoporous zeolite (MCM-41, SBA-15), ethylene carbonate (EC), propylene carbonate (PC) and $LiClO_4$ were fabricated with a simple direct evaporation method. These polymer composite electrolytes were stable up to 5.5 V (versus Li/Li^+) and the lithium ion cells assembled with these polymer electrolytes show a good performance at a discharge rate below *C*/2 (Wu et al., 2006). Composite polymer electrolyte (CPE) membranes, comprising P(VdF-HFP), aluminum oxyhydroxide ($AlO[OH]_n$) of two different sizes 7 μm/14 nm and $LiN(C_2F_5SO_2)_2$ as the lithium salt were prepared using a solution casting technique. The incorporation of nanofillers greatly enhanced the ionic conductivity and the compatibility of the composite polymer electrolyte (Stephan et al., 2006). Nano SiO_2–P(VDF-HFP) composite porous membranes were prepared as the matrix of porous polymer electrolytes through *in situ* composite method based on hydrolysis of tetraethoxysilane and phase inversion. It is found that the *in situ* prepared nano silica was homogeneously dispersed in the polymeric matrix, enhanced conductivity and electrochemical stability of porous polymer electrolytes, and improved the stability of the electrolytes against lithium metal electrodes (He et al., 2005). Composite polymer membranes comprising of P(VdF-HFP) / Al_2O_3 were prepared by phase inversion technique with poly(ethylene glycol) (PEG) as an additive. The polymer membrane prepared with a weight ratio of PVdF-HFP (40):PEG (40):Al_2O_3 (20) showed maximum protonic conductivity due to the combined effect of inert filler and its porous nature (Kumar et al., 2007). The protonic conductivity of silica polymerized *in situ* within a P(VdF-HFP) matrix has been studied. The conductivity linearly increases with the silica content (Carrière et al., 2001). Gel polymer electrolyte membranes composed of P(VdF-HFP) and surface modified aluminum or titanium oxide were prepared according to the so-called Bellcore process. The ionic conductivity of polymer membrane increased by more than one order of magnitude upon the addition of filler into polymer host (Stolarska et al., 2007). Nanocomposite polymer electrolyte (NCPE) membranes of P(VdF-HFP) matrix with ethylene carbonate and diethyl carbonate mixtures as plasticizing agents, SiO_2 nanoparticles as filler and complexed with $LiPF_3(CF_3CF_2)_3$ were prepared by solvent casting technique. NCPE membranes containing 2.5 wt% of SiO_2 exhibited enhanced conductivity of 1.13 mS cm^{-1} at ambient temperature (Arvindan & Vickraman, 2007). Various amounts of nanoscale rutile TiO_2 particle were used as fillers in the preparation of P(VdF-HFP)-based porous polymer electrolytes. Physical, electrochemical and transport properties of the electrolyte films were investigated in terms of surface morphology, thermal and crystalline properties, swelling behavior after absorbing electrolyte solution, chemical and electrochemical stabilities, ionic conductivity, and compatibility with lithium electrode (Kim et al., 2003).

In the present chapter we report novel composite polymer electrolytes by incorporating dedoped (insulating) polyaniline nanofibers instead of nanoparticles into P(VdF-HFP)-(PC+DEC)-$LiClO_4$ gel polymer electrolyte system and PEO-P(VdF-HFP)-$LiClO_4$ blend electrolyte system. The concentration of dedoped polyaniline nanofibers has been varied and its effects on ionic transport in both the systems have been investigated.

3. Synthesis techniques of nanofibers

3.1 Electrospinning

Electrospinning has been recognized as an efficient technique for the fabrication of polymer nanofibers. Various polymers have been successfully electrospun into ultrafine fibers in recent years mostly in solvent solution and some in melt form. In terms of the flexibility of the process, electrospinning is able to fabricate continuous nanofibres from a huge range of materials. Of the major classes of materials, electrospinning is able to produce nanofibres of polymers, composites, semiconductors and ceramics (Huang et al., 2003; Chronakis, 2005). The formation of nanofibers through electrospinning is based on the uniaxial stretching of a viscoelastic solution. There are basically three components in a eletrospun setup: a high voltage power supply, a capillary tube with a needle and a metal collector. A typical electrospinning set up is shown in Fig. 1. When a sufficiently high voltage is applied to a liquid droplet, the body of the liquid becomes charged, and electrostatic repulsion counteracts the surface tension and droplet is stretched, at a critical point a stream of liquid erupts from the surface. This point of eruption is known as the Taylor cone (Taylor, 1969). If the molecular cohesion of the liquid is sufficiently high, stream breakup does not occur (if it does, droplets are electrosprayed) and a charged liquid jet is formed. As the jet dries out in flight, the mode of current flow changes from ohmic to convective as the charge migrates to the surface of the fibre. The jet is then elongated by a whipping process caused by electrostatic repulsion initiated at small bends in the fibre, until it is finally deposited on the grounded collector. The elongation and thinning of the fibre resulting from this bending instability leads to the formation of uniform fibres with nanometer-scale diameters (Li & Xia, 2004). The main advantage of the electrospinning nanomanufacturing process is that it is cost effective compared to that of most bottom-up methods. The nanofibers prepared from Electrospinning process are often uniform and continuous and do not require expensive purification unlike submicrometer diameter whiskers, inorganic nanorods and

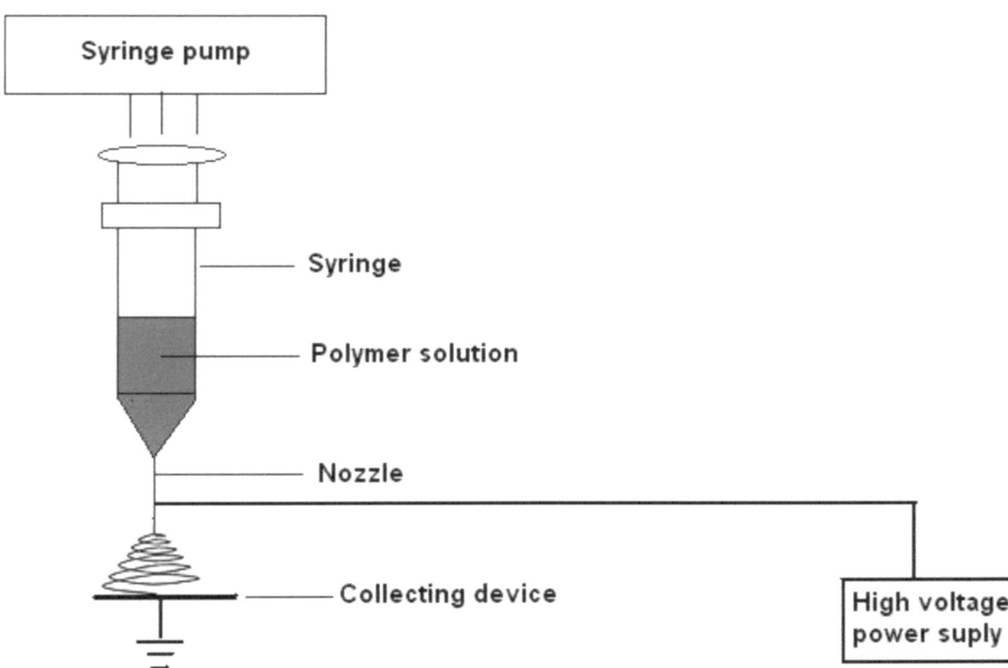

Fig. 1. Shematic diagram of electrospinning setup

carbon nanotubes (Dzenis, 2004). Polymer nanofibers mats are being considered for use in composite materials reinforcement, sensors, filtration, catalysis, protective clothing, biomedical applications, space applications such as solar cells, and micro- and nano optoelectronic device such as LEDs and photocells. Carbon nanofibers made from polymeric precursors further expand the list of possible uses of electrospun nanofibers (Li & Xia, 2004; Subbiah et al., 2005).

3.2 Template synthesis

An effective way to produce nanometer fibers (or nano-tubes) is based on the use of membrane-template techniques (Martin, 1961; Delvaux et al., 2000; Steinhart et al., 2002). Membranes, with nanochannels generated by fission-fragment tracks or by electrochemical etching of aluminum metal, are used as templates for either chemical or electrochemical deposition of conductive polymers (Pathasarathy & Martin, 1994), metal (Van de Zande et al., 1997), semiconductor (Klein et al., 1993), and other materials for the generation of nanofibers or nanotubes. Since the nanochannels on membranes are very uniform in size, the diameter and the aspect ratio of the nanofibers synthesized by the membranetemplate technique can be precisely controlled. This greatly facilitates the interpretation of optical data and the processing of these fibers (or tubes) into 2-D nanostructured materials (de Heer et al., 1995). Single-crystal semiconductor nanofibers can also be grown catalytically by metallorganic vapor phase epitaxy and laser ablation vapor-liquid-solid techniques (Morales & Lieber, 1998). The synthesis of these one-dimensional structures with diameters in the range of 3 to 15 nm holds considerable technological promise for optoelectronic device applications.

3.3 Phase separation

This method is normally used to synthesisze the polymer nanofibers. In phase separation, a polymer is first mixed with a solvent before undergoing gelation. The main mechanism in this process is the separation of phases due to physical incompatibility. One of the phase – which is that of the solvent – is then extracted, leaving behind the other other phase. In nutshell phase separation technique involves five basic steps (Ma & Zhang, 1999)

i. Dissolution of polymer.
ii. Liquid–liquid phase separation process.
iii. Polymer gelation (controls the porosity of nanoscale scaffolds at low temperature).
iv. Extraction of solvent from the gel with water.
v. Freezing and freeze-drying under vacuum.

Gelation is the most critical step that controls the porous morphology of the nanofibrous foams. The duration of gelation vary with polymer concentration and gelation temperature. At low gelation temperature nanoscale fiber networks are formed, whereas high gelation temperature results in the formation of a platelet-like structure due to the nucleation of crystals and their growth. This formation of platelet-like structure is overcome by increasing the cooling rates, which produce uniform nanofibers. However, gelation condition or polymer concentration does not affect the average diameter of fibers significantly. An increase in polymer concentration results in the decrease of porosity and increase of mechanical properties (Young's modulus and tensile strength). Other process parameters such as types of polymer and solvent, and thermal treatment also influence the morphology of the nanofibrous scaffolds (Zhang et al, 2005). The advantage of the phase separation process is that it is a relatively simple procedure and the requirements are very minimal in terms of equipment compared with the previously discussed techniques of electrospinning and template synthesis.

3.4 Interfacial polymerization

Another effective and versatile way to synthesize nanofibers of conductive polymers is the interfacial polymerization technique. Generally conductive polymers like polyaniline, polypyrrole, PEDOT etc. are synthesized by this technique. This method was originally developed by Huang (Huang, 2003), where they synthesized nanofibers of polyaniline (PAni). Interfacial polymerization does not depend on any specific template or dopant. Highquality polyaniline nanofibers are obtained even when common mineral acids such as hydrochloric, sulfuric and nitric acids are used as dopants. They found it is the nature for PANI to form nanofibrillar morphology. In order to obtain pure PANI nanofibers, the secondary growth of the initially formed nanofibers must be suppressed. The interfacial polymerization of PAni involves the polymerization of aniline at the interface between two immiscible liquids, where the newly formed PAni nanofibers diffuse away from the interface to the aqueous solution because of their hydrophilic nature. This makes more reaction sites available at the interface and avoids further growth of the PAni nanofibers. S. Goel et. al. (Goel et al., 2007) reported the synthesis of polypyrrole nanofibers in the presence of different dopants including hydrochloric acid (HCl), ferric chloride (FeCl3), ptoluene sulfonic acid (p-TSA), camphor sulfonic acid (CSA), and polystyrene sulfonic acid (PSSA) using a simple interfacial oxidative polymerization method. They observed that that the electrical conductivity of PPy nanostructures depends upon the nature of dopant (PPy-p-TSA > CSA > HCl > FeCl3 > PSSA), PPy-p-TSA nanofibers showing the highest electrical conductivity of 6×10^{-2} S/cm.

4. P(VdF-HFP)-(PC+DEC)-$LiClO_4$-Dedoped polyaniline nanofibers composite gel polymer electrolyte system:

4.1 Preparation

The host copolymer poly(vinylidenefluoride-co-hexafluoropropylene) {P(VdF-HFP)} ($M_w \approx 400000$) and salt lithium perchlorate ($LiClO_4$) were received from Aldrich, USA.
These two raw materials were heated at 50° C and 100° C respectively before use to remove the moisture. Organic solvents propylene carbonate (PC) and diethyl carbonate (DEC) were used without further treatment as obtained from EMerck. Dedoped polyaniline (PAni) nanofibers were synthesized by the interfacial polymerization technique (Huang, 2006). The interfacial polymerization reaction was carried out in 30ml glass vials. 1M amount of aniline was dissolved in 10ml of organic solvent carbon tetrachloride (CCl_4). Ammonium peroxydisulfate {$(NH_4)_2S_2O_8$} (0.25M) was dissolved in 10 ml of double distilled water and dopant acid (HCl). The polyaniline nanofibers were dedoped with 1M NaOH. The electronic conductivity of PAni nanofibers was measured with Keithley 2400 LV souecemeter. The electronic conductivity of doped nanofibers is of the order of 10^{-4} Scm^{-1}, whereas after dedoping with NaOH solution the electronic conductivity was found to be of the order of 10^{-11} Scm^{-1}. This confirms the insulating nature of dedoped polyaniline nanofibers.
P(VdF-HFP)-(PC+DEC)-$LiClO_4$ - x wt. % dedoped PAni nanofiber (x= 0, 2, 4, 6, 8,10) membranes were prepared by conventional solution casting technique. Predetermined amounts of P(VdF-HFP), $LiClO_4$ and (PC+DEC) were dissolved in acetone in the ratio of

6:1.5:1.5:1 (by weight). Subsequently the solution was stirred at 50°C for 12 hours. PC has high dielectric constant (ε = 64.6) but has high viscosity (η = 2.53), whereas DEC has low dielectric constant (ε = 2.82) but has low viscosity (η = 0.748). Combination of PC and DEC (1:1 by volume) solvent was used as optimization for high dielectric constant (ε = 33.71) and low viscosity (η = 1.639) to achieve high ionic conductivity. After 12 hours of mixing at 50°C the dedoped polyaniline nanofiber was added in the gel polymer solution and allowed to stir for another 12 hours. The viscous solution thus obtained was cast onto Petri dish and allowed to dry at room temperature. Different membranes were synthesized by varying the concentration of dedoped polyaniline nanofibers.

The ionic conductivity of the nanocomposite polymer electrolyte films was determined by ac impedance measurements using a Hioki 3532-50 LCR Hitester in the frequency range from 42 Hz to 5MHz. The temperature dependence of ionic conductivity was also measured by heating the samples from room temperature (25°C) to 60°C. The nature of conductivity of nanofibers dispersed gel polymer electrolytes was determined by transference number measurements using Wagner polarization technique with polymer electrolyte membrane between graphite blocking electrodes. The transference number was found to be = 0.98 indicating that conductivity was essentially ionic in nature. The interfacial stability of nanocomposites polymer electrolytes was studied by fabricating stainless steel/polymer electrolyte/stainless steel cells at room temperature and monitored for 20 days. X-ray diffraction patterns of the prepared membranes were obtained by Rigaku miniflex diffractometer at room temperature. Surface morphology of the composite electrolytes was studied by using Scanning Electron Microscope (SEM) (JEOL 6390 LV). The size of PAni nanofibers was determined by TEM (JEOL-TEM-100 CXII).

4.2 Results and discussion

4.2.1 TEM studies

Fig. 2 shows the TEM micrograph of PAni nanofibers. From the figure it is observed that nanofiber is composed of randomly packed polymer chains. As the PAni nanofibers are

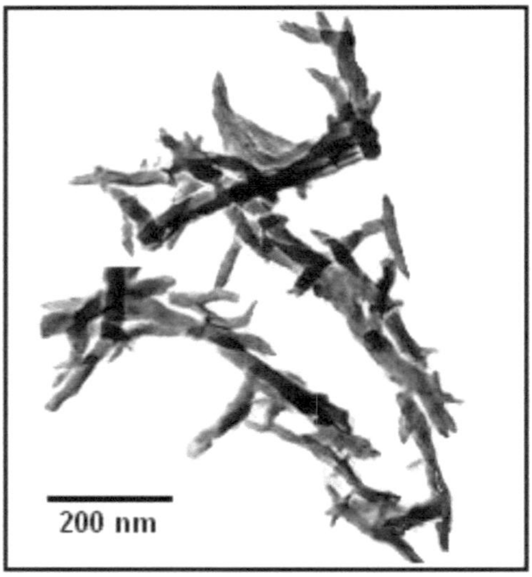

Fig. 2. TEM micrograph of dedoped Pani nanofibers

synthesized by interfacial polymerization, overgrowth of polyaniline on the nanofiber scaffolds does not take place and nanofibrillar morphological units are formed. The diameter and length of the fibers are found to be 20 nm to 30 nm and more than 1000 nm respectively. These high aspect ratio (> 50) nanofibers were used in polymer electrolytes as a filler to increase the ionic conductivity and electrochemical properties.

4.2.2 Ionic conductivity studies

Fig. 3a shows the impedance plot of P(VdF-HFP)-(PC+DEC)-$LiClO_4$ polymer electrolyte and Fig. 3b presents the complex impedance spectra of P(VdF-HFP)-(PC+DEC)-$LiClO_4$-x% dedoped polyaniline nanofiber (x = 2, 4, 6, 8 and 10) composite polymer electrolytes. The

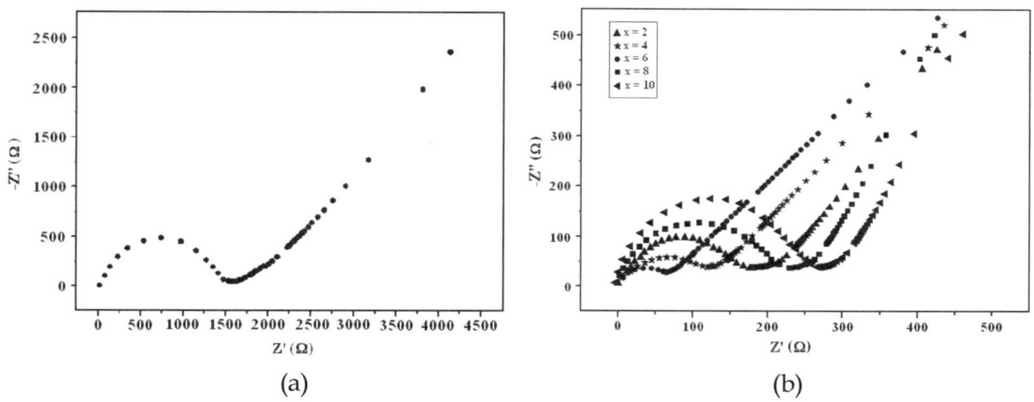

Fig. 3. (a) Complex impedance spectrum of P(VdF-HFP)-(PC+DEC)-$LiClO_4$ gel polymer electrolyte without incorporating dedoped nanofibers.
(b) Complex impedance spectra of P(VdF-HFP)-(PC+DEC)-$LiClO_4$-x% dedoped polyaniline nanofibers (x = 2, 4, 6, 8 and 10)

impedance spectra comprise a distorted semicircular arc in the high frequency region followed by a spike in the lower frequency region (Aravindan & Vickraman, 2008). The high frequency semicircle is due to the bulk properties and the low frequency spike is due to the electrolyte and electrode interfacial properties. The impedance spectra can be modeled as an equivalent circuit having a parallel combination of a capacitor and a resistor in series or parallel with a constant phase element (CPE) (Marzantowicz et al., 2005). The impedance of CPE is given by

$$Z_{CPE} = k\,(j\omega)^{-p} \quad \text{where } 0 < p < 1 \tag{1}$$

When p=0, Z is frequency independent and k is just the resistance and when p=1, $Z = k/j\omega = -jk/\omega$, the constant k_1 now corresponds to the capacitance. When p is between o and 1, the CPE acts in a way intermediate between a resistor and a capacitor. The use of series CPE terms tilts the spike and parallel CPE terms depress the semicircle. The bulk electrical resistance value (R_b) is calculated from the intercept at high frequency side on the Z' axis. The ionic conductivity is calculated from the relation $\sigma = l/R_b r^2 \pi$; where l and r are thickness of polymer electrolyte membrane and radius of the sample membrane discs and R_b is the bulk resistance obtained from complex impedance measurements. The value of σ_{ionic} of

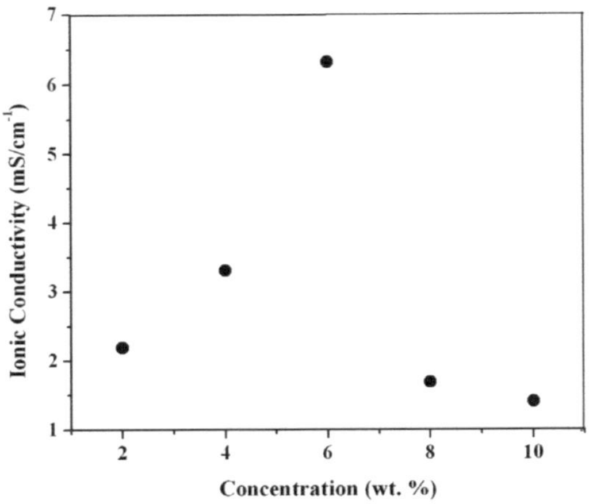

Fig. 4. Variation of ionic conductivity of P(VdF-HFP)-(PC+DEC)-$LiClO_4$-x% dedoped polyaniline nanofibers with nanofiber concentration.

plasticized nanocomposite polymer electrolytes was evaluated from complex impedance spectra and expressed as a function of nanofiber concentration at room temperature (25°C) as shown in Fig. 4. It is observed that the σ_{ionic} increases with increase in nanofiber concentration. Maximum conductivity was found to be 6.31 $\times 10^{-3}$ Scm^{-1} at room temperature for x = 6 wt. % dedoped polyaniline nanofiber filler, which is by over one order of magnitude higher as compared to that (2.5×10^{-4} Scm^{-1}) for polymer electrolyte without nanofibers. However as the filler (dedoped nanofiber) concentration increases beyond 6 wt. %, the ionic conductivity decreases. It is known that plasticized polymer electrolytes show highly porous structure created by plasticized rich phase (Aravindan & Vickraman, 2007). When dedoped polyaniline nanofiber is incorporated in the porous membrane, the movement of Li^+ ion through the pores is facilitated by the filler due to formation of conduction path resulting in higher conductivity. As the nanofiber content increases from 2 wt. % to 6 wt. % the porous structure is remarkably widened leading to entrapment of large volume of liquid electrolyte in the pores, which results in increase in ionic conductivity. Moreover the reorganization of P(VdFHFP) chains is prevented due to high aspect ratio (> 50) nanofibers leading to increase in amorphicity with increasing concentration of nanofibers which is consistent with XRD results. However, XRD results show that beyond 6 wt. % the nanofibers get phase separated out from the polymer matrix and form insulating aggregation (Wieczorek et al., 1996), which impede the Li^+ ion motion resulting in decrease in ionic conductivity. Fig. 5 shows the conductivity versus temperature inverse plots of polymer electrolyte membranes in the temperature range from 25°C to 60°C. Highest ionic conductivity of 2.2×10^{-2} Scm^{-1} has been found at 60°C with 6 wt. % of dedoped polyaniline nanofibers. The figure shows that the ionic conduction in nanocomposites polymer electrolytes obey the Vogel-Tamman-Fulcher (VTF) relation (Quartarone et al., 1998)

$$\sigma = \sigma_0 \, exp\,(-B/k\,(T-T_0)) \tag{2}$$

where B is a constant with dimensions that of the energy, T_0 is the idealized glass transition temperature at which the probability of configurational transition tends to become zero and is generally regarded as having a value between 20 to 50 K below glass transition

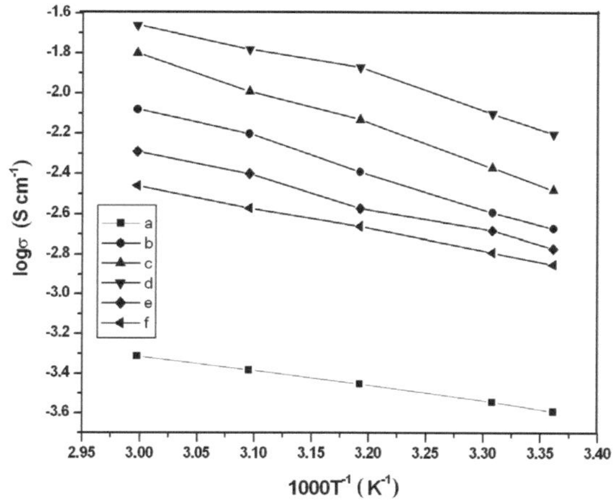

Fig. 5. Logσ vs. temperature inverse curves of P(VdF-HFP)-(PC+DEC)-$LiClO_4$-x% dedoped polyaniline nanofibers membranes (a) x=0, (b) x=2, (c) x=4, (d) x=6, (e) x=8, (f) x=10.

temperature (T_g) of the polymer. Since T_g of P(VdF-HFP) is -100°C, T_0 will be far below the temperature regions of measurements from room temperature (25°C) to 60°C. Therefore, VTF behavior (2) can be modeled as Arhenius behavior as shown lnσ vs. 1000/T plots in Fig. 5. As expected the increase in temperature leads to increase in ionic conductivity because as the temperature increases the polymer expands to produce more free volume, which leads to enhanced ionic mobility and polymer segmental mobility. The enhancement of ionic conductivity by the dedoped polyaniline nanofiber can be explained by the fact that the nanofiber inhibits the recrystallization kinetics, helping to retain the amorphous phase down to relatively low temperatures (Rhoo et al., 1997).

4.2.3 Interfacial stability

Compatibility of nanocomposites polymer electrolyte with electrode materials is an important factor for polymer battery applications. In order to improve the interfacial stability of polymer electrolytes before and after incorporating dedoped polyaniline nanofiber, the ionic conductivity was measured by fabricating stainless steel/polymer electrolyte membrane/stainless steel cells at room temperature and monitored for 20 days. Polymer electrolytes without nanofiber and containing 6 wt. % of dedoped polyaniline nanofiber have been selected to observe the effect of nanofiber on interfacial stability and the results are shown in Fig. 6.

It reveals that ionic conductivity of both the electrolytes decreased with time but decrease of ionic conductivity in the polymer electrolyte without nanofiber is much larger as compared to that of polymer electrolytes containing nanofibers. This result confirms that the interfacial stability of the polymer electrolyte containing nanofibers is better than that of without nanofibers. This can be attributed to the fact that when nanofiber is added passivation of polymer electrolyte due to reaction with electrode material decreases. High aspect ratio (> 50) nanofibers get accumulated on the surface of the electrode and effectively impede the electrode electrolyte reaction (Zhang et al., 2004). Fig. 7 schematically depicts that when dedoped (insulating) nanofibers are incorporated in the polymer electrolyte the electrolyte does not make direct contact with the electrode, which increases the interfacial stability.

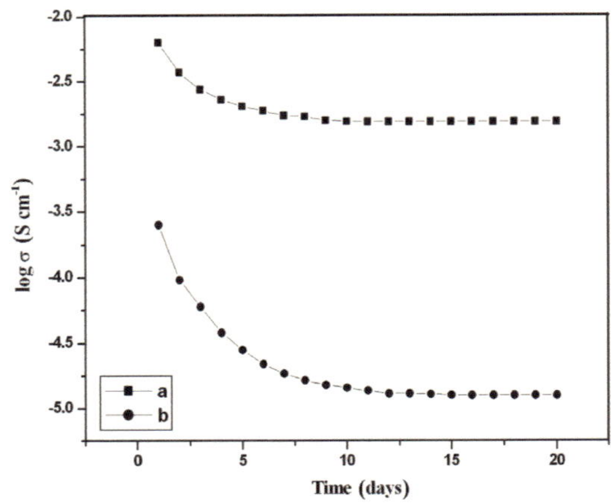

Fig. 6. Interfacial stability of (a) P(VdF-HFP)-PC+DEC-$LiClO_4$-6% dedoped polyaniline nanofibers polymer electrolyte and (b) P(VdF-HFP)-PC+DEC-$LiClO_4$ gel polymer electrolyte

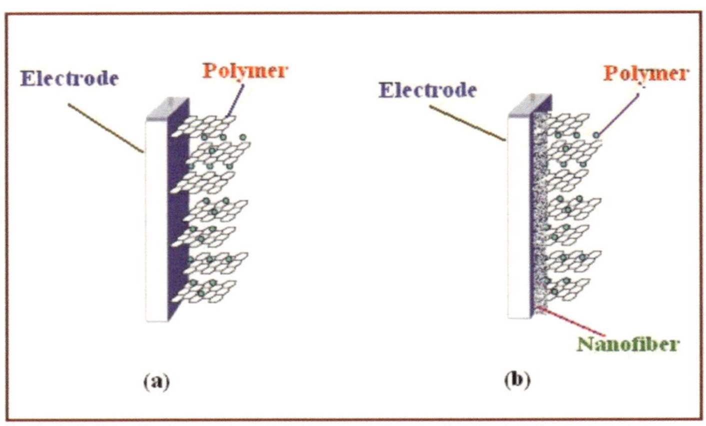

Fig. 7. Schematic representation of electrode/polymer electrolyte interfacial passivation (a) without and (b) with dedoped polyaniline nanofibers. without nanofibers

4.2.4 XRD analysis

X-ray diffraction patterns of pure P(VdF-HFP) and dedoped polyaniline nanofibers are presented in Fig. 8(a & b). In Fig. 8a the peaks at 2θ=20° and 38° correspond to (020) and (202) crystalline peaks of P(VdF) (Abbrent et al, 2001). This is a confirmation of partial crystallization of the PVdF units in the copolymer to give an overall semi-crystalline morphology for P(VdF-HFP). High intensity peaks at 2θ=20° and 2θ=23° are observed in the XRD pattern of dedoped polyaniline nanofiber (Fig. 8b). Fig. 9(a-f) shows the XRD patterns of P(VdF-HFP)-(PC+DEC)-$LiClO_4$-x% dedoped polyaniline nanofiber composite polymer electrolytes. It is observed that the addition of dedoped polyaniline nanofibers in polymer electrolytes increases the broadening and decreases the intensity of XRD peaks. This is due to fact that addition of nanofibers prevents polymer chain reorganization causing significant disorder in the polymer chains which promotes the interaction between them (Kumar et al., 2007).

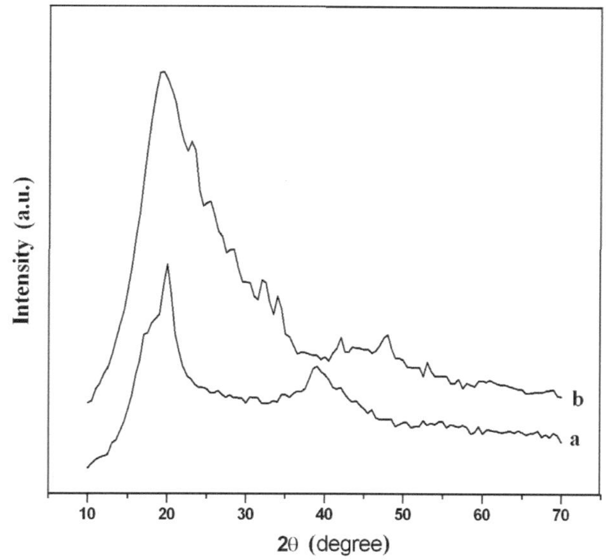

Fig. 8. XRD patterns of pure (a) P(VdF-HFP) and (b) Dedoped polyaniline nanofibers.

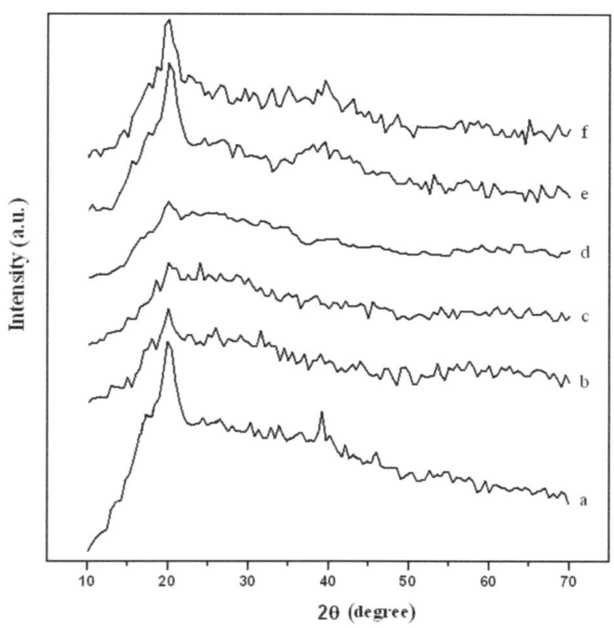

Fig. 9. XRD patterns of P(VdF-HFP)-(PC+DEC)-$LiClO_4$-x% dedoped polyaniline nanofibers membranes (a) x=0, (b) x=2, (c) x=4, (d) x=6, (e) x=8, (f) x=10.

4.2.5 Morphological study

Scanning electron micrographs of P(VdF-HFP)-(PC+DEC)-$LiClO_4$-x% dedoped polyaniline nanofiber gel composite polymer electrolytes are shown in Fig. 10(a-f). It is observed that the polymer electrolytes show highly porous structure with uniform pore distribution. Addition of fillers (dedoped polyaniline nanofibers) resulted in improved morphology, since the fillers occupied the pores along with the plasticizers. It is observed that pore distribution on the surface of the polymer electrolytes becomes denser with the increase of filler (dedoped polyaniline nanofibers) content at first, and reaches the maximum when the weight ratio of filler is about 6 wt. % (Fig. 10d), subsequently decreases when filler content increases further {Fig. 10(e-f)}. In composite gel polymer electrolytes the porous structure gives conducting pathways for Li^+ movement (Arvindan & Vickraman, 2007). The high aspect ratio of nanofibers remarkably increases the pore density and widens the porous structure of the polymer electrolytes (Xi et al., 2006). The above phenomenon is possibly due to the fact that the dedoped nanofibers try to occupy the pores in the gel polymer electrolyte and in the process pore distribution becomes denser. Highly porous structure leads to better connectivity of the liquid electrolyte through the pores accounting for the increase in ionic conductivity. Highly porous surface morphology of the polymer electrolytes is effectively formed on account of the interaction of dispersed dedoped (insulating) nanofibers with polymer component as well as the affinity with solvent molecules (Kim et al., 2003). However beyond 6 wt. % of filler content the nanofibers get phase separated from the P(VdF-HFP) matrix and form insulating clusters. These phase separated insulating percolation clusters impede ion movement and hence ionic conductivity decreases. The phenomenon of phase separation is strongly supported by XRD results as discussed in section 3.4.

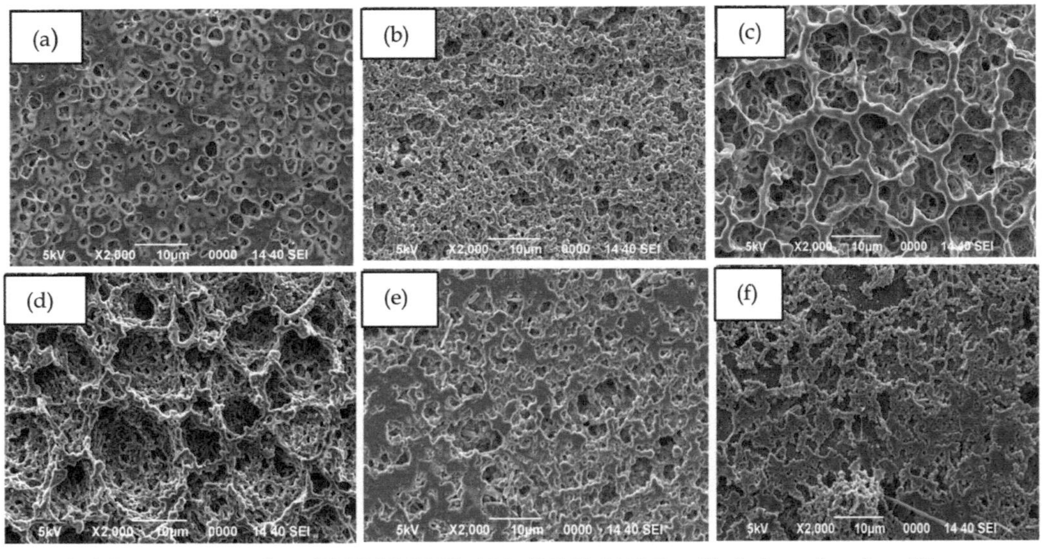

Fig. 10. SEM micrographs of P(VdF-HFP)-(PC+DEC)-$LiClO_4$-x% dedoped polyaniline nanofibers membranes (a) x=0, (b) x=2, (c) x=4, (d) x=6, (e) x=8, (f) x=10.

4.2.6 FTIR studies

FTIR is a powerful tool to characterize the chain structure of polymers and has led the way in interpreting the reactions of multifunctional monomers including rearrangements and isomerizations (Pavia et al., 2001). FTIR spectra of P(VdF-HFP), $LiClO_4$, dedoped polyaniline nanofibers and polymer complexes are shown in Fig. 11. The symmetric and asymmetric C-H stretching vibration of pure P(VdF-HFP) is observed at 3000cm^{-1}. Frequency 1633 cm^{-1} is assigned to C=O, C=C bonding. Frequencies 1486 cm^{-1} and 1404 cm^{-1} are assigned to $-CH_3$ asymmetric bending and C-O stretching vibration of plasticizer propylene carbonate and diethyl carbonate. Frequencies 12861066 cm^{-1} are assigned to -C-F- and $-CF_2-$ stretching vibration. Frequency 881 cm^{-1} is assigned to vinylidene group of polymer. The vibrational peaks of PVdF and $LiClO_4$ are shifted to (1786, 1401, 882, 837 cm^{-1}) and (1633, 1154, 626 cm^{-1}) in the polymer electrolyte respectively. The C-N stretching vibration of secondary amine in polyaniline nanofiber arises at 1289 cm^{-1}. The ammonium ion displays broad absorption in the frequency region 33503050 cm^{-1} because of N-H stretching vibration. The N-H bending vibration of secondary aromatic amine of polyaniline nanofiber occurs at 1507 cm^{-1}. The

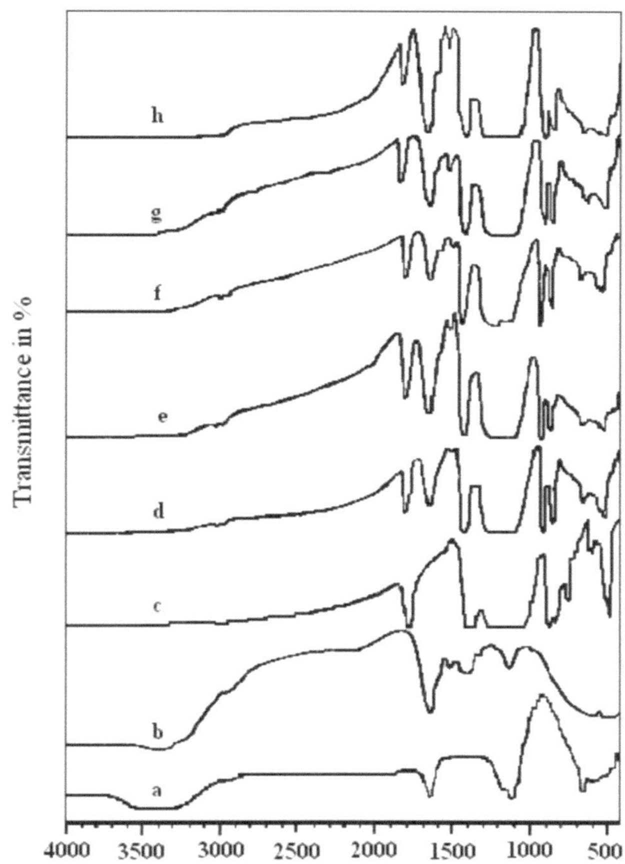

Fig. 11. FTIR spectra of (a) $LiClO_4$, (b) dedoped polyaniline nanofibers, (c) P(VdF-HFP), (d) P(VdF-HFP)-(PC+DEC)-$LiClO_4$-2% dedoped polyaniline nanofibers, (e) P(VdF-HFP)-(PC+DEC)-$LiClO_4$-4 % dedoped polyaniline nanofibers, (f) P(VdF-HFP)-(PC+DEC)-$LiClO_4$-6 % dedoped polyaniline nanofibers, (g) P(VdF-HFP)-(PC+DEC)-$LiClO_4$-8 % dedoped polyaniline nanofibers, (h) P(VdF-HFP)-(PC+DEC)-$LiClO_4$-10 % dedoped polyaniline nanofibers composite polymer electrolyte systems.

frequency 1637 cm^{-1} of polyaniline nanofiber is assigned to C=C of aromatic ring. As mentioned earlier the C-H symmetric and asymmetric bending frequencies are observed at 3000 cm^{-1} in pure P(VdF-HFP). However, after incorporating dedoped polyaniline nanofibers the corresponding bands in the electrolyte systems show a large shift (3021 cm^{-1}) in higher frequency region which is a characteristic of highly disordered conformation (Porter et al., 1987). This is in good agreement with the XRD results.

5. PEO/P(VdF-HFP)-$LiClO_4$-Dedoped polyaniline nanofibers composite solid polymer electrolyte system:

5.1 Preparation

The host polymer PEO (M_W = 6, 00,000), the copolymer P(VdF-HFP) (M_W = 4,00,000) and salt lithium perchlorate ($LiClO_4$) were received from Aldrich, USA. All the raw materials were heated at 50 °C under vacuum. Organic solvents acetonitrile and acetone were used as received from Emerk to prepare thin polymer electrolyte membranes by solution casting technique. Appropriate amount of PEO and salt $LiClO_4$ (O/Li = 8) were dissolved in acetonitrile and then mixed together, stirred and heated at 50 °C. P(VdF-HFP) was fixed at 40 wt. % of PEO for all samples, was stirred in presence of acetone at 50 °C. Subsequently both the polymer solutions were mixed, stirred and heated at 50 °C for 12-14 hours. Dedoped polyaniline nanofibers were then added in the blend polymer solutions and allowed to stir for another 7-8 hours. The viscous solution thus obtained was cast onto Petri dish and allowed to dry at room temperature. This procedure provided mechanically stable, free standing and flexible membranes. The blend based composite polymer electrolyte membranes used in this study were denoted as PEO-$LiClO_4$-P(VdF-HFP)-x% dedoped polyaniline nanofibers (x = 0, 5, 10, 15, 20, 25).

5.2 Results and discussion

5.2.1 X-Ray diffraction studies

X-ray diffraction patterns of pure PEO, P(VdF-HFP) and dedoped polyaniline nanofibers are presented in Fig. 12(a-c). High intensity peaks at $2\theta=20^{\circ}$ and $2\theta=23^{\circ}$ are observed in the XRD pattern of dedoped polyaniline nanofibers. In Fig. 12b the peaks at $2\theta=20^{\circ}$ and 38° correspond to (020) and (202) crystalline peaks of P(VdF-HFP). PEO shows a characteristic peak at $2\theta=20^{\circ}$. Fig. 13 shows the XRD patterns of PEO-P(VdF-HFP)-$LiClO_4$-x% dedoped polyaniline nanofibers composite polymer electrolytes. It is observed that when P(VdF-HFP) is blended with PEO, no additional peak appears, only the intensity of crystalline peaks decreases suggesting that the amorphicity increases (Leo et al., 2002). When dedoped polyaniline nanofibers are incorporated in the PEO-P(VdF-HFP)-$LiClO_4$ the intensity further decreases as shown in Fig. 13(b-f). The degree of crystallinity is determined by a method described elsewhere (Saikia et al., 2006). It is observed that the degree of crystallinity decreases with increasing nanofibers concentration and reaches a minimum at 15 wt. % nanofibers concentration. This reduction in crystallinity upon addition of nanofibers is attributed to the suppression of the reorganization of polymer chains by the nanofibers (Scrosati et al., 2001). However, at higher concentration of nanofibers (>15 wt.%), the degree of crystallinity increases with increasing nanofibers concentration indicating that crystalline phase starts increasing above 15 wt.% of nanofibers concentration due to reorganization of polymer chains in PEO-P(VdF-HFP)-$LiClO_4$ electrolyte system. At 20 wt.% and 25 wt.% of nanofibers concentration an additional peak appears at $2\theta=23^{\circ}$, which can be assigned to dedoped polyaniline nanofibers suggesting that above 15 wt.% polyaniline nanofibers get phase separated from the PEO-P(VdF-HFP)-$LiClO_4$ polymer electrolyte phase.

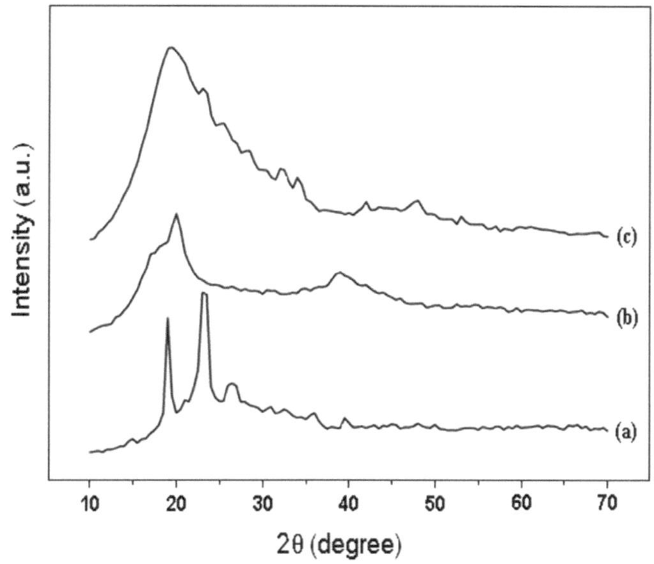

Fig. 12. XRD patterns of (a) PEO, (b) P(VdF-HFP), (c) dedoped polyaniline nanofibers.

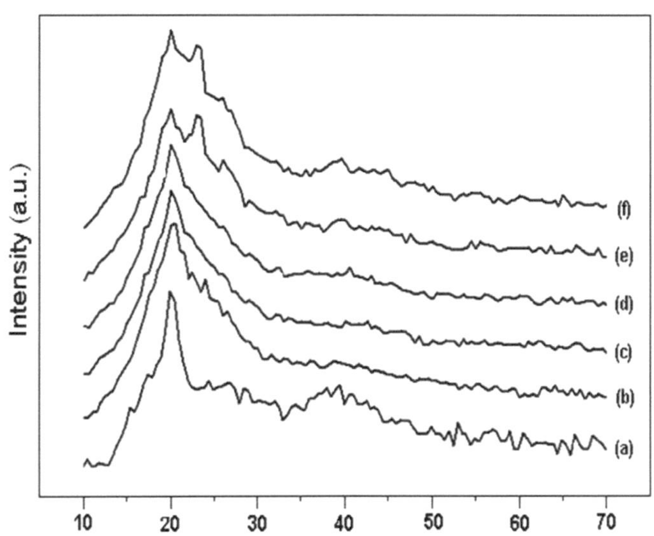

Fig. 13. XRD patterns of PEO-P(VdF-HFP)-$LiClO_4$-x% dedoped polyaniline nanofibers polymer electrolyte membranes (a) x = 0, (b) x = 5, (c) x = 10, (d) x = 15, (e) x = 20 and (f) x = 25.

5.2.2 Ionic conductivity studies

The complex impedance plots for PEO-P(VdF-HFP)-$LiClO_4$ polymer electrolyte membranes with different concentration of polyaniline nanofibers are presented in Fig. 14(a-f). The variation of ionic conductivity with increasing concentration of nanofibers is shown in Fig. 15. It is observed that the σ_{ionic} increases with increase in concentration of nanofibers. Maximum conductivity was found to be 3.1×10^{-4} Scm^{-1} at room temperature for 15 wt. % dedoped polyaniline nanofiber fillers, which is over seven times higher as compared to that (4.5×10^{-5} Scm^{-1}) for polymer electrolyte without nanofibers. However, as the filler (dedoped nanofibers) concentration increases beyond 15 wt. %, the ionic conductivity decreases. The enhancement up to 15 wt. % of nanofibers concentration seems to be correlated with the fact that the dispersion of dedoped polyaniline nanofibers to PEO-P(VdF-HFP) prevents polymer chain reorganization due to the high aspect ratio (>50) of nanofibers, resulting in reduction in polymer crystallinity, which gives rise to an increase in ionic conductivity. The increase in ionic conductivity may also result from Lewis acid-base interaction (Rajendran & Uma, 2000; Croce et al., 2001; Chung et al., 2001, Stephan & Nahm, 2006). In the present composite polymer electrolytes, the oxygen atom in PEO has two lone pair of electrons and nitrogen atom in PAni nanofibers has one lone pair of electrons, which act as strong Lewis base centers and Li^+ cations as strong Lewis acid giving rise to numerous acid-base complexes in the composite polymer electrolyte. This allows mobile ions to move more freely either on the surface of the nanofibers or through a low density polymer phase at the interface, which results in enhanced ionic conductivity. The reduction in crystallinity upon addition of polyaniline nanofibers up to 15 wt. % is consistent with XRD results. Enhancement in ionic conductivity can also be attributed to the creation of polymer-filler interface. The filler-polymer interface is a site of high defect concentration providing channels for faster ionic transport (Kumar & Scanlon, 1994) and the structure and chemistry of filler-polymer interface may have even more important role than the formation of amorphous phase in the electrolyte.

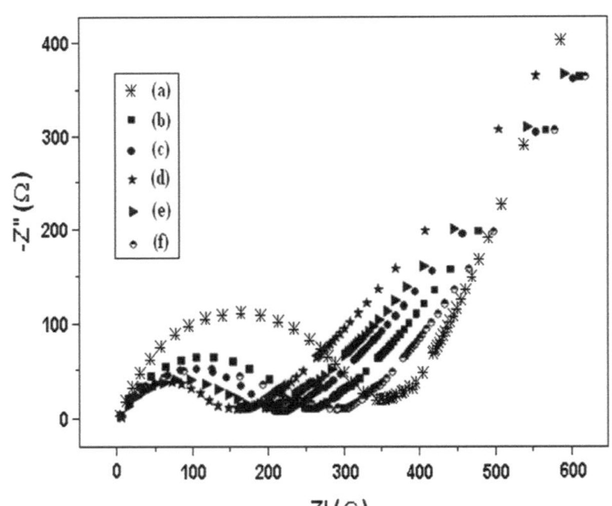

Fig. 14. Complex impedance spectra of PEO-P(VdF-HFP)-$LiClO_4$-x% dedoped polyaniline nanofibers polymer electrolyte membranes (a) x = 0 (b) x = 5, (c) x = 10, (d) x = 15, (e) x = 20 and (f) x = 25.

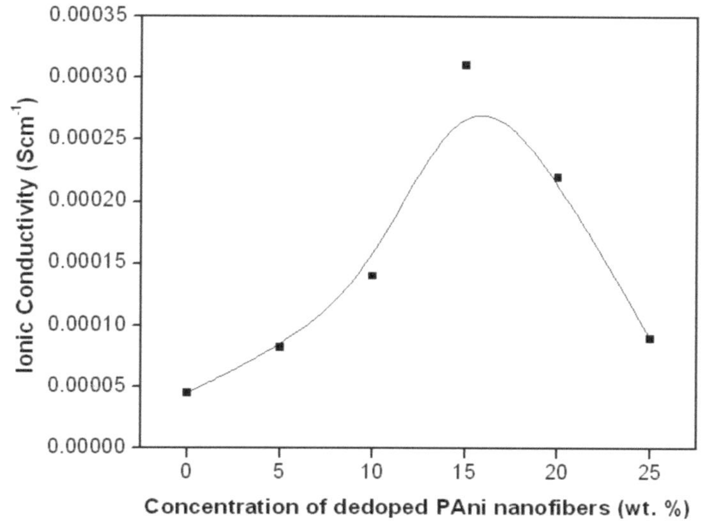

Fig. 15. Variation of ionic conductivity with different concentration of dedoped polyaniline nanofibers.

On the other hand, the decrease in ionic conductivity for concentration of nanofibers higher than 15 wt. % can be attributed to the blocking effect on the transport of charge carriers resulting from the phase separation of nanofibers (Kim & Park, 2007). Besides, above 15 wt. % of nanofibers concentration a depressed semicircle is seen in the impedance spectra, which is characteristic of a system where more than one conduction processes are present simultaneously (Kurian et al., 2005). SEM micrographs show that, at higher concentration of nanofibers (15 wt. %), a two phase microstructure is observed. This could be attributed to the fact that at higher concentration of nanofibers, uniform dispersion of nanofibers in PEO-P(VdF-HFP) matrix is difficult to achieve due to formation of phase-separated morphologies. This is expected to affect the conductivity of the system, since a large concentration of Li^+ cations are trapped in the phase separated nanofibers. Thus the decrease of ionic conductivity above 15 wt. % nanofibers content can be attributed to the effect of phase separation, which is consistent with the XRD and SEM results. Fig. 16 shows the conductivity versus temperature inverse plots of polymer electrolyte films in the temperature range from 25°C to 80°C. All the samples show a break point at around 60 °C, near the melting temperature of PEO, reflecting the well-known transition from PEO crystalline to amorphous phase.

As expected the increase in temperature leads to increase in ionic conductivity because as the temperature increases the polymer chains flex at increased rate to produce larger free volume, which leads to enhanced polymer segmental and ionic mobilities. The enhancement of ionic conductivity by the dedoped polyaniline nanofibers can be explained by the fact that the nanofibers inhibit the recrystallization kinetics, helping to retain the amorphous phase down to relatively low temperatures (Rhoo et al, 1997).

5.2.3 Morphological studies

The SEM micrographs for PEO-P(VdF-HFP)-$LiClO_4$-x% dedoped polyaniline nanofibers membranes are presented in Fig. 17(a-f). In general three-four phases are known to coexist

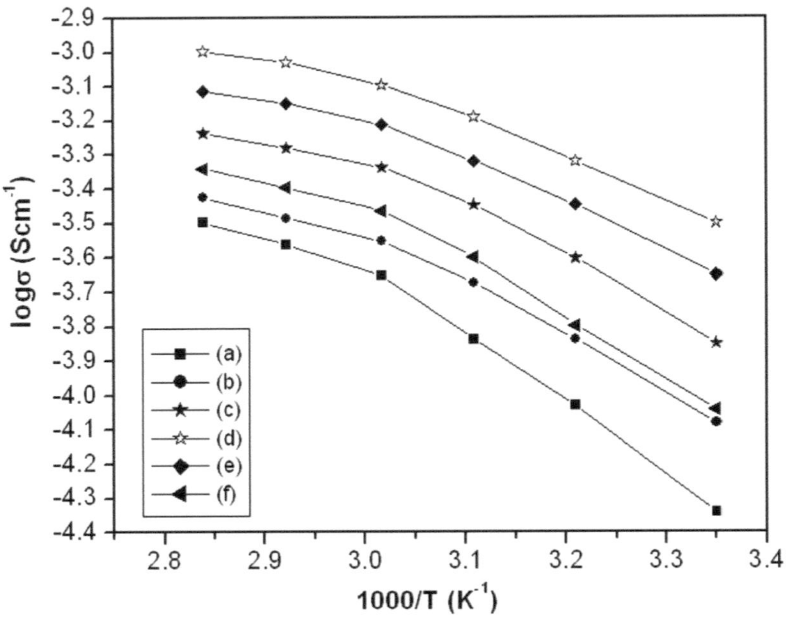

Fig. 16. logσ vs. temperature inverse curve PEO-P(VdF-HFP)-$LiClO_4$-x% dedoped polyaniline nanofibers polymer electrolyte membranes (a) x = 0, (b) x = 5, (c) x = 10, (d) x = 15, (e) x = 20 and (f) x = 25

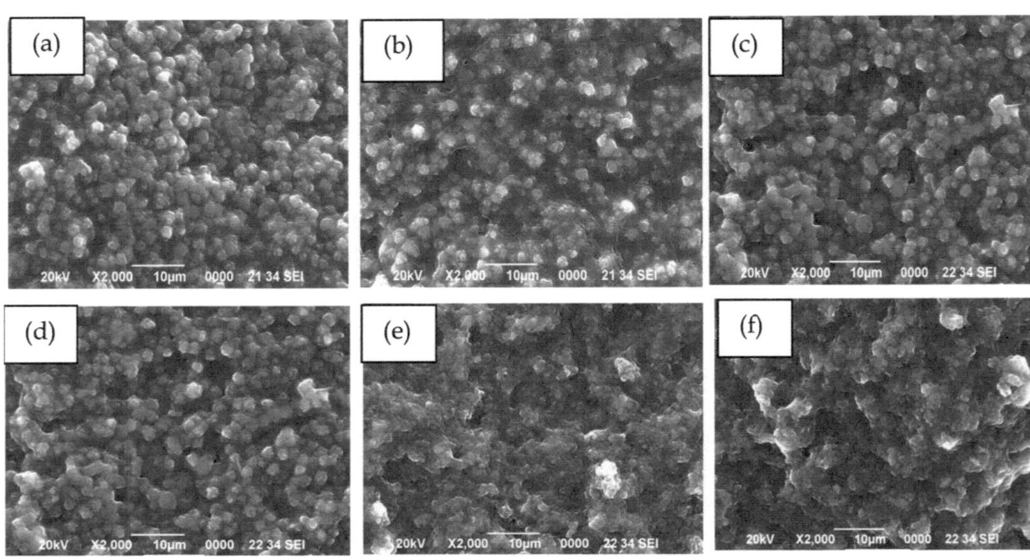

Fig. 17. SEM micrographs of PEO-P(VdF-HFP)-$LiClO_4$-x% dedoped polyaniline nanofibers polymer electrolytes (a) x = 0, (b) x = 5, (c) x = 10, (d) x = 15, (e) x = 20 and (f) x = 25.

in the PEO based polymer electrolytes viz. crystalline PEO phase, crystalline PEO-Li salt complex phase and amorphous PEO phase. It is observed that below 15 wt. % nanofibers concentration (Fig. 17a-c), the surface morphology is granular and smooth, which could be attributed to the reduction of PEO crystallinity due to complexation with lithium salt and polyaniline nanofibers. At 15 wt% concentration of nanofibers, rough morphology and sharp interfaces are observed (Fig. 17d) which may facilitate lithium ion conduction along the interface (Saikia & Kumar, 2005).

Fig. 17e shows that at 20 wt% of nanofiber concentration a two phase microstructure is observed due to phase segregation of nanofibers. Phase separation becomes more prominent at 25 wt% of nanofibers as shown in Fig. 17f. The nanofibers get phase separated from the PEO-P(VdF-HFP) polymer matrix and form domain like regions, which may act as physical barriers to the effective motion of the ions leading to decrease in ionic conductivity.

6. Conclusions

Stable dedoped PAni nanofiber-P(VdF-HFP) composites can be readily prepared by interfacial polymerization followed by solution casting method. The TEM result shows that polyaniline nanofibers of diameter 20-30 nm in size are formed by interfacial polymerization. The ionic conductivity of dedoped PAni nanofiber-P(VdF-HFP) based composite electrolyte was influenced by increase in PAni content. It obeys the Arrhenius Law of conductivity. The dedoped PAni nanofiber-P(VdF-HFP) based composite electrolyte has the highest ionic conductivity of 6.3×10^{-3} Scm^{-1} at room temperature. The crystallinity is found to decrease in case of composites. Increase in ionic conductivity can be attributed to the fact that dispersed nanofibers improve the porous structure of the gel polymer electrolytes forming better connectivity for ion motion through the liquid electrolyte. This is confirmed by SEM results. However at higher filler content conductivity decreases because the dedoped nanofibers start forming insulating clusters that encumber ion movement. The interfacial stability of the nanofibers dispersed polymer gel electrolyte membranes is observed to be better than that of gel polymer electrolytes without nanofibers. XRD analysis reveals that amorphicity increases upon addition of nanofibers which could be attributed to the reduction in chain reorganization of polymer by nanofibers.

In the PEO-P(VdF-HFP)-$LiClO_4$ polymer electrolyte the XRD, SEM and conductivity results show that the conductivity of increases when dedoped polyaniline nanofibers are added as a filler upto a concentration of 15 wt%. At higher concentration (> 15 wt%) the polyaniline nanofibers get phase separated from the polymer matrix. The three moieties PEO, P(VdF-HFP) and dedoped polyaniline nanofiber no longer remain a miscible uniform phase but nanofibers get phase separated.

Both the electrolyte systems show that by using dedoped (insulating) polyaniline nanofibers as fillers the ionic conductivity can be enhanced. It is also observed that there is a certain critical concentration above which the dedoped polyaniline nanofiber phase gets separated out from the electrolyte which thereby reduces the ionic conductivity. A probable explaination for this effect can be stated as when the concentration of the dedoped nanofibers is increased agglomeration occurs, the nanofiber phase gets separated out from the electrolyte and some domain like structures are formed which is evident from the SEM images. These domain like structures caused due to the increase in concentration of the dedoped nanofibers create barriers in the conducting path and hinders the ionic movement thereby reducing the ionic conductivity.

7. References

Abbrent, S.; Plestil, J.; Hlavata, D.; Lindgren, J.; Tegenfeldt, J. & Wendsjo, A. (2001). Crystallinity and morphology of PVdF-HFP-based gel electrolytes. *Polymer*, 42, 1407-1416.

Abraham, K.M. (1993). Highly conductive polymer electrolytes, In: *Applications of electroactive polymers*, Scrosati, B. (Ed.), 75-112, Chapman and Hall, London.

Ahn, J.-H. ; Wang, G.X.; Liu, H.K. & Dou, S.X. (2003). Nanoparticle-dispersed PEO polymer electrolytes for Li batteries. *J. Power Sources*, 119121, 422-426.

Algamir, M. & Abraham, K.M. (1994). Lithium batteries, new materials, developement and prospectives, In: *Industrial chemistry library, vol. 5*, Pistoia, G. (Ed.), 93-136, Elsevier, Amsterdam.

Appetecchi, G.B.; Crose, F. & Scrosati, B. (1995). Kinetics and stability of the lithium electrode in poly(methylmethacrylate)-based gel electrolytes. *Electrochim. Acta*, 40, 991-997.

Aravindan, V. & Vickraman, P. (2007). Polyvinylidenefluoride-hexafluoropropylene based nanocomposite polymer electrolytes(NCPE) complexed with $LiPF_3(CF_3CF_2)_3$. *Eur. Polym. J.* 43, 5121-5127.

Aravindan, V. & Vickraman, P. (2008). Characterzation of SiO_2 and Al_2O_3 incorporated PVdF-HFP based composite polymer electrolytes with $LiPF_3(CF_3CF_2)_3$. *J.Appl. Polym. Sci.*, 108, 1314-1322.

Bhattarai, S.R.; Bhattarai, N.; Yi, H.K.; Hwang, P.H., Cha, D.I. & Kim, H.Y. (2004). Novel biodegradable electrospun membrane: scaffold for tissue engineering. *Biomaterials*, 25, 2595–2602.

Carrière, D.; Barboux, P.; Chaput, F.; Spalla, O. & Boilot J. P. (2001). Enhanced connectivity in hybrid polymers. *Solid State Ionics*, 145,141-147.

Chatterjee, A. & Deopura, B.L. (2002). Carbon nanotubes and nanofibre: an overview. *Fiber Polym.*, 3(4), 134–139.

Chen, H. W.; Chiu, C.Y.; Wu, H.D.; Shen, I.W. &. Chang, F.C. (2002). Solid state electrolyte nanocomposites based on poly (ethylene oxide), poly (oxypropelene) diamine, mineral clay and lithium perchlorate. *Polymer*, 43, 5011-5016.

Choe, H.S.; Giaccami, J.; Alamgir, M. & Abraham, K.M. (1995). Preparation and characterization of poly(vinyl sulfone)- and poly(vinylidene fluoride)- based electrolytes. *Electrochim. Acta*, 40, 2289-2293.

Chronakis, I. S. (2005). Novel nanocomposites and nanoceramics based on polymer nanofibers using electrospinning process—a review. *J. Mater. Process. Technol.* 167, 283-293.

Chung, S.H.; Wang, Y.; Persi, L.; Croce, F.; Greenbaum, S.G.; Scrosati, B. & Plichta, E. (2001). Enhancement of ion transport in polymer electrolytes by addition of nano-scale inorganic oxides. *J. Power Sources* 97-98, 644-648.

Croce, F.; Appetecchi, G.B.; Persi, L. & Scrosati, B. (1998). Nanocomposite polymer electrolytes for lithium batteries. Nature, 394, 456-458.

Croce, F.; Persi, L.; Scrosati, B.; Serraino-Fiory, F.; Plichta, E. & Hendrickson, M.A. (2001). Role of the ceramic fillers in enhancing the transport properties of composite polymer electrolytes. *Electrochim. Acta*, 46, 2457-2461.

De Heer, W.A.; Bacsa, W.S.; Chatelain, A.; Gerfin, T.; Humphrey-Baker, R.; Forro, L. & Ugarte, D. (1995). Aligned nanotube films: production and optical and electronic properties. *Science*, 268:845-847.

Delvaux, M.; Duchet, J.; Stavaux, P.-Y.; Legras, R. & Demoustier-Champagne, S. (2000). Chemical and electrochemical synthesis of polyaniline micro- and nano-tubules. *Synth. Met.*, 113, 275280.

Dzenis, Y. (2004). Spinning continuous fibers for nanotechnology. *Science*, 304, 1917-1919.

Fenton, B.E.; Parker, J.M. & Wright, P.V. (1973). Complexes of alkali metal ions with poly(ethylene oxide). *Polymer*, 14, 589.

Feuillade, G. & Perche, Ph. (1975). Ion-conductive macromolecular gels and membranes for solid lithium cells. *J. Appl. Electrochem.*, 5, 6369.

Goel, S.; Mazumdar, N.A. & Gupta A. (2009). Synthesis and characterization of polypyrrole nanofibers with different dopants. *Polymers for Advanced Technologies*, DOI: 10.1002/pat.1516.

Golodnitsky, D.; Ardel, G. & Peled, E. (2002). Ion-transport phenomena in concentrated PEO-based composite polymer electrolytes. Solid State Ion. 147, 141-155.

Gray, F.M. (1991). *Solid polymer electrolytese fundamental and technological applications*, VCH, London, NewYork.

He, X.; Shi, Q.; Zhou, X.; Wan, C. & Jiang, C. (2005). In situ composite of nano SiO2 -P(VDF-HFP) porous polymer electrolytes for Li-ion batteries; *Electrochim. Acta*, 51,1069-1075.

Huang, J. (2006). Syntheses and applications of conducting polymer polyaniline nanofibers. *Pure Appl. Chem.*, 78, 15-27.

Huang, Z. -M.; Zhang, Y. Z.; Kotaki, M. & Ramakrishna, S. (2003). A review on polymer nanofibers by electrospinning and their applications in nanocomposites. *Compos. Sci. Technol.* 63, 2223-2253.

Jacob, M.M.E.; Hackett, E. & Giannelis, E.P. (2003). From nanocomposite to nanogel polymer electrolytes. *J. Mater. Chem.* 13, 1-5.

Kalyana Sundaram, N.T.; Vasudevan, T. & Subramania A. (2007). Synthesis of ZrO_2 nanoparticles in microwave hydrolysis of Zr (IV) salt solutions—Ionic conductivity of PVdF-co-HFP-based polymer electrolyte by the inclusion of ZrO_2 nanoparticles. *Journal of Physics and Chemistry of Solids*, 68, 264-271.

Kim, K.M.; Ko, J. M.; Park, N. -G.; Ryu, K. S. & Chang, S. H. (2003). Characterization of poly(vinylidenefluoride-co-hexafluoropropylene)-based polymer electrolyte filled with rutile TiO_2 nanoparticles. *Solid State Ionics*, 161(1-2), 121-131.

Kim, K.M.; Park, N.-G. ; Ryu, K.S. & Chang, S.H. (2002). Characterization of poly(vinylidenefluoride-co-hexafluoropropylene)-based polymer electrolyte filled with TiO_2 nanoparticles. *Polymer*, 43, 3951-3957.

Kim, K.M.; Ryu, K.S.; Kang, S. -G.; Chang, S. H. & Chung, I. J. (2001). The effect of silica addition on the properties of poly(vinylidene fluoride)-co-hexafluoropropylene)-based polymereElectrolytes. *Macromol. Chem. Phys.*, 202(6), 866-872.

Kim, S. & Park, S.-J. (2007). Preparation and electrochemical behaviors of polymeric composite electrolytes containing mesoporous silicate fillers. *Electrochim. Acta* 52, 3477-3484.

Klein, J.D.; Herrick II, R.D.; Palmer, D.; Sailor, M.J.; Brumlik, C.J. and Martin, C.R. (1993). Electrochemical fabrication of cadmium chalcogenide microdiode arrays. *Chem. Mater.*, 5, 902-904.

Kovac, M.; Gaberscek, M. & Grdadolnik, J. (1998). The effect of plasticizer on the microstructural and electrochemical properties of a $(PEO)_nLiAl(SO_3 Cl)_4$ system. Electrochim. Acta, 44, 863-870.

Kumar, B. & Scanlon, L.G. (1994). Polymerceramic composite electrolytes. *J. Power Sources*, 52, 261-268.

Kumar, G. G.; Kim, P.; Nahm, K. S. & Elizabeth, R. N. (2007). Structural characterization of PVdF-HFP/ PEG/Al_2O_3 proton conducting membranes for fuel cells. *J. Membr. Sci.*, 303,126-131.

Kumar, R.; Subramania, A.; Kalyana Sundaram, N.T.; Vijaya Kumar, G. & Bhaskaran, I. (2007). Effect of MgO nanoparticles on ionic conductivity and electrochemical properties of nanocomposite polymer electrolyte. *J. Membr. Sci.*, 300, 104-110.

Kurian, M.; Galvin, M.E.; Trapa, P.E.; Sadoway, D.R. & Mayes, A.M. (2005). Singleion conducting polymer–silicate nanocomposite electrolytes for lithium battery applications. *Electrochim. Acta*, 50, 2125-2134.

Leo, C.J.; Rao, G.V.S. & Chowwdari, B.V.R. (2002). Studies on plasticized PEO–lithium triflate-ceramic filler composite electrolyte system. *Solid State Ionics*, 148, 159-171.

Lobitz, P.; Fullbier, H.; Reich, A. & Ambrahsat, K. (1992). Polymer solid electrolytes-PEO and alkali halides: A modified preparative technique. *Solid State Ionics*, 58, 49-54.

Ma, P.X. & Zhang, R. (1999). Synthetic nano-scale fibrous extracellular matrix. Synthetic nano-scale fibrous extracellular matrix. *J. Biomed. Mater. Res.*, 46, 60–72.

Martin, C. R. (1994). Nanomaterials: A membrane-based synthetic approach. *Science* 266, 1961-1966.

Marzantowicz, M.; Dygas, J.R.; Jenninger, W. & Alig, I. (2005). Equivalent circuit analysis of impedance spectra of semicrystalline polymer. *Solid State Ionics*, 176, 2115-2121.

Michael, M.S. & Prabaharan, S.R.S. (2004). Rechargeable lithium battery employing a new ambient temperature hybrid polymer electrolyte based on PVK+PVdF–HFP (copolymer). *J. Power Sources*, 136, 408-415.

Morales, A.M. & Lieber, C.M. (1998). A laser ablation method for the synthesis of crystalline semiconductor nanowires. *Science*, 279, 208-211.

Pathasarathy, R.V. & Martin, C.R.. (1994). Synthesis of polymeric microcapsule arrays and their use for enzyme immobilization. *Nature*, 369:298-301.

Pavia, D.L.; Lampman, G.M. & Kriz, G.S. (2001). *Introduction to spectroscopy*, third ed., Harcourt College Publ., USA.

Porter, M.D.; Bright, T.B.; Allara, D.L. & Chidsey, C.E.D. (1987). Spontaneously organized molecular assemblies. 4. Structural characterization of n-alkyl thiol monolayers on gold by optical ellipsometry, infrared spectroscopy, and electrochemistry. *J.Am. Chem.Soc.*, 109, 3559-3568.

Quartarone, E.; Mustarelli, P. & Magistris, A. (1998). PEO-based composite polymer electrolytes. Solid State Ionics, 110, 1-14.

Rajendran, S.; Kannan, R. & Mohendran O. (2001). An electrochemical investigation on PMMA/PVdF blend-based polymer electrolytes. *Mater. Lett.*, 49, 172-179.

Rajendran, S. & Uma, T. (2000). Conductivity studies on PVC/PMMA polymer blend electrolyte. *Mater. Lett.*, 44, 242-247.

Rhoo, H.J.; Kim, H.T.; Park, J. M. & Hwang, T. S. (1997). Ionic conduction in plasticized PVC/PMMA blend polymer electrolytes. *Electrochim. Acta*, 42, 1571-1579.

Saikia, D.; Hussain, A.M.P.; Kumar, A.; Singh, F. and Avasthi, D.K. (2006). Ionic conduction studies in Li $^{3+}$ ion irradiated (PVDF-HFP)-(PC+DEC)-$LiCF_3SO_3$ gel polymer electrolytes. *NIMB*, 244, 230-234.

Saikia, D. & Kumar, A. (2005). Ionic transport in P(VDF-HFP)-PMMA-$LiCF_3SO_3$-(PC+DEC)-SiO_2 composite gel polymer electrolyte. *Eur. Polym. J.*, 41, 563-568.

Scrosati, B.; Croce, F. & Panero, S. (2001). Progress in lithium polymer battery R&D. *J. Power Sources*, 100, 93-100.

Scrosati, B.; Croce, F. & Persi L. (2000). Impedance spectroscopy study of PEObased nanocomposite polymer electrolytes. *J. Electrochim Soc*, 147, 1718-1721.

Song, J.Y.; Wang, Y.Y. & Wan, C.C. (2000). Conductivity study of porous plasticized polymer electrolytes based on poly(vinylidene fluoride). A comparison with polypropylene separators. *J. Electrochem. Soc.*, 147, 3219-3225.

Steinhart, M.; Wendolff, J. H.; Greiner, A., Wehrspohn, R. B.; Nielsch, K.; Choi, J. & Gosele, U. (2002). Polymer Nanotubes by Wetting of Ordered Porous Templates. *Science*, 296, 1997.

Stephan, A.M. & Nahm, K.S. (2006). Review on composite polymer electrolytes for lithium batteries. Polymer, 47, 5952-5964.

Stephan, A.M.; Nahm, K.S.; Kumar, T.P.; Kulandainathan, M.A.; Ravi, G. & Wilson, J. (2006). Nanofiller incorporated poly(vinylidene fluoride–hexafluoropropylene) (PVdF–HFP) composite electrolytes for lithium batteries. J. Power Sources, 159, 1316-1321.

Stephan, A. M.; Nahm, K. S.; Prem Kumar, T.; Kulandainathan, M. A.; Ravi, G. & Wilson, J. (2006). Nanofiller incorporated poly(vinylidene fluoride–hexafluoropropylene) (PVdF–HFP) composite electrolytes for lithium batteries; *J. Power Sources*, 159,1316-1321.

Stephan, A.M.; Renganathan, N.G.; Kumar, T.P.; Pitchumani, R.; Shrisudersan, J. & Muniyandi, N. (2000). Ionic conductivity studies on plasticized PVC/PMMA blend polymer electrolyte containing $LiBF_4$ and $LiCF_3SO_3$. *Solid State Ionics*, 130, 123-132.

Stolarska, M.; Niedzicki, L.; Borkowska, R.; Zalewska, A. & Wieczorek, W. (2007) Structure, transport properties and interfacial stability of PVdF/HFP electrolytes containing modified inorganic filler. *Electrochim. Acta*, 53,1512-1517.

Subbiah, T.; Bhat, G.S.; Tock, R.W. Parameswaran, S. & Ramkumar, S.S. (2005). Electrospinning of nanofibers. *J. Appl. Polym. Sci.* 96, 557–569.

Sundararajan, S.; Bhushan, B., Namazu, T. & Isono, Y. (2002). Mechanical property measurements of nanoscale structures using an atomic forcemicroscope. *Ultramicroscopy*, 91(1–4):111–118.

Suthar, A. & Chase, G. (2001). Nanofibres in filter media. *Chem. Eng.*, 726, 26–28.

Tarascon, J.-M.; Gozdz, A.S.; Schmutz, C.; Shokoohi, F. & Warren, P.C. (1994). Performance of Bellcore's plastic rechargeable Li-ion batteries. *Solid State Ionics*, 86, 49-54.

Taylor, G.I. (1969). Electrically driven jets. *Proc R. Soc. London*, Ser A, 313, 453-475.

Li, D. & Xia, Y. (2004). Electrospinning of nanofibers: Reinventing the wheel?. *Adv. Matter.*, 16, 1151-1170.

Van de Zande, B.M.I.; Bohmer, M.R.; Fokkink, L.G.J. & Shonenberger. C. (1997). Aqueous gold sols of rod-shaped particles. *J. Phys. Chem.B*, 101, 852-854.

Wang, Z. & Tang, Z. (2004). A novel polymer electrolyte based on PMAML/PVDF-HFP blend. Electrochim. Acta, 49,1063-1068.

Wieczorek, W.; Stevens, J.R. & Florjanczyk, Z. (1996). Composite polymer based solid electrolytes. The Lewis acidbased approach. *Solid State Ionics*, 85, 67-72.

Wu, C.-G.; Lu, M.I.; Tsai, C.-C. & Chuang, H.-J. (2006). PVdF-HFP/ metal oxide nanocomposites: The matrices for highconducting, low-leakage porous polymer electrolytes. *J. Power Sources*, 159, 295-300.

Xi, J.; Qiu, X.; Li, J.; Tang, X.; Zhu, W. & Chen, L. (2006). PVDFPEO blends based microporous polymer electrolyte: effect of PEO on pore configurations and ionic conductivity. *J. Power Sources*, 157, 501-506.

Zhang, S.S.; Ervin, M.H.; Xu, K. & Jow, T.R. (2004). Microporous poly (acrylonitrile-methyl methacrylate) membrane as a separator of rechargeable lithium battery. *Electrochim. Acta*, 49, 3339-3345.

Zhang, Y.; Ouyang, H.; Lim, C.T.; Ramakrishnan, S. & Huang, Z.M. (2005). Electrospinning of gelatin fibers and gelatin/PCL composite fibrous scaffolds Electrospinning of gelatin fibers and gelatin/PCL composite fibrous scaffolds. *J. Biomed. Mater. Res. Part B: Appl. Biomater.*, 72B, 156–165.

3

Nanoreinforced Adhesives

Silvia G. Prolongo, María R. Gude and Alejandro Ureña
University Rey Juan Carlos
Spain

1. Introduction

1.1 Adhesive joints

An adhesive may be defined as a material which when applied to surfaces of materials can join them together and resist separation. The term adhesion is used when referring to the attraction between the substances, while the materials being joined are commonly referred to as substrates or adherends (Kinloch, 1987).

The adhesive properties of some substances have been used for thousands years. During Prehistory, for example, man has employed several plant resins as adhesives, either neat or with other materials to improve their properties (Regert, 2004; Wadley, 2005). However, the science and technology of adhesion and adhesives has not progressed significantly until the middle of 1940s (Kinloch, 1987), when the II World War promoted the development of different technologies, between them, the polymer science, closely related to the adhesives. Since then, great advances have been carried out in all aspects of adhesives technology. Nowadays, the practical demands upon adhesives have changed. The main aim is no longer to achieve simply strong bonds; that has been mastered. The targets are durability under a variety of harsh environments, together with enhanced toughness and, in some specialised cases, adequate performance at relatively high temperatures (Allen, 2003).

Adhesive joints offer many advantages, with regard to other ways of joining materials, like welding, brazing, riveting or bolting:

- The ability to join any type of material, even to make dissimilar joints.
- An improved stress distribution in the joint, reducing the stress concentrations caused by rivets or bolts.
- Adhesive boding can potentially reduce the weight of a structure, mainly because of the ability to join thin-sheet materials efficiently.
- An improvement in the corrosion resistance between dissimilar materials, and also with regard to the use of mechanical fasteners.
- Adhesive joints can be used for sealing, insulating (heat and electricity) and damping vibrations.
- Less expensive than other types of joints, especially when bonding large areas.

They have some drawbacks:

- Require careful surface preparation of adherends, especially in order to attain a long service-life from adhesive joints in hostile environments.
- Limitation on upper service temperature.
- The strength of adhesive joints is relatively low compared to mechanical joints.

- Disassembly cannot be carried out without incurring significant damage to the joint.
- Non-destructive test methods for adhesive joints are relatively limited compared to those used with other fastening methods.
- Heat and pressure may be required for assembly.
- Jigs and fixtures may be required for assembly.

1.2 Mechanisms of adhesion

The mechanisms or theories of adhesion try to explain how an adhesive bond is formed between two materials. There is not a universal mechanism to explain all bonds. In fact, in adhesive joints there is usually more than one mechanism contributing to the adhesive strength.

Mechanical interlocking: This theory proposes that mechanical interlocking of the adhesive into the irregularities of the substrate surface is the major source of intrinsic adhesion (Kinloch, 1987). This means that the adhesion is directly related to the substrate roughness. Mechanical, like grit blasting, or chemical roughening, like anodizing, generate different values of surface roughness and also different features. The size and shape of these features has an influence on the adhesion, providing a tortuous path which prevents the separation of the adhesive from the adherend (Fisher, 2005). However, this theory is not able to explain the good adhesion strength attained in some cases between smooth surfaces.

Diffusion theory: The diffusion of segments and chain ends of polymers was suggested as a mechanism for the adhesion of similar polymers. This requires that the macromolecules or chain segments of polymers (adhesive and substrate) possess sufficient mobility and are mutually soluble (Kinloch, 1987). Two polymers, or a polymer and a solvent, are miscible when they have similar solubility parameters. This theory demonstrates the autohesion of plastics using hot or solvent welding, and also explains why polymers with very different solubility parameters do not present good adhesion between them.

Electronic theory: In adhesive joints of metallic substrates, the different nature of the materials (metal and polymer) facilitates the transfer of electrons from the metal to the adhesive, in order to equilibrate the Fermi levels of both metal and polymer. The result is the creation of an electric double layer at the interface (Allen, 2003). The existence of that layer is easy to demonstrate. For example, it causes the flashes of light and noise which occur when an adhesive tape is stripped form a solid surface (Allen, 2003). However, it is not clear if such electrostatic forces promote an increase of the joint strength or they are a result of that increase (Kinloch, 1987).

Adsorption theory: The adsorption theory of adhesion is the most widely applicable theory and proposes that, provided sufficiently intimate molecular contact is achieved at the interface, the materials will adhere because of the interatomic and intermolecular forces which are established between the atoms and molecules in the surfaces of the adhesive and substrate. This means that the adhesive has to spread over the solid surface. A liquid wets a solid when the contact angle between a liquid drop and a solid surface is lower than 90°; in other words, when the surface free energy of the surface is high than the surface tension of the liquid. The Young equation (Young, 1805) describes the relationship between surface free energy, γ_{SV}, and contact angle, θ, in the three-phase contact point (Figure 1):

$$\gamma_{SV} = \gamma_{SL} + \gamma_{LV} \cdot \cos\theta \quad (1)$$

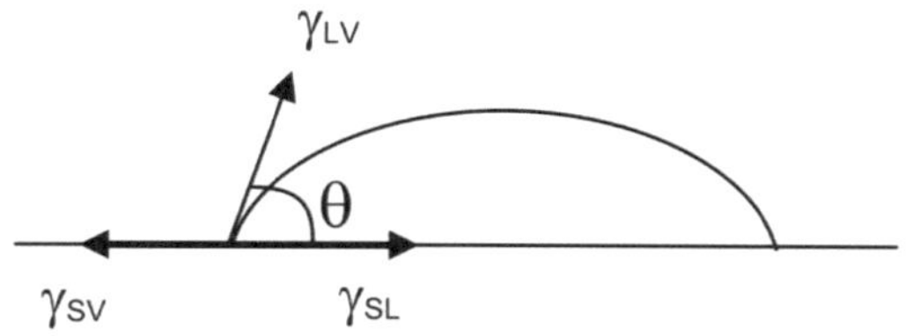

Fig. 1. Schematic representation of the interactions existing when a liquid wets a surface.

The work of adhesion, W_A, is defined as the difference between the sum of the surface free energies of the solid and liquid phases and the interfacial free energy:

$$W_A = \gamma_{SV} + \gamma_{LV} - \gamma_{SL} \tag{2}$$

Combining equations (1) and (2) the relationship between the work of adhesion and the contact angle can be obtained:

$$W_A = \gamma_{LV}(1 + \cos\theta) \tag{3}$$

This equation is very useful to estimate the strength of an adhesive joint, taking into account that the higher the work of adhesion, the higher the adhesive strength. The surface energy of a liquid, γ_{LV}, is a known parameter, and the contact angle is very easy to determine.

Weak boundary layer theory: This is not strictly a mechanism of adhesion, but it is a theory which allows explaining the lack of adhesion in many cases. Oxide layers, low molecular weight species, oils and other contaminants are weak boundary layers, poorly adhered to the substrate. If any of such layers is present on the surface of the adherend when bonding, the joint will fail between this layer and the substrate with low strength.

1.3 Surface preparation

In order to achieve good adhesive properties, one requirement is that the adherends must present adequate surface properties. As it was mentioned above, the surface free energy of the substrate should be higher than the surface tension of the adhesive. In some cases, especially when the adherends are polymers or polymer matrix composites, the surfaces have to be modified to increase their surface free energy. There are other reasons to apply surface treatments before adhesive bonding, like removing weak boundary layers, increasing the surface roughness, creating specific chemical groups or homogenizing the surfaces to improve the reproducibility of the results.

There are several types of surface treatments available: mechanical (grit blasting), chemical (anodizing, acid etching) or energetic (plasma, laser). In every case the most suitable treatment has to be selected, taking into account the material, shape and size of the adherend, the adhesive and other circumstances related to the manufacturing.

2. Nanoreinforced adhesives: potential advantages

Polymer nanocomposites manufactured from an effective dispersion of nanofillers (nanoparticles, nanofibres, nanotubes, etc) into a polymeric matrix (thermoplastic or thermosetting) have been proposed as a powerful tool for generating new multifunctional materials with improved mechanical, physical and chemical properties. Due to their small size and large surface area, nanoreinforcements would be able to provide unique combination of properties, which are not possible to be reached for conventional fillers with sizes in the micrometer range. Of particular importance, it is the requirement of achieving a good distribution of the nanofiller in the polymer, in order to obtain the pursued increases in properties, without loss of other characteristics of the nanocomposite (i.e. processability) because of the high tendency to particle aggregation.

The development and commercialization of nanoparticles such as nanoclays, carbon nanotubes (CNT) or nanofibers (CNF), inorganic nanoparticles and other, offer new possibilities to tailor adhesives in the nanoscale range. Due to the large surface area of the nanosized particles only small amounts are needed to cause significant changes in the resulting properties of the nanocomposite adhesives. It could provide a new generation of structural adhesives with combination of thermal, electrical or thermomechanical properties which also provide higher environmental durability because of their lower water absorption and enhaced ageing properties.

The potential of nanofillers for adhesive formulations is promising, and their effects, most of them based on the chemical and physical interactions developed between the nanoparticle surface and the resin at the reinforcement-matrix interfaces, can be classified on the following groups:

a) Mechanical properties. Many of the new applications of structural adhesives (i.e. transportation application such as aircraft industry) require stable materials under service conditions which imply high temperature environments, beside to be resistant to failure resulting from vibration and fatigue loading. The addition of nanofillers to base adhesive formulations generally increases their modulus and mechanical strength. However, the main objective in these cases is to increase fracture toughness without loss of adhesive characteristics. Research in improving the fracture toughness of brittle polymers (i.e. thermosets) using nanoreinforcements holds great promise. Although the toughness of these brittle resins is usually increased by means of the addition of rubber fillers, other mechanical properties are usually degraded. For example, the improvement of the toughness of epoxy resins by incorporating nanofillers (i.e CNTs) in the resin system has been reported by numerous researchers. The participation of new mechanisms of fracture energy consumption generated by the interaction between cracks and nanofillers (crack deflection, crack bridging, fiber pull-out, etc) is considered responsible of the toughening effect associated to the nanoreinforcement addition.

Gojny et al. (2005) have published an overview in *Composites Science and Technology* over the influence of nanofiller on the fracture toughness of brittle epoxy resins and the related micromechanical mechanisms. These authors consider toughening mechanisms participate at two different dimensional levels: 1) micro-mechanical mechanisms, such as crack deflection at agglomerates, crack pinning, crack blunting and the extension of the plastic deformation zone and 2) nano-mechanical mechanisms, such as interfacial debonding, pull-out and crack bridging with participation of the nano-sized structure of CNTs.

Improvements in toughness with addition of low contents of nanofiller have been reported for numerous authors, not only in the case of nanoreinforced polymers but also in situations in which the nanoreinforced matrix is included in a more complex system such as continuous fiber reinforced composites. The manufacture of multiscale composites by incorporation of nanofiller inside the matrix composite is also considerate as a potential method to improve those properties which are highly depended on the matrix (among them, thoughness). Manufacture of these composites requires that nanoreinforced resins keep their rheological and wetting characteristics to make possible the infiltration of fibre performs. Both types of properties are also required by nanoreinforced adhesives. In this research line, R. Sadeghian et al. (2006) have manufactured by Vaccum Resin Transfer Moulding (VARTM) hybrid composites constituted by CNF nanoreinforced polyester/glass fiber, improving the mode-I delamination resistance G_{IC} about 100 %when CNF concentration up to 1 wt% is incorporated in the polyester matrix. These authors characterized also the viscosity dependence on the CNF concentration noticing a notable increase in resin viscosity when we CNF concentration raised from 1 to 1.5 wt%. This problem, which limits the processability of multiscale composites by infiltration methods, must be considered also in the case of nanoreinforced adhesives.

b) Electrical properties. In relation with the electrical properties, one of the most interesting fields of application is the incorporation of carbon nanotubes or carbon nanofiber as fillers in electrical conductive adhesives. The aim is to improve the performance of conductive adhesives in comparison to common products. An increase of electrical conductivity is observed in these kinds of nanocomposites with increasing CNT or CNF contents, showing clear percolation behaviour. The conductivities of the many of the developed composites show magnitudes below materials like copper. The percolation threshold values depend on the type of nanoreinforcement, being lower in the case of CNT than for CNF. The method of dispersion also has a dramatic influence on the conductivities of the nanocomposites, both for the effectiveness of the dispersion and for the effect of the applied dispersion method (mechanic stirring, ultrasonication, calandering, etc) on the nanorinforcement integrity. High energetic dispersion processes may damage the nanofillers decreasing their aspect ratio, which affect to the percolation behaviour.

The electrical conductivity is usually detrimentally influenced by the application of functionalization treatments to the carbon nanoreinforcement. Although these kinds of treatments (oxidation, amination, fluoridation, etc) usually improve the nanofiller dispersion and favour the formation of covalent bond with the polymer matrix, they are always connected to structural changes (i.e rupture of the CNTs, resulting in a reduced aspect ratio) and, therefore, to a reduction of the electrical conductivity. Figure 2 shows the change in specific conductivity with the percentage of nanofiller for two epoxy nanocomposites reinforced with double wall CNT (untreated and aminofuntionalizated), compared with the effect of the addition of carbon black. The lowest percolation threshold value is reached for the Epoxy/DWCNT; an increase in this value is observed in the case of the Epoxy/DWCNT-NH_2 because of the damage of the nanofillers during the funtionalization treatment.

c) Thermal Properties and Thermal Stability. Thermal stability is one of the most important properties of polymer nanocomposites for potential applications as functional or structural components at elevated temperatures. Thermal stability and degradation behaviour of nanocomposites have been studied by several researchers. For example, Sarathi et al. (2007) showed that the addition of nanoclays (i.e organo-montmorillonites) in epoxy increases the heat deflection temperature up to a critical percentage of nanoclay in epoxy, about 5 wt % above which it reaches a steady state. Addition of nanoclays also improves the thermal stability reducing, in relation with unreinforced epoxy, the loss of weight measured during a thermogravimetric analysis. Decomposition temperatures of nanocomposites generally increased with increasing nanofiller contents, indicating that the thermal decomposition of the matrix is retarded by the presence of the nanoreinforcement. These results may be

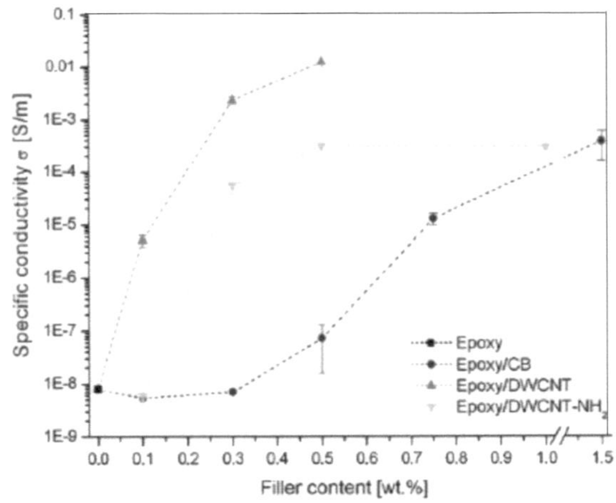

Fig. 2. Electrical conductivity of the nanocomposites as a function of filler type and content (Gojny et al., 2005).

attributed to the physical barrier effect, having experimental proofs that not only nanoclays but also CNTs impede the propagation of decomposition reactions in the nanocomposites (Kim & Kim, 2006).

Other thermal property that can be controlled by the addition of low amount of nanoparticles is the coefficient of thermal expansion (CTE). In the specific case of thermosetting resins, the CTE values can be differenced below and above the glass transition temperature (T_g). Considering the application of these resins as adhesive, the most useful CTE concerns the temperature below T_g, since adhesive would lose most of its mechanical properties at temperatures higher than T_g. Since CNT shows negative CTEs values (longitudinal CTE of SWNTs has been estimated to be -12×10^{-6} K^{-1} while a transverse CTE was predicted to be -1.5×10^{-6} K^{-1}) (Kwon et al., 2004; Jiang et al., 2004), the aditions of SWCNTs could lead to a lower CTE in SWNT nanocomposites. This effect will be so much remarkable when dispersion of nanoreinforcement is more effective. For example, S. Wang et al. have shown that the CTE of the functionalized SWCNTs-epoxy composites below T_g could be diminished by 52 and 42% by the incorporation of 1% by weight of nanotubes which were subjected to simple functionalization treatments (mechanical chopping and oxidization) to improve their dispersion (Wang et al., 2007).

The addition of some kind of nanofillers (i.e. CNT) can also increase the thermal conductivity of nanocomposites. Heimann et al. (2008), have shown that the thermal conductivity rises almost linearly with rising content of CNT in the polymer matrix (epoxy matrix). The composite with the highest portion of CNT tested (10 wt %) points out an enhancement nearly 4.4 times compared to the matrix without CNT (Figure 3); no influence of the method of dispersion could be observed.

d) Gas and Liquid Barrier Properties. The barrier properties of the nanocomposites are considerably improved as compared to that of pure or macroscopically filled polymers. The reason for the dramatic drop in permeability has been attributed to the existence of well-dispersed nanoreinforcements with a large aspect ratio (nanoclays, CNT, CNF). Most studies on polymer nanocomposite barrier properties are based on the tortuous pathway concept (Nielsen, 1967), where the nanofiller phase is assumed to be impermeable for gas

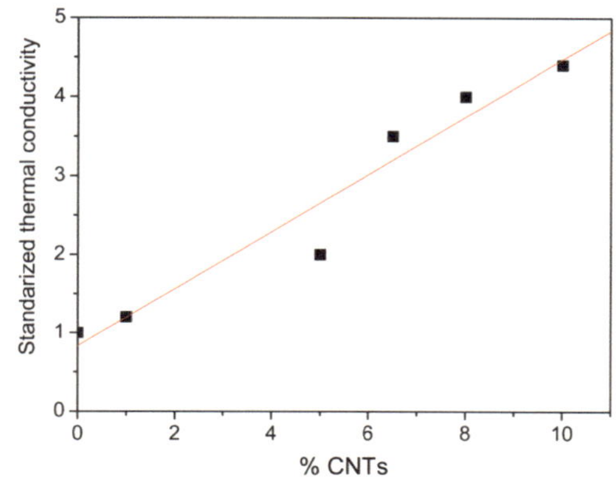

Fig. 3. Standardized thermal conductivity of epoxy and epoxy CNT reinforced nanocomposites (Heimann et al., 2008).

and liquid molecules, which forces the gas molecules to follow a tortuous path thereby increasing the effective path length for diffusion.

One of the potential advantages of nanoreinforced adhesives related with these barrier properties is use as a moisture barrier. Moisture permeation is a measure of the ability of a material to resist moisture to penetrate through its thickness. Several important parameters must be considered, including the volume fraction (V_f) and the aspect ratio of the nanoparticles. Higher aspect ratios provide greater barrier improvement according to the equation:

$$\frac{P_c}{P_0} = \frac{1}{1+(L/2W)V_f} \tag{4}$$

where P_c and P_0 are the permeability coefficients of the nanocomposite and the neat polymer, respectively; *L/W* is the aspect ratio of the nanofiller, defining the term $(1 + (L/2W)Vf)$ as the tortuosity factor. Reductions in moisture permeability in the range of 57-86% for epoxy resins nanoreinforced with nanoclays have been determined, deducing that the very large aspect ratio of the clay platelets is the main factor to reach an effectively increased the moisture penetration path, which is responsible for the reduced permeability. (Kim et al., 2005).

Although nanoplatelets have been shown as very effective gas and liquid barriers in polymeric matrices, recent studies on the transport properties, sorption and diffusion of water vapour carried out on epoxy resin filled with multi-walled carbon nanotubes, have also showed the improved effect of the barrier properties with increasing MWCNT concentration (Guadagno et al., 2009)

Water absorption is other of the properties of polymer which can be improved by the dispersion of nanofillers. This improvement can be significant for resins which preset pourer behaviour under prolonged water exposure, such as epoxy. The substantial decrease of permeability brought about by nanocomposite structures is a major advantage of polymer–clay nanocomposites, due to the tortuous path presented by high aspect ratio clay. The Toyota researchers determined that the rate of water absorption in their polyamide 6-clay nanocomposite was reduced by 40% compared with the pristine polymer. However, these results are more contradictory in the case epoxy matrix nanocomposites where only the rate of absorption is reduced, while the equilibrium water uptake is relatively unaffected.

In spite of those potential advantages of the nanoreinforced adhesives, the incorporation of nanofillers into the adhesive may originate problems associated to the increase in viscosity and the modification of the wetting behaviour with regard to the neat adhesive. It is enough shown that the addition of nanoparticles into liquid resins increases their viscosity; and for the particular case of CNTs, it has been found that increase in the viscosity of the nanocomposites filled with CNTs was much higher than increase in the viscosity of polymer composites filled with carbon fibers (CF) or carbon black (CB). Beside, nanocomposites filled with functionalized CNTs, that have better dispersion of the CNT, show a complex viscosity at low frequency.

3. Adhesives reinforced with inorganic nanoparticles

Due to the novelty of the nanocomposites, there are not much scientific researches which analyse the viability for the use of nanoreinforced polymers as adhesives. Further, the most of the found publications about nanoreinforced adhesives are centred in the reinforcement of epoxy adhesives with different kinds of carbon nanotubes. This is probably associated to the fact that the epoxy resins reinforced with carbon nanotubes are being currently the most studied nanocomposites by the scientific community. Even so, several researches have been found about the reinforcement of adhesives with inorganic nanoparticles. The nature of the added nanofiller is varied, being nano-sized particles of silica and alumina some of the most used. Also, the published results are varied. Among other reasons, the study of the adhesive ability of a resin, modified or not, depends on several factors, such as the nature of the adherends and the applied surface treatments on them, the geometry of the joints (single lap, butt, T-joing, etc) and the type of test carried out to determine the strength of the joints (lap shear, peel, pull off, wedge, etc). Besides all these variables, new ones are added, which are associated with the own manufacture of nanocomposites, like the dispersion techniques and methods applied, the previous chemical treatments carried out over nanofiller surfaces and the geometry, structure and other characteristics own of nanoreinforcements, among others. Despite of this, this section of the chapter contains a summary of some of the most interesting published results on adhesives reinforced with nano-sized particles of inorganic nature.

The most of bibliography found about the addition of nanofiller into the adhesives is mainly based on epoxy adhesives. Compared with other adhesives, epoxy ones produce joints with high shear strength and excellent creep properties. The delamination resistance and impact of the epoxy joints are, however, relatively low. Due to their good properties, these adhesives are frequently used in high responsibility applications where their relative high cost is not as relevant. It is expected that the advantages obtained by the addition of the relative expensive nano-scale filler compensate the increase of price of the adhesive. In fact, the addition of nanofiller into epoxy adhesives could enhance the main debilities of the epoxy joints, such as their strength and toughness. Moreover, it should increase the electrical conductivity of these resins, becoming from isolator to conductive materials. This is especially interesting because of the epoxy adhesives are frequently used for joining metals and carbon fiber reinforced composites. Both are electrical conductors and it will be very profitable that their joint remains this electrical behaviour, using an electrical conductive adhesive.

Lanlan Zhai and collaborators have published several researches on the effect of the addition of alumina nanoparticles in epoxy adhesives (Zhai et al., 2006; Zhai et al., 2008), analysing their pull-off strength over steel. Some of the extracted results are shown in Figure 4a, together with a scanning electron micrograph of the alumina nanoparticles added (Figure 4b) and image of the nanoreinforced epoxy surface captured by transmission electron microscopy (Figure 4c).

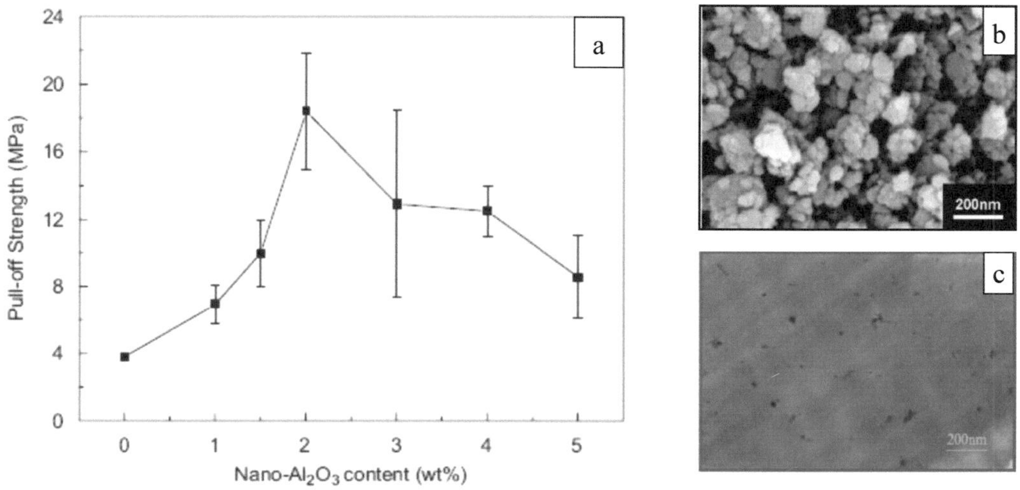

Fig. 4. (a) Effect of nano-alumina additive content on the adhesion strength, (b) SEM micrograph of alumina nanoparticles, (c) TEM micrograph of the nanoreinforced epoxy surface (Zhai et al., 2008).

It is probed that the addition of alumina nanoparticles causes a drastic increase of the adhesive strength, which reaches the maximum value when the nanofiller content is 2 wt %. The pull-off strength of this nanoreinforced adhesive is almost five times higher than that of pure epoxy adhesive. This increase is intimately associated with a change in failure mode, which becomes from interfacial failure for non modified adhesive to a mixed cohesive-interfacial failure mode for the joints bonded with nanoreinforced adhesives. At high nano-alumina contents, the adhesive strength falls because the surface wetting ability of the adhesive is reduced by the increase of its viscosity.

The modification of adhesives by the addition of alumina nanoparticles has been also studied in epoxy-based film adhesives, which are incrementing their use for joining aluminium and polymer composite parts in the aircraft industry. These applications typically require the modification of epoxy formulations to increase the adhesion, toughness and peel strength of the joints, because they are usually subjected to vibration and fatigue loads besides high service temperature environments. The most widely used modifiers of epoxy-based film adhesives consist of reactive liquid elastomers, which increase the toughness of the joints but limit their modulus, thermal stability and hot-wet performance. Also, the phase separation of the rubber can imply a reduction in shear strength. Gilbert et al. (2003) confirmed that the addition of 5 wt % nano-alumina into an epoxy formulation that was filmed on polyester random mat scrim achieved increases in the peel strength of almost 50% and in shear strength of 15% for joints of aluminium substrates. Contradictory results were obtained in the measurements of mode I and mode II fracture toughness of nanoreinforced epoxy adhesives when the nature of substrate was carbon fiber/epoxy laminates. They were strongly dependent on whether the composite systems were cocured or bonded. Mode I fracture toughness of the cocured composites increased, while values for the bonded systems drastically fell down by the addition of the nanofiller.

Other kinds of inorganic nanoparticles have been added to adhesives in order to enhance their properties and behaviour. In particular, several researches have been published using nanosized particles of silica. It is well known that thermophysical and thermomechanical properties, such as thermal conductivity, coefficient of thermal expansion, tensile and breaking strength of epoxy resins improve considerably due to the silicate nanopowder into the matrix polymer. Also, the addition of low concentrations of nanosilica particles to a typical rubber toughened epoxy adhesive leads to very significant increases in the toughness and single lap shear strength of the joints (Klug & Seferis, 1999; Kinloch et al., 2003). This increase is related to the enhancing of the plastic deformation of the epoxy matrix due to the appearance of different toughness mechanisms, such as crack deflection and crack twisting around the nanoparticles. On the other hand, Bhowmik et al. (2009) probed that the exposure under high-energy radiation of a nanosilica reinforced epoxy resin causes an increase of its crosslinking density, essentially affecting the overall behaviour and mechanical properties of the nanoreinforced polymer. In fact, they report an increase of more than 100% of the lap shear strength of the titanium joints when the adhesive was reinforced with nano-silicate particles and exposed to high energy radiation regard to the adhesive strength of non-modified adhesive. Patel et al. (2006) analysed the strength of acrylic-silica hybrid adhesives, prepared in situ by sol-gel, through both peel and lap shear tests using aluminium, biaxially oriented polypropylene (PP) and wood as substrates. They found enhancements on the joint strengths with hybrid nanoreinforced adhesives compared to neat acrylic ones, which were associated to changes in the failure mode from interfacial failure for neat acrylic adhesive to slip-stick failure in the case of the hybrid composites. As with alumina nanoparticles, the joint strength increases with increase in nanofiller loading up to certain content due to the higher cohesive strength and higher interaction between the substrates and the nanoreinforced adhesives. High contents of nanoreinforcements imply the fall of adhesive and mechanical properties because of both an increase of the adhesive viscosity and problems to disperse rightly the nanofiller, appearing micro-sized agglomerations.

Lanlan Zhai et al. (2006) published a comparative study about the effectiveness of different kind of inorganic nanoparticles on the stickiness of epoxy adhesives. In particular, they used nanoparticles of Al_2O_3 (whose average diameter was 80 nm), nano-$CaCO_3$ (with 40 - 80 nm of diameter) and nano-SiO_2 (whose size was 10 – 20 nm in diameter). These nanofillers were added in 2 wt % regard to the epoxy adhesive mass. Low carbon steel sheets were used as adherends, which were abraded with different silicon carbide paper, polished to an optical flatness and finally degreased and dried. As shown in Figure 5, the adhesion strength, measured through pull-off adhesion test, of the epoxy adhesives incorporating three kinds of nanoparticles was greatly improved compared with pure epoxy adhesive. The highest increase is obtained by the adhesive reinforced with nanoparticles of alumina, from 3.4 to 18.4 MPa, while the strength of the nano-$CaCO_3$ modified epoxy adhesive was as much as that of nano-SiO_2 modified system, no more than 12 MPa.

The increase of adhesive strength by the addition of nanofillers into the adhesives implies a stronger anchoring associated to changes on the physical and chemical properties of the modified adhesives. The different enhancements found as function of nanoparticle nature may be attributed to the chemical properties of nanoparticles, which may have influence in

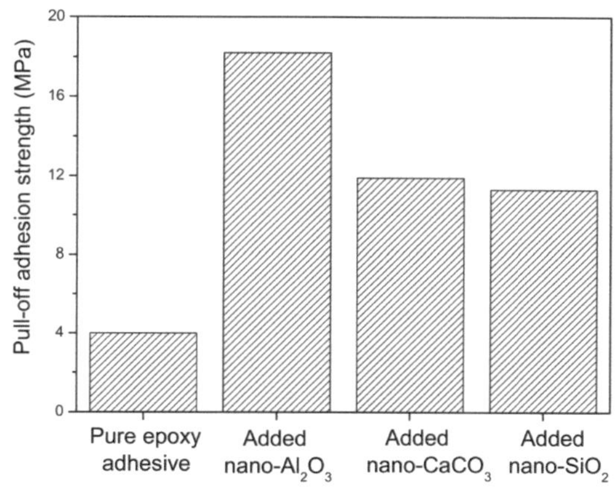

Fig. 5. The pull-off adhesion strength of epoxy adhesives reinforced with different inorganic nanoparticles (Zhai et al., 2006).

the chemical interaction of the surfaces of steel and epoxy adhesives, producing some chemical bonds on the interface and therefore enhancing the adhesion strength. The formation of bridges between the adhesive and the adherends was confirmed by the analysis of the interface morphology through scanning electron microscopy. Figure 6a shows the morphology of the interface of steel and epoxy adhesive reinforced with 2% nano-Al_2O_3. It is possible to observe some epoxy fibers connecting with the steel substrate, which implied that both surfaces had contacted closely. The gap between steel and epoxy adhesive was likely to result from abrading and polishing of the metallographic specimen. Figure 6b shows the morphology of the surface boundary of steel and epoxy adhesive without nanoparticles, which is radically different. The gap is evidently wider than that of nanoreinforced joint, implying a weaker adhesion. For this reason, it was easy for the contamination to fall into the gap in the process of preparing the metallographic specimen.

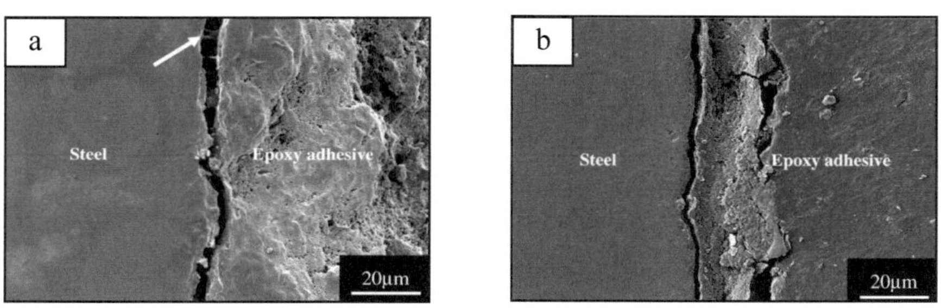

Fig. 6. SEM micrographs of steel - epoxy interface morphology with adhesive nanoreinforced with nanoparticles of alumina (a) and with neat epoxy adhesive (b) (Zhai et al., 2006).

Polyhedral-oligomeric-silsesquioxanes (POSS) are other kind of inorganic silica particles which are actually commercialized, being nanocages of 1.5 nm in size with organic substituents. The substituents can be inactive, physically compatibles with the matrix, or reactive, which promote curing or grafting reactions with the polymer. The effect of the addition of low amounts of POSS into epoxy adhesives is strongly dependent on the nanostructure of the epoxy/POSS network, which in turn depends on the functional groups (reactive or nonreactive) of the POSS (Dodiuk et al., 2005). The highest values of shear and peel strength are obtained when the crosslinking degree of nanoreinforced adhesive is high. Due to the large surface area of POSS, only relatively small amounts (< 4 wt %) are needed to cause significant changes on the properties of the epoxy resin. In fact, excess of POSS amount implies the plasticization of the matrix, decreasing the joint strength.

Finally, Patel et al. (Patel et al., 2006) analysed the effect of the addition of an organically modified montmorillonite nanoclay, commonly named Cloisite 10A, on the joint strength bonded with a very soft acrylic adhesive. With a high surface energy adherend, like aluminium, clay nanoreinforced adhesives displayed gradual increment in peel strength with the increase of filler content, measuring enhancements of up 45% regard to neat adhesive. However, the observed improvement with low surface energy substrate, polypropylene, was lower. This indicates a favourable interaction between the silicates and aluminium substrate. The lap shear strength spectacularly increases with the nanoclay addition, up to 146, 130 and 142% in joints of aluminium-aluminium, wood-wood and polypropylene-polypropylene, respectively. Besides the adhesive properties, the addition of nanoclay into the adhesives enhances their barrier performance. This is especially interesting in the use of polyurethane adhesives (Osman et al., 2003). They are commonly used in producing laminates for food packing due to their flexibility and wide application temperature range. However, their use is limited due to their low barrier performance, as oxygen and humidity barriers. The inclusion of small volume fractions of montmorillonite in polyurethane adhesives decreases their gas transmission rate due to the impermeability of the inorganic nanoparticles.

4. Adhesives reinforced with carbon nanotubes

One of the nanosized filler which has generated higher expectation are the carbon nanotubes (CNTs). CNTs, composed of one grapheme layer (SWCNT) or many grapheme layers (MWCNT), are a novel crystalline carbon form. The growing interest of these materials is associated with their spectacular and new properties theoretically expected. Independently of nanotube type and its diameter, the value of the plane elastic modulus should reach the reported one for the graphite, which is 1.06 TPa (Nelly, 1981). Its yield strength is still unknown, although it must be also similar to that calculated for the graphite, which is estimated around 130 GPa (Perepelkin, 1972). The mechanical strength of MWCNTs has could be measured by Atomic Force Microscopy, giving values around 14 GPa (Wong et al., 1997). Due to its very low diameter and in spite of its high stiffness, the carbon nanotubes present very high flexibility, bending fully reversible up to 110° critical angle for SWCNT (Salvetat et al., 1999). In addition to their mechanical properties, the nanotubes present very interesting physical properties. They have metallic and semiconducting electrical character, field emission properties and high thermal conductivity, among others. Therefore, these materials have been widely researched as nanofiller in the manufacturing of composites, using different matrix materials, polymers, ceramics and metals. As it was said above, research in improving the fracture toughness of brittle thermosets using nanotechnology holds great promise.

As it is well known, in order to reach the best properties of nanocomposite, CNTs must be totally dispersed into the composite matrix. For it, numerous alternatives have been proposed (Xie et al., 2005; Vaisman et al., 2006; Prolongo et al., 2008) such as the use of solvents and surfactants to disperse rightly the CNTs with the epoxy monomer. Other proposals are based on the application of high mixing forces, using high shear mechanical mixers or ultrasonic. One of the last proposals with higher success is the use of a three roll mini-calander (Gojny et al., 2005). This procedure consists of passing the CNT/epoxy mixture through several rotating cylinders with a very small gap between them, around 50 - 5μm.

Among other difficulties, the increase of the viscosity of epoxy monomer by the addition of CNTs is high due to the very high specific area of these nanofillers (200 - 700 g/m^2). This usually hinders the manufacturing of the nanocomposite. S. G. Prolongo et al., in *Journal of Adhesion Science and Technology*, (2009) analysed the rheological behaviour of mixtures of epoxy monomer with different contents of MWCNTs. The results are shown in the Figure 7. The used MWCNTs were partially functionalized with amino groups (0.5 wt% NH_2). For this reason, the nanoreinforced mixtures were thermally treated in order to enhance the chemical reaction between oxirane rings of epoxy monomer and amino groups of the nanotubes.

In spite of the viscosity increase, the shear rheological behaviour seems remaining constant. Both neat and nanoreinforced epoxy resins show Newtonian behaviour at the high shear rate applied and relative low temperature (< 70 ºC). This is explained by the preferential orientation of the nanoreinforcement in the flow rate at high shear rates. Other authors (Hyun et al., 2001) reported that the addition of a very small amount of nanotubes induce non-Newtonian behaviour at very low shear rate, which is probably associated with the non-orientation of nanotubes. On the other hand, at relative high contents of CNTs, the application of a thermal pre-curing treatment to the nanoreinforced mixtures implies an increase of their viscosity, indicating that the chemical reaction between epoxy monomer and amino-functionalized carbon nanotubes occurs. The more CNTs content is added, more amine groups anchored to nanotube react, decreasing the mobility of epoxy molecules and therefore increasing their shear viscosity.

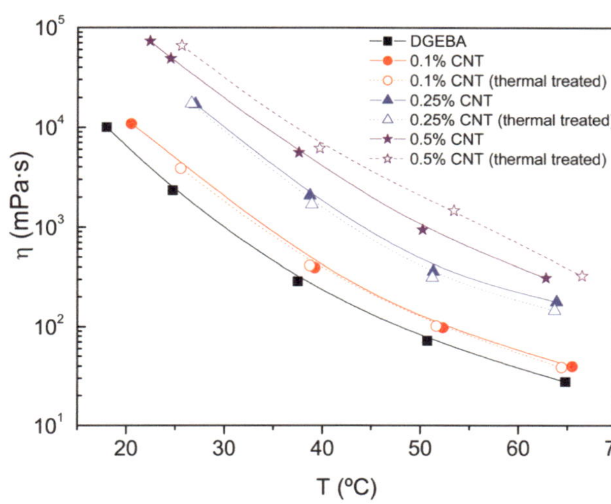

Fig. 7. Shear viscosity versus temperature plots for non-cured epoxy monomer (DGEBA, squares) and the epoxy mixtures with 0.1 (circles), 0.25 (triangles) and 0.5 wt % (stars) amino-functionalized MWCNTs non-thermal treated (solid lines) and precured at 130 ºC for 1 hour (dot lines) (Prolongo et al., 2009).

The addition of CNTs to epoxy resins modifies many of their properties, but this chapter is centred in the application of these nanoreinforced composites as adhesives. Several works have been found about the addition of CNTs on epoxy adhesives in order to enhance the mechanical strength and toughness of the bonded joints. Suzhu Yu et al. (2009) studied the mechanical behaviour and durability in humid environments of the A2024-T3 aluminium joints bonded with an epoxy adhesive reinforced with MWCNTs. Figure 8 shows some of the obtained results.

The wedge test has a relatively high stress concentration at or near the interface of the joint and is sensible to environmental attack. Therefore, it is usually used to provide quantitative durability data for a joint. As it can be observed in the figure, the crack developed in three steps for almost all the specimens: initial crack, crack propagation in about 3 - 8 h of immersion, and crack propagation after the first 3 - 8 h of immersion. The initial crack length was reduced up 70.3% for the epoxy resins randomly reinforced with 0.5 wt % CNTs compared to the obtained for the joint with neat epoxy adhesive. The decrease of the initial crack length occurs from 0.5 to 1 wt % CNTs, then the length increases with increasing CNTs fraction from 1 to 5 wt %. Obviously, the addition of CNTs into the epoxy adhesive causes a significant improvement in the bond strength of the joints, which is attributed to the excellent properties of the nanofiller. From a theoretical point of view, the strength of the adhesives should be monotonously increased with increasing CNT loading; thus, the initial crack length of the specimen would have monotonously decreased with the nanotube fractions. This is true at low CNTs contents (< 1 wt %). At higher contents, adverse effects of CNTs might have resulted from aggregation and poor dispersion of the nanofiller into epoxy matrix. The agglomerates can act as defects and reduce the strength of the adhesives. The propagation crack rate at first 3 - 8 h of immersion is also reduced by the addition of CNTs. In fact, the joint bonded with neat epoxy adhesive failed and broke after 3 h of immersion in water. This suggests that the water resistance of the adhesive increases with the CNT loading. The nanoreinforced epoxies must be able to resist water attack, so its adverse effects on the strength and durability of joints are not so significant. This is explained because carbon nanotubes are hydrophobic in nature and therefore their addition into the adhesive enhances the water resistance of the joints. The failure mode of the joint with neat epoxy adhesive is cohesive failure, referring to crack propagation on the adhesive (figure 8b). Interestingly, for the joints with CNT filled epoxy adhesives, more interfacial failure, referring to crack propagation on the adhesive-adherend interface, is developed with increasing nanofiller content. In fact, for epoxy adhesive reinforced with 5 wt % CNTs, only one surface is covered with adhesive in most areas; the other one mainly showed the metal surface.

Several works (Hsiao et al., 2003; Meguid & Sun, 2004) have studied the mechanical strength of CNT reinforced epoxy adhesives to join carbon fiber reinforced polymer (CFRP) composites. The shear strength increased by 31.2 and 45.6% when 1 and 5 wt % MWCNTs, respectively, were added in the epoxy system (Hsiao et al., 2003). These increments are associated with the enhanced mechanical properties of the nanoreinforced adhesives and the change of the failure mode of joints. The fracture of joints bonded with non-modified epoxy adhesive occurred at the epoxy along the bonding interface. In fact, no significant damages were observed on the composite adherends. In contrast, the failure observed for nanoreinforced joints was cohesive in the adherends. The nanotubes effectively transferred the load to the graphite fibers in the adherends and the failure was in the composite. For this reason, the graphite fibers of the composite adherends were highly damaged after the test.

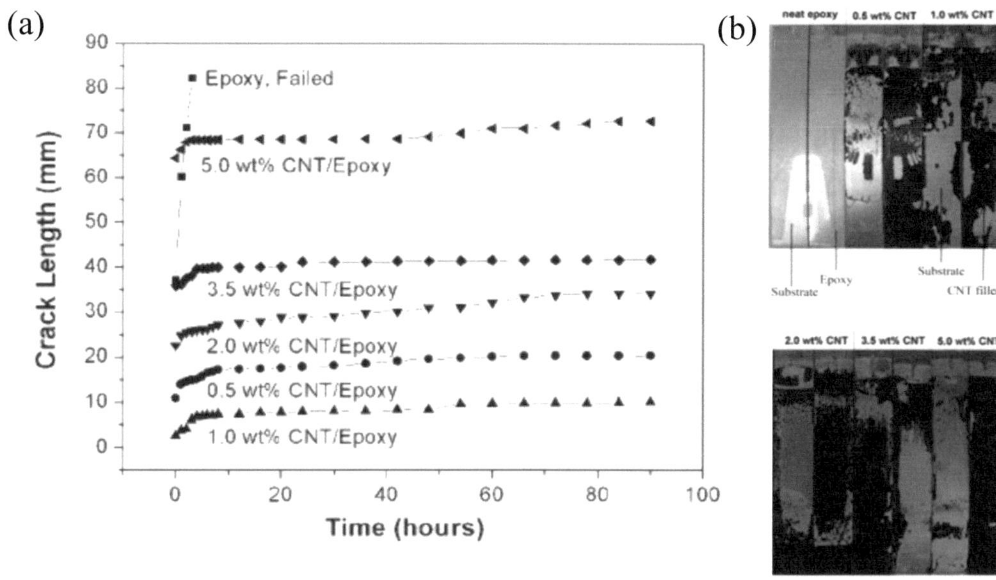

Fig. 8. (a) Crack propagation of the CNT filled epoxy adhesive joints as a function of immersion time into water at 60 °C, (b) photomicrographs of the wedge specimens after inmersion into water at 60 °C for 90 h (Yu et al., 2009).

Meguid and Sun studied the adhesive properties of nanoreinforced epoxy adhesive using dissimilar joints, formed by carbon fiber/epoxy laminate and aluminium alloy 6061-T6. The results reveal that the presence of uniformly dispersed carbon nanotubes causes an increase of the bonding strength. A remarkable improvement in Young´s modulus as well as ultimate tensile strength of the nanoreinforced adhesives is also appreciated. The increase continues with the increase in the weight percentage of nanofiller. However, as other authors have already observed, there is an optimum content of nanofiller. At very high carbon nanotubes contents, above 10 wt %, the properties degrade to below the ones of the neat epoxy adhesive. These results indicate the sensibility of the shear and tensile properties of the adhesive to the concentration of the nanofiller. Taking into account the fracture study of the tested specimens, this behaviour is attributed to the following. The nanotubes are characterised by large areas per unit gram. As the number of adhesively joined points increases, the adhesive strength of the epoxy increases leading to a higher strength of the joint. However, it seems that there is a limit to the number of dispersed nanotubes beyond which a drop in the properties is observed. Once the CNTs fully fill the gaps and porosities and all contact points are established, the addition nanotubes could not interact effectively within the epoxy adhesive and consequently poor matrix infiltration occurs. The additional nanotubes may force the polymer molecules to take up a strained conformation and thereby considerably modify molecular structures of polymer and interfaces that can be easily debond. Also, the agglomeration of CNTs could act as failure initiation sites, which could results in lowering the strength and stiffness of the adhesive.

Finally, Saeed & Zhan (2007) analysed the adhesive properties of several thermoplastic polyimides filled with MWCNTs, using steel as adherends. They confirmed that the addition of CNTs to this kind of adhesives also enhance their adhesive properties. In particular, they measured the lap shear strength and the adhesive energy of the joints. Also, according to other authors, they found a maximum content of CNTs (0.5 - 1.0 wt %) from which the joint strength decrease, due to a change in the failure mode. The joint with high percentages of CNTs failed in adhesive mode, showing poor wetting of adherend surfaces. These authors also probed that the increase in the lap shear strength by the CNTs addition remains even up to 200 °C.

In addition to the improvements of the mechanical and adhesive properties, the addition of carbon nanotubes into epoxy adhesives implies other important physical change of the resin. Their electrical conductivity radically changes. In fact, the epoxy thermosets are typically electrical insulator. In contrast, the nanotubes have metallic or semiconducting behaviour depending on their structural configuration. S. G. Prolongo at., in a work published in the *Journal of Nanoscience and Nanotechnology* (2009) studied the reduction of the electrical resistivity of the epoxy resins by the addition of MWCNTs, which was measured in 14 magnitude orders, from 10^{17} Ω·cm for non-modified epoxy resin to 10^{3} Ω·cm for the reinforced resin with 0.25 wt % CNTs. In fact, it was probed that the percolation threshold should be lower than a content of CNTs of 0.1 wt %. Tao Wang et al. (2006) also measured the modification of the electrical conductivity of pressure-sensitive adhesives by the CNTs addition. This property changes from 10^{-11} S/m for the neat adhesive to 1 S/m for nanofilled ones. In this system, they determined that the percolation threshold reaches 0.3 wt %.

5. Adhesives reinforced with alumina nanofibers

The nanoscale alumina fiber powder is usually produced by the electron-explosion of metal wire, appearing linear insulate nanofibers together with co-mingled in a bundle. These nanofibers are usually thin (2 - 4 nm in diameter) with a very high aspect ratio (20 - 80) and therefore a high surface area (300 - 700 m^2/g). The Young's modulus of the alumina nanofibers is 300 GPa and their tensile strength is 2 GPa (Meguid & Sun, 2004). The fibers have unique sorption properties, cationic and anionic chemisorption properties, such as scavenging precious and heavy metals from water.

The addition of alumina nanofibers into epoxy adhesives causes a light increase of the peel and strength of the joints of aluminium substrates (Gilbert et al., 2003; Meguid & Sun, 2004). However, the effect of nanoreinforcement of epoxy adhesives with nano-alumina fibers on the toughness of the carbon fiber/epoxy composite joints significantly varies as a function of the manufacturing method applied, depending on whether the adhesive had been bonded to the composite or cocured with the prepreg. Gilbert et al. (2003) published an interesting work about the effect of alumina nanofibers in these systems. Figure 9 shows some of the obtained results.

For the unmodified system, the bonded adhesive performed almost twice as effectively as the cocured adhesive. Nevertheless, the fracture toughness of the cocured samples tended to increase with the addition of nanoscale modifiers, while the values resulting from bonded samples tended to decrease substantially with the addition of the nano-modifiers. This behaviour is difficult to explain. The reduction of mode I fracture toughness may have been associated with an increase in the matrix stiffness by the addition of the nanofiller, causing the adhesive crack propagation. In contrast, the increase of toughness on cocured joints is usually attributed to increased crack tortuosity, blunting of the crack tip or increased matrix strength. These mechanisms seem to be enhanced by the presence of nanofibers in the cocured adhesive because of the modified adhesive may have formed covalent chemical

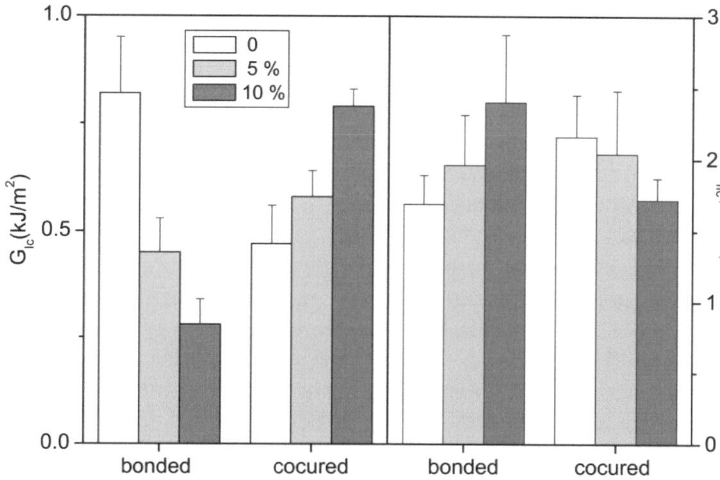

Fig. 9. Effect of alumina nanofibers content on the mode I and mode II fracture toughness of adhesives cocured with and bonded to high performance carbon fiber/epoxy prepregs (Gilbert et al., 2003).

bonds and undergone molecular entanglements with the prepreg matrix when the adhesive system was cocured with the prepreg. It is worthy to note that the effect of the nanofiber introduction into the epoxy adhesive on the mode II fracture toughness is exactly opposite. For cocured samples, the mode II fracture toughness decreases with the addition of nanofibers. It is difficult to state the cause. Gilbert et al. (2003) have observed that in these samples the crack propagated out of the adhesive layer, indicating that the shear strength of the adhesive had exceeded that of the composite material. In contrast, results of the nano-modified adhesives in the bonded system could indicate that the addition of alumina nanofibers increase the strength of the matrix.

6. Adhesives reinforced with carbon nanofibers

The carbon nanofibers (CNFs), grown through vapour carbon deposition, usually present diameters in the order to 20 - 200 nm and very different lengths from 10 to 100 µm. Their estimated axial Young´s modulus is in the range of 100-1000 GPa, depending on the nanofiber configuration. This parameter is particularly sensible to the shell tilt angle. The nanofibers with small tilt angles form the axial direction present much higher stiffness than the ones with large tilt angles. The mechanical strength of CNFs is usually around 2.5 and 3.5 GPa (Tibbetts & Beetz, 1987). It is known that the expected mechanical properties of carbon nanofibers are lower than those corresponding to carbon nanotubes. However, the nanofibers usually have higher length and they are also less expensive than CNTs. The high length of nanofibers is an important aspect to use them as structural nanofillers. Several researches (Bucknall et al., 1994; Zerda & Lesser, 2001) have demonstrated that the effective toughening may not be energetically favourable at nano-length scale, being generally necessary filler lengths greater than 100 nm. It has been probed (Odegard et al., 2003) that long fiber reinforced composite can effectively arrest the crack propagation, which determines the material strength and fracture toughness. However, short fillers might not have this positive effect on the mechanical properties of the composite. Other interesting property of the carbon nanofibers is their high electrical conductivity, which is about 4 x 10^3 S/cm (Al-Saleh & Sundararaj, 2009). This value is similar to the reported for other traditional fillers, such as carbon fibers or carbon black, which are also electrical conductive materials, the electrical conductivity of traditional long carbon fibers is 1.7 x 10^3 S/cm (Al-Saleh & Sundararaj, 2009). However, the main advantage of the nanofibers is their high aspect ratio and their high specific surface area, which allow manufacturing composites with high electrical conductivity at very low filler content, meaning a very low increase of density.

It is widely known that a "good adhesive" must have a high wettability over the adherend. This behaviour is frequently determined by the measurement of the contact angle. The smaller the contact angle, higher wettability on the substrate is obtained. S. G. Prolongo et al. (2009) have published the effect of the addition of CNFs into an epoxy adhesive on its contact angle, using carbon fiber epoxy laminate as adherend. As it is expected, the measurement of contact angle depends on the adhesive nature and substrate but it is also dependent on the characteristics of the surface of the adherends, such as their surface energy, roughness, etc. Therefore, several surface treatments, commonly used for composites, were tested. The obtained results are shown in Figure 10.

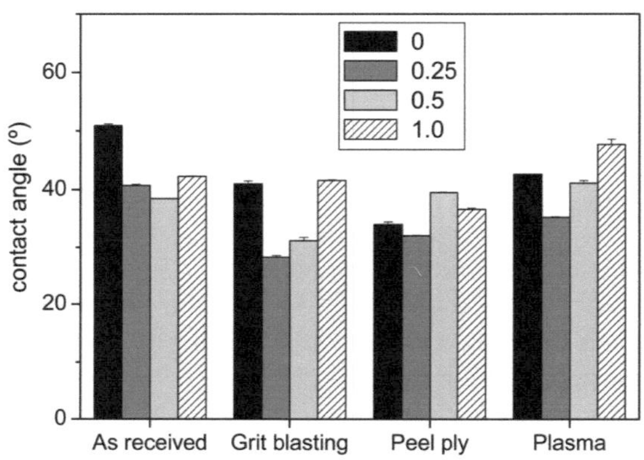

Fig. 10. Contact angle of neat epoxy adhesive and modified adhesives reinforced with 0.25, 0.5 and 1 wt % CNFs on carbon fiber/epoxy laminates (non-treated and surface treated with grit blasting, peel ply and plasma) (Prolongo et al., 2009).

The contact angle generally decreases by the addition of nanofibers, meaning an increase of the wettability of the adhesive. This could be explained by the nano-scale size of the nanofiller and the higher chemical compatibility between the carbon/epoxy composite and the nanoreinforced epoxy adhesive with carbon nanofibers. The nano-scale size of the nanofibers could enhance their permeation on the porous and grooves of the adherend surface, which would increase the wettability of the adhesive. The increase of the CNFs content implies an increase of the contact angle although the measured value is lower than that of neat epoxy adhesive in most cases, except to the epoxy adhesive reinforced with 1 wt % CNFs. This increase seems be associated with the worse dispersion of the nanofiller into the adhesive. It was demonstrated, in works published in *Composites Science and Technology* (Prolongo et al., 2008; Prolongo et al., 2009), that the epoxy nanocomposites with 0.25 wt % presented a suitable dispersion of filler, although at high magnifications, it was possible to observe that the nanofibers tend to be tangled. However, at relative higher nanofibers contents, it was observed the appearance of large agglomerates, with one or more microns of diameters. These agglomerates increase the effective size of used filler, causing a decrease of the adhesive wettability. In these works, the technique of dispersion used to manufacture the nanoreinforced adhesives is based on the use of chloroform as solvent and ultrasonic and high shear mechanical stirring as mixing techniques.

The rheological behaviour of the adhesives is especially interesting to analyse their primer ability. In order to define the application conditions, it is necessary to determine the variation of the viscosity as function of the temperature and its dependence with the shear rate. It is well known that the addition of nanofillers into the non-cured resins causes an important increase of their viscosity. In this chapter, it was demonstrated that the viscosity of an epoxy monomer nanoreinforced with carbon nanotubes is higher than the one of neat monomer (Figure 7). However, for the same content of carbon nanofiller, the viscosity of the reinforced epoxy resin is much higher with CNTs than with CNFs (Prolongo et al., 2009, *Journal of Adhesion Science and Technology*), as it is shown in Figure 11. This figure shows the dependence of the viscosity with the temperature for an epoxy monomer commonly used in formulations of epoxy adhesives, dyglycidyl ether of Bisphenol A (DGEBA), which is reinforced with different contents of carbon nanotubes and nanofibers. It can draw attention that the amount of nanofiller added is different but it was chosen with the criterion of not greatly increase the viscosity.

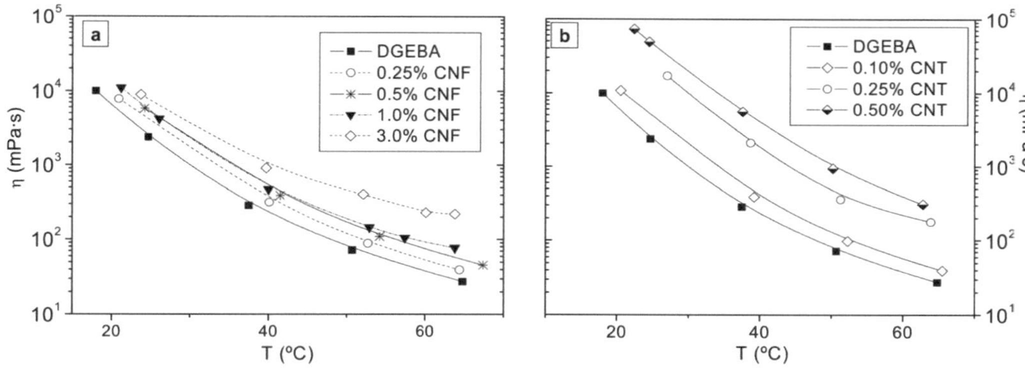

Fig. 11. Viscosity as function of the temperature of epoxy monomer (DGEBA, diglycidyl ether of Bisphenol A) reinforced with different contents of CNFs (a) and CNTs (b) (Prolongo et al., 2009).

The higher viscosity of the mixtures with CNTs could be associated to the higher specific surface of the added nanotubes, close to 300 m^2/g, than the one of nanofibers, in the range of 150 - 200 m^2/g. Other reason is the better dispersion degree observed for the nanocompose reinforced with CNFs regard to the resin with nanotubes. In spite of the increase of viscosity, the rheological behaviour is not affected by the addition of nanofillers, remaining constant their dependence with the temperature. Due to an increase of the adhesive viscosity could cause difficulties in the manufacture of the joints associated to the decrease in the wettability of the adhesive, the study of the rheological behaviour seems a suitable method to determine the optimum content of the nanofiller to add into the epoxy resin. In principle, a higher content of nanoreinforcement would imply better mechanical properties of the adhesive, provided the dispersion of the nanofiller is right. However, high contents of nanofibers usually originate the appearance of agglomerates, which commonly cause a decrease of the mechanical properties of the adhesive.

There are several works published concerning to the effect of the addition of carbon nanofibers into adhesives in order to enhance the strength and toughness of the joints. Xu et al. (2007) studied the tensile strength of joints bonded with a commercial epoxy adhesive reinforced with carbon nanofibers. The nature of adherends varied from aluminium to poly(methylmetacrilate) (PMMA). The bonding surfaces of both were sand-blasted before the adhesive was applied to these bonding areas. The CNFs were directly added into epoxy adhesive, which was later treated at high temperature and sonication. The obtained results are shown in Figure 12.

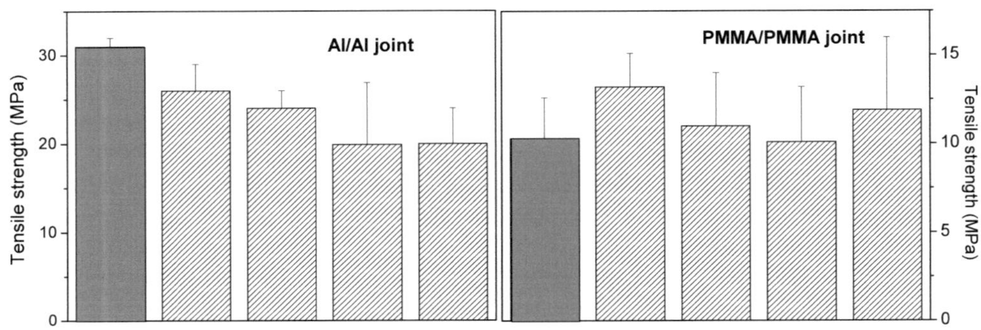

Fig. 12. Tensile strength of Al/Al (a) and PMMA/PMMA (b) joints bonded with an epoxy adhesive reinforced with different fiber weight percents. Left dark bar corresponds with the strength of joints bonded with pure epoxy adhesive (Xu et al., 2007).

All the tensile strengths of Al/Al joints bonded with nanofiber reinforced adhesives are below the tensile strength featuring neat epoxy. In contrast, the strength of PMMA/PMMA joints bonded featuring nanoreinforced adhesives with different fiber weight contents, from 0.3 to 0.8%, exceed the value measured for the joint bonded with non-modified adhesive. The maximum increase observed was up to 30%. Seeing the discrete results obtained, the authors of this work postulated that discontinuous nanofibers or nanotubes with high strength and stiffness, when they are added as reinforcement to matrices, could generate high stress at the fiber-matrix interface and an inefficient interfacial shear stress transfer could be occur. Thus the strong nanofibers can not carry high load. It is necessary continuous forms of nanofibers or nanotubes without finite ends, precluding the presence of extra matrix material at the end of nanofiller, in order to eliminate stress concentrations. Therefore, the nanofibers should be very long or at least being aligned to reach important increases of strength or fracture toughness.

Prolongo et al., in *The Journal of Adhesion* (2009), analysed the effect of the addition of carbon nanofibers into epoxy adhesives on the lap shear strength of joints of carbon fiber/epoxy laminates. Besides the enhancement of the mechanical properties of the nanoreinforced adhesives, the very small size of the filler could enhance the adhesion on the substrate, generating new anchor-points on them. This mechanism could be enhanced when the adherend is a composite of epoxy matrix reinforced with long fibers due to the high chemical compatibility with the epoxy adhesive reinforced with carbon nanofibers. Figure 13 shows the lap shear strength of the joints bonded featuring neat epoxy adhesive and the ones reinforced with different CNFs contents. The adherends were treated with different surface treatments such as plasma, grit-blasting and peel ply in order to increase the wettability of the adhesive and therefore to increase the adhesive strength. It is observed that the addition of carbon nanofibers scarcely affects the joint strength in spite of the nanofilled adhesives showed lower contact angles (see Figure 10), which implies an enhancement of the wettability. In contrast, as it is expected, the lap shear strength strongly depends on the surface treated applied to the composite. The highest strength is obtained for the laminate treated by plasma due to the higher wettability of the adhesives on these surfaces. Grit blasted joints also present high strength due to the increase of the roughness of the adherends, enhancing the mechanical adhesion.

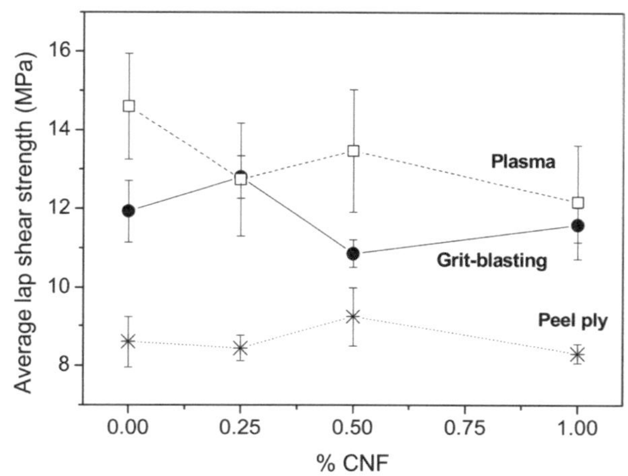

Fig. 13. Average lap shear strength of the joints bonded with neat epoxy adhesive and modified adhesives reinforced with 0.25, 0.5 and 1 wt % CNFs using carbon fiber/epoxy laminates as adherends, treated with grit blasting, peel ply and plasma (Prolongo et al., 2009).

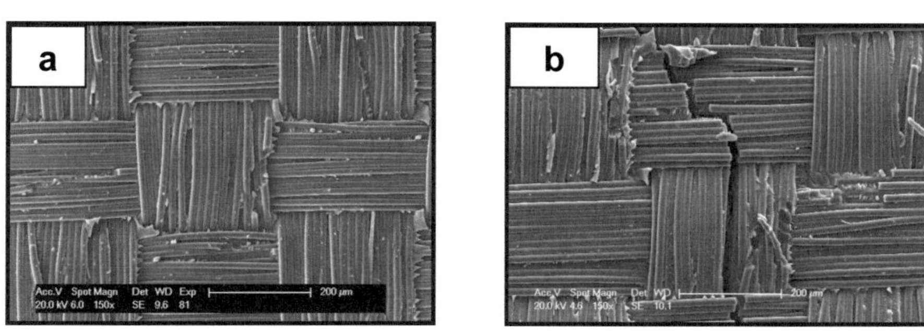

Fig. 14. Scanning electron micrographs of the fracture surface of tested joints with peel ply treated adherends, whose failure mode was adhesive at the interface: (a) non-modified epoxy adhesive and (b) epoxy adhesive reinforced with 0.5 wt% CNFs (Prolongo et al., 2009).

Despite the fact that the measured values of lap shear strength are similar, the fracture surfaces generated by the tested joints with the neat epoxy adhesive and the reinforced ones presented significant differences. For example, Figure 14 shows the fracture surfaces of the joints whose adherends were treated by peel ply. While the surface of the non-reinforced adhesives scarcely showed deformation, the surface of the epoxy adhesives reinforced with CNFs present long cracks on the peel ply texture and even micro-scale zones of cohesive failure, meaning small pieces of pulled out epoxy matrix of the laminates. These evidences indicate higher adhesion ability of the reinforced adhesives.

Figure 15 shows some micrographs obtained at higher magnification for fracture surfaces of epoxy adhesive reinforced with 0.5 wt % CNFs, whose adherends were treated by plasma. Figures 15a and 15b correspond to the cohesive failure zones while Figures 15c and 15d show the adhesive failure zones. This detailed study of fracture surfaces shows some interesting points. The cohesive failure zone (Figures 15a and 15b) in the composite adherend can be distinguished by the presence of the fiber imprints. Within the imprints of the carbon fibers, it is possible to observe striations, which are the bright bands. These striations, running along the fiber axis, correspond to the characteristic surface roughness of PAN-based carbon fibers (intermediate modulus, IM7) used to manufacture the laminate. These marks on the epoxy matrix of the composite are generated during the test and provide a clear indication of the adhesive shear failure mode at the matrix/fiber interface. The presence of shear forces at the crack tip causes the delamination of the interface, sliding the fiber surface over the matrix. It is known that the fracture energy in mode II is higher than in mode I for thermosetting carbon fiber/epoxy composites. On the other hand, the epoxy matrix of the composite present the typical pattern of shear cusps or *hackles* (Figure 15b) characteristic of mode II shear failure observed by other authors both in epoxy carbon fiber laminates and in adhesively bonded CFRP joints during shear testing. The cusps are

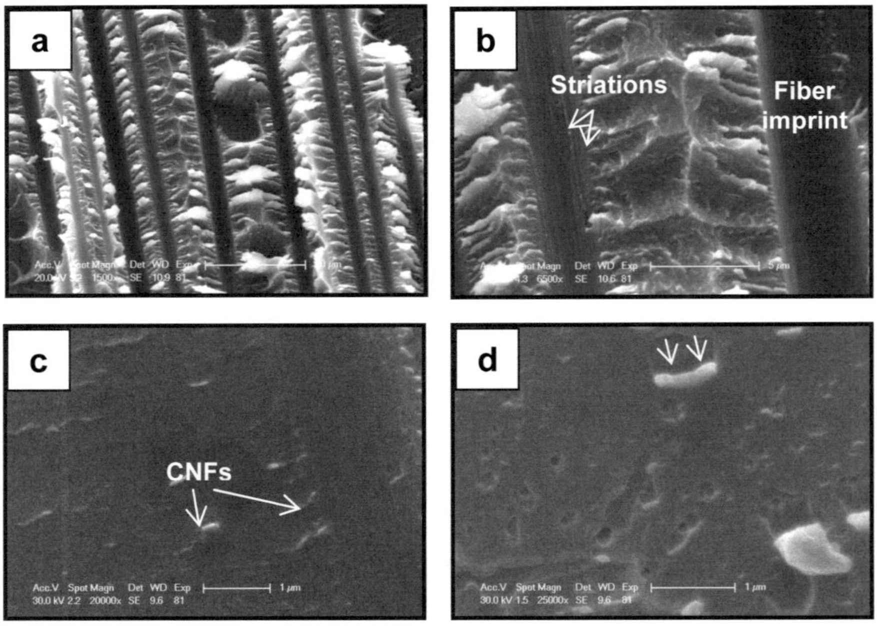

Fig. 15. Scanning electron micrographs of the fracture surface of tested joints whose adehernds was treated by plasma and whose adhesive was reinforced with 0.5 wt % CNFs. Its failure mode was mixed adhesive-cohesive (Prolongo et al., 2009).

oriented perpendicular to the fibers, bent over along them with a width approximately equal to the distance between the fibers. Figures 15c and 15d correspond to the adhesive failure zone. In particular, they are SEM micrographs at very high magnifications of the face with nanoreinforced epoxy adhesive, which show the roll played by CNFs in the crack propagation and final formation of shear cusps. Small void with sizes in the rage of the nanofiber diameters found on surfaces of the cusps could show the participation of these carbon nanofibers in pull-out mechanisms from the adhesive matrix. The shear sliding of those CNFs oriented on the fracture plane (white arrow in Figure 15d) favour the matrix deformation in mode II.

Finally, although the shear and tensile strength of the joints seem not to be widely increased by the addition of carbon nanofibers, it is known that the electrical resistivity of the epoxy resins markedly decreases. Table 1 shows the decrease of the electrical resistivity of epoxy resin when different contents of carbon nanofibers are added. It is seen that the studied epoxy nanocomposites follow typical percolation behaviour. The percolation threshold of the electrical resistivity is the sharp jump by several orders of magnitude which is attributed to the formation of a three-dimensional conductive network of the fillers within the matrix. In the studied system, an epoxy resin reinforced with CNFs, the electrical percolation threshold seems to occur between 0.25 and 0.50 wt %. In fact, the resulting nanocomposites are electrically conductive while the neat epoxy resin is an insulating material. The low percolation threshold of the nanocomposites is justified by the large aspect ratio of the nanoreinforcements, forming a percolating network throughout the epoxy matrix. The decrease of the electrical resistivity with an increase in reinforcement content is attributed to the probably of reinforcement contact.

wt % CNF	ρ ($\Omega \cdot$cm)
0	10^{17}
0.25	$1.8 \cdot 10^{7}$
0.5	$1.2 \cdot 10^{5}$
1.0	$3.2 \cdot 10^{4}$

Table 1. Electrical resistivity of epoxy resins reinforced with different contents of CNFs (Prolongo et al., 2009)

Similar conclusions was drawn by Thao Gibson et al. (2005), who studied the development of epoxy based adhesives formulated with coated and uncoated vapour-grown carbon nanofibers. They confirmed that the shear strength of metal-metal and composite-metal joints remained constant with the addition of CNFs into the adhesive. However, this modification caused an important decrease of the electrical resistivity and an increase of thermal conductivity, from 5.10^{10} $\Omega \cdot$cm and 0.8 W/mK for neat epoxy to 0.2 $\Omega \cdot$cm and 2.8 W/mK for the adhesive reinforced with 20 wt % CNFs. It is worthy to note that the desired properties for a high electrically/thermally conductive adhesive in the aerospace industry are an electrical resistivity lower than 10^{8} $\Omega \cdot$cm and a thermal conductivity higher than 1.0 W/mK.

7. Summary of main results

The following tables collect a summary of the most relevant results published about the mechanical properties of the joints bonded with neat and nanoreinforced epoxy adhesives, which were determined by lap shear (Table 2), peel (Table 3) and double cantilever beam tests (Table 4). Table 5 shows the same properties for other kinds of adhesives.

Filler	Adherend	Neat adhesive	Nanoreinforced adhesive	Variation (%)	Reference
Al_2O_3 nanoparticles	Al	237 MPa	273 MPa	+15	Gilbert, 2003
SiO_2 nanoparticles	Al	25.5 MPa	28.5 MPa	+12	Klug, 1999
	Al	20.8 MPa	23.0 MPa	+11	Kinloch, 2003
	Ti	25 MPa	40 MPa	+60	Bhowmik, 2009
POSS	Al	21 MPa	24 MPa	+14	Dodiuk, 2005
MWCNT	CF/epoxy	-	-	+46	Hsiao, 2003
Al_2O_3 nanofibers	Al	237 MPa	265 MPa	+12	Gilbert, 2003
	Al-CF/epoxy	-	-	+30	Meguid, 2004
CNF	PMMA	28 MPa	32.5 MPa	+16	Xu, 2007
	CF/epoxy	11.9 MPa	12.8 MPa	+8	Prolongo, 2009

Table 2. Lap shear strength of different adhesive joints: comparison between neat and nanoreinforced epoxy adhesive

Filler	Adherend	Neat adhesive	Nanoreinforced adhesive	Variation (%)	Reference
Al_2O_3 nanoparticles	Al	87 N	130 N	+49	Gilbert, 2003
SiO_2 nanoparticles	Al	3.1 N/mm	5.5 N/mm	+77	Kinloch, 2003
POSS	Al	0.19 N/mm	0.49 N/mm	+158	Dodiuk, 2005
Al_2O_3 nanofibers	Al	87 N	119 N	+37	Gilbert, 2003

Table 3. Peel strength of different adhesive joints: comparison between neat and nanoreinforced epoxy adhesive

Filler	Adherend	Neat adhesive	Nanoreinforced adhesive	Variation (%)	Reference
Al_2O_3 nanoparticles	CF/epoxy	0.47 kJ/m^2	0.85 kJ/m^2	+81	Gilbert, 2003
SiO_2 nanoparticles	CF/epoxy	0.59 kJ/m^2	0.74 kJ/m^2	+25	Klug, 1999
	Al	1.2 kJ/m^2	2.3 kJ/m^2	+92	Kinloch, 2003
Al_2O_3 nanofibers	CF/epoxy	0.47 kJ/m^2	0.79 kJ/m^2	+68	Gilbert, 2003

Table 4. Mode I fracture toughness (G_{IC}) of different adhesive joints: comparison between neat and nanoreinforced epoxy adhesive

Filler	Adhesive	Adherend	Test	Neat adhesive	Nanoreinforced adhesive	Variation (%)	Reference
SiO_2 nanoparticles	Acrylic	Al	Lap shear	106.3 kPa	339.6 kPa	+219	Patel, 2006
		Wood		40.7 kPa	128.9 kPa	+217	
		PP		34.8 kPa	185.6 kPa	+433	
		Al	Peel	1.36 kN/m	1.98 kN/m	+46	
		PP		0.16 kN/m	0.25 kN/m	+56	
Nanoclay	Acrylic	Al	Lap shear	131.7 kPa	323.6 kPa	+146	
		Wood		147.9 kPa	339.5 kPa	+130	
		PP		86.0 kPa	208.5 kPa	+142	
		Al	Peel	1.49 kN/m	1.98 kN/m	+33	
		PP		0.20 kN/m	0.27 kN/m	+35	
MWCNT	Polyimide	Steel	Lap shear	18 MPa	22.5 MPa	+25	Saeed, 2007

Table 5. Mechanical properties of different adhesive joints: comparison between neat and nanoreinforced adhesive

8. Concluding remarks

The addition of nanofillers causes an important increase of the mechanical properties of different adhesives, although the obtained enhancements depend on numerous factors, such as the nature of the adhesive and adherends, the applied surface treatment or the tested property. Also, they depend on the nature and content of nanofiller. It seems that the best results were obtained with the addition of nano-sized silica particles and carbon nanotubes. Specially, the effect of these nanofillers is more noticeable in the peel strength and mode I fracture toughness. In general, it is observed that there is an optimum content of nanofiller for which the adhesive properties measured are the maximum. At higher contents, the properties fall back. This fall is usually accompanied with a change of the failure mode of the joints. Frequently, the joints bonded with neat adhesives present failure at the interface while the failure shown for the joints bonded with nanoreinforced adhesives is cohesive. Finally, at relative high contents of nanoreinforcement, the failure mode of the joints is interfacial again. The improvement of the adhesive properties by the addition of nano-sized filler has been associated to different phenomena. Between them, the nanoparticles and nanofibers can fill the gaps and porosities of the adherend, establishing new contact points and enhancing the interfacial strength due to the mechanical anchoring mechanism. On the other hand, it was probed that the nanoreinforced adhesives present higher wettability than the neat epoxy resins, which justifies a high adhesive strength of the joints. Some authors affirm the formation of chemical bonds between nanoreinforced adhesive and the surface of the substrates. Other works justify the increase of the strength and toughness of the joints by the enhancement of mechanical properties of the adhesive. The worsening of the joint properties at relative high nanofiller contents can be also explained by different mechanisms, such as the increase of the adhesive viscosity or the appearance of agglomerations due to dispersion problems.

Finally, the addition of nanofillers into adhesives can improve other interesting properties, like the gas permeability, thermal conductivity and electrical conductivity. It is worthy to note that the addition of carbon nanotubes or nanofiber implies an increase of the electrical conductivity of the adhesives, becoming from insulate to semiconductor or electrical conductor material, which is the special relevance in the joint of electrical conductive substrates.

9. References

Allen, K.W. (2003). "At forty cometh understanding": A review of some basics of adhesion over the past four decades. *International Journal of Adhesion and Adhesives*, 23, 2, (March 2003) 87-93, ISSN 0143-7496

Al-Saleh, M.H. & Sundararaj, U. (2009). A review of vapor grown carbon nanofiber/polymer conductive composites. *Carbon*, 47, 1, (January 2009) 2-22, ISSN 0008-6223

Bhowmik, S.; Benedictus, R.; Poulis, J.A.; Bonin, H.W. & Bui, V.T. (2009). High-performance nanoadhesive bonding of titanium for aerospace and space applications. *International Journal of Adhesion and Adhesives*, 29, 3, (April 2009) 259-267, ISSN 0143-7496

Bucknall, C.G.; Karpodinis, A. & Zhang, X.C. (1994). A model for particle cavitation in rubber-toughened plastics. *Journal of Materials Science*, 29, 13, (July 1994) 3377-3383, ISSN 0022-2461

Dodiuk, H.; Kenig, S.; Blinsky, I.; Dotan, A. & Buchman, A. (2005). Nanotailoring of epoxy adhesives by polyhedral-oligomeric-sil-sesquioxanes (POSS). *International Journal of Adhesion and Adhesives*, 25, 3, (June 2005) 211-218, ISSN 0143-7496

Fisher, L.W. (2005). *Selection of engineering materials and adhesives*, Taylor & Francis, ISBN 0824740475, Boca Raton

Gibson, T.; Rice, B.; Ragland, W.; Silverman, E.M.; Peng, H-H.; Strong, K.L. & Moon, D. (2005). Formulation and evaluation of carbon nanofiber-based conductive adhesives, *Proceedings of 50th International SAMPE Symposium*, Long Beach, May 2005

Gilbert, E.N.; Hayes, B.S. & Seferis, J.C. (2003). Nano-alumina modified epoxy based film adhesives. *Polymer Engineering and Science*, 43, 5, (May 2003) 1096-1104, ISSN 0032-3888

Gojny, F.H.; Wichmann, M.H.G.; Fiedler, B.; Bauhofer, W. & Schulte, K. (2005). Influence of nano-modification on the mechanical and electrical properties of conventional fibre-reinforced composites. *Composites Part A: Applied Science and Manufacturing*, 36, 11, (November 2005) 1525-1535, ISSN 1359-835X

Gojny, F.H.; Wichmann, M.H.G.; Fiedler, B. & Schulte, K. (2005). Influence of different carbon nanotubes on the mechanical properties of epoxy matrix composites – A comparative study. *Composites Science and Technology*, 65, 15-16, (December 2005) 2300-2313, ISSN 0266-3538

Guadagno, L.; Vertuccio, L.; Sorrentino, A.; Raimondo, M.; Naddeo, C.; Vittoria, V.; Iannuzzo, G.; Calvi, E. & Russo, S. (2009). Mechanical and barrier properties of epoxy resin filled with multi-walled carbon nanotubes. *Carbon*, 47, 10, (August 2009) 2419-2430, ISSN 0008-6223

Heimann, M.; Wirts-Ruetters, M.; Boehme, B. & Wolter, K.J. (2008). Investigations of carbon nanotubes epoxy composites for electronics packaging. *Proceedings of 58th Electronic Components & Technology Conference*, pp. 1731-1736, ISBN 9781424422302, Orlando, May 2008, Institute of Electrical and Electronics Engineers

Hsiao, K.-T.; Alms, J. & Advani, S.G. (2003). Use of epoxy/multiwalled carbon nanotubes as adhesives to join graphite fibre reinforced polymer composites. *Nanotechnology*, 14, 7, (July 2003) 791-793, ISSN 0957-4484

Hyun, Y.H.; Lim, S.T.; Choi, H.J. & John, M.S. (2001). Rheology of poly(ethylene oxide)/organoclay nanocomposites. *Macromolecules*, 34, 23, (November 2001) 8084-8093, ISSN 0024-9297

Jiang, H.; Liu, B.; Huang, Y. & Hwang, K.C. (2004). Thermal expansion of single wall carbon nanotubes. *Journal of Engineering Materials and Technology*, 126, 3, (July 2004) 265-270, ISSN 0094-4289

Kim, J-K.; Hu, C.; Woo, R.S.C. & Sham, M-L. (2005). Moisture barrier characteristics of organoclay-epoxy nanocomposites. *Composites Science and Technology*, 65, 5, (April 2005) 805-813, ISSN 0266-3538

Kim, J.Y. & Kim, S.H. (2006). Influence of multiwall carbon nanotube on physical properties of poly(ethylene 2,6-naphthalate) nanocomposites. *Journal of Polymer Science Part B-Polymer Physics*, 44, 7, (April 2006) 1062-1071, ISSN 0887-6266

Kinloch, A.J. (1987). *Adhesion and Adhesives: Science and Technology*, Chapman and Hall, ISBN 041227440 X, London

Kinloch, A.J.; Lee, J.H.; Taylor, A.C.; Sprenger, S.; Eger, C. & Egan, D. (2003). Toughening structural adhesives via nano- and micro-phase inclusions. *The Journal of Adhesion*, 79, 8-9, (August-September 2003) 867-873, ISSN 0021-8464

Klug, J.H. & Seferis, J.C. (1999). Phase separation influence on the performance of CTBN-toughened epoxy adhesives. *Polymer Engineering and Science*, 39, 10, (October 1999) 1837-1848, ISSN 0032-3888

Kwon, Y-K.; Berber, S. & Tomanek, D. (2004). Thermal contraction of carbon fullerenes and nanotubes. *Physical Review Letters*, 92, 1, (January 2004) 015901, ISSN 0031-9007

Meguid, S.A. & Sun, Y. (2004). On the tensile and shear strength of nano-reinforced composite interfaces. *Materials and Design*, 25, 4, (June 2004) 289-296, ISSN 0261-3069

Nelly, B.T. (1981). *Physics of graphite*, Applied Science, ISBN 0853349606, London

Nielsen, L.E. (1967). Models for the permeability of filled polymer systems. *Journal of Macromolecular Science, Part A*, 1, 5, (August 1967) 929-942, ISSN 1060-1325

Odegard, G.M.; Gates, T.S.; Wise, K.E.; Parck, C. & Siochi, E.J. (2003). Constitutive modelling of nanotube-reinforced polymer composites. *Composites Science and Technology*, 63, 11, (August 2003) 1671-1687, ISSN 0266-3538

Osman, M.A.; Mittal, V.; Morbidelli, M. & Suter, U.W. (2003). Polyurethane adhesive nanocomposites as gas permeation barrier. *Macromolecules*, 36, 26, (December 2003) 9851-9858, ISSN 0024-9297

Patel, S.; Bandyopadhyay, A.; Ganguly, A. & Bhowmick, A.K. (2006). Synthesis and properties of nanocomposite adhesives. *Journal of Adhesion Science and Technology*, 20, 4, (2006) 371-385, ISSN 0169-4243

Perepelkin, K.E. (1972). Comparative estimate of the theoretical, highest attainable strength and rigidity of oriented layer structures. *Soviet Materials Science*, 8, 2, (March 1974) 198-202, ISSN 0038-5565

Prolongo, S.G.; Burón, M.; Gude, M.R.; Chaos-Morán, R.; Campo, M. & Ureña, A. (2008). Effects of dispersión techniques of carbon nanofibers on the thermo-physical properties of epoxy nanocomposites. *Composites Science and Technology*, 68, 13, (October 2008) 2722-2730, ISSN 0266-3538

Prolongo, S.G.; Campo, M.; Gude, M.R.; Chaos-Morán, R. & Ureña, A. (2009). Thermo-physical characterisation of epoxy resin reinforced by amino-functionalized carbon nanofibers. *Composites Science and Technology*, 69, 3-4, (March 2009) 349-357, ISSN 0266-3538

Prolongo, S.G.; Gude, M.R.; Sánchez, J. & Ureña, A. (2009). Nanoreinforced epoxy adhesives for aerospace industry. *Journal of Adhesion*, 85, 4-5, (April 2009) 180-199, ISSN 0021-8464

Prolongo, S.G.; Gude, M.R. & Ureña, A. (2009). Synthesis and characterisation of epoxy resins reinforced with carbon nanotubes and nanofibers. *Journal of Nanoscience and Nanotechnology*, 9, 10, (October 2009) 6181-6187, ISSN 1533-4880

Prolongo, S.G.; Gude, M.R. & Ureña, A. (2009). Rheological behaviour of nanoreinforced epoxy adhesives of low electrical resistivity for joining of carbon fiber/ epoxy laminates. *Journal of Adhesion Science and Technology*, accepted, ISSN 0169-4243

Regert, M. (2004). Investigating the history of prehistoric glues by gas chromatography-mass spectrometry. *Journal of Separation Science*, 27, 3, (February 2004) 244-254, ISSN 1615-9314

Sadeghian, R.; Gangireddy, S.; Minaie, B. & Hsiao, K.T. (2006). Manufacturing carbon nanofibers toughened polyester/glass fiber composites using vacuum assisted resin transfer molding for enhancing the mode-I delamination resistance. *Composites Part A: Applied Science and Manufacturing*, 37, 10, (October 2006) 1787-1795, ISSN 1359-835X

Saeed, M.B. & Zhan, M-S. (2007). Adhesive strength of nano-size particles filled thermoplastic polyimides. Part-I: Multi-walled carbon nano-tubes (MWNT)-polyimide composite films. *International Journal of Adhesion and Adhesives*, 27, 4, (June 2007) 306-318, ISSN 0143-7496

Salvetat, J.P.; Bonard, J.; Thomson, N.; Kulik, A.; Forro, L. & Benoit, W. (1999). Mechanical properties of carbon nanotubes. *Applied Plysics A*, 69, 3, (September 1999) 255-260, ISSN 0947-8396

Sarathi, R.; Sahu, R.K. & Rajeshkumar, P. (2007). Understanding the thermal, mechanical and electrical properties of epoxy nanocomposites. *Materials Science and Engineering A*, 445-446, (February 2007) 567-578, ISSN

Tibbetts, G.G. & Beetz, C.P. (1987). Mechanical properties of vapour grown carbon fibers. *Journal of Physics D-Applied Physics*, 20, 3, (March 1987) 292-297, ISSN 0022-3727

Vaisman, L.; Wagner, H.D. & Marom, G. (2006). The role of surfactants in dispersion of carbon nanotubes. *Advances in Colloid and Interface Science*, 128-130, (December 2006) 37-46, ISSN 0001-8686

Wadley, L. (2005). Putting ochre to the test: replication studies of adhesives that may have been used for hafting tools in the Middle Stone Age. *Journal of Human Evolution*, 49, 5, (November 2005), 587-601, ISSN 0047-2484

Wang, S.R.; Liang, Z.Y.; Gonnet, P.; Liao, Y.H.; Wang, B. & Zhang, C. (2007). Effect of nanotube functionalization on the coefficient of thermal expansion of nanocomposites. *Advanced Functional Materials*, 17, 1, (January 2007) 87-92, ISSN 1616-301X

Wang, T.; Lei, C-H.; Dalton, A.B.; Creton, C.; Lin, Y.; Fernando, K.A.S.; Sun, Y-P.; Manea, M.; Asua, J.M. & Keddie, J.L. (2006). Waterborne, nanocomposite pressure-sensitive adhesives with high tack energy, optical transparency, and electrical conductivity. *Advanced Materials*, 18, 20, (October 2006) 2730-2734, ISSN 0935-9648

Wong, E.W.; Sheehan, P.E. & Lieber, C.M. (1997). Nanobeam mechanics: elasticity, strength, and toughness of nanorods and nanotubes. *Science*, 277, 5334, (September 1997) 1971-1975, ISSN 0036-8075

Xie, X-L.; Mai, Y-W. & Zhou, X-P. (2005). Dispersion and alignment of carbon nanotubes in polymer matrix: A review. *Materials Science and Engineering R*, 49, 4, (May 2005) 89-112, ISSN 0927-796X

Xu, L.R.; Li, L; Lukehart, C.M. & Kuai, H.C. (2007). Mechanical characterization of nanofiber-reinforced composite adhesives. *Journal of Nanoscience and Nanotechnology*, 7, 7, (July 2007) 2546-2548, ISSN 1533-4880

Young, T. (1805). An essay on the cohesion of fluids. *Philosophical Transactions of the Royal Society of London*, 95, (January 1805) 65-87, ISSN 0261-0523

Yu, S.; Tong, M.N. & Critchlow G. (2009). Wedge test of carbon-nanotube-reinforced epoxy adhesive joints. *Journal of Applied Polymer Science*, 111, 6, (March 2009) 2957-2962, ISSN 0021-8995

Zerda, A.S. & Lesser, A.J. (2001). Intercalated clay nanocomposites: morphology, mechanics and fracture behavior. *Journal of Polymer Science Part B-Polymer Physics*, 39, 11, (June 2001) 1137-1146, ISSN 0887-6266

Zhai, L.; Ling, G.; Li, J. & Wang, Y. (2006). The effect of nanoparticles on the adhesion of epoxy adhesive. *Materials Letters*, 60, 25-26, (November 2006) 3031-3033, ISSN 0167-577X

Zhai, L.; Ling, G. & Wang, Y. (2008). Effect of nano-Al_2O_3 on adhesion strength of epoxy adhesive and steel. *International Journal of Adhesion and Adhesives*, 28, 1-2, (January 2008) 23-28, ISSN 0143-7496

4

Fabrication of Bio-nanocomposite Nanofibers Mimicking the Mineralized Hard Tissues via Electrospinning Process

Gyeong-Man Kim
Fraunhofer Institute for Cell Therapy and Immunology, Perlickstraße 1, 04103 Leipzig, Germany

1. Introduction

From physicochemical point of view, e.g, the architecture and the chemistry, mineralized hard tissues such as bone and dentin can be considered as unique "living nanocomposites". They are nature-made materials, which common exhibit highly complex and strongly hierarchical structures on different length scales [1-3]. Microscopically, in the lowest level of such mineralized hard tissue the minerals are deposited in parallel fashion into a structured organic matrix, yielding a highly anisotropic three-dimensional structure [4]. The mineral components are calcium phosphates known as mainly hydroxyapatite (HAp) nanocrystals with formula $Ca_{10}(PO_4)_6(OH)_2$ [5,6], having a plate-like shape with highly variable length (20-40 nm), width (~20nm) and thickness (~1.5-5 nm) dimensions. While the inorganic components are predominantly responsible for the mechanical properties of hard tissues, the organic matrix not only gives them their toughness but also contains the osteoblasts, odontoblasts and mixtures of type I collagen and specialized non-collagenous proteins, including growth factors necessary for continual growth and repair [7].

In spite of high demand in clinical medicine, nature's ability to self-organize inorganic component with preferred orientation in the bioorganic matrix is still not reproducible by synthetic procedures because of their complex nature. Therefore, in fields ranging from biology and chemistry to materials science and bioengineering a big challenge is being required to facilitate fabrication of bone and dentin-like biocomposite materials, which may allow the ingrowth of hard tissues with improving the mechanical properties with respect to the hard tissue regeneration. Several major requirements, which are critical in the development of biomaterials for hard tissue regeneration purpose, have to be taken into account: (1) the material must have a positive influence on the amount and quality of newly formed hard tissue, (2) the material must shorten the period of hard tissue formation and healing process of damaged tissues, (3) the material must not cause any adverse effects, namely inflammatory or toxic response, and (4) it must obtain mechanical properties that match with the natural hard tissues [8,9].

In recent years, particular attention has been paid to the biomimetic approaches, which allow us to mimic such nature-made bio-inorganic and bio-organic composite materials [10]. The main idea in biomimetic approaches is to control and fabricate the morphology and the composition of developed biomaterials, in which the nanocrystallites of inorganic compounds are being dispersed with preferential orientation in the organic matrices. Owing to its large potential in biomedical applications, many studies have been reported the preparation of bone-like biocomposites of HAp and bioactive organic components such as collagen, chondroitin sulfate, chitosan and amphiphilic peptide by direct precipitation [11-13], poly(lactic acid) through a solvent-cast technique [14,15], and polyamide by solution method [16,17]. Some studies showed a similar nanostructure to bone, in which the crystallographic *c*-axes of the HAp were regularly aligned along the collagen fibers [18]. Additional studies have also demonstrated the preparation of collagen/HAp composites by anchoring of HAp particles in the collagen matrix [19,20]. Although these nanocomposites exhibited high biocompatibility and improved mechanical properties with similar chemical composition to natural bone, they did not show any complex hierarchical nanostructures as bone.

Recently, electrospinning technique has received a growing attention because polymer fibers prepared by this technique achieve fiber diameters in the range from micrometers down to a few nanometers straightforwardly and cost-effectively [21,22]. In a typical process, a polymer solution is forced through a capillary, forming a pendent drop at the tip of capillary. Then a high voltage is applied between the capillary and a grounded collection target. When the electric field strength overcomes the surface tension of the droplet, a polymer solution jet is initiated and accelerated toward the collection target. As the jet travels through the air, the solvent evaporates and a non-woven polymeric fabric is formed on the target. Because the resulting non-woven fabrics often resemble the superstructure features of natural extra cellular matrix, they have gained a great interest in tissue engineering as scaffold materials for tissue regeneration, immobilized enzymes and catalyst systems, and wound dressing articles. In addition, the high specific surface area and highly porous three-dimensional structure enables their use in high density cell and tissue cultures. To date, it has been well established that the ES process allows easy incorporation of particles with different habits, such as 1-dimensional carbon nanotubes [23,24], 2-dimensional layered silicates [25-27] or 3-dimensional SiO_2 nanoparticles [28] and many others [29], into the nanofibers.

Although pure HAp rarely occurs in biological systems, the synthetic HAp has been proven to possess excellent biocompatibility, bioactivity and osteoconduction [30,31]. These characteristics are most probably attributed to its chemical composition and crystallographic structure similar to those of the mineral part of natural biominerals occurring in bone, dentine, cartilage, tendons and ligaments [32,33]. Therefore, to date, the synthetic HAp has been used with a great attention in widespread biomedical applications such as fillings and/or substitutes for bulk bone defects or voids in the form of powders, porous blocks or granules, and coatings for orthopedic (e.g. hip-joint prosthesis) and dental implants [34-36].

Based on the above mentioned considerations, in the present work we successfully mimicked bio-nanocomposite fibers to the lowest level of hierarchical organization similar to bone. Our approach comprised using the electrospinning technique on the basis of PVA as matrix and HAp nanoparticles as the inorganic phase for hard tissue regeneration purpose [37]. To our best knowledge, such kind of bone-like electrospun fibers based on the biocompatible polymer with HAp has not yet been documented. The idea to use PVA is supported by several factors: PVA is one of the well-known water-soluble synthetic polymers and the backbone chains are highly interconnected by hydrogen bonding because of the presence of abundant hydroxyl groups. These characteristics contribute to an excellent chemical resistance and mechanical properties, including high tensile and impact strength or creep resistance. On the other hand, the hydroxyl groups can be used to securely incorporate molecules of biological species, such as collagens, hyaluronan [38] or deoxyribonucleic acid [39]. PVA is truly biodegradable with nontoxic byproducts being water and carbon dioxide and it is highly biocompatible [40] with living organs and cells, which leads to broad biomedical and pharmaceutical applications [41,42]. In addition, PVA/HAp nanocomposites (NC) were chosen for two reasons: (1) PVA can strongly interact with the surface of HAp, because PVA is hydrophilic and HAp has hydroxyl groups on the surface, and thus (2) PVA/HAp NC can conveniently be electrospun from an aqueous medium.

The capability and feasibility of the technique presented in this work provides a promising alternative approach for the fabrication of bio-nanocomposite fibers. These fibers provide widespread future applications related to hard tissue (bone and dentin) replacement and coating implants, because of their resemblance to the nanostructure of living bone from the physiochemical point of view.

2. Results and discussion

2.1 Morphology and crystal structure of HAp

Figure 1a shows a TEM micrograph of HAp particles, in which the morphology of the individual HAp particles is clearly revealed to be rod-like shape. By semi-quantitative analysis, the particles had width of 10 to 30 nm (with an average of ~18 nm) and lengths of 20 to 120 nm (average of ~50 nm) (see Figure 1b) [44]. To get more detail of surface and crystal structure of HAp we performed HR-TEM. Figure 2a shows an overview image of agglomerates of HAp nanoparticles. The image indicates that HAp nanoparticles consist of well defined crystalline rods. Hence, hereafter, we denote HAp nanoparticles to HAp nanorods.

Most striking feature obtained from TEM investigation is that all HAp nanorods exhibit nanoporous morphology on the surface. Such electron-lucent nanopores were reported by several authors, suggesting that the voids were sublimation holes caused by electron beam damage [45]. However, at least, based on our results we exclude this argument, because the nanopores always appeared, even if the duration and dose of electron beam on the HAp during TEM were rigorously changed. Therefore, it could be concluded that such nanoporous morphology is an intrinsic nature of our HAp nanorods [46]. High porosity on the HAp surface may provide cell anchorage sites and structural guidance. In addition, and within an *in vivo* milieu, they provide the interfaces to respond to the physiological and biological clues and to remodel the extracellular matrix integrating with the surrounding native tissue.

In Figure 2b a HR-TEM image of a single isolated HAp nanorod along the [010] zone axis together with its corresponding fast Fourier transform (FFT) pattern is shown. A calculated inter-planar spacing between vertical dark and white lines (d-spacing) of a prevailing HR-TEM fringe is 0.342 nm, normal to the long axis of the HAp nanocrystal, i.e., the c-lattice planes corresponding to the (002) reflection. In addition, FFT pattern showed several sets of lattice fringes corresponding to spacing of 0.827 nm, 0.463 nm, 0.386 nm, 0.317 nm and 0.278 nm, which are attributed to the (100), (110), (111), (102) and (112) planes, respectively (discussed as following in XRD).

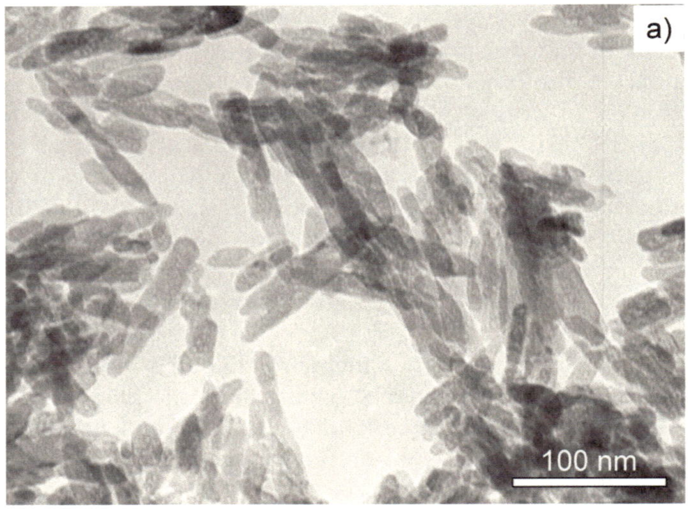

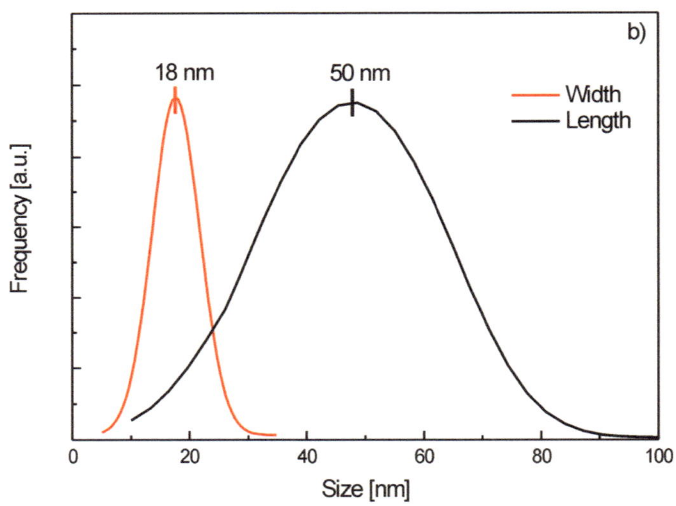

Fig. 1. a) TEM micrograph of HAp nanoparticles and b) particle size distribution.

To identify the crystal structure and lattice fringes of HAp nanorods compared with HR-TEM data we carried out the XRD measurements. The X-ray powder diffraction pattern of pure HAp nanoparticles is shown in Figure 3. The observed positions (2θ) and their corresponding $d_{2\theta}$ and indices (hkl) from the XRD pattern of HAp nanoparticles are listed in table 1. The XRD profile of the HAp is indexed with the cell dimensions a = 0.941 nm, c = 0.69 nm, and the structural refinements are carried out in the $P6_3/m$ space group [47]. No noticeable diffraction peak appears other than that of HAp. These results suggest that the *d* spacings of the corresponding planes of our HAp nanoparticles are in good agreement with those on the JCPDS card (No 9-432).

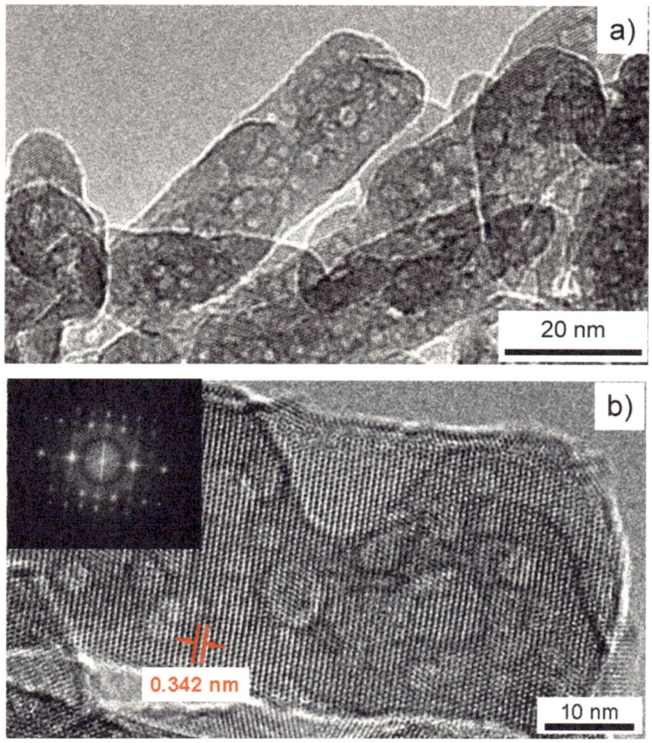

Fig. 2. a) HR-TEM overview image of agglomerates of Hap nanoparticles and b) micrograph of overview HAp nanoparticles, showing the nanoporous morphology and b) HR-TEM image of a single HAp nanorod with its corresponding FFT pattern (as inset).

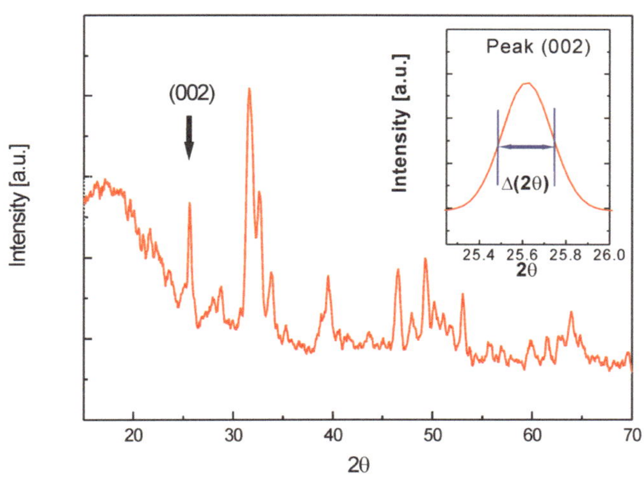

Fig. 3. XRD pattern of pure HAp nanorods. Inset indicates the enlarged (002) reflection (denoted with an arrow) used for calculation of the crystal size.

To estimate the crystal size of the HAp nanoparticles we applied the Scherrer's equation (eq. (1)) for the broadening of the corresponding X-ray spectral peaks [48]:

$$L = \frac{K\lambda}{B(2\theta)\cos\theta} \tag{1}$$

where L is the crystallite size, K is a constant taken as the default value of 0.89, λ is the wavelength of the x-ray, B(2θ) is the full width at the half maximum intensity (FWHM) and θ is the Bragg angle. The width of the diffraction peak with the intensity (002) was selected for the calculation (see inset in Figure 3). The estimated crystallite size L of Hap is 18.1 nm. This is well consistent with that from TEM measurements (average 17.7 nm).
Based on the analysis of TEM, HR-TEM and XRD, it can be concluded that the HAp nanorods are high quality crystals containing nanopores. The rods exhibit a growing direction along [001] with a d-spacing of 0.68 nm along its long axis.

2θ	$d_{2\theta}(nm)$	*hkl*
25.8	0.34	002
31.8	0.28	211
32.8	0.27	300
34.0	0.26	202
39.8	0.23	310
46.5	0.20	222
48.1	0.19	312
49.4	0.18	213
53.1	0.17	004

Table I. 2θ-values observed from XRD pattern of pure HAp nanorods, d-values determined from 2θ- values ($d_{2\theta}$) and corresponding indices (*hkl*)

2.2 Morphology and crystalline structure of ES-PVA fibers with and without HAp

Figure 4a shows a representative SEM micrograph for the electrospun fibers of pure PVA. Using water as solvent and combination of optimal electrospinning conditions (6 cm of working distance, 10 kV of operating voltage), the average diameter of ES fibers is approximately 160 nm without any sign of bead formation. The structure of electrospun fibers from PVA/HAp nanocomposite is just identical to those from pure PVA, except for minor difference in their average size. The average diameter of ES-fibers shifts from 160 nm to 170 nm upon addition of HAp (see Figure 4b). Nevertheless, the majority of diameter in both cases lies in the range of 100 to 250 nm.
In order to discern any variation in crystalline content as a result of HAp addition to the PVA the XRD measurements on the electrospun fibers PVA with and without HAp were carried out and the resulting XRD patterns were analyzed in more detail (see Figure 5). In the case of the electrospun pure PVA fibers, the diffraction peaks reveal at 2θ = 19.5° (d-spacing = 0.454 nm) and 40.5° (d-spacing = 0.223 nm) [49]. The peak corresponds to a typical doublet reflection of (101) and (10-1) planes of the semicrystalline atactic PVA, which contain the extended planar zigzag chain conformation, while the latter belongs to the (220) reflection [50]. This diffraction profile indicates that the electrospun fibers are still preserved

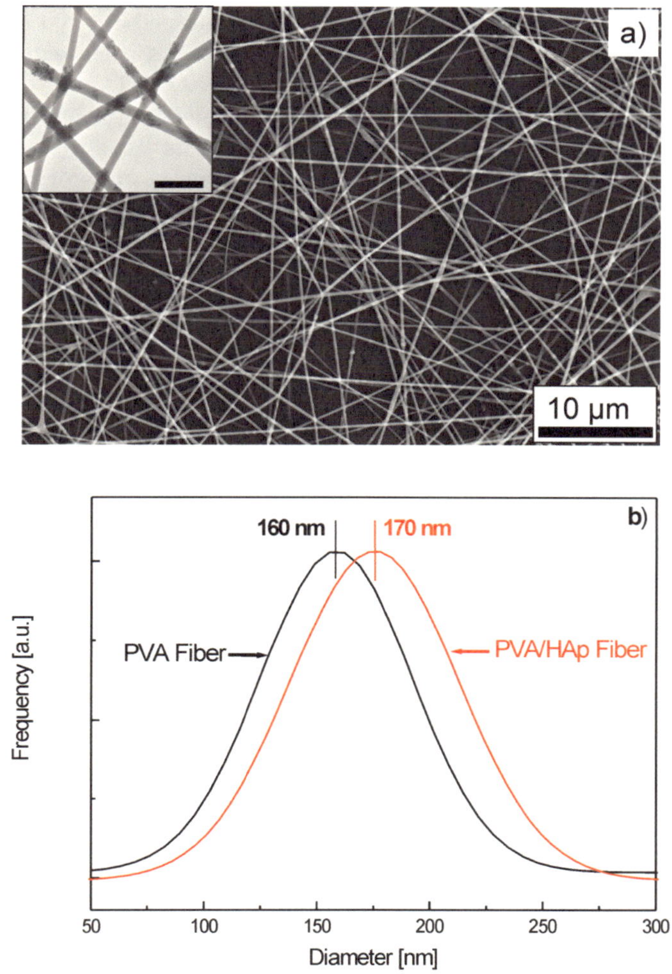

Fig. 4. a) SEM and TEM (as inset, scale bar indicates 100 nm) micrographs of as-electrospun PVA/HAp nanocomposite fibers and b) diameter distribution of ES-PVA and -PVA/HAp nanocomposite fibers.

in its semicrystalline nature associated with the trans-planar chain conformation. This could be attributed to the fact that due to the intense intermolecular hydrogen bonding the PVA chains retain their trans-planar chain conformation even under the ultra-high shear stress during the electrospinning process.

In contrast, after integrating HAp nanorods into PVA, as seen in Figure 5, the position of the doublet peak slightly is shifted to a lower value, implying an increase in the size of the PVA crystallites. In addition, the peak area is broadened to some extent, indicating that the crystallinity of PVA is suppressed by the introduction of HAp into the PVA matrix. From close inspection of the doublet peak (see inset in Figure 5) this result is clearly evidenced by an increase of d-spacing from 0.454 nm to 0.462 nm corresponding to the peak position at 2θ = 19.5° and 19.2° for ES-PVA and ES-PVA/HAp, respectively. The increase in the crystalline thickness might be attributed to the high packing efficiency of the tie molecules in the amorphous regions within the electrospun fibers. As a consequence, the overall diameter of electrospun fibers increased by the introduction of HAp into PVA fiber matrix. The suppression of the crystallinity, i.e., broadening of peak area, should be attributed to the structural change from the lamellar structure to the microfibrillar structure associated with the lamellar break-up by strong elongational stress during electrospinning. Applying the Debye-Scherer Equation (1) with regard to the broadening of the doublet peak we have estimated the average crystallite size, L, corresponding to the microfibril diameter of PVA crystallite; 3.2 nm and 2.7 nm for the ES-PVA fibers and ES-PVA/HAp fibers, respectively.

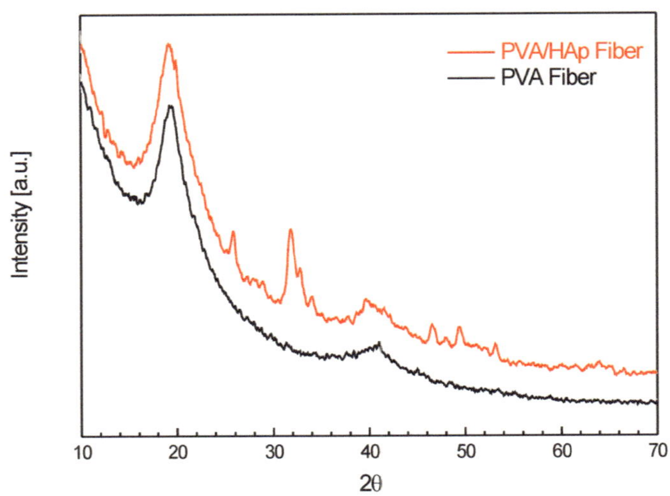

Fig. 5. XRD patterns of electrospun PVA and PVA/HAp nanocomposite fibers. The scans are shifted along the y-axis for clarity.

It should be noted that the suppression of crystallinity for the electrospun fibers compared with the bulk polymer is a general occurrence. The electrospinning process is associated with a high shear stress and a very rapid structure formation of the polymer material, thus this method gives rise to a less appropriate environment to form perfect crystallites as in the bulk. Further suppression of PVA crystallinity in the electrospun fibers upon introducing HAp is caused by destruction of the orientation order of the macromolecule chains and even by formation of amorphous bound layers around the HAp nanoparticles. The first reason means that the HAp nanorods act as steric hindrances within the electrospun fibers, and thus the crystallization is restricted by a decreasing amount of intact crystalline region. The latter could be envisaged as a fact that the strong interaction caused by hydrogen bonding leads to trap PVA molecules onto the HAp nanorods coupled with their large specific surface areas. As a consequence, the regular planar zigzag conformation collapses and converts into an amorphous structure on the interfacial region of Hap. Thus, the overall crystallinity of PVA is decreasing when the HAp nanorods are incorporated into the PVA matrix. This result is well consistent with the DSC measurements as described later.

The typical internal morphology of the electrospun fibers containing HAp is shown in Figure 6a. It is clearly seen that the HAp nanorods are well distributed within the electrospun fiber and, of most interest, a large number of HAp nanorods are preferentially oriented parallel to the longitudinal direction of the electrospun PVA fibers. A HR-TEM image of a single isolated HAp nanoparticle with the corresponding fast Fourier transform (FFT) pattern is shown in Figure 6b. It can be clearly confirmed that the d-spacing of (001) plane of HAp, i.e. the crystallographic *c*-axis, oriented normal to the long axes of the PVA fiber. As mentioned above, the ultra-high shear stress upon electrospinning not only induces an orientation of macromolecules but also properly initiates the alignment of nanorods of HAp to a large extent.

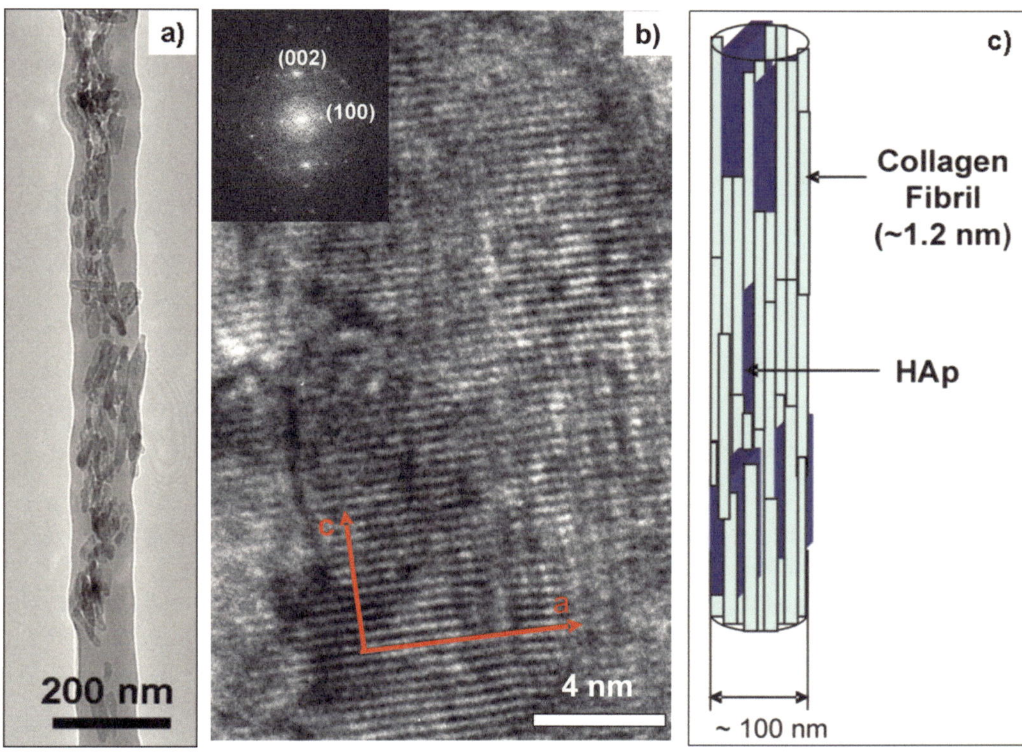

Fig. 6. a) TEM image of fiber morphology taken from a single electrospun PVA/Hap nanocomposite fiber. b) HR-TEM image and its corresponding FFT pattern (c is the fiber axis). c) schematic illustration of mineralized collagen fibrils.

3.3 Thermal analysis: DSC and TGA

In order to investigate the thermal behavior, such as melting, crystallization and formation of crystalline structure, we performed DSC measurements. Since DSC is sensitive to the thermal history of the samples, the first heating run of DSC at a rate of 10K/ min was taken into account to characterize thermal properties of all samples in the present work. The results from DSC and the characteristics observed for all systems are presented in Figure 7 and summarized in Table II.

	T_g	T_m	T_c	ΔH_m	Crystallinity %
PVA powder	41.7	218	179	65.3	47
PVA fiber	48.3	219	190	61.4	44
PVA/HAp fiber	50.3	216	187	57.8	39*

* Since the HAp nanorods do not contribute to the enthalpy of fusion for the PVA matrix, the weight only fraction of PVA is taken into consideration.

Table II. Glass transition, melting and crystallization temperatures, melting enthalpies and calculated crystallinities

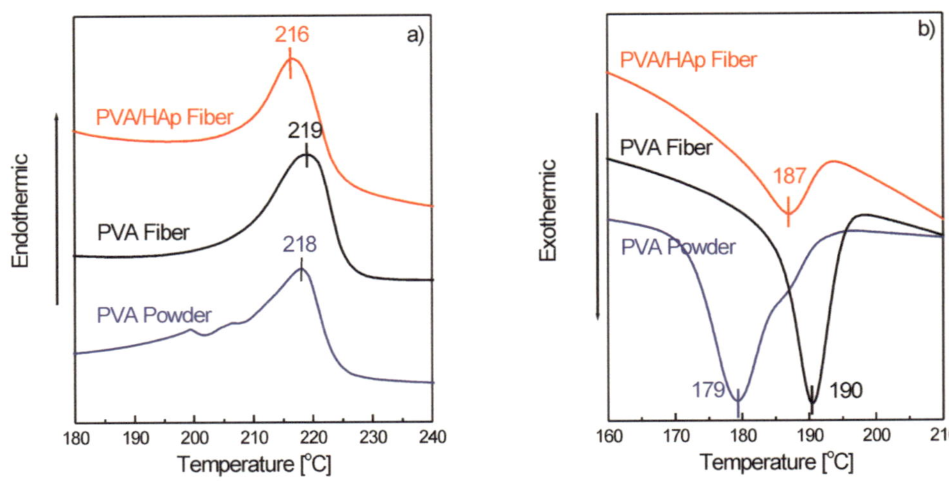

Fig. 7. DSC thermograms for the PVA powder, electrospun PVA fibers with and without HAp nanorods: a) endothermic and b) exothermic traces.

The endothermic peaks during the heating stage, which are centered at 216 -218 °C, are shown in Figure 7a. This is attributed to the melting of the crystalline PVA phase.

While the melting temperature in the electrospun fibers of pure PVA is almost unchanged compared with that in the pure PVA powder, after adding the HAp nanorods the melting temperature of the electrospun PVA/HAp NC shifts slightly to a lower value compared with others. In contrast, the melting enthalpy - the degree of crystallinity - decreases by the electrospinning, and this decrement is even more pronounced by the incorporation of HAp nanorods. A decrease in the enthalpy of fusion and the melting temperature suggests that the crystallinity and perfection of the crystal structure are reduced by HAp loading.

The lower crystallinity after electrospinning can be readily explained by the fact that the jet being ejected from a capillary under the strong electrostatic potential is highly experienced by an extensive elongational flow to stretch the polymer chains in the electrostatic filed direction and simultaneously evaporation of the solvent is followed within a very short time scale that generally leads to a lowering of the temperature (like quenching). Due to such an extremely high elongational flow rate as well as the very limited time scale during the electrospinning, the polymer chains are forced to align parallel to the fiber direction. As a consequence, the molecules in the electrospun fibers exhibit a less favorable packing.

A further decrease in crystallinity after adding HAp nanorods onto the electrospun fibers indicates that there are interactions between PVA and HAp nanoparticles, i.e., the hydrogen bonding between them. The intense hydrogen bonding can also change the chain configuration around the interfacial region of HAp nanorods, induced most likely by the amorphous layer around them. In other words, the well dispersed HAp nanorods within the electrospun fibers readily generate defects in the crystalline phase of PVA. As a consequence, less heat is needed to destroy the hydrogen bonding or to set the PVA chains free to melt, thus resulting in a decreasing melting temperature.

To determine the crystallization temperature (T_c), the samples were heated up to 250°C, kept at this temperature for 5 min and then cooled down at a rate of 10°C/min. Figure 7b shows the exotherms upon cooling from the melt. The crystallization temperature significantly increases upon electrospinning (ΔT_c= +11°C, comparing the PVA powder with the electrospun fibers of PVA) and then decreases with incorporation of the HAp nanorods (ΔT_c= -3 °C compared with the electrospun fibers of PVA/HAp NC to those of the electrospun PVA fibers). However, it should be noted that T_c for the electrospun fibers with and without HAp is still significantly higher than that of the PVA powder. All these changes can be attributed to the formation of nuclei at an earlier stage and at a higher temperature during the cooling process. The shifting to higher temperatures upon electrospinning corresponds to a promoted rapid crystallization. This is the intrinsic nature of very rapid structure formation in the electrospinning process, and thus insufficient time is provided to form perfect crystallites in the PVA matrix than in the bulk.

The decrease in T_c in the electrospun fibers by addition of HAp nanorods may be attributed to the facts that due to higher thermal conductivity of HAp nanorods compared to that of PVA, the heat will be more diversely distributed in the samples containing the HAp nanorods. Thus, HAp nanorods provide a high degree of supercooling during the electrospinning, which leads to a rapid formation of heterogeneous crystallization nuclei. On the other hand, the hydroxyl groups presented on the surface of HAp nanorods preferentially serve as heterogeneous nucleation sites for PVA crystallization, however, due to the strong interaction between HAp nanorods and PVA molecules via hydrogen bonding the crystallization process is further confined in its dimensional reduction of the PVA molecules, thereby the crystallization process will be shortened [51]. Upon introduction of HAp into ES-fibers the T_c shifts towards a higher temperature. However, a decrement of the crystallization enthalpy is observed, which is most likely attributed to the decrease of conformation entropy due to the preferential orientation of the polymer chains.

PVA exhibits an other endotherm at 41.7 °C corresponding to the glass transition temperature (T_g) of PVA [data not shown here] [52]. The overall glass transition temperatures of the PVA matrix of the electrospun fibers with and without HAp were found to have higher temperatures compared to the PVA powder, as shown in Table 2. These results indicate that the oriented PVA chains are highly confined in the electrospun fibers and upon adding HAp to the electrospun fibers more constraints are imposed on the polymer chains due to the strong interactions (such as hydrogen bonding) between them, thereby the segmental motions of the polymer chains are greatly constricted.

Figure 8 shows typical TGA thermograms with the corresponding first order derivatives (DTGA) of weight loss as a function of temperature for ES-fibers with and without HAP along with pure PVA powder. The samples were measured in the temperature range from 30 °C up to 1000 °C with a constant rate of 10 °C/ min under nitrogen atmosphere. The TGA and DTGA show that all samples exhibited three distinct weight loss stages at 30 - 210 °C (5 wt.-% loss of weakly physisorbed water), 210 – 400 °C (decomposition of side chain of PVA) and 400 – 540 °C (decomposition of main chain of PVA) [53]. Nevertheless, major weight losses are observed about 75 wt.-% in the range of 210-540 °C for all samples, which are corresponding to the structural decomposition of PVA.

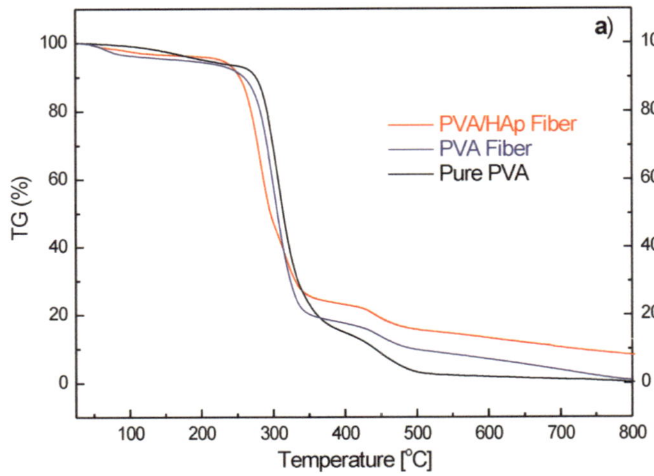

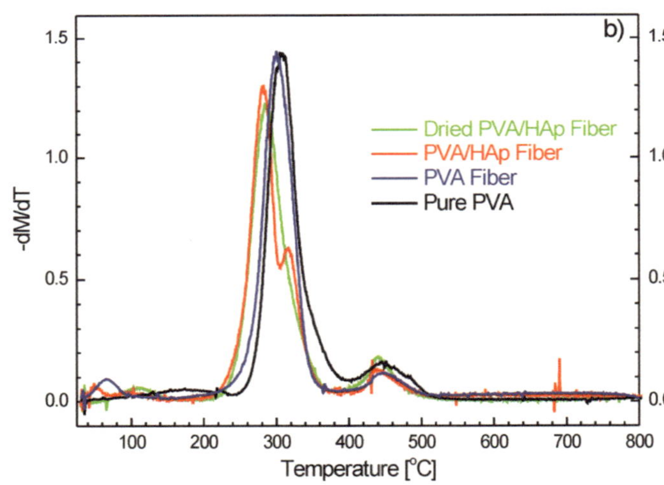

Fig. 8. a) TGA curves for the PVA powder, electrospun PVA fibers with and without HAp nanorods and b) the first order derivatives of TGA (the dried PVA/HAp nanocomposite fibers are included for comparison, as green curve).

Figure 8b shows the first order derivatives of TGA curves, revealing the temperatures at which the maximum decreases of mass occur. The temperature at the maximum mass loss rate is 308 °C for the PVA powder, 286 °C for the electrospun fibers of PVA, and 281 °C for the electrospun fibers of PVA/HAp NC. DTGA data clearly show that the side chains of PVA are readily decomposed prior to the main chains. It can be also recognized that the thermal decomposition behavior of the electrospun PVA fibers is similar to that of the pure PVA powder. However, this process is accompanied with a decrease from 308 to 286 °C in its maximum mass loss rate upon electrospinning.

The maximum mass loss rate of the electrospun fibers of PVA/HAp NC further shifts to a lower temperature compared with that of the electrospun pure PVA. It is interesting to note here that the first derivative of the electrospun fibers of PVA/HAp NC exhibits a distinctive peak at 317 °C, which is not appearing in the TGA curve. To identify this peak we dried the electrospun PVA/HAp NC fibers in air at 80 °C for 8h, then cooled in a desiccator and TGA was carried out again. This peak completely disappears in the DTGA curve of the dried electrospun PVA/HAp NC fibers. This result clearly indicates that the peak at 312 °C arises from the loss of lattice water which may be called "structural" water or water trapped within HAp nanorods, consisting with data of Rootare and Craig [54] for the chemisorbed water layer. Over about 600 °C, all TGA diagrams become flat and mainly the inorganic residue (i.e. HAp nanorods) remains. From the amounts of the residue at 800 °C, the inorganic contents of the electrospun PHA/HAp NC are estimated of about 7.5 wt.-%.

Evidently, the thermal decomposition of both electrospun fibers of PVA and PVA/HAp NC shifts slightly toward the lower temperature range compared with bulk PVA powder. This implies that the electrospinning suppresses the thermal stability of PVA and the addition of HAp to the PVA fibers even enhances this effect.

3.4 Fourier transform infrared spectroscopy

FTIR spectroscopy was carried out to analyze any complex structural changes upon electrospinning as well as interactions between HAp and PVA. The results are summarized in Table III. The representative FTIR absorption spectrum of the HAp nanorods in the 4000 cm^{-1} to 500 cm^{-1} range is shown in Figure 9 [55]. The phosphate ions, PO_4^{3-}, are the principal

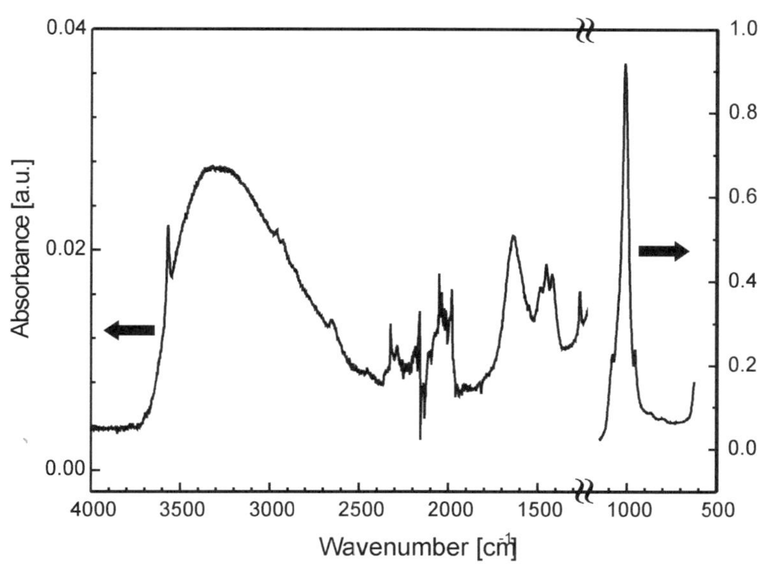

Fig. 9. FTIR spectrum of pure HAp nanorods.

HAp	PVA	PVA fiber	PVA/HAp fiber	Assignment
3567				stretching of OH
~3315				Water
	3278	3278	3296	stretching of OH
	2935	shoulder	2939	asymmetric stretching of CH_2
	2906	2914	2911	symmetric stretching of CH_2
	1710	1711	1717	stretching of CO
	1654	1657	1674	O=H, C=C
1638				Water
1485				
1453				
1421			1421	CO_2^{-3}
	1417	1418	1418	O-H, C-H bending, $\gamma(CH_2)$, $\delta(OH)$
	1376	1374	1378	CH_2 wagging
	1325	1326	1326	δ(O-H) with CH wagging
	1238	1241	1240	C-O-C
	1143	shoulder	shoulder	Stretching of CC (crystalline sequence of PVA)
	1088	1087		stretching of CC and bending of OH (amorphous sequence of PVA)
1088			1092	symmetric stretching of PO_4^{3-}
1019			1033	asymmetric stretching of PO_4^{3-}
962			962	symmetric stretching of PO_4^{3-}
	919	917	920	Bending of CH_2
	839	831	838	Rocking of CH_2
			756	
			632	bending of OH
			599	PO_4^{3-} deformation vibration

Table III. Assignments of FTIR absorption bands of pure HAp nanorods, pure PVA powder, the electrospun fibers of PVA and PVA/HAp NC.

molecular components of HAp giving to the IR absorbance in the 550-1200 cm^{-1} region. The characteristic peaks at 962, 1019 and 1088 cm^{-1} correspond to the stretching vibration of PO_3^{-4} and at 599 cm^{-1} to the deformation vibrations of PO_3^{-4}. As other major components, OH^- ions are identified by observation of the broad band from about 3700 to 2500 cm^{-1}. The peak of this band is centered at about 3297 cm^{-1}, which is a typical assignment of the stretching mode of OH^- ions. In addition, the two distinct bands are observed at 3576 and 1639 cm^{-1}, which arise from the stretching mode and bending mode of H_2O molecules, respectively. Furthermore, some other impurity ions are also identified by the peaks

appearing in the 1490 to 1410 cm^{-1} range. These are attributed to the stretching mode of a trace amount of CO_3^{2-} groups presented in HAp.
Typical FTIR spectra of the pure PVA powder and the electrospun fibers of PVA and PVA/HAp NC along with pure HAp nanorods are shown in Figure 10. As shown in Figure 10 (curve a), the characteristic absorption bands of PVA occur at 3278 cm^{-1} (stretching of OH), 2935 cm^{-1} (asymmetric stretching of CH_2), 2906 cm^{-1} (symmetric stretching of CH_2), 1417 cm^{-1} (wagging of CH_2 and bending of OH), 1143 cm^{-1} (stretching of CO from crystalline sequence of PVA), 1088 (stretching of CO and bending of OH from amorphous sequence of PVA), 919 cm^{-1} (bending of CH_2) and 838 cm^{-1} (rocking of CH) [56].

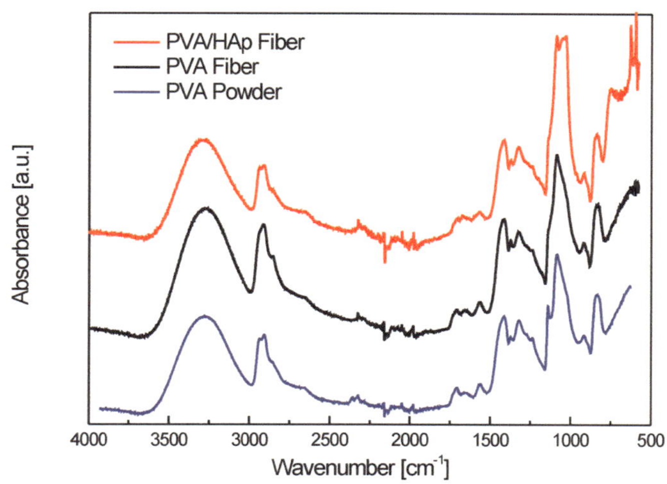

Fig. 10. FTIR spectra for the PVA powder, electrospun PVA fibers and PVA/HAp nanocomposite fibers.

The electrospun fibers of PVA show spectral features similar to that for the PVA powder, but with some extent of changes in their relative intensity and peak position (curve b in Figure 10). While the OH stretching of electrospun PVA appearing at 3278 cm^{-1} is unaltered in its position when compared to the pure PVA powder, absorption bands position changes for other peaks are observed in the electrospun PVA fibers. Minor changes occur in the carbonyl stretching in the 2900 to 2950 cm^{-1} region and a major shift appears in the CC stretching band at 1143 cm^{-1}. The asymmetric stretching band of CH_2 at 2935 cm^{-1} becomes a shoulder and its symmetric stretching band shifts to a higher value at 2914 cm^{-1} from 2906 cm^{-1}. This suggests that the electrospinning has a strong influence on the wavelength and consequently the absorption maxima of stretching vibration shifts toward higher frequencies in comparison with the pure PVA. It is important to emphasize here that the intensity of the 1143cm^{-1} band, which is associated with the CC stretching mode of PVA, is crystalline sensitive and thus is generally used for qualitative and/or quantitative analysis of PVA crystallinity [57]. Accordingly, it can be clearly recognized that the intensity of the 1143cm^{-1} band in the electrospun PVA fibers compared with pure PVA powder appears significantly changed in form of a shoulder. This result indicates that the degree of crystallinity of PVA decreases upon electrospinning, consisting with the trends observed from DSC measurements.

The FTIR spectrum of the electrospun PVA/HAp NC fibers (curve c in Figure 10) exhibits the characteristic absorption bands of PAV along with additional bands corresponding to the phosphate groups of HAp. After incorporating HAp nanorods into the PVA fibers the predominant broad absorption band associated with the OH stretching of PVA centered at 3278 cm^{-1} shifts to a higher wave number region (ν_{OH} = 3296 cm^{-1}). This result suggests that the hydrogen bonding becomes stronger in the electrospun PVA/HAp NC than in pure PVA as well as in the electrospun PVA fibers due to the increase in the number of OH groups by adding of HAp.

In addition, the major absorption band of PO_4^{3-} stretching appearing at 1019 cm^{-1} in the HAp spectrum, moves to 1033 cm^{-1}. This red shift might be attributed to the interactions between PVA molecules and HAp particles. Also the spectra of PVA adsorbed on HAp influence clearly the recovering of the carbonyl stretching region (see in the red circled area). Furthermore, the overall PVA crystallinity decreases by loading of HAp within the electrospun fibers, which is clearly evidenced by weakening of the crystalline band at 1143 cm^{-1}.

It is of interest to note here that the absorption bands associated with H_2O molecules distinctly appear at 3576 cm^{-1} and 1639 cm^{-1} in the spectrum of HAp. The higher peak at 3576 cm^{-1} is completely missing in the electrospun fibers of PVA/HAp NC, while the peak at 1639 cm^{-1} still exists in the spectrum despite of a large extent of reduction in its intensity. In conjunction with the TGA results we confirmed that the former band at 3576 cm^{-1} is identified with the loosely physisorbed water at the surface of HAp nanorods, whereas the latter arises from the chemisorbed molecular water within the HAp lattice.

4. Conclusions

In the present work we demonstrate a powerful technique for fabricating biocompatible and biodegradable PVA/HAp nanocomposite fibers in order to mimic mineralized hard tissues for bone regeneration purpose by applying the electro spinning process. Various techniques, including TEM, HR-TEM, SEM, XRD, DSC, TGA, FTIR spectroscopy were performed to characterize the resulting electrospun PVA/HAp composite nanofibers in comparison with pure PVA and PVA/HAp nanocomposites before electrospinning. Morphological investigation showed that the HAp nanoparticles exhibit nanoporous morphology, which provides enlarged interfaces being a prerequisite for physiological and biological responses and remodeling to integrate with the surrounding native tissue. The most striking physiochemical feature of the electrospun PVA/HAp composite nanofibers is that the HAp nanorods are preferentially oriented parallel to the longitudinal direction of the electrospun PVA fibers as confirmed by electron microscopy and XRD. This feature bears strong resemblance to the nanostructure of mineralized hard tissues serving as building block of bone. Furthermore, the PVA as matrix and HAp nanorods as inorganic phases strongly interact through hydrogen bonds within the electrospun PVA/HAp nanocomposite fibers. The strong bonding due to the presence of a great extent of OH groups in the PVA polymer and the HAp nanorods leads improved thermal properties. The hybrid electrospinning shown in the present work provides great potential as a convenient and straightforward technique for the fabrication of biomimicked mineralized hard tissues suitable for bone and dentin replacement and regeneration. However, a great challenge still exists mainly in how to stabilize the electrospun PVA/Hap nanofibers when in contact with an aqueous medium. Such deficiency could be resolved by proper chemical and/or physical treatments of the material that are currently under investigation.

5. Experimental procedure

5.1 Materials and electrospinning process

All experiments utilized commercial grade PVA (Aldrich) with an average molecular weight, M_w, of about 115.000, and a degree of hydrolysis of 98-99%. The HAp nanoparticles were of commercial grade Ostim® from *aap* Implantate AG, Germany. At first, an aqueous solution of 7 wt.-% pure PVA was prepared by dissolving the PVA polymer in distilled water at 80 °C with vigorous stirring for about 8h. To obtain electrospinnable solution, the prepared 7 wt.-% aqueous solution of PVA was slowly cooled to room temperature, sonicated for 30 min and kept at this temperature for one night, to eliminate air bubbles. In addition, to prepare the solution of PVA/HAp nanocomposite the 7.5 wt.-% HAp was added into the 7 wt.-% aqueous solution of PVA. This solution was also vigorously stirred with a magnetic stir for at least 8h at room temperature, followed by sonication for 2h to ensure homogeneity. Electrospinning was carried out at room temperature in a vertical spinning configuration, using a 1 mm inner diameter flat-end needle with a 5 cm working distance. The applied voltages were in the range from 3 kV to 20 kV, driven by a high voltage power supply (Knürr-Heizinger PNC, Germany). The electrospun fibers were collected either directly on Cu-grids or glide glasses.

5.2 Characterization

5.2.1 Electron microscopy

The morphology and particle size of the as-received HAp were investigated by conventional transmission electron microscopy (TEM) (JEOL 200CX operated at 200 kV) as well as by high-resolution TEM (Philips CM200 FEG\ST Lorentz electron microscope with a field emission gun and at an acceleration voltage of 200 kV). Micrographs were recorded with a 1k*k, 14 bit CCD-camera (model multiscan, Gatan company, USA) and fed to a computer for on-line image processing (Digital Micrograph 3.3.1, Gatan company, USA). Samples for electron microscopy were prepared by dropping aqueous solution of HAp and direct electrospinning of PVA/HAp NC on Cu-grids covered with an ultra thin carbon layer for characterization of HAp particles and their spatial dispersion within the electrospun fibers, respectively. The size and morphology of the electrospun fibers were investigated by field emission gun environmental scanning electron microscopy (FEG-ESEM, Philips ESEM XL 30 FEG). The diameter of the fibers and their size distribution were analyzed by measuring over 200 fibers in the randomly recorded FEG-ESEM micrographs, using image analysis software (Analysis, Soft Imaging System Co., Germany). Samples for FEG-ESEM were directly electrospun on the slide glasses and followed by Au sputtering.

5.2.2 Wide-Angle X-Ray Diffraction (WAXD)

To analyze the crystalline structure of the PVA powder, the electrospun (ES) PVA fiber and the ES-PVA/HAp NC fibers wide-angle X-ray scattering (WAXS) was performed with a Seifert XRD 3000 using CuKα radiation (λ = 1.54056 Å). The tube source was operated at 40 kV and 30 mA. Scans were run over a 2θ range of 10-60° with a step of 0.05° and a dwell time of 1 s.

5.2.3 Differential scanning calorimetry (DSC) and Thermogravimetric analysis (TGA)

DSC measurements were conducted to measure the glass-transition temperatures (T_g), the melting and crystallization behavior with a Mettler-Toledo DSC 820 under a nitrogen atmosphere. The samples were sealed in aluminum pans and were heated and cooled in the temperature range 25 to 250 °C in the DSC instrument with a rate of 10 °C/min. The weight of each sample was approximately 0.5 mg. The DSC temperature and heat flow values were calibrated with indium as standard. The degree of relative crystallinity X_c, was estimated from the endothermic area by the following equation (1):

$$X_c = \Delta H_f \,/\, \Delta H_f^o \tag{1}$$

where ΔH_f is the measured enthalpy of fusion from DSC thermograms and ΔH_f^o is the enthalpy of fusion for 100% crystalline PVA (ΔH_m = 138.6 J/g from literature) [43].
Thermo gravimetric analysis (TGA) was performed using a Perkin-Elmer TGA7 by heating from room temperature to 750°C at a heating rate of 20°C/min under a nitrogen flow.

5.2.4 Fourier Transform Infrared spectroscopy (FTIR)

An FTIR spectrometer (FTIR-Spectrometer S2000, Perkin-Elmer) equipped with a fixed 100 μm diameter aperture and a mercury-cadmium-telluride (MCT) detector was used to analyze the absorbance in the wave number range of 500-4000 cm^{-1} with a resolution of 2 cm^{-1}.

6. References

[1] Weiner S, Traub W. Bone structure: from angstroms to microns. FASEB 1992;6:879–85.
[2] Rho JY. Kuhn-Spearing L, Zioupos P. Mechanical properties and the hierarchical structure of bone. Med. Eng. Phys.1998; 20: 92-102.
[3] Fratzl P. Characterizing natural fibre composites with hierarchical structure. Fibre Diffraction Rev 2002;210:32-9
[4] Park JB, Lakes RS. Biomaterials, 2nd ed. New York: Plenum Publishing Co, 1992. p. 194.
[5] Ravaglioli A, Krajewski A, Bioceramics: Materials, Properties, Applications. Chapman and Hall, London, UK, 1992.
[6] Elliott JC. Structure and Chemistry of the Apatites and Other Calcium Orthophosphated; Elsevier; Amsterdam, 1994.
[7] Butler, W.T. & Ritchie, H. The nature and functional significance of dentin extracellular matrix proteins. Int. J. Dev. Biol. 39, 169–179 (1995).
[8] Park J, Lutz R, Felszeghy E, Wiltfang J, Nkenke E, Neukam FW, Schlegel KA. The effect on bone regeneration of a liposomal vector to deliver BMP-2 gene to bone grafts in peri-implant bone defects. Biomaterials 2007; 28: 2772-82.
[9] A. Kocialkowski, W.A. Wallace and H.G. Prince, Clinical experience with a new artificial bone graft: preliminary results of a prospective study. Injury 21 (1990), pp. 142–144.
[10] Dujardin E, Mann S. Bio-inspired materials chemistry. Adv Mater 2002; 14: 775-88
[11] Kikuchi M, Itoh S, Ichinose S, Shinomiya K, Tanaka J. Self-organization mechanism in a bone-like hydroxyapatite/collagen nanocomposite synthesized in vitro and its biological reaction in vivo. Biomaterials 2001; 22: 1705-11.
[12] Yamaguchi I, Tokuchi K, Fukuzaki H, Koyama Y, Takakuda K, Monma H, Tanaka J. Preparation and mechanical properties of chitosan/hydroxyapate nanocomposites. Bioceramics 2000;192-1:673-76.
[13] Chen F. Wang ZC, Lin CJ. Preparation and characterization of nano-sized hydroxyapate particles and hydroxyapatite/chitosan nanocomposite for use in biomedical materials, Mater. Lett. 2002; 57: 658-62.
[14] Liao SS, Cui FZ, Zhang W, Feng QL. Hierarchically Biomimetic Bone Scaffold Materials: Nano-HA/Collagen/PLA Composite. J. Biomed. Mater. Res. Part B: Appl. Biomater. 2004; 69B: 158-65.

[15] Deng XM, Hao JY, Wang CS. Preparation and mechanical properties of nanocomposites of poly(D,L.lactide) with Ca-defient hydroxyapatite nanocrystals. Biomaterials 2002; 22: 2867-73.

[16] Wei J, Li YB, Chen WQ, Zuo Y. A study on nanocomposite of hydroxyapatite and polyamide. J Mater Sci 2003;38:3303-06.

[17] Hung M, Feng JQ, Wang JX, Zhang XD, Li YB, Yan YG. Synthesis and characterization of nano-HA/PA66 composites. J Mater Sci: Mater Med 2003;14:655-60.

[18] Hartegerink JD, Beniash E, Stupp SI. Self-Assembly and Mineralization of Peptide-Amphiphilie Nanofibers. Science 2001;294:1684-88.

[19] Marouf H A, Quayle A A, Sloan P. In vitro and in vivo studies with collagen/hydroxyapatite implants. Int J Oral Maxillofac Implants 1990;5:148–54.

[20] C. Du, F.Z. Cui, X.D. Zhu and K. de Groot, Three-dimensional nano-HAp/collagen matrix loading with osteogenic cells in organ culture. J Biomed Mater Res 1999;44:407–15.

[21] Reneker, D. H.; Chun, I. Nanometre diameter fibres of polymer, produced by electrospinning. Nanotechnology 1996;7:216-23.

[22] Kim GM, Wutzler A, Radusch HJ, Michler GH, Simon P, Sperling RA, Parak WJ. One-Dimensional Arrangement of Gold Nanoparticles by Electrospinning. Chem. Mater. 2005;17:4949-57.

[23] Dror Y, Salalha W, Khalfin RL, Cohen Y, Yarin AL, Zussman E. Carbon Nanotubes Embedded in Oriented Polymer Nanofibers by Electrospinning. Langmuir 2003;19:7012-20.

[24] Kim GM, Michler GH, Pötschke P. Deformation processes of ultrahigh porous multiwalled carbon nanotubes/polycarbonate composite fibers prepared by electrospinning. Polymer 2005;46:7346-51

[25] Fong H, Liu W, Wang CS, Vaia RA. Generation of electrospun fibers of nylon 6 and nylon 6-montmorillonite nanocomposite. Polymer 2002;43:775-80.

[26] Kim GM, Lach R, Michler GH, Chang CH. The Micromechanical Deformation Process of Electrospun Polymer Nanocomposite Fibers. Macromol. Rapid Commun. 2005;26:728-33.

[27] Kim Michler GH, Ania F, Balta Calleja FJ. Temperature-dependence of polymorphism in electrospun nanofibres of PA6 and PA6/clay nanocomposite, Polymer 2007;48:4814-23.

[28] Kim GM, Lach R, Michler GH, Pötschke P, Albrecht K. Relationships between phase morphology and deformation mechanisms in polymer nanocomposite nanofibres prepared by electrospinning process. Nanotechnology 2006;17:963-72.

[29] Wang M, Singh H, HattonTA, Rutledge GC. Field-responsive superparamagnetic composite nanofibers by electrospinning. Polymer 2004;45:5505-14.

[30] de Bruijn JD, van Blitterswijk CA, Davies JE. Initial bone matrix formation at the hydroxyapatite interface in vitro. J Biomed Mater Res 1995;29:89–99.

[31] Hamadouche M, Sedel L. Ceramics in orthopaedics. J Bone Joint Surg Br 2000;82:1095–1099.

[32] Traub W, Arad T, Weiner S. Three-dimensional ordered distribution of crystals in turkey tendon collagen fibers. Proc. Natl. Acad. Sci. USA. 1989;86:9822-26.

[33] Hench LL. Bioceramics: From Concept to Clinic. J. Am. Ceram. Soc. 1991, 74, 1487-510.

[34] W. Suchanek, M. Yoshimura, Processing and Properties of Hydroxyapatite-Based Biomaterials For Use as Hard Tissue Replacement Implants. J. Mater. Res. 1998, 13, 94-117.

[35] Hench LL. Bioceramics. J. Am. Ceram. Soc. 1998;81:1705-28.

[36] LeGeros RZ. Calcium phosphates in oral biology and medicine. Monogr Oral Sci 1991;15:1-21.

[37] Kobayashi H, PVA-HAp nanocomposites for artificial cornea. Adv in Sci & Tech 2006;53:9-16.
[38] Scotchford CA, Cascone MG, Downes S, Giusti P, Osteoblast responses to collagen-PVA bioartificial polymers in vitro: the effects of cross-linking method and collagen content. Biomaterials 1998;19:1–11.
[39] Aoi K, Takasu A, Okada M. DNA-based polymer hybrids. Part I. Compatibility and physical properties of poly(vinyl alcohol)/DNA sodium salt blend. Polymer 2000;41:2847–53.
[40] Tamada Y, Ikada Y. In: E. Chiellini, P. Giusti, C. Migliaresi and L. Nicolais, Editors, Polymers in medicine 2, Plenum Publishing Company, New York (1986), p. 101.
[41] Stejskal, J.; Kratochvil, P.; Helmstedt, M. Polyaniline Dispersions. 5. Poly(vinyl alcohol) and Poly(N-vinylpyrrolidone) as Steric Stabilizers. Langmuir 1996;12:3389-92.
[42] Finch, C. A. Poly(vinyl alcohol) Development; Wiley: Chichester, England, 1992; pp 18-32.
[43] Peppas NA, Merrill EW, Differential scanning calorimetry of crystallized PVA hydrogels. J. Appl. Polym. Sci. 1976;20:1457-65.
[44] Rusu VM, Ng CH, Wilke M, Tiersch B, Fratzl P, Peter MG, Size-controlled hydroxyapatite nanoparticles as self-organized organic-inorganic composite materials, Biomaterials 2005;26:5414-26.
[45] Sato K, Kogure T, Iwai H, Tanaka J, Atomic-Scale {10-10} Interfacial Structure in Hydroxyapatite Determined by High-Resolution Transmission Electron Microscopy, J Am Ceram Soc 2002; 85:3054–58.
[46] Sutter B, Ming DW, Clearfield A, Hossner LP. Mineralogical and Chemical Characterization of Iro-, Manganese-, and Copper-Containing Synthetic Hydroxyapatites. Soil Sci Soc Am J 2003;67:1935-1942.
[47] Astala R, Stott MJ. First Principles Investigation of Mineral Component of Bone:CO_3 Substitution in Hydroxyapate. Chem. Mater. 2005;17:4125-33.
[48] Klug HP, Alexander LE. X-ray diffraction precedures for polycrystallite and amorphous materials. New York: Wiley; 1974.
[49] Ricciardi, R.; Auriemma F, Rosa CD, Laupretre F. X-ray Diffraction Analysis of Poly(vinyl alcohol) Hydrogels, Obtained by Freezing and Thawing Techniques. Macromolecules 2004;37;1921-27.
[50] Bunn CW, Crystal structure of polyvinyl alcohol. Nature 1948, 161, 929-30.
[51] Pritchard, J. G. Poly(vinyl alcohol)--Basic Properties and Uses, Polymer Monographs; Gordon and Breach Science: New York, 1970; Vol 4.
[52] Strawhecker KE, Manias E. Structure and Properties of Poly(vinyl alcohol)/Na+ Montmorillonite Nanocomposites. Chem Mater 2000;12:2943-49.
[53] Gimenez, V.; Mantecom, A.; Cadiz, V. Modification of poly(vinyl alcohol) with acid chlorides and crosslinking with difunctional hardeners. J Polym Sci Part A:Polym Chem 1996;34:925-34.
[54] Rootare HM, Craig RG. Vapor phase adsorption of water on hydroxyapatite, J Dental Res 1977;56:1437-48.
[55] Rey C, Shimizu M, Collins B, Glimcher MJ, Resolution-enhanced Fourier transform infrared spectroscopy study of the environment of phosphate ion in the early deposits of a solid phase calcium phosphate in bone and enamel and their evolution with age: 2. Investigations in the v_3 PO_4 domain, Calcif. Tissue Int., 1990, 46,384.
[56] Ding B, Kimura E, Sato T, Fujita S, Shiratori S, Fabrication of blend biodegradible nanofibrous nonwoven mats via multi-jet electrospinning, Polymer 2004;45:1895-1902.
[57] Tadokoro H, Seki S, Nitta I. Some information on the infrared absorption spectrum of polyvinyl alcohol from deuteration and pleochroism. J. Polym. Sci. 1956;22:563-66.

5

Diversity of Nanofibers from Electrospinning: from Graphitic Carbons to Ternary Oxides.

Yu Wang, Idalia Ramos and Jorge J. Santiago-Aviles
University of Pennsylvania and University of Puerto Rico at Humacao.
Philadelphia, PA, USA and Humacao, PR.
USA

1. Introduction

Electrospinning is a simple method of obtaining polymer fibers with nanoscopic diameter. It uses electrical forces to produce polymer fibers with nanometer scale diameters. Electrospinning occurs when the electrical forces at the surface of an organic solution or melt overcome the surface tension and cause an electrically charged jet to be ejected. As the the solvent evaporates, an electrically charged fiber remains. This charged fiber can be directed by electrical forces and then collected in sheets or other useful geometrically forms. In this monograph we are exploring the use of electrospinning in the generation of nanoscopic and microscopic fibers of conductors such as graphitic carbons, semiconductors such as SnO_2, and insulators such as the Perovskite PZT. The discussion will center mostly in the fibers electrical properties and it applications.

Carbon Nanofibers: Carbon nanofibers, like other quasi-one-dimensional nanostructures such as nanowires, nanotubes and molecular wires have potential application in a multiplicity of fields, such as high-temperature catalysis, heat-management materials in aircraft, and filters for separation of small particles from gas or liquid. Of more importance to us, there is a possible use as building blocks for bottom-up assembly applications in nanoelectronics and photonics [Mrozowski, 1979; Hu et al, 1999; Duan et al, 2001] Carbon fibers are usually produced by spinning from organic precursor fibers or by chemical vapor deposition (CVD). While the spinning method can only produce microscale carbon fibers, CVD can synthesize carbon fibers with diameters from several microns down to less than 100 nm [Bahl et al, 1998; Endo et al, 2001]. However, CVD involves a complicated process and high cost. Electrostatic generation, or electrospinning technique, invented in the 1930s [Formhals,1934], recently gained renewed interest because it can spin a variety of ultrafine polymer fibers in a micro- or even nanoscale at low cost [Doshi & Reneker, 1995]. By simply pyrolyzing electrospun ultrafine polymer fibers, with a subsequent heat treatment, Chun et al. [Reneker & Chun, 1996] and the authors [Wang et al, 2003] have obtained carbon nanofibers. In general, carbons may include classic carbons such as soot, charcoal, graphite, and "new" carbons. Among the new variants we can mention carbon fibers derived from polyacrylonitrile (PAN), and glass-like carbons derived from nongraphitizable precursors, i.e., various types of more or less crystallized polycrystalline graphites [Iganaki & Radovic, 2002]. These building techniques could overcome fundamental limitations of conventional microfabrication based on lithography [Hu et al, 1999; Duan et al, 2001].

Carbon fibers have wide applications in structural materials such as composites, and potentially in a multiplicity of nonstructural applications such as sensors [Rebouillat et al, 1999]. The recent "rediscovery" of electrostatistic deposition has enabled one to spin a variety of ultra-fine polymer fibers in a simple way, which can be heat treated into carbon fibers with diameter in the nanoscale range [Doshi & Reneker, 1995; Reneker & Chung, 1996; Wang et al, 2002]. The application of carbon nanofibers as sensing elements relies on their electronic transport properties being modulated by the sensing element physico-chemical interaction with the analyte. The authors recently evaluated the size of single electrospun polyacrylonitrile (PAN)-derived carbon nanofibers using a scanning probe microscope for measuring their conductivity at room temperature, and found that the conductivity depends on annealing temperature and time [Wang et al, 2003]

It is well known that the electrical conductivity of pyrolytic graphite increases with temperature. Such temperature dependence was at first explained by the simple two-band (STB) model [Klein, 1964]. The STB model also predicts a level off of the conductivity at a very low temperature. However, recent experimental results show that the conductivity of carbon fibers is very sensitive to temperature at very low values (K) [Bright & Singer,1979; Spain et al, 1983; Koike & Fukase,1987]. Such anomaly has been attributed to weak electron localization [Koike & Fukase,1987], electron-electron interaction [Koike & Fukase,1987], the Kondo effect [Koike & Fukase,1987], and hopping mechanism [Baker & Bragg, 1983], all of which show very weak effects unless evaluated at very low temperatures. As to the overall temperature dependence of conductivity, two-dimensional (2-D) weak localization, hopping and tunneling [Abeles et al, 1975] mechanisms have been put forward as possible explanations.

Although classical electron transport theory predicts an increase of electrical resistance in the presence of a magnetic field [Putley, 1960], Mrozowski and Chaberski found a decrease of resistance with magnetic field, or negative magnetoresistance, in partially ordered (pregraphitic) carbons [Mrozowski & Chaberski, 1956]. Since then, negative magnetoresistance has been found not only in poorly graphitized bulk carbon [Hishiyama, 1996] and carbon thin film [Faist & Lohneysen, 2002], but also in carbon fibers, irrespective of whether the carbon fibers were derived from PAN [Koike & Fukase,1987], benzene [Endo et al, 1982], pitch-derived [Bright & Singer, 1979], or CVD [Fuji et al, 2002]. The most commonly accepted model accounting for the negative magnetoresistance was Bright's model [Bright, 1979]. This model attributes the resistance decrease to the increase of the density of the states and carrier density with magnetic field, arising from the formation of Landau levels. However, the Bright model cannot account for all of the observed phenomena, including the strong temperature dependence of magnetoresistance below liquid-helium temperature, and the absence of magneto-resistance saturation at high-magnetic field. Then, Bayot et al. [Bayot et al, 1984, 1990] explained the effect using a weak-localization mechanism, which results as a consequence of any small disorder in the electronic system. The weak-localization effects in pregraphitic carbon fibers are due to their turbostatic phase structure, in prior 2-D.

It is noteworthy that the anomalous temperature and magnetic field dependence of conductivity have been found in carbon fibers with diameters larger than 10 mm. It is interesting to evaluate the scaling of such effects, that is, whether similar effects exist in carbon fibers with a diameter of nanoscale although such evaluation becomes increasingly difficult as the diameter is reduced. Note that most of the investigated carbon fibers were heat treatment temperature was higher than 1000 °C, with their observed negative magnetoresistance of a few percents in magnitude. The weak localization in carbon fiber originates from its disordered nature. By lowering the carbonization temperature, we can probe a lower degree of order in the carbon fiber, and may observe a stronger weak localization effect. In this chapter we comment on the temperature dependence of the electrical conductivity of carbon nanofiber pyrolyzed at a lower temperature of 1273 K, its large negative magnetoresistance at low temperature, and attempt to explain such properties within the frame work of STB and 2D weak-localization models.

Oxide fibers

Binary oxides fibers: Semiconducting tin oxide (stannic oxide, SnO_2), with a rutile structure and a wide bandgap (E_g = 3.6 eV), is chemically inert, mechanically hard, thermally heat-resistant and has a wide variety of existing and potential applications in sensors and optoelectronics such as solar cells, displays and electrochromic devices [Chopra et al, 1983; Williams, 1987]. While the optoelectronics applications of the oxide are mostly due to its wide bandgap, which makes it transparent up to ultraviolet light, its sensor applications are derived from its conductivity modulation by species chemisorbed on its surface and their interaction with non-stoichiometric oxygen vacancies in its lattice. Although the two kinds of applications require SnO_2 with different nature and levels of crystal defects, such as dopants and oxygen vacancies, both have taken advantage of the thin film morphology, and therefore SnO_2 thin film has been a research focus. So far, the thin film has been synthesized by various methods, such as evaporation [Seal & Shukla,2002], sputtering [Shuah & Fun, 1986], spray pyrolysis [Sinclair et al, 1965], chemical vapor deposition [Santhi et al, 1980] and the sol-gel process [Davazoglou,1997], and its synthetic processes have been characterized and correlated to its final stoichiometry, phase constituents and crystal defects. The preference for thin films in sensor applications is due to its higher surface-to volume ratio than that of the bulk shape and its restriction to the grain growth perpendicular to the substrate. The ratio is even higher and the grain growth is further confined for a fibrous shape. Unfortunately, SnO_2 fiber has been synthesized in only a limited number of ways, such as by laser ablation [Mishra et al, 2002], thermal decomposition [Liu et al, 2003], oxidizing electrodeposition of a template [Xu et al, 2002] and electrospinning [Kolmakov, 2003]. Of these methods, electrospinning is especially interesting in that it is easy, inexpensive, versatile and flexible. The technique was invented as early as the 1930s [Li et al, 2003] and was recently revitalized to synthesize ultra-fine polymer fibers. We were the first to report the synthesis of micro- and nanoscopic inorganic (lead zirconate titanate) fibers using electrospinning [Wang et al, 2004, Wang & Santiago-Aviles, 2002] and we also developed two recipes for the electrospinning of SnO_2 fibers: one was modified from that of SnO_2 thin film fabrication through the sol-gel route [Wang & Santiago-Aviles, 2004] and the other, developed independently, thermally decomposes a single metal-organic, dimethyldineodecanoate tin
($C_{22}H_{44}O_4Sn$), mixed with a solution of poly (ethylene oxide) (PEO, HO−[CH_2-CH_2-O−]$_n$-H) in chloroform ($CHCl_3$). The $C_{22}H_{44}O_4Sn$ compound was chosen because it is inexpensive, commercially available, neither too toxic nor too harmful to the environment and, most importantly, has appropriate rheological properties, especially when it is mixed with PEO/$CHCl_3$ solution for electrospinning [Wang et al, 2005]. The second recipie can further introduce pores to fibres as to enhance their ratio of surface area to volume. Such porous SnO_2 fibres have electrical properties highly sensitive to their environment [Wang et al, 2004]. Since the precursor solution ($C_{22}H_{44}O_4Sn$ /PEO/$CHCl_3$) contains organic groups and Sn–C and Sn–O bonds that are infrared-active, Fourier-transform infrared (FTIR)

spectroscopy will be an effective way to reveal changes in their structure and atomic bonding. This paper uses FTIR, together with thermogravimetric (TG) and differential thermal (DT) analysis (TGA and DTA) and x-ray diffraction (XRD) to further identify the synthesized fibres and to reveal a series of changes that lead to the conversion of the starting chemicals into the final product of porous ultra-fine SnO_2 fibres. This information will help us to control and tailor the micro/nanostructure, porosity and lattice defects of the final SnO_2 fibres so as to meet different specific application requirements.

Transparent conductive oxides (TCOs) have received extensive attention because of their important optoelectronic applications such as electrochromic devices, heat mirrors, and transparent electrodes and antireflection coatings in solar cells. Usually such oxides are semiconductors and their transparency is due to their wide band gap [Wang et al, 2007]. Tin oxide or stannic oxide (SnO_2) is a typical TCO. With a wide band gap of around 3.6eV makes it transparent up to the ultra-violet (UV) light. Although intrinsic stoichiometric single crystal SnO_2 is an insulator, its conductivity can be greatly increased either by impurity doping or by the introduction of oxygen vacancies in its lattice, which donate electrons [Chopra et al, 1983]. Since most optoelectronic, as well as sensing, applications prefer a thin film shape [Chopra et al, 1983; Jarzebski & Marton, 1976], SnO_2 thin films have been synthesized by numerous methods, such as chemical vapor deposition [Davazoglou, 1997], sol-gel [Terrier et al, 1997], spray pyrolysis [Shanti et al, 1999] and polymeric precursor [Giraldi et al, 2006], and their electrical and optical properties have been well characterized [Davazoglou, 1997; Giraldi et al, 2006]. However, for many applications such as a line light source, a fibrous shape is required. Unfortunately, so far only a few methods, namely, thermal decomposition [Xu et al, 2002], laser ablation [Liu et al, 2003], template oxidization [Kolmakov et al, 2003], vapor deposition [Mathur et al, 2005] and electrospinning [Wang et al, 2004,2005], have been developed to fabricate SnO_2 nanofibers or ribbons and their optical properties have been only barely touched [Liu et al, 2004; Dharmaraj et al, 2006] even though such properties characterization is indispensable for their applications. To our knowledge, no report has been made on characterizing their optical band gap, the most important parameter for their optoelectronic applications. The authors of this article have developed two recipes for electrospinning SnO_2 fibers [Wang et al, 2004,2005] and characterized their electrical properties. We investigate their optical and photoconductive properties, with the emphasis on the determination of their optical band gap and conductance response to UV light.

Binary oxide semiconductors have important sensing and optoelectronic applications [Seal & Chukla, 2002; Batzil & Diebold, 2005]. Usually, such oxides have a wide band gap and are good insulators in their undoped and stoichiometric state. However, oxygen vacancies leading to nonstoichiometry can easily be formed in their lattice, donate electrons, and greatly increase their conductivity. Their conductivity is also modulated by species chemisorbed on their surface and the subsequent interaction between the chemisorbed species and the nonstoichiometric oxygen vacancies. We fabricated nanofibers using electrospinning and metallorganics decomposition (MOD) techniques [Wang et al, 2004, 2007]. The conductivity of our synthesized fiber is highly sensitive to its environment, suggesting promising sensing applications. In this letter, we fabricated a gas sensor based on a single electrospun SnO_2 nanofiber and used it to detect moisture, and methanol gases. Although nanowires fabricated in other ways have been used to detect CO, and other gases [Kolmakov et al, 2003], this might be one of the earliest such use using electrospun fibers.

Civilian and industrial safety control, environmental protection and homeland security have stimulated great demand for novel chemical sensors, including gas sensors that can monitor a small amount of toxic, inflammable and/or explosive gases such as hydrogen (H_2) and carbon monoxide (CO) and odorous components such as O_3 and NO_x. The core element in a gas sensor is its sensing material. Binary oxide semiconductors constitute a promising family of sensing materials used in gas sensors because they are cost-effective, chemically inert, mechanically hard, and thermally heat-resistant, and therefore can be used in a harsh environment and are reliable over a long term [Williams, 1987; Seal & Chukla, 2002; Batzill & Diebold, 2005]. Electrically, they have a wide band gap and are good insulators if they are pure and stoichiometric. However, point defects, such as oxygen vacancies, can easily form in their lattice leading to non-stoichiometry, and act as donors as depicted below.

$$V^{=}_{o} = V^{-}_{o} + e, \quad (1a)$$

$$V^{-}_{o} = V_{o} + e, \quad (1b)$$

Such interactions greatly increase their conductivity. On the other hand, environmental oxygen atoms can be adsorbed on their surface, pick up electrons from the conduction band,

$$O_2\,(g) + 2e = 2O^{-}\,(ad), \quad (2a)$$

$$O_2\,(g) + 4e = 2O^{=}\,(ad), \quad (2b)$$

This chemisorption leads to a positively charged layer due to electron depletion immediately below the negatively charged surface and greatly decreasing their conductivity. When the surface is exposed to a reductive gas such as CO, however, the surface-adsorbed O^- or $O^=$ ions will react with the reducing gas molecules, release electrons,

$$CO\,(g) + O^{-}\,(ad) = CO_2\,(g) + e, \quad (3a)$$

$$CO\,(g) + O^{=}\,(ad) = CO_2\,(g) + e, \quad (3b)$$

Leading to a decrease of the depletion zone, and greatly increasing the conductivity. Such conductivity modulation by surface chemisorbed species and their interaction with the point defects provides a reliable gas detecting mechanism for binary oxide gas sensors [Williams, 1987; Seal & Chukla, 2002; Batzill & Diebold, 2005]

Tin oxide (SnO_2) sensors represent some of the early-commercialized chemical sensors (the Taguchi sensor [Naoyoshi, 1975]). Usually the SnO_2 sensing element is used in the shape of a thin film [Capone et al, 2001;Mandayo et al, 2003; Korotchenkov et al, 1999] because of its inherent higher surface-to-volume ratio than bulk. A fibrous or ribbon shape is more favorable for surface sensing than bulk and thin films in that it has an even higher surface-to-volume ratio. The ratio can also be increased by the introduction of pores into the SnO_2 thin film [Jin et al, 1998]. So, porous nanofiber/nanoribbon will be doubly favorable for surface sensing. Unfortunately, these had not been synthesized until recently, when we fabricated porous SnO_2 ribbons, with horizontal and vertical dimensions of 100 nm–20 μm and 10–100 nm, respectively, from the metallo-organic precursor dimethyldineodecanoate tin ($C_{22}H_{44}O_4Sn$) using electrospinning and thermal decomposition techniques [Wang et al, 2004, 2007]. As SnO_2 gas sensors are usually used in atmosphere above room temperature for maximum sensitivity, it is essential to evaluate the electrical conductance (G) of our SnO_2 nanoribbons in analyte gas atmosphere. We also want to find its temperature (T) dependence between room temperature and its probable elevated operating temperature, which is usually 473 to 773 K [Williams, 1987; Seal & Chukla, 2002; Batzill & Diebold, 2005]. Previous studies on bulk and thin-film SnO_2 has demonstrated that, due to the temperature-sensitive thermodynamics of surface chemisorption/desorption and surface– lattice interaction, conductance is dictated by the temperature to a large extent [Jarzebski & Marton, 1976; Ryzhikov et al, 2003]. In this chapter we briefly explored the electrical conductance of our single electrospun porous nanoribbon and its T -dependence in ambient air, and hopefully paves the way for potential sensing applications.

Ternary and more complex oxide fibers: Lead zirconate titanate, Pb (Zr_xTi_{1-x}) O_3 (PZT), is a well-known ferroelectric material with significant technological importance [Yoshikawa et al, 1995]. PZT fibers have potential for utilization in high performance hydrophones and ultrasonic transducer applications [Yoshikawa et al, 1995]. The author's group synthesized PZT fibers with diameter from 100 nm to 20□m by means of electrospinning and metallo-organic decomposition [Wang et al, 2002; Wang & Santiago-Aviles, 2004]. Since then, it has been a challenge for us to evaluate their ferro electric properties. Usually, PZT is evaluated as dielectric media in a sandwiched metal/PZT/ metal capacitor [Trolier-Mckinstry & Muralt, 2004]. Such evaluation does not work for a single PZT fiber with diameter less than 10μm because of its ultra low capacitance. Recently, piezoresponse imaging (PRI) has been developed to probe polarization domains in ferro-electric thin films □(including PZT thin film), measure their properties in micro- or nano-scale, and correlate domain polarization and local properties directly with topography and morphology [Birk et al, 1991;Harnagea, 2007]. We will explore the use of such techniques in elucidating PZT properties.

2. Experimental details

Carbon Nanofibers: The fibers precursor is a commercial polyacrylonitrile (PAN) and N, N-dimethyl formamide (DMF) solution, in a ratio of 600 mg PAN to 10 ml DMF. The solution was deposited on silicon wafers with a 150-nm-thick film of silicon oxide and patterened with 1 X 1 mm2 gold contact array. A homemade electrospinning setup [Wang et al, 2003], was used to spin single precursor fibers between two isolated gold contacts (Fig. b). The as-spun PAN fibers were pyrolyzed and heat treated at 1273 K for 30 min in a vacuum of 10^{-6} torr. The processed fibers were characterized using Raman scattering at room temperature with a green laser (wavelength = 514 nm) as the exciting radiation. The cross-sectional dimensions and area (S) were evaluated using a scanning probes microscopy (SPM) [Wang et al, 2002, 2003].

A two point probe setup was used to continuously monitor the conductance (G) in the temperature range between 300 and 1.9 K, back and forth, without any applied magnetic field. Its resistance (R) was measured from 295 K down to 15 K with a sampling interval around 0.02 K. To suppress the possible heating effect during the measurement, the constant dc current passing through the fiber was kept at 1 μA or below, and the temperature was controlled automatically. Conductance was also measured at 1.9, 3.0, 5.0, and 10.0 K while the applied magnetic field, perpendicular to the fiber, was increased or decreased continuously between -9 and 9 Tesla twice. According to a previous analysis [Rebbouillat et al, 1998], the contact resistance is much less than that of the nanofiber itself. The length L and cross section area S of the fibers were measured using an optical microscope and a scanning probe microscope (SPM) operated in tapping mode. The conductivity was finally determined using $\sigma = GL/S$.

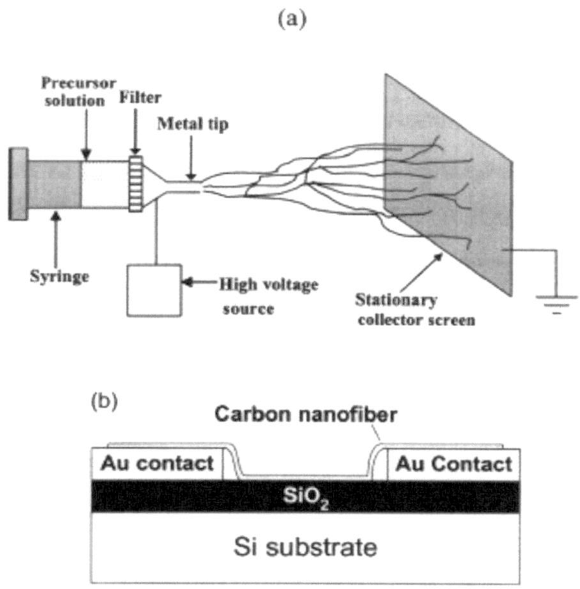

Fig. 1. Schematic of (a) homemade electrospinning setup (b) carbon nanofiber deposited on substrate

Binary Oxides: The SnO_2 fiber fabrication process was characterized by a Fourier Transform Infrared Spectrometer, equipped with a semi-demountable liquid cell with rectangular potassium bromide (KBr) windows and a universal attenuated total reflectance (UATR) sampling accessory with composite zinc selenide and diamond crystals on its top plate. The PEO powder was sampled using the UATR; the pristine $C_{22}H_{44}O_4Sn$ liquid and the polyethylene oxide/$CHCl_3$ and precursor ($C_{22}H_{44}O_4Sn$/PEO/$CHCl_3$) solutions were sampled using the liquid cell. Since our liquids have strong IR absorption, the spacer was withdrawn from the liquid cell and the liquid samples were pressed between two KBr window plates. The produced film is estimated to be no thicker than 10 μm, thinner than the thinnest spacer commercially available (50 μm). The as-deposited and heat-treated mats were scratched off their substrates, ground into powder and sampled using the UATR. To compensate for the high cutoff wave number (650 cm^{-1}) caused by the UATR and Si substrate, the ground powders were then sampled as mulls using the liquid cell with its spacer thickness = 50 μm. Mulls were prepared by dispersing ground powders in the Nujol® (Perkin-Elmer) oil and stirring the mixture electromagnetically until a homogeneous liquid suspension was formed.

Tin Oxide Nanofibers: The synthetic procedure, is briefly recounted here to facilitate the analysis of the results in this chapter. Commercially available PEO (molecular weight 900 000) and chloroform were mixed in the ratio of 10 mg PEO/1 ml $CHCl_3$ and the mixture was stirred using a magnet until the homogeneous solution formed and commercial alkoxide $C_{22}H_{44}O_4Sn$ was added to the solution in a volume ratio of 2:1. The new mixture was stirred until it finally became a homogeneous solution with the appropriate viscosity. TGA and DTA on the precursor solution was conducted using platinum (Pt) pans in a simultaneous differential technique module from room temperature to 700 °C at a heating rate of 10 °C min^{-1}. The precursor solution, after loading into the pan, was left in the air for its solvent to evaporate until its weight became relatively stable for the thermal analysis to start. Precursor fibres and mats were electrospun using a homemade set-up [84–88] onto thermally oxidized single-crystal silicon wafers with (111) orientation, and heat-treated subsequently in air for 2 h at 200, 300, 400, 500 and 600 °C, respectively. The silicon substrates used were thermally oxidized until their surface oxide thin film had a thickness of around 180 nm. The as-deposited precursor fibres were observed under an optical microscope equipped with a digital camera. The microscope, when in its differential interference contrast (DIC) mode, can form an image which appears distinct and three-dimensional. The fibres after heat treatment were observed under an scanning electron microscope (SEM), operated at an accelerating voltage of 3–5 kV, and characterized using an x-ray diffractometer equipped with a Cu Kα x-ray source and a graphite monochromator. Gold contacts of 50 μm x 50 μm x 200 nm with neighboring distance 50 μm were thermally evaporated along one single ribbon using nickel TEM grids as masks, which were fixed by the magnetic force from a flat magnet sheet on the backside of the substrate (figures 2(a) and 2(b)). Electrical properties of single porous nanoribbons were characterized using a self-assembled system (figure 2(c)) in ambient air with a relative humidity of 90%. The sample was heated using a hot plate. A thermocouple was directly attached to the substrate and used to monitor the temperature. To create a uniform local temperature distribution around the sample, an aluminum case was used to enclose the sample. The sample was heated from 300 to 660 K and, immediately after the measurement at 660K, cooled down to 300K with the measuring temperature isochronally varied around every 10K. At each measuring temperature, the measurement was not performed until the temperature was stable for 5 min. I–V characteristics were measured using the two probe method and a semiconductor characterization system. The measuring voltage was swept cyclically, that is, from 0 →1V→−1V→0V with ±0.01V steps.

Fig. 2. Deposition of Au contacts along a single SnO2 ribbon using TEM grids as masks: (a) schematics, (b) optical micrograph of resulting sample, (c), schematics of the testing system.

Ternary and more complex oxides:

3. Comments on the experimental results:

Carbon Nanofibers: Fig. 3 shows an SEM image of the heat-treated carbon fiber. Its horizontal diameter was measured to be approximately 120 nm. Scanning probe microscopy height image analysis revealed an elliptical cross-sectional profile with approximately the same horizontal diameter of 120 nm, a vertical diameter of only 75 nm [1,Wang et al, 2003], and its area S = 7068 ± 200 nm^2. The Raman microscattering spectrum shows two strong peaks centered on 1371 and 1588 cm^{-1}, respectively, indicating disordered and graphitic carbons in the nanofiber. The in-plane graphitic crystallite size La was estimated to be around 2.5 nm [2, Wang et al, 2003].

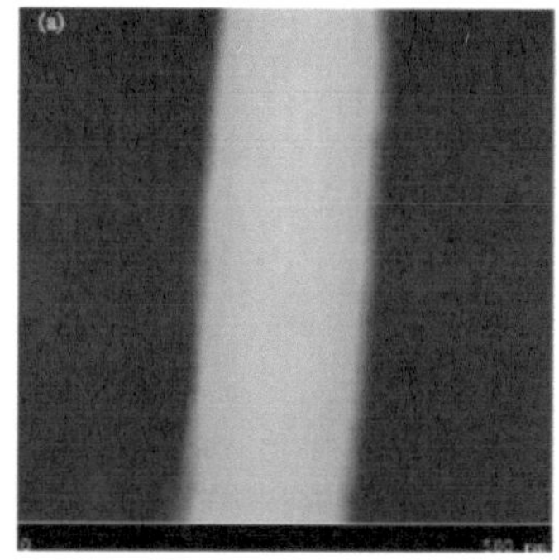

Fig. 3. SEM micrograph of carbon nanofiber

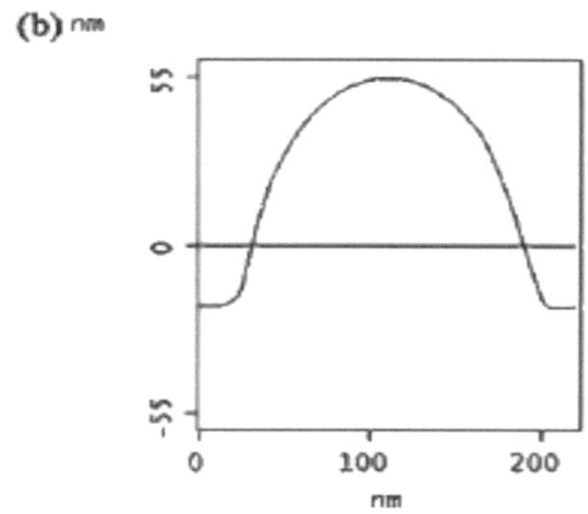

Fig. 4. (a) Scanning probe micrograph of a carbon nanofiber and (b) its average cross-section profile.

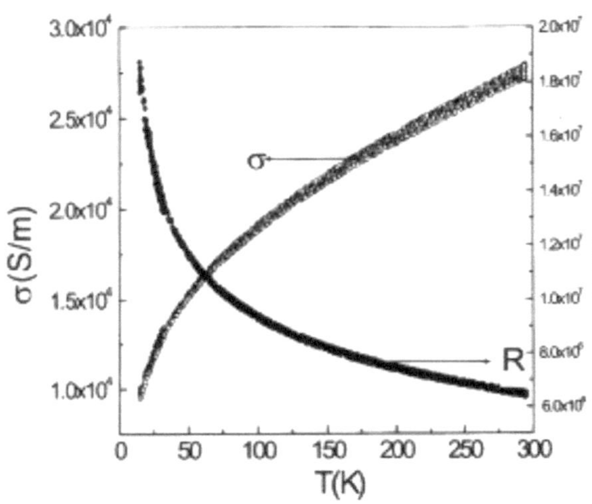

Fig. 5. Temperature dependence of R and σ

The figure above (5), shows a plot of R and σ versus temperature (T) in the range from 15 to 295 K. Note that σ increases monotonically and smoothly from 1.0 X 10^4 S/m at 15 K to 2.75

x 10^4 S/m at 295 K, indicating the semiconducting nature of the fiber. Similar temperature dependence of conductivity was also found in carbon microfibers [Spain & Volin, 1983].

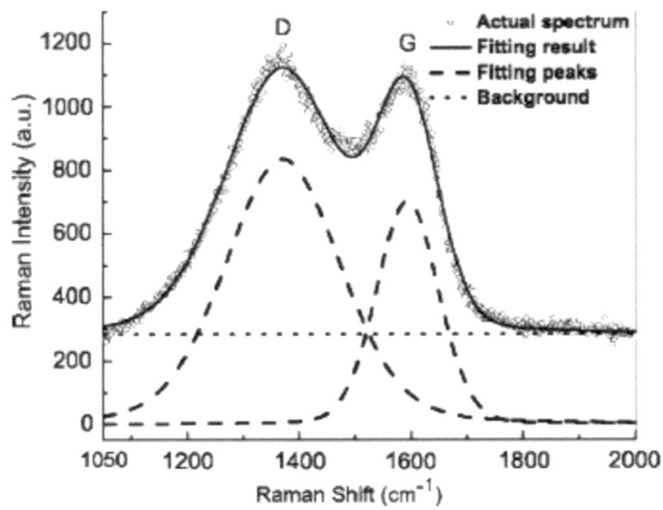

Fig. 6. Raman spectrum of the pyrolized fiber

Fig. 6 above, shows the Raman spectrum of the same carbon nanofibers whose G and D peaks, centered at 1371 and 1588 cm^{-1}, attest to the coexistence of disorder and graphitic carbon in the nanofibers. From the ratio of the integrated intensity of D peak to G peak, the inplane graphitic crystallite size La was estimated to be 2.47 ± 0.08 nm [Mott & Davies, 1979]. The bands were fitted to a Gaussian-Lorentzian shape, with their integrated intensity ratio R=2.1. Since MC is positive for HTT = 1173 K, the carrier mobility cannot be measured from the negative parabolic field dependence of MC as in the case of HTT = 1273 K [146] and the n value is unknown for the time being. However, n and kFl values can still be estimated. At room temperature, σ value for HTT = 1173 K is only half of that for HTT = 1273 K [Fig. 5(c)]. If the ratio is attributed exclusively to the difference in n, n value for HTT = 1173 k will be half of that for HTT = 1273 K [146], i.e., 10^{25} m^{-3}, $k_F \approx (3\pi^2 n)^{1/3} \approx 6\text{x}10^8$ m^{-1}, and its $k_F l \approx k_F L_a$ is 1.2. While the fiber treated at 1273 K between 10 and 200 K, manifesting a WL regime that is not the case for the fiber treated at 1173 K, where the linear relation between ln σ(T) and T1/2 [Fig. 5(d)] confirms the VRH regime between 300 and 5 K. Although such a relation cannot identify whether the VRH regime is of the Mott type (with d=1) or the E-S type, the turbostratic phase structure of our pregraphitic carbon fibers demands d=2 [Mrozowski & Chaberski, 1956], and σ(T) with TES = 187 K. Using $\xi \approx L_a \approx 2$ nm, we can estimate $\varepsilon_r \approx 200$. This value seems reasonable in that it lies between the metallic and insulating limits of $\varepsilon_r \approx \infty$and $\varepsilon_r \approx 10$, respectively.

Fig. 7(a) shows the magnetoresistance MR, defined as MR = $\rho(B)/\rho(0)$ - 1, and the magnetoconductivity MC = $\sigma(B)/\sigma(0)$ -1 of the fibers at temperatures of 1.9, 3.0, 5.0 and 10 K with magnetic field B from -9 to 9 T. At all four investigated temperatures, the MR is negative. Its magnitude increases with an increase in B and a decrease in T. It is noteworthy that MR = -0.75 at T = 1.9 K and B = 9 T, one of the largest negative magnetoresistance known to the authors. Since |MR| is quite large, MR ≈ $(MC)^{-1}$ is not always valid. The

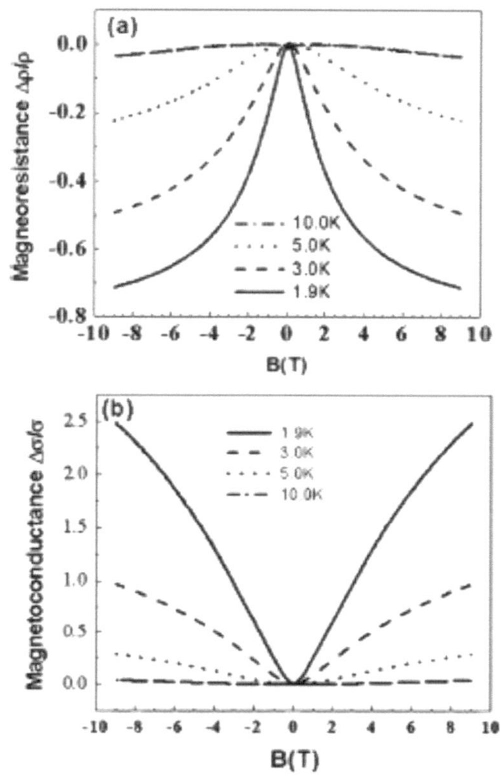

Fig. 7. (a) Large negative magnetoresistance and (b) positive magnetoconductance of carbon fibers with magnetic field from -9 to 9 T.

magnetic field dependence of magnetoresistance for most electronic systems exhibiting 2-D weak localization is relatively easy to understand. In carbon fibers, however the origin is *a priori* not obvious. Bright attributed the origin to the turbostatic nature of the samples, which should have an electronic structure nearly the same as that of 2-D graphite. In the 2-D regime, the correction to the sheet conductance G produced with the magnetic field B perpendicular to the plane of the 2-D carrier system is given by the following expression [Bayot et al, 1989, 1990], [Rosenbaum, 1985]:

$$G(B,T)=G_\infty+\frac{e^2}{\pi h}\left[\frac{3}{2}\Psi\left(\frac{1}{2}+\frac{B_2}{B}\right)-\Psi\left(\frac{1}{2}+\frac{B_1}{B}\right)-\frac{1}{2}\Psi\left(\frac{1}{2}+\frac{B_3}{B}\right)\right],$$

$$B_1=B_0+B_{s.o.}+B_s,$$

$$B_2=B_i(T)+\tfrac{4}{3}B_{s.o.}+\tfrac{2}{3}B_s,\qquad (4)$$

$$B_3=B_i(T)+2B_s,$$

G_∞ is the sheet conductance at infinite magnetic field, or as calculated in the classical Boltzmann formulation of the transport theory; ψ is digamma function, and B_k (k=0, I, s, s.o.) represents the characteristic field associated with the scattering mechanism κk standing for elastic scattering (o), inelastic scattering (i), magnetic impurity scattering (s) and spin-orbit coupling (s.o.). Therefore, the magnetoconductance takes the form

$$\mathrm{MC} = \eta(T)\left[\frac{3}{2}\Psi\left(\frac{1}{2}+\frac{B_2}{B}\right)-\Psi\left(\frac{1}{2}+\frac{B_1}{B}\right)-\frac{1}{2}\Psi\left(\frac{1}{2}+\frac{B_3}{B}\right)-\ln\frac{B_2^{3/2}}{B_1 B_3^{1/2}}\right], \quad \text{where} \quad \eta(T)=\frac{e^2}{\pi h G(0,T)} \tag{5}$$

Given a temperature T, (5) contains only four unknown parameters: η(T), B_1, B_2, and B_3. They can be derived from a nonlinear fitting. The results show that B_2 and B_3 have very close fitting values. In fact, their difference is less than their respective fitting errors. This indicates that both $B_{s.o}$ and B_s are very small, i.e., both magnetic impurity scattering and spin-orbit coupling are very weak. For simplicity, we assume $B_{s.o} \approx B_s \approx 0$, and $B_1 = B_0$, $B_2 = B_3 = B_i(T)$. Then (5) can be simplified as

$$\mathrm{MC} = \eta(T)\left[\Psi\left(\frac{1}{2}+\frac{B_2}{B}\right)-\Psi\left(\frac{1}{2}+\frac{B_1}{B}\right)-\ln\frac{B_2}{B_1}\right]. \tag{6}$$

Nonlinear curve fitting using (6) and Mathematica software (shown in Fig. 8) showed that B_i increases with T, i.e., the inelastic scattering intensifies when the temperature increases.
The temperature dependence of zero field conductivity can be considered in parallel to the modified STB model [Wang & Santiago-Aviles, 2003], which accounts for the conductivity phenomenological, several other models can explain the temperature dependence of the conductivity, namely: 1) 2-D weak localization model: According to the 2-D weak localization model [Lee & Ramakrishnan, 1985; Langer et al, 1996]

$$\sigma(T) = \sigma(0) + \frac{e^2}{2\pi^2\hbar}\ln\left[1+\left(\frac{T}{T_C\,(B,\tau_s)}\right)^p\right] \tag{7}$$

where σ(0) and TC (B, τs) are two constants. If (7) is used to fit the temperature dependence of the zero field conductivity, we obtain, σ(0) = 0, and p = 1.15 (Fig. 9). The value p=1.15 is in agreement with the previous results of p ≈ 1.00 in carbon microfibers [Bayot et al, 1989] and multiwalled carbon nanotube [Langer et al, 1996]. It indicates that the dominating inelastic scattering mechanism is likely to be disorder enhanced electron-electron scattering in 2-D system.
2) Variable range hopping model: This phenomenon occurs in highly disordered materials because the distribution of energy states makes it more favorable to hop to a distant empty state of nearly the appropriate energy than to a nearby empty state that has a much higher energy level. Because of the high resistivity and lack of long-range order of the fiber, one

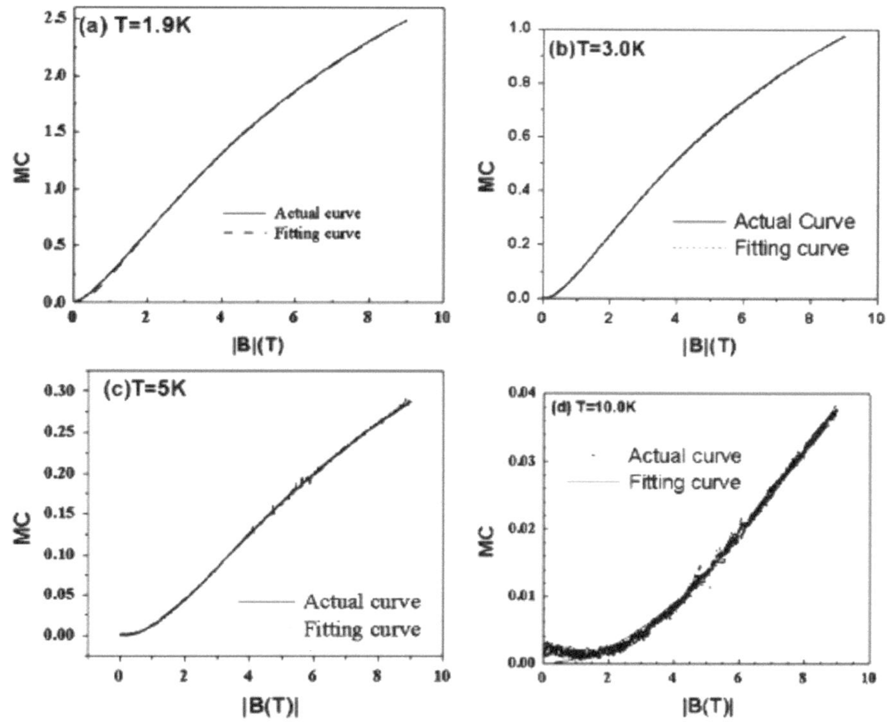

Fig. 8. Comparison of magnetoconductance fitting curves [using (6)] with experimental curves.

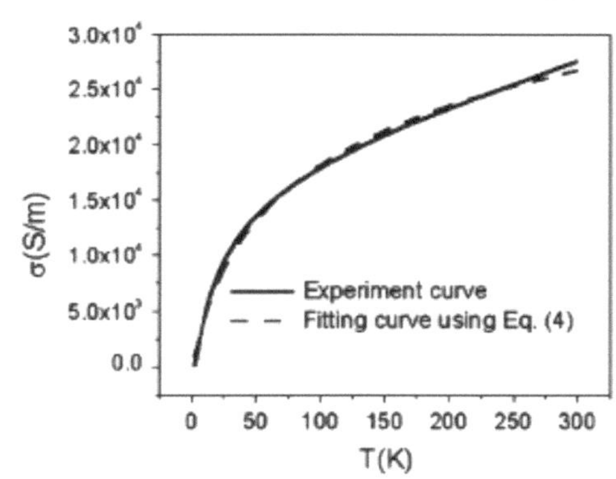

Fig. 9. Fitting of conductivity curve to (7).

would expect that the electrical conduction would result from a hopping mechanism. The dimensionality enters the equation during the summing of the available states. For a d-dimensional system [Mott & Davies, 1979]

$$\sigma(T) = A \exp\left[-\frac{B}{T^{1/(d+1)}}\right] \tag{8}$$

where d = 1, 2, 3, and A, B are two constants. Since there is controversy about the dimensionality of the carbon fibers with respect to variable range hopping, a good three-parameter fit to all the data was found by least square fitting of the data to (8)(Fig. 11). The fitting results d = 0.295 indicates that the dimensionality of hopping lies between 2 and 3, or that d = 2 and d = 3 coexist.

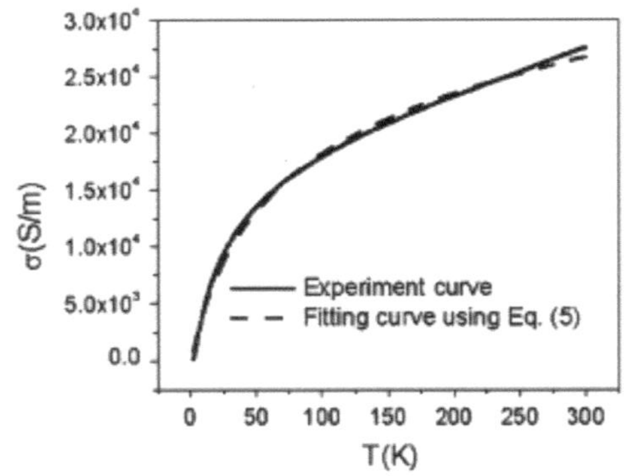

Fig. 10. Fitting of conductivy curve to (5).

Tunneling can be considered as part of transport, the tunneling model was developed for metallic particles imbedded in a highly resistive matrix but the only essential feature of the metallic particles is that the conduction electrons therein are delocalized. As revealed by XRD and Raman spectra, the graphite domains within our carbon nanofibers have a size of 1 to 2 nm [3, Wang et al, 2003]. Since electrons on the small fully carbonized basic units fit this criterion, it is not unreasonable that the model may apply to the investigated fibers. For dc conduction, electron tunnel between the charging centers imbedded in a highly resistive matrix of totally disordered carbon might be plausible. When the electric field is low, the conduction is ohmic resulting in thermally activated charge carriers hopping to the nearest neighbor-charging center. For high-electric fields, the conduction is highly nonohmic and is the result of field-induced tunneling. The conductivity in this model has the same form as 1-d variable range hopping [Abeles et al, 1975]

$$\sigma(T) = A_1 \exp\left[-\frac{B_1}{T^{1/2}}\right] \tag{9}$$

Where A_1 and B_1 are two constants. If the σ(T) versus T curve is fit to (9) [Fig. 11], we obtain A_1 = 41172 S/m, B_1 = 7.92617 $K^{1/2}$. It seems reasonable to try to fit the data to a model in which 3-D variable range hopping and tunneling between domains, in which the electrons are delocalized, coexist. The conductivity for such a case is

$$\sigma(T) = A \exp\left[-\frac{B}{T^{1/(d+1)}}\right] + A_1 \exp\left[-\frac{B_1}{T^{1/2}}\right]. \tag{10}$$

Fig. 12 shows the results of fitting the data to (10). The fitting is excellent. The figure also shows the curves for the two components. They indicate that the conduction is mostly undertaken by a tunneling mechanism. It increases the convexly with the temperature, the hopping mechanism accounting for only a small fraction of the total (its contribution to the conductivity increases concavely with the temperature). As such, the hopping mechanism can be excluded from the main transport mechanism(s) in the investigated carbon fiber.

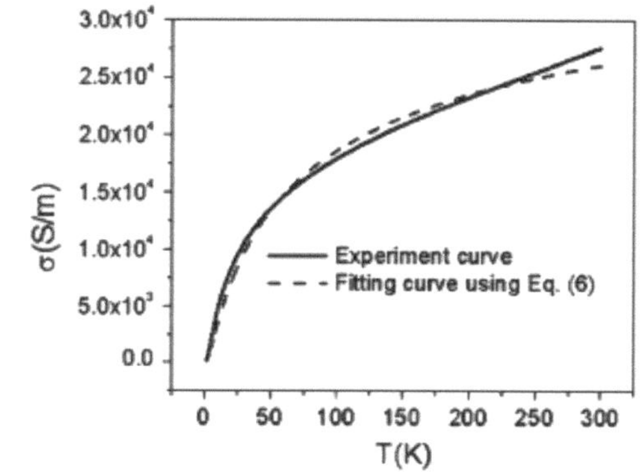

Fig. 11. Fitting of conductivity curve to (9)

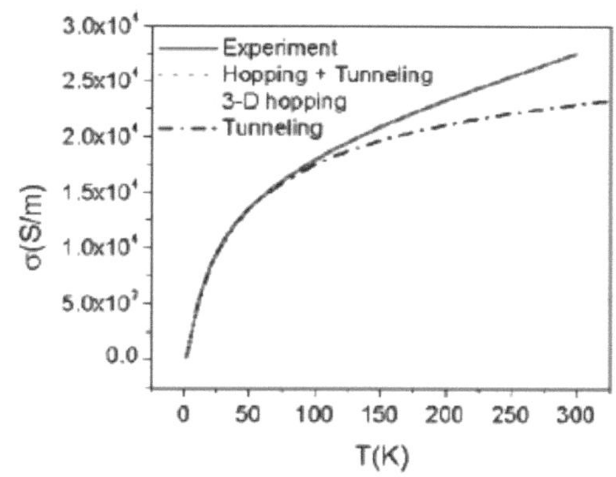

Fig. 12. Fitting of conductivity to (7)

Since both (7) and (9) fit the experimental σ versus T curve quite well, the fitting alone seems not enough to determine whether the main transport mechanism is 2-D weak localization effect, or the tunneling mechanism. However, the good description of the T- and B-dependence of the large MR that can be done using 2-D weak localization effect indicates that the same effect is mainly responsible for the T-dependence of the conductivity. This harmonizes with the low-electrical field setup during the conductance measurement. The voltage applied between the two conducting pads, separated by a distance of 1 mm, is 0.03-0.6 V. So the average electrical field in the carbon nanofiber between the two pads is 30-600 V/m, not strong enough for the tunneling mechanism to dominate.

Binary oxides: Previously we discussed the precursor mixing process for SnO_2. Here, Figure 13 shows thermal analysis curves of the precursor solution. As the temperature increases from room temperature, the weight decreases steadily at first and then slows between 100 and 150 °C, beyond which the weight decreases drastically to 5.2% around 270 °C, where the weight-loss rate drops sharply to a very low level until 385 °C. Above this value the weight remains practically unchanged. The corresponding derivative TG (DTG) curve also indicates three temperature regions in term of the weight loss, peaked around 60, 246 and 301 °C, respectively. The DTA curve, on the other hand, reveals two partially overlapping processes with their exothermic effect ranging from 220 °C to 280 °C (peaked around 263 °C) and from 280 °C to 390 °C (peaked around 345 °C), respectively. Combining the results, one can see that the TG, DTG and DT curves point out four distinct temperature regions with their respective overall changes: $CHCl_3$ evaporates from room temperature to 100 °C (note that 60 °C is its boiling point in air); PEO and $C_{22}H_{44}O_4Sn$ decompose rapidly into an intermediate product between 220 and 300 °C; the intermediate product undergoes a minor weight loss between 300 and 385 °C; and no major chemical reaction occurs with weight change above 400 °C.

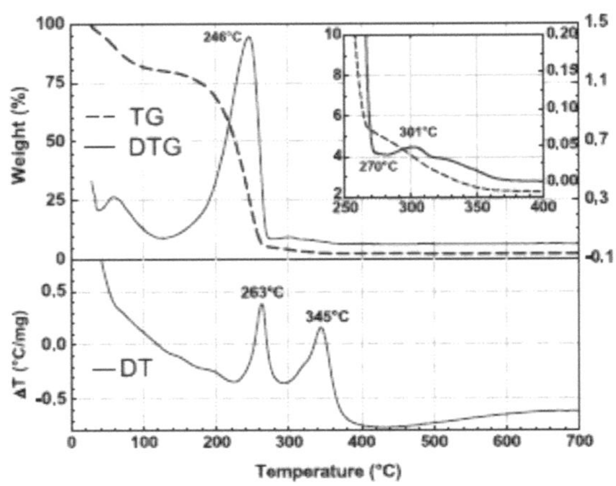

Fig. 13. Thermal analysis (TG, DTG and DT) curves of the precursor solution. The inset highlights TG and DTG curves between 250 and 400 °C

The XRD spectra shown in figure 14 indicate that the samples were amorphous and the 100, 101, 200 and 211 characteristic peaks of the rutile-structured SnO_2 [JCPDS cards] do not emerge until 400 °C. The peaks become more distinct as the heat treatment temperature rises. Additional peaks indexed to the 220 and 310 planes, indiscernible after the heat treatment at and below 500 °C, became evident at 600 °C although they are not as strong and distinct as the others. Except for the Si (111) peak [JCPDS cards] from the substrate in the spectrum of 500 °C, all peaks are characteristic of rutile-structured SnO_2. The multiple diffraction peaks reflect the polycrystalline nature of the sample. Their relative intensity, similar to that in the JCPDF card of rutile-structured SnO_2, indicates no obvious texture (preferential orientation) of crystallites within the samples. The combination of thermal analysis and XRD results gives us a clearer picture of the process: while the exothermic process between 220 and 300 °C does not produce any crystallization, the other process, between 300 and 385 °C, results in the incipient rutile lattice. Above 400 °C, the major change is the further development of the incipient lattice, i.e. phase transformation without obvious weight or stoichiometric change. Optical micrographs show precursor fibres of smooth surfaces with indent belts (figures 15(a) and (b)) along their length axis, whose formation mechanism is under investigation. The equal-thickness interference fringes in figure 21(a) clearly indicate that the height/thickness of the precursor fiber varies along the direction of the horizontal diameter. Optical microscopic observation also revealed that the final SnO_2 fibres have typical lengths of up to a few millimeters. SEM observation further revealed that the fibres are porous and their diameters range from 100 nm to 40 μm (figures 15(c)–(f)).

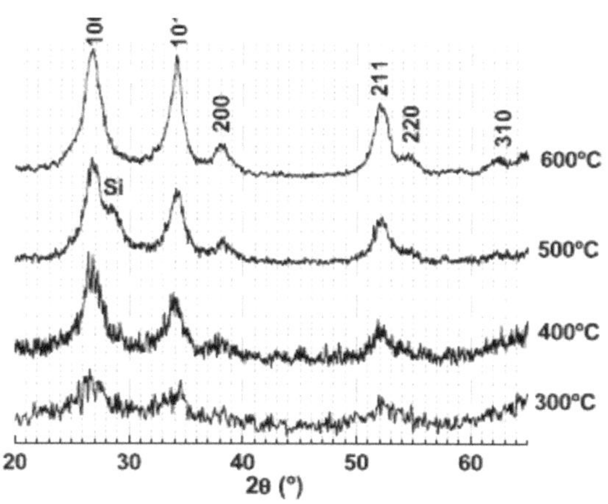

Fig. 14. X-ray diffraction spectra of mat samples heat treated from 300 to 600 °C. The indexes were assigned as the rutile structure of SnO_2 according to [JCPDS cards]. The extra peak at 500 °C is due to (111) diffraction of the single-crystal Si substrate according to [JCPDS cards]. The mat sample heat treated at 500 °C was not dense and thick enough for the diffracted x-ray by the Si substrate not to be detected.

The most porous regions in the final fibres lie in the indent bands of the precursor fibres. We are investigating how the former evolve from the latter and trying to quantitatively measure the surface area of our porous fibres using the BET or any other pertinent method [Brunauer et al, 1938]. The difficulty of the measurement lies in synthesizing enough of the material to do the isotherms and separating our fibres from their substrates without changing their porous morphology. Figure 15(a) shows FTIR spectra of the starting chemicals PEO, $CHCl_3$ and $C_{22}H_{44}O_4Sn$ with their major absorption bands tentatively assigned. The assigning for $CHCl_3$ was easy as it and its IR spectrum are well known and widely published [Silverstein & Webster, 1998]. The assigning for PEO was helped by [Ratna et al, 2006] and [Deng et al, 2006]. The work for $C_{22}H_{44}O_4Sn$ was the most difficult: it is rarely reported and its IR spectrum is unavailable in the literature.

In the FTIR spectrum of the as-deposited precursor mats (figure 16(a)), there is no chloroform absorption bands, indicating a total evaporation of the volatile solvent during the electrospinning. In fact, the as-deposited spectrum is similar to that of pure $C_{22}H_{44}O_4Sn$ except for the sharp bands centred around 1553 and 1411 cm^{-1}. The former 'extra' band was weakened and shifted to 1530 cm^{-1} by the heat treatment at 200 °C, which, however, did not change other major characteristic bands except for the shift of the bands below 800 cm^{-1} toward lower wavenumber. Significant changes occur after the heat treatment at 300 °C, where the bands of such organic groups as methyl (−CH3), methylene (−CH2), methylidyne (-CH), carbonyl (−C=O) and hydroxyl (-O-H) almost disappear (figure 16(b)), suggesting their decomposition. Meanwhile, there appears a band around 760–730 cm^{-1}, which can be attributed to the asymmetric Sn-O-Sn stretching [Almaric-Popescu, 2001], and the Sn-C band merges with the band previously around 634 cm^{-1} into a broad band centred around

540 cm^{-1}. These changes are accompanied by an overall exothermic effect and drastic weight loss (figure 13), characteristic of thermal decomposition of organics. All these strongly suggest that the organic groups in the precursor fibres were thermally decomposed between 200 and 300 °C. The product of the thermal decomposition, as previously revealed by its XRD spectrum, is an amorphous, rather than rutile-structured, SnO_2.

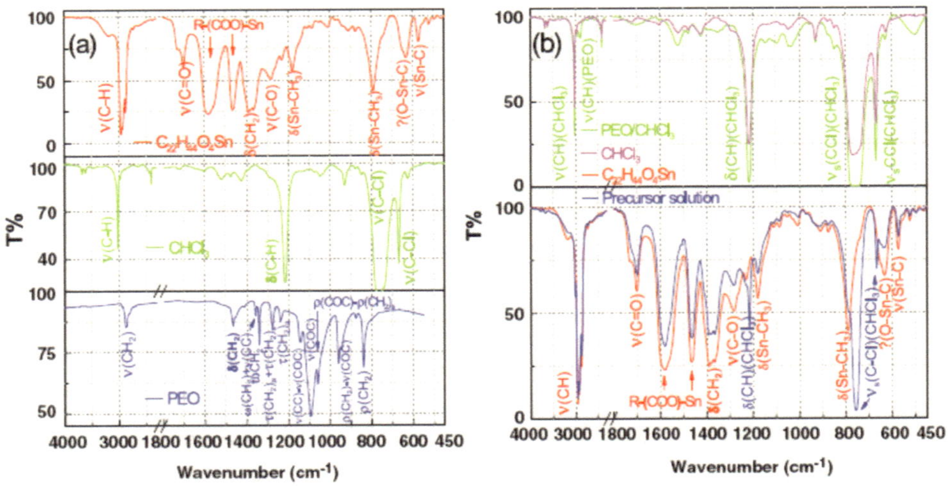

Fig. 15. The FTIR spectra and their absorption band assignment of the starting (a) chemicals: PEO, $CHCl_3$ and $C_{22}H_{44}O_4Sn$, and (b) solutions: PEO/$CHCl_3$ and $C_{22}H_{44}O_4Sn$/PEO/$CHCl_3$. The $C_{22}H_{44}O_4Sn$ bands were tentatively assigned with the help of neodecanoic acid and tetramethyltin spectra; its band around 634 cm^{-1} may be due to the O-Sn-C bridging. The spectrum of the PEO/$CHCl_3$ solution is mainly that of $CHCl_3$ except for the weak C-H stretching band around 2976 cm^{-1}, which is from PEO. The $C_{22}H_{44}O_4Sn$/PEO/$CHCl_3$ precursor solution spectrum, however, is mainly the superposition of those of PEO/$CHCl_3$ and $C_{22}H_{44}O_4Sn$. Fortunately, we found IR spectra of neodecanoic acid ($C_9H_{19}COOH$) [Ratna et al, 2006] and tetramethyltin ($(CH_3)_4Sn$) [Japanese spectral data base], which have the same major groups and structural chains as $C_{22}H_{44}O_4Sn$. The major bands in the two spectra, tentatively assigned with the help of [Kuag, 2002; Fiaz et al, 2002; Hori et al, 1986; Jing et al, 1995], are also predominant in the $C_{22}H_{44}O_4Sn$ spectrum except for the band around 634 cm^{-1}, which may reflect the bridging of O-Sn-C (figure 16(a)). Note that the band around 574 cm^{-1} is assigned to the Sn–C stretching [Kuag, 2002; Fiaz et al, 2002; Hori et al, 1986; Jing et al, 1995], those around 792 and 1180 cm^{-1} are assigned to the Sn-CH_3 bending and that around 1580 cm^{-1} is assigned to R-(COO)- Sn [Hori et al, 1986]. The assignments are listed in table 1. The major IR absorption bands of the PEO/$CHCl_3$ solution mainly come from $CHCl_3$ except for the C-H stretching (υC-H) band around 2890 cm^{-1}, characteristic of the C-H stretching in the chain of HO–[CH_2–CH_2–O–]$_n$–H. The FTIR spectrum of the precursor solution, however, appears to be the combination of $C_{22}H_{44}O_4Sn$ and the PEO/$CHCl_3$ spectra. The C–Cl band of $CHCl_3$ between 840 and 700 cm^{-1} merges with the Sn–CH_3 band of $C_{22}H_{44}O_4Sn$ between 850 and 700 cm^{-1} to form a broader band of the precursor solution from 850 to 700 cm^{-1}, peaked at 760 cm^{-1} and with a shoulder at 780 cm^{-1} (figure 16(b)). It is noteworthy that many bands overlap or combined with other bands and their assignments may not be unique. However, that will not affect our later discussion about general chemical changes during the synthetic process.

PEO		$C_{22}H_{44}O_4Sn$	
Band position (cm^{-1})	Assignments	Band position (cm^{-1})	Assignments
3000–2750	$\nu(CH_2)_a$	2960	$\nu(CH_3)_a$
2952, 2800, 2739, 2693	$\nu(CH_2)_a$	2935	$\nu(CH_2)_a$
1467	$\delta(CH_2)_a - \delta(CH_2)_a$	2875	$\nu(CH_3)_s$
1359	$\omega(CH_2)_s + \nu(CC)$	1730, 1700	$\nu(C{=}O)$
1341	$\omega(CH_2)_a$	1580, 1485	R–(COO)–Sn
1280	$\tau(CH_2)_a + \tau(CH_2)_s$	1465,1390–1360	$\delta(CH_2)$
1242, 1234	$\tau(CH_2)_a$	1280	$\nu(C\text{–}O)$
1145	$\nu(CC) - \nu(COC)$	1230	$\tau(C\text{–}H)$
1095	$\nu(COC)$	1180	$\delta(Sn\text{–}CH_3)$
1060	$\rho(COC) + \rho(CH_2)_s$	793	$\delta(Sn\text{–}CH_3)$
960, 946	$\rho(CH_2)_s - \nu(COC)_a$	634	?(unassigned)
840	$\rho(CH_2)_a$	574	$\nu(Sn\text{–}C)$

Table 1. Infrared band assignments for PEO and $C_{22}H_{44}O_4Sn$. (Note: δ bending; ω wagging; ν stretching; ρ rocking; τ twisting; a anti-symmetric; and s symmetric; signs + and – denote phase relations of coupled coordinates.)

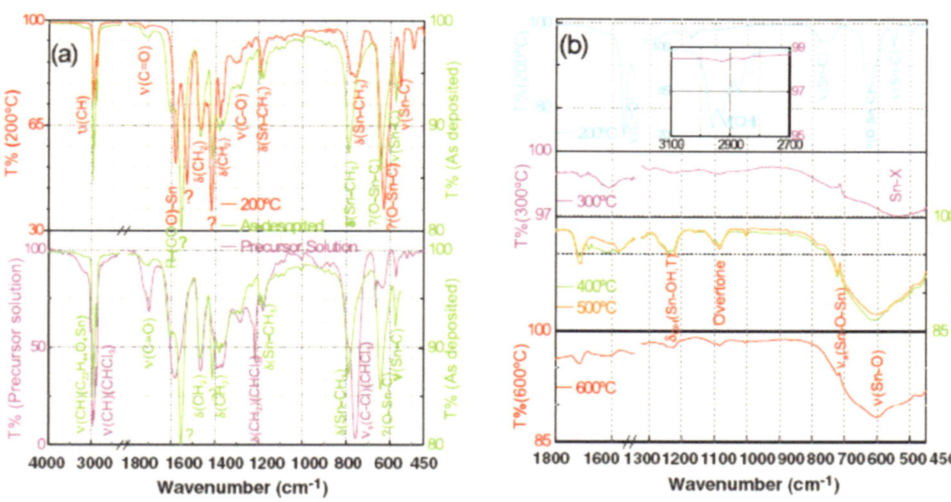

Fig. 16. The FTIR spectra and their absorption band assignment of the precursor solution PEO, $CHCl_3$ and $C_{22}H_{44}O_4Sn$, and mats as-deposited and heat-treated at temperatures between 200 and 600°C. Note that the $CHCl_3$ bands disappeared immediately after the electrospinning. Bands of other organic groups survive the heat treatment at 200 °C and mostly disappear by the heat treatment at 300 °C, above which spectra do not vary so sensitively with the heat-treating temperature. The inset highlights the disappearance of the C-H stretching band between 3100 and 2700 cm^{-1} of the mats heat-treated at 200 and 300 °C.

FTIR spectra (Figure 17(a)) show that organic groups in the precursor fibers decompose mostly between 200°C and 300°C, and transforms into incipient polycrystalline rutile structure around 400°C. On the other hand, the fibers were transparent from near IR (wavelength 1100nm) to its UV absorption edge. The edge shifts toward the longer wavelength when HTT increases from 100°C to 300°C, above which the shift is not apparent (Figure 17(b)).

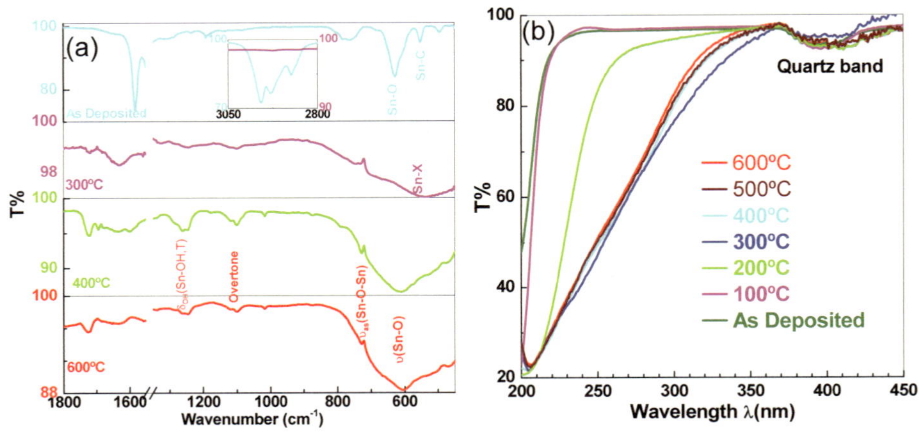

Fig. 17. (a) FTIR and (b) UV/VIS spectra of as-deposited and heat-treated fibers

It is well accepted that the band gap is direct in the rutile-structured SnO_2 [Chopra el al, 1983]. If direct transition is allowed at zero momentum, the optical absorption coefficient α follows, on the long-wave limit,

$$\alpha(E) = [A(E - E_g)]^{1/2}, \tag{11a}$$

or

$$[\alpha(E)]^2 = A(E - E_g), \tag{11b}$$

as a function of photon energy (E), where E_g is the optical band gap and A is a constant [Bardeen et al, 1956]. Since α and transmittance T are correlated as

$$T = e^{-\alpha L}, \tag{12a}$$

or

$$\alpha = -(\ln T)/L, \tag{12b}$$

where L is thickness of sample, we have

$$(\ln T)^2 = AL^2(E - E_g), \tag{13}$$

Heat treating temperature (°C)	300°C	400°C	500°C	600°C
Optical band gap E_g (eV)	3.985~0.004	3.997~0.001	3.952~0.002	4.087~0.002

Table 2. Optical band gap determined from UV absorption edge

which predicts a linear relation between (ln T)² and E. Such a linear relation is confirmed in the photon energy range of 4.3-5.2eV), and its extrapolation to the E-axis gives us E_g values for different Heat treatment temperatures (Table 2). Overall, the E_g value increases with HTT but the variation is very small. The values lie within the reported range of 3.8-4.2 eV for SnO_2 thin film using the same method [Terrier et al, 1997; Shanti et al, 1999], and is higher than the better accepted electrical band gap (E_{g0}) of stoichiometric single crystal SnO_2. The possible reasons for the difference include the quantum confinement caused by the porosity [Lehemann & Gosele, 1991] and the Burnstein-Moss effect [Burnstein, 1954; Moss, 1954] (or the free carrier concentration [Ye et al, 2005]), as well as the defect effect [Terrier et al, 1997]. It is noteworthy that, curves of αi (i=1/2, 1) and (αE)j (j=1/2, 1 and 2) vs E have also been linearly fitted and extrapolated to obtain optical E_g. Such ways of data processing sometimes result in approximately the same Eg values as those listed in Table I, or values closer to the E_{g0} value. However, they lack of convincing physical grounds in our case, and their results are not presented here. The four-probe measurement shows that the conductance (G) of fibers is insensitive to the visible light, which is consistent with its transparency in the visible light range However, G is sensitive to UV light (Figure 19). Upon the UV illumination, G increases sharply and then asymptotically to its steady value by 58% (the on-to-off ratio is 1.58); when the UV light is turned off, G decreases asymptotically by 32% (the on-to-off ratio is 1.32) to its new steady values. The 90%- rise and fall times of the conductance are 89 and 119 seconds, respectively. The high optical band gap and photoconductance imply promising optoelectronic applications of the nanofibers. The relatively long response times are attributed to the surface adsorption and desorption processes triggered by the generation and recombination of photo electron-hole pairs. As the experiment was conducted in the air, various species had been adsorbed on the fiber surface and their adsorption had been in equilibrium [Jarzebski & Marton, 1976].

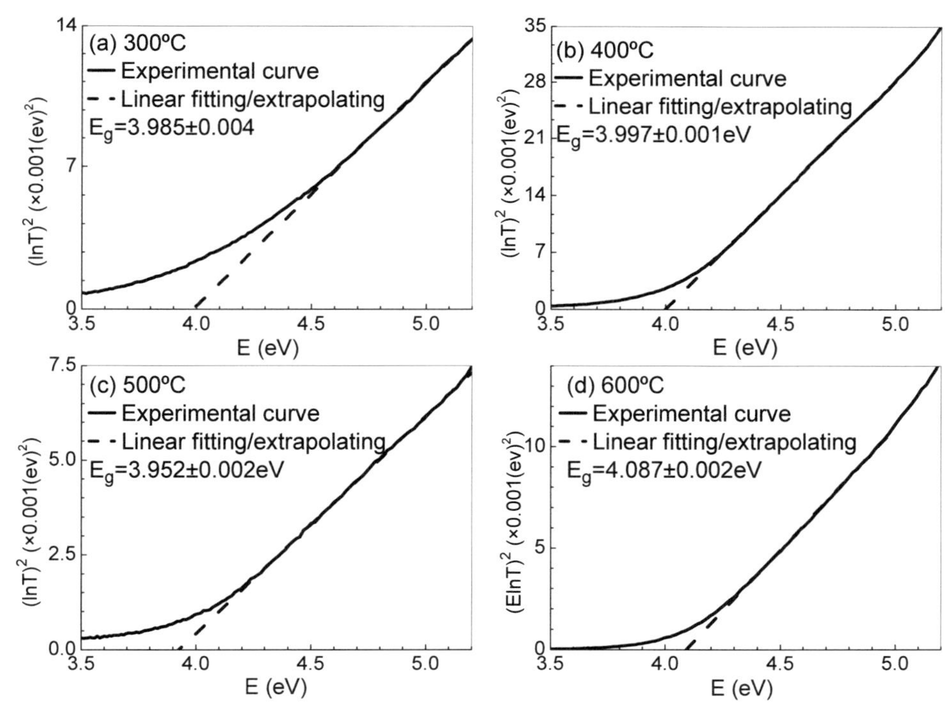

Fig. 18. Determination of optical band gap E_g from the UV absorption edge.

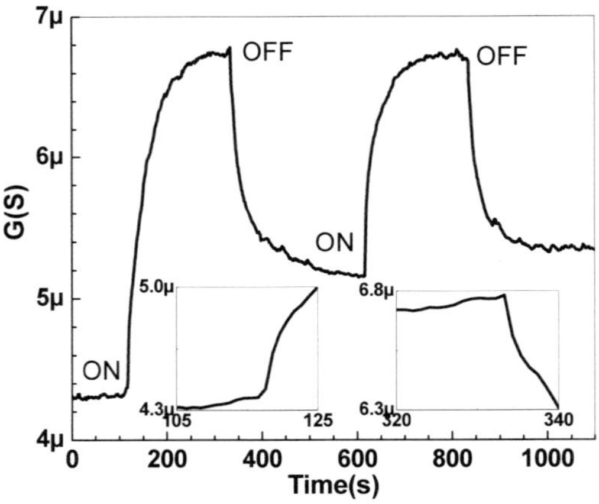

Fig. 19. Transient conductance responses to 254 nm UV light.

The UV-generated electron-hole pairs disturb equilibrium and species are adsorbed or desorbed to attain a new steady surface state. When the UV light is off, the photo electrons and holes recombine with each other, species were adsorbed or desorbed to return to the equilibrium. Since the processes involve surface migration and diffusion of species, the overall stabilization takes time. It is also interesting that G does not return to its initial value after the first UV on-and-off cycle, implying an irreversible surface change by the UV illumination. However, G returns to its starting value on the second cycle.

The sensor's typical time response, Fig.20, show the turn-on and off of gases with their main response parameters listed in Table 3. Upon exposure to the nitrogen carrier gas at a higher flow rate, the sensor showed a higher relative conductance (G) change and shorter response times, defined as the times it takes G to change by 90% of the measured change. Such response is reasonable in that a higher flow rate generates a higher moisture concentration in the chamber, accumulates and ventilates it in a shorter time. In fact, the measured response times are those of the whole testing system, and are affected not only by the sensor's intrinsic response time but also by gas generation, transport and venting times. A higher flow rate decreases the latter three times. Additionally, G increased after the first

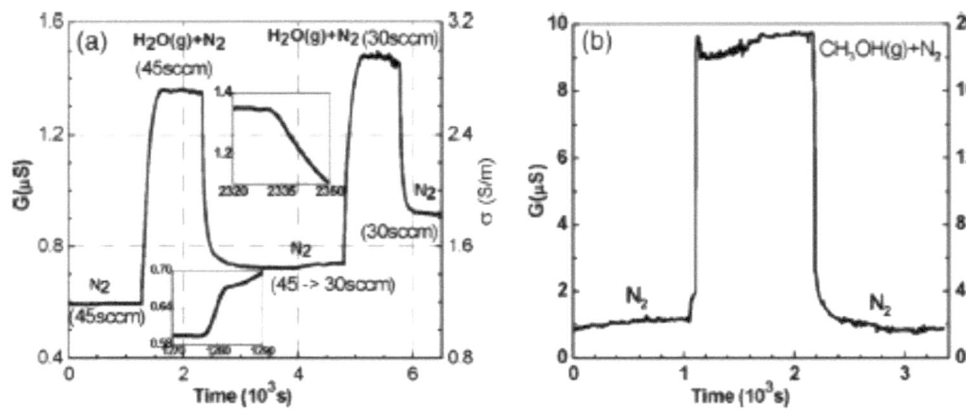

Fig. 20. Response of the sensor to (a) moisture and (b) methanol gas.

exposure to the moisture, suggesting a not totally reversible change (such as adsorption) during that exposure. This may have contributed to the higher G level during the second exposure than during the first.
Such response times, although close to the lower limit of the response time range of sensors based on thin films [Seal & Chukla, 2002], are still dictated by surface adsorption and desorption processes, which, after triggered by the introduction and exhaustion of reducing or oxidizing analytes, involve atomic diffusion and surface species migration

Analyte Gas	N_2 flow rate (sccm)	G(high)/G(low) – 1 Rise	Fall	90% Response time (s) Rise	Fall
CH_3OH	45	845%	845%	38	10
H_2O	45	229%	88%	108	150
H_2O	30	105%	62%	218	187

Table 3. Main response characteristics of the sensor

With the same flow rate of 45 sccm, and increased G by 229% and 845%, respectively. The rise and fall times are 108 and 150 s for water vapor, and 38 and 10 s for ethanol, respectively. The higher relative G change and the shorter response times with do not necessarily mean higher sensitivity to than to since their concentration in the chamber could not be measured. It could, at least partially, be attributed to the higher volatility of , which may have accumulated a higher concentration in the chamber within a shorter time. The effect is consistent with that of the flow rate on the sensor's responses to moisture. Although the sensor's response times are not exactly the same as the measured system response times, they cannot be much less than the latter and are expected to be several to tens of seconds.
All measured I-V characteristics are linear, as typically shown in figure 21(a), and their linear slope (conductance) varies with the measuring temperature as shown in figure 21(b). Apparently both the heating and cooling curves show that G increases slowly with T until 420–480 K, and rises sharply beyond that T range. A careful look into the curves (see the inset in figure 21(b)), however, finds noteworthy differences between the heating and cooling curves. Whereas G does not increase until 400 K during the heating, it decreases steadily during cooling down to 300 K; at the same temperature from 300 to 580 K, G is higher during the cooling than the heating, and, after the heating and cooling cycle, G returns to a higher value than its initial value. Such differences and hysteresis strongly

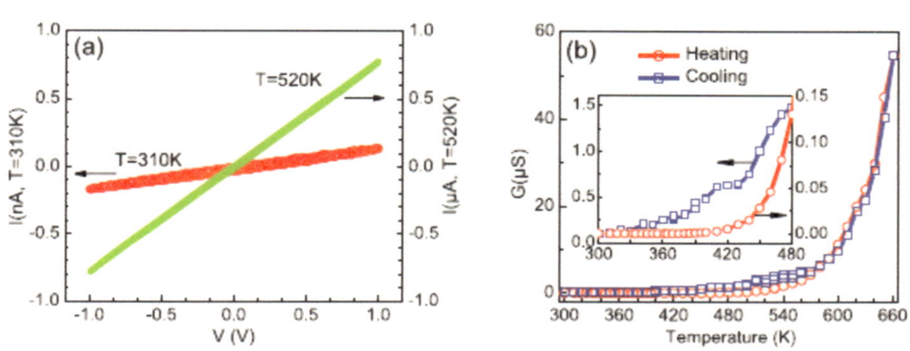

Fig. 21. (a) I-V characteristics and (b) temperature dependence of the conductance

suggest irreversible changes during the cycle, as is the case in SnO_2 thin films [Williams, 1987; Chopra et al, 1983; Jarzebski &Marton, 1976Ryzhikov et al, 2003]

To look into the mechanism that controls the behaviors shown in figure 21(b), ln G is plotted against 1/T in figure 22(a), where the hysteresis between 300 and 570 K is more obvious. Moreover, during heating from 380 to 660 K, cooling from 660 to 570 K, and cooling from 570 to 300 K, there exist Arrhenius type relations

$$\text{Ln } G = \ln G_0 - \Delta E/kT, \tag{14}$$

where k, G_0 and ΔE are the Boltzmann constant, the pre-exponential factor, and the thermal activation energy, respectively, and their fitting values are listed in table 4.

Temperature range (K)	300–380	380–570	570–660
Heating	Non-Arrhenius relation	$\Delta E = 0.918 \pm 0.004$ eV,	
Cooling	$\Delta E = 0.259 \pm 0.006$ eV, $G_0 = 569 \pm 7$	$G_0 = (9.2 \pm 0.2) \times 10^{-4}$	

Table 4. Fitting values for thermal activation energy ΔE and pre-exponential factor G_0.

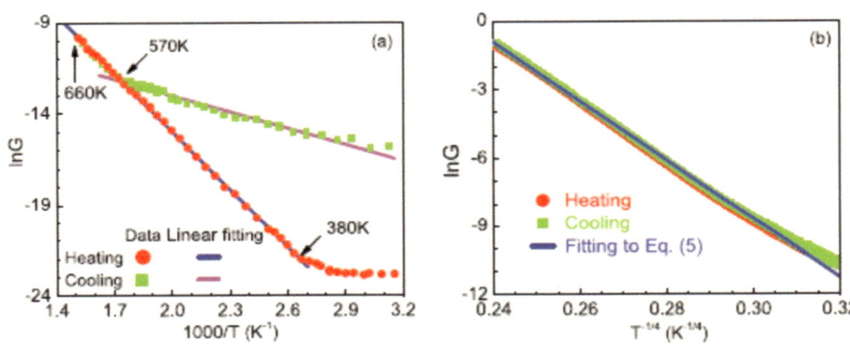

Fig. 22. (a) Arrehnius relation in air and (b) VRH relation in vacuum

For comparison, the T-dependence of conductance in a similar SnO_2 nanoribbon between 300 and 90 K in a vacuum is shown in figure 22(b). Obviously, there is almost no G hysteresis in a vacuum. Although figures 22(a) and (b) correspond to different temperature ranges, their trends around 300 K are comparable. While G is not sensitive to T in air during cooling, σ or equivalently G in vacuum strongly depends on T according to the mechanism of variable-range hopping (VRH) in three dimensions,

$$\ln \sigma(T) = \ln \sigma_0 - (T_0/T)^{1/4}, \tag{15}$$

where σ0 and T0 are constants. The comparison clearly indicates that G was controlled by surface-related processes during the heating in air. In fact, the weak G hysteresis in figure 22(b) may be attributed to the partially reversible adsorption and desorption of residual gases in the chamber whose vacuum was only 10^{-3} Torr. A similar bifurcate linear ln G versus 1/T relation also exists in SnO_2 thin films and has been attributed to adsorption and desorption of multiple kinds of molecules and species on SnO_2 surface [Williams, 1987; Chopra et al, 1983; Jarzebski & Marton, 1976; Ryzhikov et al, 2003]. However, the specific temperature ranges and ΔE values for SnO_2 thin films differ from our results. The difference is due to themorphology, impurities and non-stoichiometric defects in SnO_2, resulting from different preparation methods and conditions, as well as in the testing environment and procedures. Since our measurement was conducted in air with a relatively high humidity, the effect of moisture, in addition to oxygen, must be taken into account. Although the detailed intermediate mechanism is still ambiguous and perhaps controversial, it is accepted that moisture, after chemisorption and reaction, donates electrons and increases G like a reducing gas following the overall path

$$H_2O\ (g) + S = H_2O{+}(ads) + e, \qquad (16)$$

where S stands for the SnO_2 surface .Note that moisture and oxygen have opposite effects on G and their adsorption and desorption dynamics are different. It is plausible that the surface-chemisorbed H_2O molecules are desorbed during heating from 300 to 380 K and that the desorption consumes conducting electrons (according to the reverse of equation (23)). However, the consumption is compensated for by the increase of bulk conducting electron concentration due to T rises, and the overall G is not sensitive to the temperature in this range. At around 380 K, oxygen starts to be desorbed from the surface and the desorption contributes to the overall net G increase as described by the Arrhenius relation equation (21). Upon cooling, oxygen is adsorbed reversibly and G follows the same Arrhenius relation until 570 K, where the H_2O molecules start to be additionally adsorbed, slowing down the overall G decreasing rate and reducing the activation energy ΔE because of its opposite effect to that of oxygen adsorption. The G versus T hysteresis seems to be a consequence of the partial replacing of the adsorbed oxygen by moisture on the SnO_2 surface after the heating and cooling cycle. Such irreversible replacement in turn can be attributed to the numerous pores within the ribbon, whose capillary effects additionally restrict adsorption, desorption, diffusion, and evaporation around and inside them [Ryzhikov et al, 2003].

The SEM and SPM analysis indicates that the $SnCl_4$ precursor led to better results in terms of uniformity/continuity of the fibers. Figure 23 shows the XRD spectra for fibers electrospun using the $SnCl_4$ precursor. After sintering at 400 °C, 110, 101 and 211 XRD peaks of rutile structure SnO_2 appear, indicating the formation of its recipient lattice. These peaks become more distinct and an additional 200, 220 and 310 peaks showed up after sintering at 600 °C,

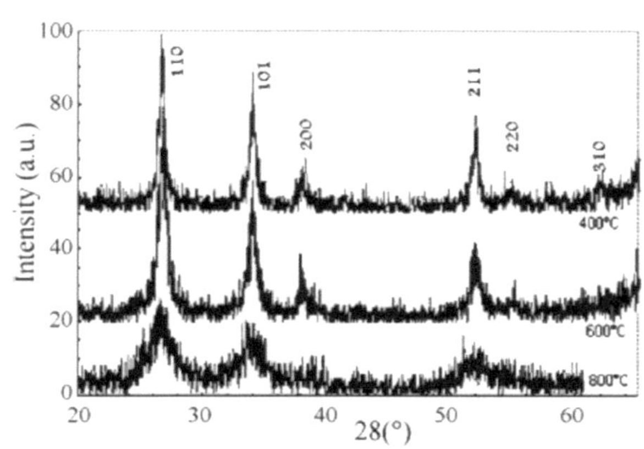

Fig. 23. XRD spectra for $SnCl_4$-based fibers at different temperatures.

indicating the development of a more integral rutile lattice between 400 °C and 600 °C. Up to 800 °C, all peaks are still identified as SnO_2 peaks, suggesting the following overall chemical reactions:

$$Sn(OH)_nCl_{4-n} + (4 - n)H_2O \rightarrow SnO_2 + 2H_2O + (4 - n)HCl, \quad (17)$$

$$[-CH_2CH_2O-]_n + (5n/2)O_2 \rightarrow 2nCO_2 + 2nH_2O, \quad (18)$$

and n = 0, 1, 2, 3, 4.

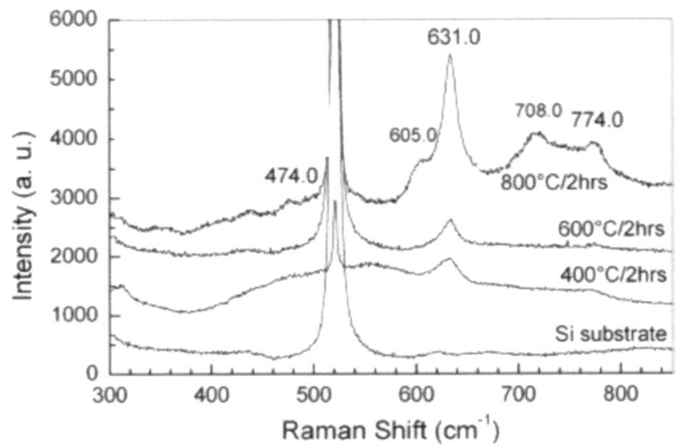

Fig. 24. Raman micro-spectra of Si substrate and $SnCl_4$-based mats sintered at different temperatures.

For sintering temperatures of 400, 600 and 800 °C, the relative intensities of the diffraction peaks are consistent with those reported in Ref. [JCPDS cards]. Typical Raman micro-scattering spectra are presented in Fig. 24 After sintering at 400°C, a peak around 631cm^{-1} shows up, whereas another weak peak centered around 774cm^{-1} begins to appear. The two peaks become more distinct after sintering between 600 and 800°C, whereas the third peak appears around 474 cm^{-1} after sintering at 800°C. The SnO_2 rutile structure belongs to the space group $P4_2/mnm$ [Scott, 1970]. Using X- ray photoelectron spectroscopy atomic sensitivity factors, the relative atomic concentrations of Sn, O, C and C were analyzed semi-quantitatively and their dependence on the sintering temperature is shown in Fig. 25.

Upon sintering at 400°C, Cl almost disappears; C concentration decreases sharply from 36.4% to 5-8%, whereas O concentration increases to more than 50%, and Sn concentration increases to 38%. The drastic decrease of carbon concentration and the regular shift of the XPS C1s peak caused by sintering suggest that the carbon is a residual element rather than an adventitious extrinsic impurity.

Ternary and more complex oxides: For the PZT Perovskite, the starting metallo-organics were still zirconium n-propoxide (dissolved in free alcohol), titanium isopropoxide, and lead 2-ethylhexanoate. The synthesis process was the same as above, except for the two refinements: (1) The viscosity was increased by reducing the fraction of xylene; (2) A syringe filter was added to the tip. Single crystal silicon wafers, coated or uncoated with a 150 nm-thick platinum film, were used as substrates. The as-electrospun precursor fibers or mats were sintered in air for two hours at 300, 400, 500, 600, 700, 800, and 850°C, respectively. Samples of the thick precursor solution, uniformly spread over a silicon wafer by spin-coating, as-electrospun and as-sintered fibers or mats were characterized using a Rigaku x-

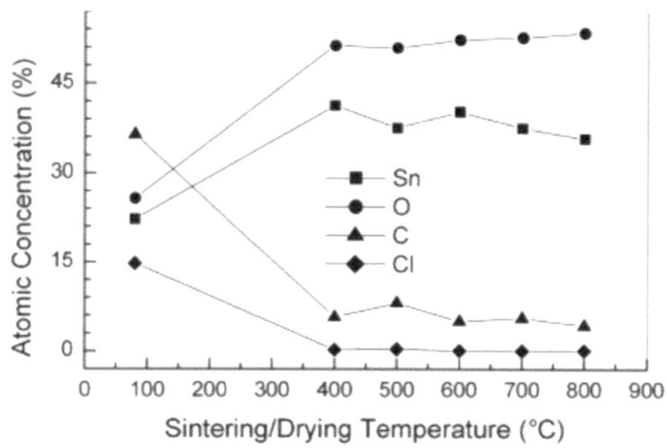

Fig. 25. Sintering temperature dependence of the atomic concentration of $SnCl_4$-based fibers.

ray diffractometer and a JEOL JSM-6300FV scanning electron microscope, equipped with an energy dispersion spectroscopy (EDS) attachment. The precursor solution and the mats were additionally characterized using an FTIR spectrometer. While the transmittance spectra of the precursor solution were obtained using a liquid cell with spacing of 0.05mm, specular reflectance spectra of mats samples were obtained using a specular reflectance accessory. In latter case, the same substrates, without any deposit and sintered in the same batch with the tested samples, were used for background scanning. The sintered samples were imaged and analyzed by piezoresponse imaging techniques (PRI). This technique is implemented through the combination of scanning probe microscopy and lock-in electrical measurement techniques [Birk et al, 1991; Harnagea, 2007]. It makes use of a sandwiched conductive SPM tip/PZT thin film/bottom metal capacitor, where the vertical SPM tip acts as the top electrode that can be precisely positioned and moved in a program-controlled way, dc and ac voltages [V_{dc} and $V_{ac}\cos(\omega t)$] are applied simultaneously between the tip and the bottom metal,

$$V_{tip}=V_{dc}+V_{ac}\cos(\omega t), \tag{19}$$

set up electric field (E) and induce piezo electric oscillation (δ) on PZT surface,

$$\delta = \delta_0 + \delta_1 \cos(\omega t + \phi_1) + \delta_2 \cos(2\omega t + \phi_2) + \ldots\ldots \tag{20}$$

which is actually detected by SPM as a voltage signal (V),

$$V = V_0 + V_1 \cos(\omega t+\varphi_1)+ V_2 \cos(2\omega t+\varphi_2) + \ldots\ldots \tag{21}$$

Where the subscript i (=0,1,2,...) stands for the ith harmonic component, ϕ_i - φ_i is the ith harmonic phase shift of the SPM detecting system, phase ϕ_1=0 and π when the polarization vector P is up and down in the local domain. Of the harmonic components of various orders, the dc or static components reflect the surface topography if V_{dc}=0; the 1st harmonic component reflects local piezo electric properties,

$$\delta_1 = d_{33}V_{ac}, \tag{22}$$

where d_{33} is the local piezoelectric constant, V_1, δ_1, and d_{33} are proportional to the magnitude of polarization (P) and their sign reflects the P direction. With V_{dc} = 0, P is the spontaneous polarization P_s. For each tip position, if V_{dc} is swept and d_{33} is measured, d_{33} can be plotted against V_{dc} as a local piezoelectric curve [68].

The local piezoelectric curve is not the same as the better-known polarization (P versus E) curve in macroscale. Usually, E is considered to be more pertinent than Vdc for properties characterization. However, d_{33} has not been plotted against E, mainly due to the difficulties in calculating E from V_{dc}. The actual interaction between the SPM tip and PZT thin film has been modeled using the layer capacitor model, the method of image charges, and the finite element analysis (FEA) [Harnagea, 2007]. In these cases, the approximate azimuthal symmetry reduces the problem into two dimensions [Lanyi et al, 1997].

PRI was conducted on electrospun PZT microfibers using a Dimension™ 3100 SPM by Digital Instruments under the conditions: V_{dc} = 0V, V_{ac} =10 V, ω=140πKHz. Figure 26 show typical SPM topography and PRI micrographs of a single fiber about 5 μm in diameter and an area within that fiber. Despite of the continuous surface topography [Fig. 26 (a)], PRI micrographs reveal numerous spontaneous polarization domains of size from 100 to 1000 nm [Figs. 26(b), 26(c)]. Their bright, dark, and grey colors correspond to up, down, and horizontal polarization directions, respectively.

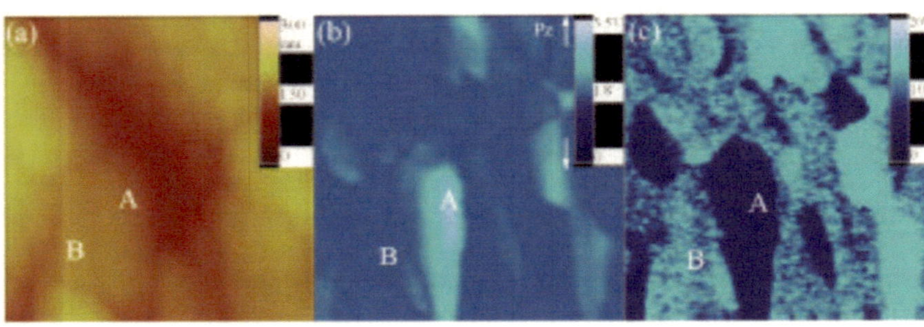

Fig. 26. (Color) (a) SPM topography, (b) PRI amplitude, and (c) PRI phase images of a single PZT microfiber (the area size is 5X5 μm^2).

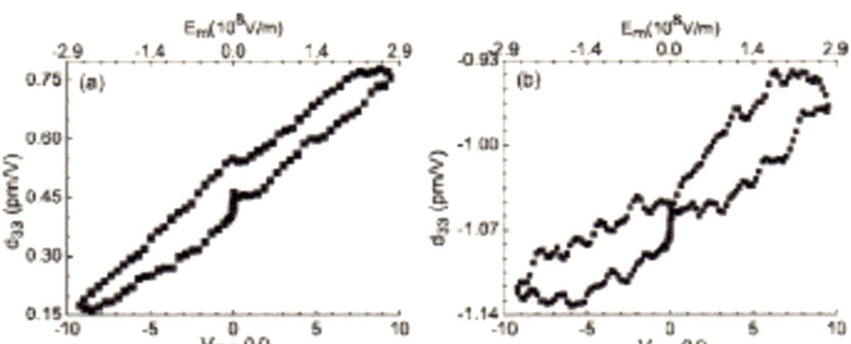

Fig. 27. Local d_{33} vs. V_{dc} curves of domains (a) A and (b) B in Fig. 26

During our local measurement, the SPM tip was positioned on two neighboring domains with opposite polarization direction (Fig. 26) and V_{dc} was swept in a cyclic way of 0 V → 10

V → -10 V → 0 V with a step of ±0.1 V. Figure 27 presents local piezoelectric curves as d_{33} against V_{dc} and E_{max} using Eq. (17) with r=35 nm. Note that d_{33} keeps its sign in both curves, implying that the polarization was not switched in two domains. Although E_{max} is as high as $2.9X10^8$ V/m. E diminishes rapidly toward the bottom metal.

4. References

Abeles B., Sheng P., Coutts M. D., and Arie Y. (1975). Structural and electrical properties of granular metal- films. *Adv. Phys*, vol. 24, no. 3, 407–16.

Amalric-Popescu D and Bozon-Verduraz F(2001), *Catal. Today*, 70, 139–154

Bahl O. P., Shen Z, Lavin J. G., and. Ross R. A (1998). Manufacture of carbon fibers, *Carbon Fibers, 3rd ed*, Donnet J.-B., Ed, Marcel Dekker, New York, 1–84

Baker D. F. and Bragg R. H. (1983). One-dimensionality in glassy carbon, *Phys. Rev. B*, vol. 28, no. 4, 2219– 2221

Barsan N. and Weimar U. (2003), *Journal of Physics: Condensed Matter*, 15, R813–R839

Bardeen J., Blatt F. J. and. Hall L. H (1954) *Photoconductivity Conference, Atlantic City*, Breckenridge R. G.,. Russell B. R and Hahn E. E., Ed, John Wiley, New York,

Batzill M and Diebold U (2005) *Prog. Surf. Sci*1956. 79, 47–154

Bayot V., Piraux L., Michenaud J.-P., and Issi J.-P. (1989), *Phys. Rev. B* 40, 3514-3519

Bayot V, Piraux L., Michenaud J.-P., and Issi J.-P. (1989), Weak localization in pregraphitic carbon fibers, *Phys. Rev. B*, vol. 40, no. 6, 3514–3523

Bayot V., Piraux L., Michenaud J. P., Issi J. P., Lelaurain M., andMoore A. (1990), Two-dimensional weak localization in partially graphitic carbons, *Phys. Rev. B*, vol. 41, no. 17, 11 770–11 778

Bergmann G. (1984), Weak localization in thin films a time of flight experiment with conduction electrons, *Phys. Rep* 107, 1, 1–58

Birk H., Glatz-Reichenbach J., Jie L, Schreck E., and Dransfeld K., J. (1991), *Vac. Sci. Technol.* B 9, 1162-1166.

Bright A. (1979) Negative magnetoresistance of pregraphitic carbons *Phys. Rev. B*, vol. 20, no. 12,. 5142–5149

Bright A. A and Singer L. S. (1979) Electronic and structural characteristics of carbon fibers from mesophase pitch. *Carbon*, vol. 17, 59–69.

Bright A. A. and Singer, L. S. (1979), *Carbon* 17, 59-64

Brinker C., Hurd A., Schunk P., Frye G. and Ashley C. (1992), *J. Non-Cryst. Solids* 147/148, 424-429

Brunauer S., Emmett P. H. and Teller E. (1938) *J. Am. Chem. Soc.* 60, 309–319

Burstein E. (1954) *Physl Rev*, 93,632-637

Candeloro P., (2005), *Microelectronic Engineering*, 78, 178-185.

Capone S., Siciliano P., Quaranta F., Rella R., Epifani M. and Vasanelli L. (2001), *Sensors Actuators B* 77, 503–511

Comini E.., (2002) *Applied Physics Letters*, 81, 1869-1875.

Chopra K. L., Major S., andPanda D. K.(1983), Transparent Conductors- A status Review, *Thin Solid Films*, 102, 1-46.

Chuah D. S. and Fun H. (1986), *Mater. Lett.* 4, 274-277

Davazoglou D. (1997), *Thin Solid Films* 302, 204-213

Deng Y., Dixon J. B. and White G. N. (2006), *Colloid Polym. Sci.*, 284, 347–356

Dharmaraj N., Kim C. H., Kim K. W., Kim H. Y., Suh E. K. (2006), *Spectrochimica Acta Part A* 64, 136-140

Doshi J. and Reneker D. H. (1995) Electrospinning process and application of electrospun fibers, *J. Electr.*, 35, 151–156

Duan X., Huang Y., Wang J., and Lieber C. M. (2001), Indium phosphid nanowires as building blocks for nanoscale electronic and optoelectronic devices, *Nature*, 409, 66–69

Duan X., Huang Y., Wangand J., and Lieber C. M. (2001)., *Nature (London)* 409, 66-72

Endo M., Kim Y. A., Takeda T., Hong S. H., Matusita T., Hayashi T., and Dresselhaus M. S.(2001), Structural characterization of carbon nanofibers obtained by hydrocarbon pyrolysis, *Carbon*,. 39, 2003–2010.

Endo M., Hishiyama Y., and Koyama T. (1982), Magnetoresistance effect in graphitizing carbon fibers prepared by benzene decomposition, *J. Phys. D, Appl. Phys.*, 15, 353–363.

Faißt A. and Löhneysen H. V. (2002), Electrical resistance and magnetoresistance of pyrocarbon at Low temperatures in fields up to 14 tesla, *Carbon*, vol. 40, 321–327.

Fiaz A., Ali S., Parvez M., Munir A., Khan K. M., Shah T. A. and Mazhar M. (2002), *Heteroatom Chem.* 13 638–649

Formhals A. (1934), *Process and Apparatus for Preparing Artificial Threads*, U.S. Pat. 1 975 504

Fuji M., Matsui M., Motojima S., and Hishikawa Y. (2002), Magnetoresistance in carbon micro-coils obtained by chemical vapor deposition, *Thin Solid Films*, 409, 78–81.

Giraldi T. R.,. Escote M. T., Maciel A. P, Longo E., Leite E. R, Varela J. A. (2006), *Thin Solid Films*, 515, 2678-2685.

Gopel W. and Schierbaum K.D. (1995), *Sensors and Actuators B*, 26, 1-8

Harnagea C. (2001), *Local piezoelectric response and domain structures in ferroelectric thin films investigated by voltage-modulated force microscopy*, Ph.D. thesis; http://sundoc.bibliothek.uni-halle.de/diss-online/01/ 01H318, June 23, 2007.

Hishiyama Y. (1976), Negative magnetoresistance in soft carbons and graphite, *Carbon*,. 8, 259–269

Hori M., Hattori S., Yoneda T. and Morita S. (1986), *J. Vac. Sci. Technol. B*, 4, 500–504

Hu J., Odom T. W., and Lieber C. M. (1999), Chemistry and physics in one dimension: Synthesis and properties of nanowires and nanotubes, *Acc. Chem. Res.*, 32, 435–445

Hu J., Odom T. W., and Lieber C. M, (1999), *Acc. Chem. Res.* 32, 435-441

Inagaki M and Radovic L. R. (2002)., *Carbon* 40, 2279-2287

Ishioka M., Okada T., and Matsubara K. (1993), Electrical resistivity, magnetoresistance, and morphology of vapor-grown carbon fibers prepared in a mixture of benzene and Linz–Donawitz converter gas by floating catalyst method, *J. Mater. Res.*, 8, 8, 1866–1874.

Jackson J. D. (1999), *Classical Electrodynamics*, 3rd ed., Wiley, New York

JCPDS Card No. 41-1445, 27-1402, 41-1445

Japanese National Institute of Advanced Industrial Science and Technology. (2006), *Spectral Database for Organic Compounds* available at http://www.aist.go.jp/RIODB/ SDBS/cgi-bin/cre index.cgi

Jarzebski Z. M. and Marton J. P. (1976), *Journal of Electrochemical Society*, 123 199C-205C, 123 (1976): 299C-310C, 123 (1976): 333C-346C.

Jensen J. O. (2004), *Spectrochim. Acta A*, 60, 2561-2570

Jin Z., Zhou H-J., Jin Z-L., Savinell R. F. and Liu C-C. (1998), *Sensors Actuators B*, 52, 188–194

Klein C. A. (1964), STB model and transport properties of pyrolytic graphites, *J. Appl. Phys.*, 35, 10, 2947–2957.

Koike Y. and Fukase T. (1987), Anomalous electrical conduction in carbon fibers at low temperatures, *Solid-State Commun.*, 62, 6, 499–502

Kolmakov A., Zhang Y., Cheng G., and Moskovits M. (2003), *Adv. Mater.* 15 (12), 997-1003

Korotchenkov G., Brynzari V. and Dmitriev S. (1999), *Sensors Actuators B* 54, 197–201

Kolmakov A. and Moskovits M. (2004), *Annu. Rev. Mater. Res.*, 34, 151-170

Kwag G. (2002), *Macromolecules* 35, 4875–4879

Langer L., Bayot V., Grivei E., Issi J.-P., Heremans J. P., Olk C. H., Stockman L., Van Haesendonck C., and Bruynseraede Y. (1996), Quantum transport in a multiwalled carbon nanotube, *Phys. Rev. Lett.*, 76, 3, 479–482

Lanyi S., Torok J., and Rehurek P. (1996), *J. Vac. Sci. Technol. B*, 14, 892-897

Law M., Goldberger J. and Yang P. (2004), *Annu. Rev. Mater. Res.*, 34, 83-88.

Lee P. A. and. Ramakrishnan T. V.(1985), Disordered electronic system, *Rev. Modern Phys.*, 57, 2, 287–337.

Lehmann V., Gosele U. (1991), *Appl. Phys. Lett*, 58, 856-860

Liu Z., Zhang D., Han S., Li C., Tang T., Jin W., Liu X., Lei B. and Zhou C. (2003), *Adv. Mater.* 15, 1754-1762

Li D.,Wang Y. and Xia Y. (2003), *Nano Lett.*, 3, 1167-1173

Mandayo G. G., Castano E., Gracia F. J., Cirera A., Cornet A. and Morante J. R. (2003), *Sensors Actuators B* 95, 90–96

Mathur S., Barth S., Shen H., Pyun J. and Werner U. (2005), *Small*, 1, 7713-7717.

Mishra S., Ghanshyam C., Ram N., Singh S., Bajpai R. P. and Bedi R. K. (2002), *Bull. Mater. Sci.*, 25, 231-245

Moss T. S. (1954), *Proceedings of Physics Society London B*, 67,775-791.

Mott N. F. and Davis E. A. (1979), *Electronic Processes in Non-Crystalline Materials*. Oxford, Clarendon Press, U.K.

Moulder J.F. et al. (1992), *Handbook of X-Ray Photoelectron Spectroscopy*, Jill Chastain, (by Perkin-Elmer Corporation, Physical Electronics Division, Minnesota, USA,).

Mrozowski S. (1979), Specific-heat anomalies and spin-spin interactions in carbons—Review, *J. Low Temp Phys.*, 35, 3/4, 231–297.

Mrozowski S. and Chaberski A. (1956), Hall effect and magnetoresistivity in carbons and polycrystalline graphites, *Phys. Rev.*, 104, 74, 74–83

Naoyoshi T. (1975), Japanese Patent No. S50-30480

NIST Chemistry Webbook, NIST Standard Reference Database Number 69, (2005), Release available at http://webbook.nist.gov/chemistry Nov 5, 2006

Pouchert C. J. (1997), *The Aldrich Library of FT-IR Spectra 2nd edn*, vol 1–4 (Wisconsin, USA: Sigma-Aldrich Co.)

Putley E. H. (1960), *The Hall Effect and Related Phenomena*, 66–97 , Butterworth, London, U.K.

Ratna D.,Divekar S., Samui A. B., Chakraborty B. C. and Banthia A. K. (2006), *Polymer* ,47, 4068–4074

Rebbouillat S., Peng J. C. M., Donnet J.-B., and Ryu S.-K. (1998), *Carbon Fibers, 3rd ed*, Donnet J.-B., Wang T. K., and Peng J. C. M., Eds, 463–542, Marcel Dekker, , New York

Renker D. H. and Chun I. (1996), Nanometer diameter fibers of polymer, produced by electrospinning, *Nanotechnology*, 7, 216–223

Rosenbaum R. (1985), Superconducting fluctuations and magnetoconductance measurements of thin films in parallel magnetic fields, *Phys. Rev. B*, 32, 4, 2190–2199

Ryzhikov I. A., Pukhov A. A., Il'in A. S., Glukhova N. P., Afanasiev K. N. and Ryzhikov A. S. (2003) *Microelectron. Eng.*, 69, 270–273

Sacher R. E., Davidsohn W. and Miller F. A. (1970), *Spectrochim. Acta* ,26, 1011–1022

Santhi E., Dutta V., Banjeree A. and Chopra K. L. (1980), *J. Appl. Phys.*, 51, 6243-6249

Santiago-Aviles J.J. and Wang Y. (2003), *Carbon*, 41, 2665-2670.

Scott J.F. (1970), *J. Chem Phys*, 53, 852-855

Seal S. and Shukla S. (2002), Nanocrystalline SnO gas sensors in view of surface reactions and modifications, *Journal of Metals*, 54, 9, 35–60

Shaw T. M., Trolier-McKinstry S., and McIntryre P. C. (2000), *Annu. Rev. Mater. Sci.*, 30, 263-284
Shanthi S., Subramanian C., Ramasamy P. (1999), *Crystal Research and Technology*, 34, 8, 1037-1046
Sinclair W., Peters F., Stillinger D. and Koonce S. (1965), *J. Electrochem. Soc.*, 112, 1096-1100
Silverstein R. M. and Webster F. X. (1998), *Spectrometric Identification of Organic Compound 6th* edn Wiley, New York
Spain I. L., Volin K. J., Goldberg H. A., and Kalnin I. (1983), Electronic properties of pan-based carbon fibers, *J. Phys. Chem. Solids*, 44, 8, 839–849.
Tang A. (1997), *Sensors and Actuators B*, 43, 161-172.
Terrier C., Chatelon J. P. and Roger J. A. (1997), *Thin Solid Films*, 295, 95-100
Trolier-Mckinstry S. and Muralt P. (2004), *J. Electroceram.*, 12, 7-16
Wang Y., Serrano S., and Santiago-Aviles J. J. (2002), Conductivity measurement of electrospun PAN-based carbon nanofiber, *J. Mater. Sci. Lett.*, 21, 1055–1057.
Wang Y., Serrano S. and Santiago-Avilés J. J. (2002), *Materials Research Society, Symposia Proceedings Vol. 702* Materials Research Society, Pittsburghp, 359-362.
Wang Y. and Santiago-Aviles J. J. (2002), *Mater. Res. Soc. Symp. Proc. 702,* 235–340
[a],Wang Y., Santiago-Aviles J. J., Furlan R., and Ramos I. (2003), Pyrolysis temperature and time dependence of electrical conductivity evolution for electrostatically generated carbon nanofibers, *IEEE Trans. Nanotechnol.*, 2, 39–43
Wang Y. and Santiago-Avilés J. J. (2003), Large negative magnetoresistance and two-dimensional weak localization in carbon nanofiber fabricated using electrospinning, *J. Appl. Phys.*, 94, 3, 1721–1727
[b], Wang Y., Serrano S., and Santiago-Aviles J. J.(2003), Raman characterization of carbon nanofibers prepared using electrospinning, *Synthetic Metals*, 138, 423–427
Wang Y. and Santiago-Avilés J. J. (2004), *Nanotechnology*, 15, 32-37
[a], Wang Y., Furlan R., Ramos I. and Santiago-Avieles J. J. (2004), *Appl. Phys. A*, 78, 1043-1047
[b],Wang Y., Aponte M., Leon N., Ramos I., Furlan R., Evoy S. and Santiago-Avilés J.J. (2004), *Semicond. Sci. Technol.*, 19, 1057-1061
[c],Wang Y., Aponte M., Leon N., Ramos I. (2004), Synthesis and characterization of tin oxide microfibers electrospun from a simple precursor solution, *Semicond. Sci. Tech.*, 19, 1057-1062.
[a],Wang Y., Aponte M., Leon N., Ramos I., Furlan R., Pinto N., Evoy S. and Santiago-Avil´es J. J. (2005), *J. Am. Ceram. Soc.*, 88, 2059-2063
[b],Wang Y., Aponte M., Leon N., Ramos I. (2005), J. *Am. Ceram. Soc. 88* ,8, 2059-2064.
Wang Y., Ramos I., and Santiago-Avilés J. (2007), Synthesis of ultra-fine porous tin oxide fibers and its process characterization, *Nanotech.*, 18, 7, 295601–295607
Xu C., Xu G., Liu Y., Zhao X. and Wang G. (2002), *Scr. Mater.*, 46, 789-793
Ye J. D., Gu S. L., Zhu S. M., Liu S. M. Zhang., Y. D, Zhang R., Shi Y. (2005), *Appl. Phys. Lett.*, 86,19,2111-2117
Ying T. H., Sarmadi A. M., Hop C. E. C. A. and Denes F .(1995), *J. Appl. Polym. Sci.*, 55, 1537–1551
Yoshikawa S., Selvaraj U., Moses P., Withams J., Meyer R, and Shrout T. (1995)., *J. Intell. Mater. Syst. Struct.*, 6, 152-158
Watson J. (1984), *Sensors and Actuators,* 5, 29-35.
Williams D. (1987), *Solid State Gas Sensors*, 71-87, P Moseley and B Tofield, Ed, Hilger Bristol U.K.

6

Preparation of Functionalized Nanofibers and Their Applications

Young-Seak Lee and Ji Sun Im
Chungnam National University
Republic of Korea

1. Introduction

Recently, nano-scaled materials have been investigated with amazing increased interest due to their many advantages, such as large surface area and many active surface sites. Among different nano-scaled materials, nanofibers have been widely applied in industry due to the ease in production processes compared to other nano-materials. Nanofibers can be identified as fibers having diameters between tens and hundreds of nanometers. This nano-scaled diameter of fibers can give an enormous surface area per unit mass.
Nanofibers are usually fabricated by the electrospinning method as a non-woven mat. Non-woven fabrics are textile materials consisting of randomly oriented fibers connected together by physical entanglements or bonds between individual fibers, without any knitting or stitching. These non-woven nanofibers are the primary alternative for traditional textiles as filtration media, energy storage media, hygienic and health/personal care textiles, thermal and sound insulating materials, ecological materials, building materials, geo-textiles and automotive textiles.

2. Formation of nanofibers by electrospinning

2.1 Theory of electrospinning

New applications require fibers with increasingly smaller diameters. Since the surface area is proportional to the fiber diameter and the volume is proportional to the square of the diameter, the specific surface area is inversely proportional to the fiber diameter, leading to high specific surface areas for small fibers. In addition, pore size depends on the fiber diameter; therefore, small fibers produce non-wovens with a small pore size. There are several methods for producing small diameter fibers using high-volume production methods, such as fibrillation, island-in-sea, and the novel melt-blowing system, in addition there are highly accurate methods such as nano-lithography and self-assembly. However, the usefulness of above methods is restricted by combinations of narrow material ranges, high costs and low production rates. In comparison, electrospinning is a simple and low cost process and has an intermediate production rate. Electrospinning is a process for sub-micron scale polymer-based filament production (usually called nanofibers) by means of an electrostatic field. Due to these forces, the meniscus of a liquid flowing out of a capillary nozzle elongates, forming a fine jet, that is later atomized into fine droplets. The droplets obtained by this method are electrically charged. A basic electrospinning setup consists of three elements: an electrical generator (high voltage supply), a capillary (jet source) and a metal collector (target). Depending on the flow rate and potential of the capillary, the droplets can be of submicron size, with a narrow size distribution. The solution is usually electrically charged by the generator, and the collector is grounded, but it is also possible to invert the process by electrically charging the collector and grounding the solution. When the electrostatic forces exceed the surface tension force, the pendent droplet at the capillary tip is stretched into a cone (called Taylor cone), and a solution stream is ejected (Dzenis, Y.A., 2004). Whether the jet will form a continuous fiber or disperse into droplets depends on polymer molecular weight, polymer chain entanglement, and the solvent applied to the process (specifically, its evaporation rate). It is known from the literature that smooth fibers are produced when the product of intrinsic viscosity (η) and polymer concentration (c), known as the Berry's number, Be=ηc, is greater than a certain critical value Becr, which is characteristic of the polymer. Specific viscosity of a polymer solution is determined as the ratio:

$$\eta_{sp}=(\eta_0-\eta_s)/\eta_s \quad (1)$$

where η_0 is the zero shear rate viscosity of the polymer solution at concentration (c), and η_s is the solvent viscosity. From this equation, the intrinsic viscosity (η) of the polymer is determined as a linear extrapolation of specific viscosity η_{sp} measured for various concentrations to the concentration at c=0:

(2)

The intrinsic viscosity (η) is also related to the molecular weight (M_W) of a linear polymer by the Mark–Houwink equation:

$$\eta=KM_W{}^a \quad (3)$$

where the constants (*K*) and (*a*) depend on the polymer, solvent and temperature.

2.2 Electrospinning setups

A basic electrospinning setup consists of three elements: an electrical generator (high voltage supply), a capillary (jet source) and a metal collector (target) (Fig. 1).

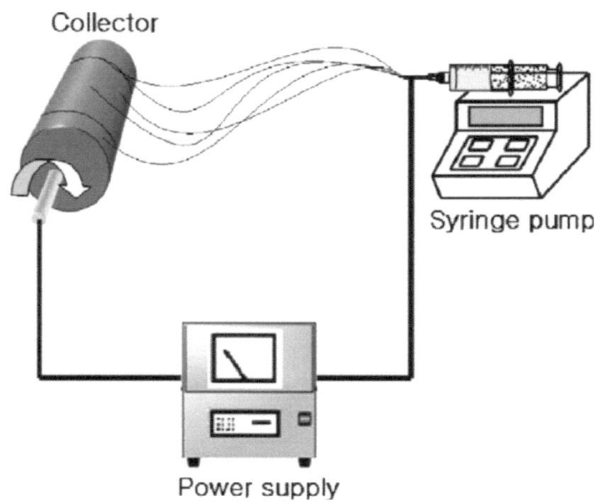

Fig. 1. Diagram of the electrospinning apparatus (J.S. Im et al., 2008).

In our group, polyacrylonitrle (PAN)-based nanofibers were obtained by the electrospinning method. The electrospinning was carried out under the following conditions: [10 wt% of PAN/DMF polymer solution, 15 kV] (J.S. Im et al., 2008).

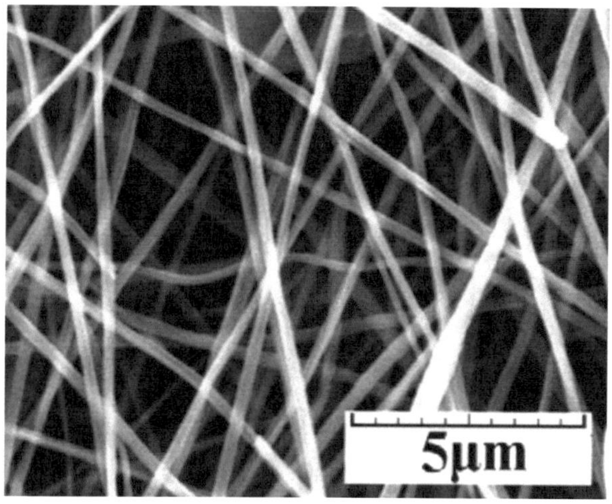

Fig. 2. PAN-based electrospun carbon fibers.

3. Functionalization of nanofibers

3.1 Activation process for porous nanofibers

Nanofibers have many benefits because of their large surface area for active reaction sites. Activation processes for improving these active sites have been applied by chemical/physical activation. As a representative case of physical activation, the activation agent was embedded into the fibers and then removed by physically removing the agents. The process for PAN-based porous nanofibers physically activated by silica is presented in Fig. 3 (J.S. Im et al., 2009). The silica-activated carbon nanofibers are shown in Fig. 4 (J.S. Im et al., 2009). The pores generated by physical activation are clearly observed.

3.2 Heat treatment effect for physical properties

Heat treatment has been carried out widely because it can change physical properties. The electrical enhancement of PAN-based carbon fibers was presented by (J.H. Huang et al., 2006) in Fig. 5. A stabilization treatment was carried out at 280 °C for a duration of 4 h, and samples with the same stabilization process parameters were used as the starting material of the carbonization treatment. Cyclic voltammetry (CV) measurements were carried out in a potential window of 0 to 1 V with different scan rates ranging from 100 to 1000 mV/s. It is noted that the curve for electrospun PAN-based carbon fibers shows a zigzag-like behavior, which indicates the hydrogen ions in the electrolytes are not easily adsorbed on the surface pores of PAN-CFs. In contrast, after thermal treatments (especially after the carbonization process), the capacitive performance is still stable even at the scan rate of 1000 mV/s. This result reveals that the capacitive performance of PAN-based carbon fibers is improved by the carbonization process.

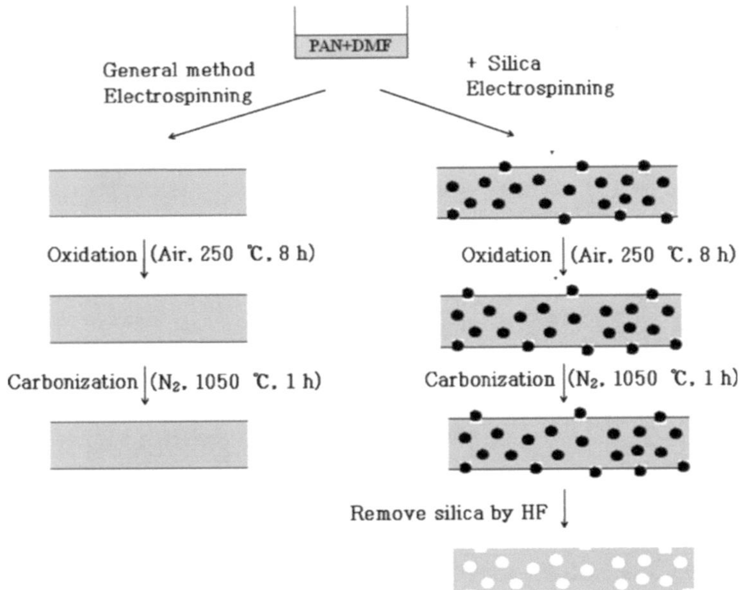

Fig. 3. The procedure of manufacturing carbon fibers and silica-activated carbon fibers.

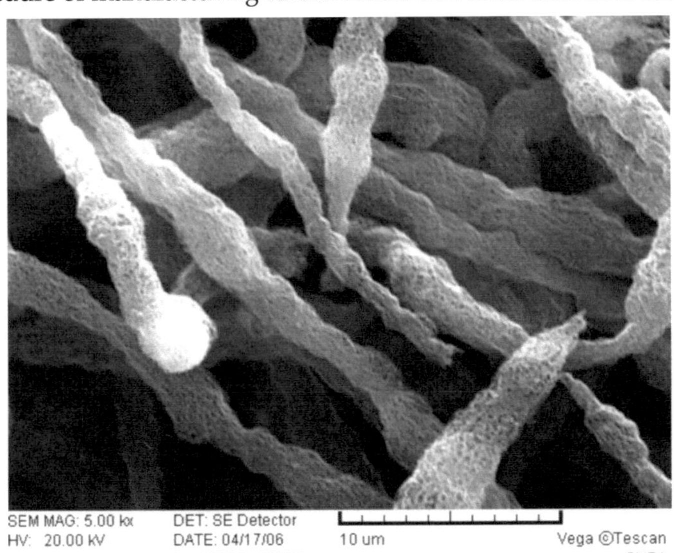

Fig. 4. Silica-activated carbon nanofibers.

This enhancement of electrical properties has been attributed to the improved electrical conductivity caused by the orientation of carbon structures. It was presented that the electrical conductivity can be improved significantly by well-oriented carbon structures through heat treatment (J.S. Im et al., In Press). The test was carried out on a variety of heat treatment temperatures and amounts of carbon nanotube additives, as shown in Fig. 6. The electrical conductivity was improved over three fold by the effects of heat treatment.

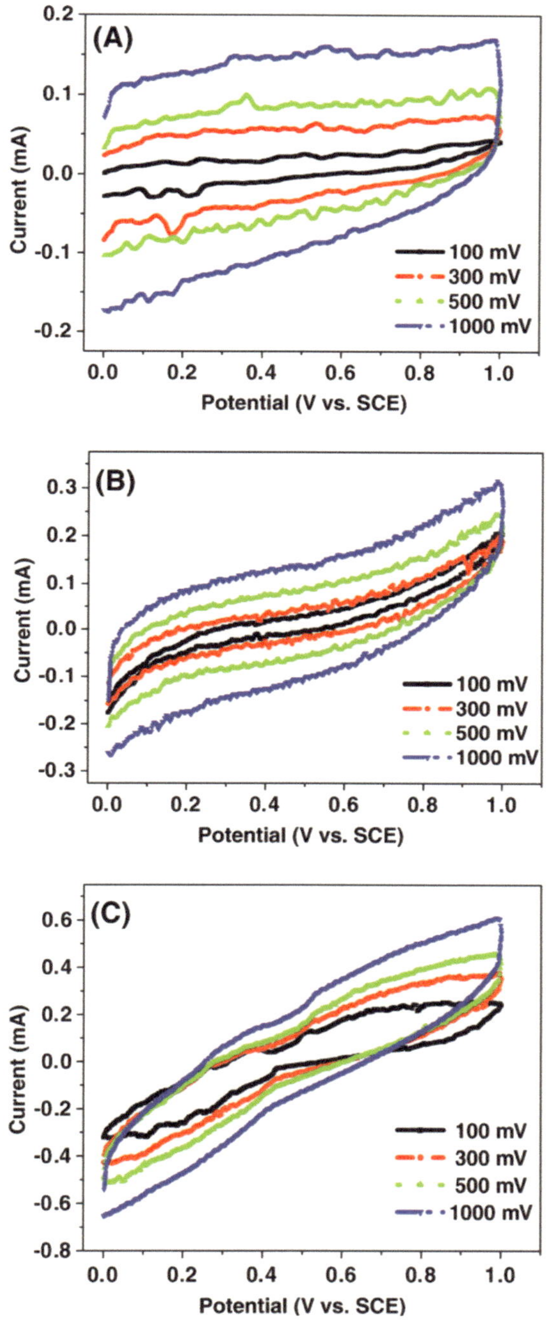

Fig. 5. Capacitive behavior of (A) electrospun, (B) stabilized, (C) carbonized PAN-based carbon fibers.

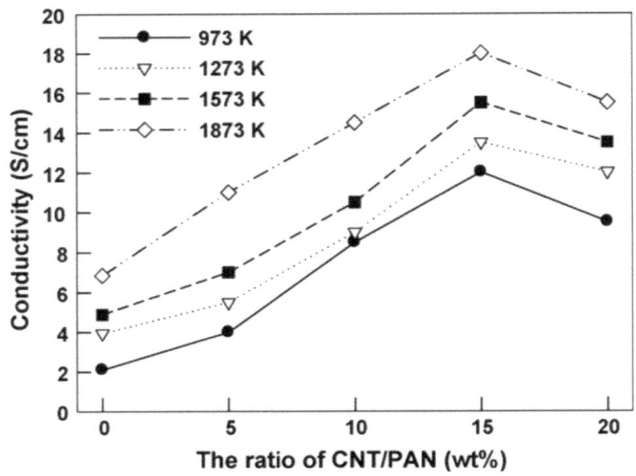

Fig. 6. Electrical conductivity of carbon fiber/carbon nanotubes composite.

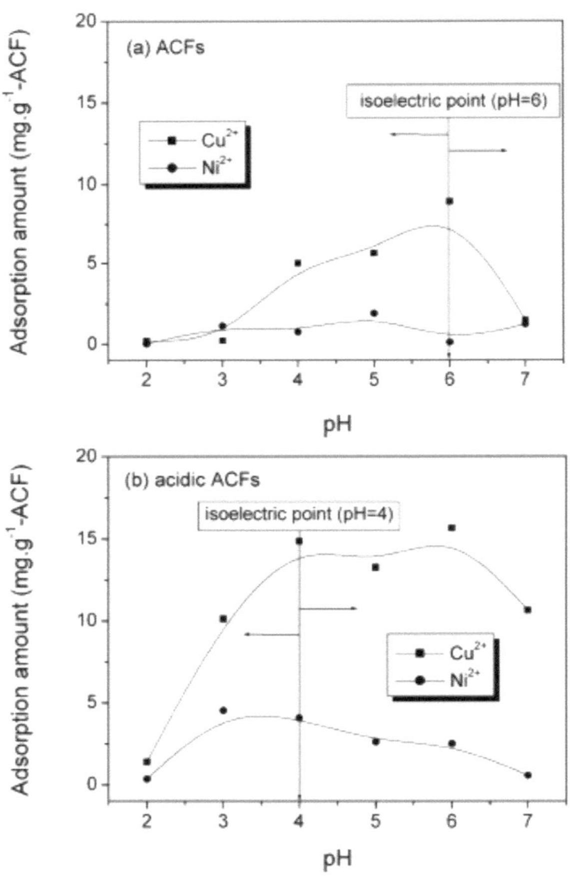

Fig. 7. The removal of multi-metal ions on (a) activated carbon fibers and (b) acid-treated activated carbon fibers as a function of pH at 25 °C.

3.3 Modification of nanofibers

The modification of nanofibers has been applied widely to give them improved properties. The result, which shows that the removal of multi-metal ions was improved by surface modification using acid treatment, is shown in Fig. 7 (S.J. Park et al., 2004). They explained that the surface functional groups containing oxygen content by acid treatment played an important role for the removal of metal ions.

A plasma treatment was carried out to enhance the surface energy of carbon nanofibers for good bonding to the matrix (V. Bruser et al., 2003). The gas-state plasma treatment was carried out by using the fluidized bed plasma reactor depicted in Fig. 8. The surface energy increased over two fold after 5 min of plasma treatment (V. Bruser et al., 2003).

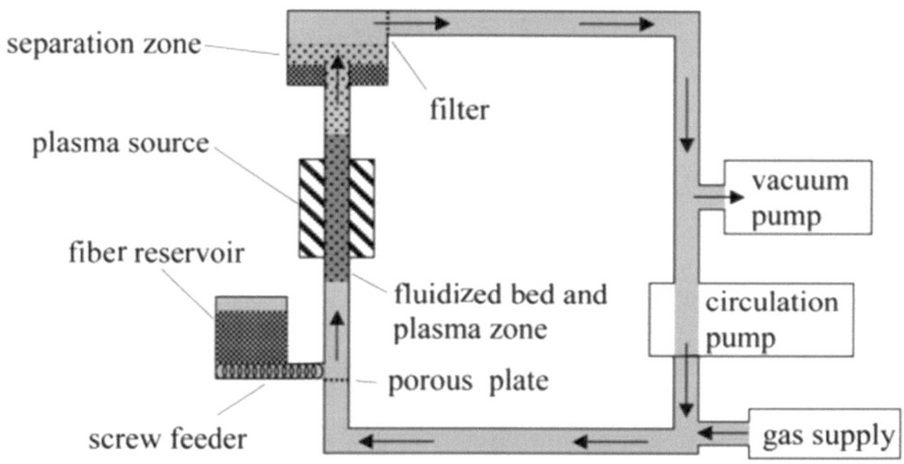

Fig. 8. Fluidized bed plasma reactor (V. Bruser et al., 2003).

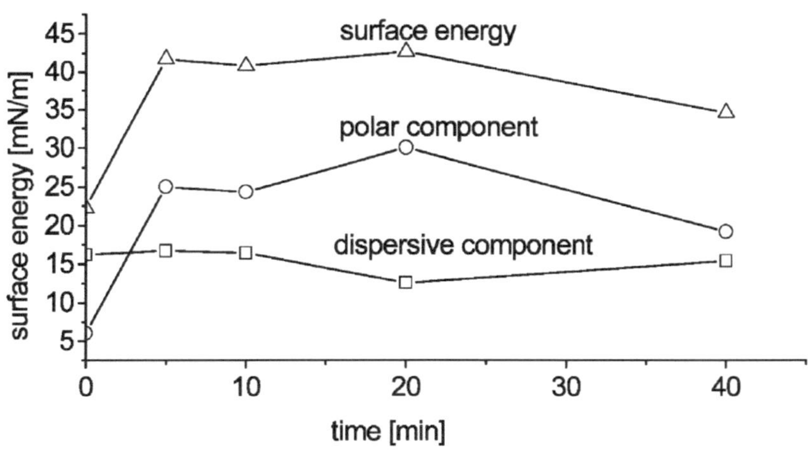

Fig. 9. Surface energy at 1.2 mbar and 80 W; gas composition: $Ar:O_2=1:1$ (V. Bruser et al., 2003).

The effects of functionalized surface groups on nanofibers were investigated based on improved gas adsorption by our group (J.S. Im et al., 2009). The fluorination treatment of porous nanofibers was carried out in the gas state for 5 min. The capacity of methane storage was increased by attraction effects of fluorine on porous nanofibers.

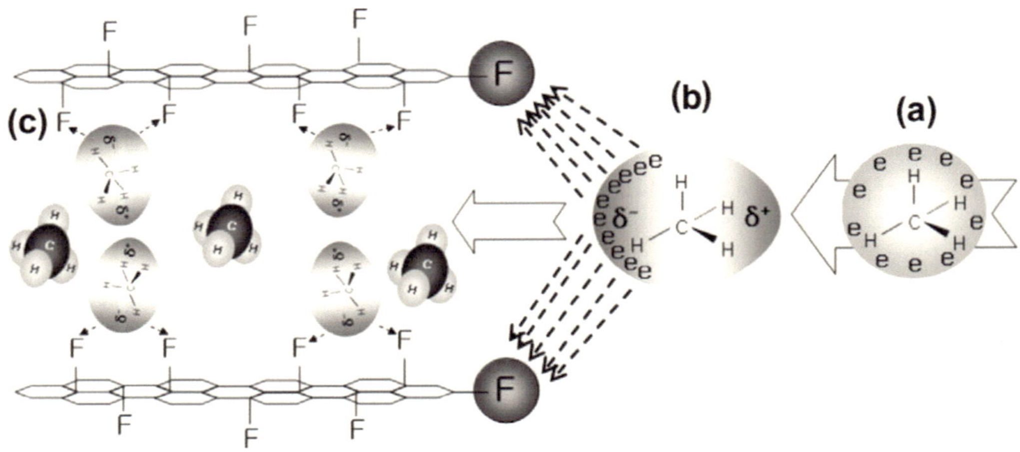

Fig. 10. Mechanism of methane storage; (a) methane molecule, (b) electron-attracted molecule and (c) adsorbed methane molecule in the carbon silt pore (J.S. Im et al., 2009).

3.4 Nanofibers complex

The use of nanofibers as the reinforcing filler and conducting additive in polymers to improve their mechanical and electrical properties is generally encountered in polymer technology. Nanofibers, such as carbon and glass fibers, are routinely used in composites of a range of different polymers. Improvement in modulus and strength, achieved by using nanofibers in a composite, has been presented by many researchers. The improved storage modulus of epoxy resin by addition of carbon nanofibers is presented in Fig. 11 (Derrick R.

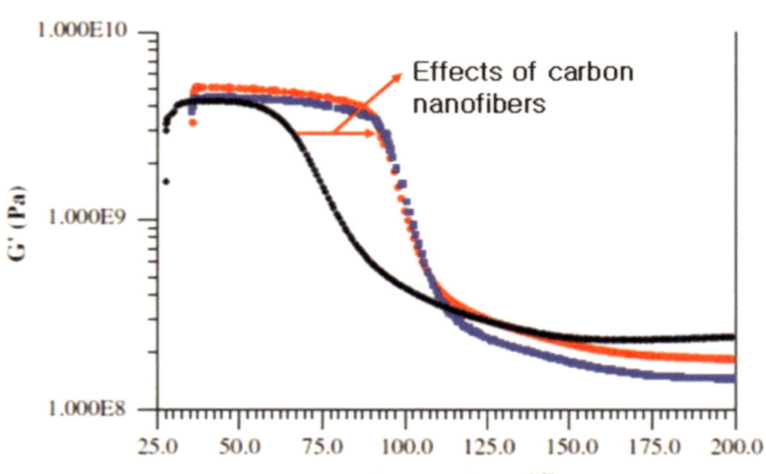

Fig. 11. Storage modulus of epoxy resin based on the effects of carbon nanofibers (Derrick R. Dean et al., 2009).

Dean et al., 2009). The transition temperature shifted significantly higher, indicating that the epoxy/carbon nanofiber composite can be stained up to higher temperatures.
This mechanical improvement is attributed to the properties at the fiber/matrix interface and are therefore dependent on the surface area of the interface. S.G. Prolongo's group synthesized Epoxy/carbon nanofiber composites by suing functionalized nanofibers with amino groups (S.G. Prolongo et al., 2009). They found that the dispersion of carbon nanofibers is improved by the functionalization process up to a nanofiber content of 1 wt%. They also found that improved dispersion of carbon nanofibers markedly affects the physical and thermo-dynamical mechanical properties of the epoxy nanocomposites. The addition of functionalized carbon nanofibers causes an important increase in the coefficient of thermal expansion and glassy storage modulus of nanocomposites.
The use of carbon nanofibers for enhanced electrical properties has also been reported widely. The group of Torsten presented results showing that a reduced specific resistance of epoxy resin was obtained by the addition of carbon nanofibers (Torsten Prasse et al., 2003). They also found that the orientation of carbon nanofibers were important for the electrical properties of the epoxy complex. When the carbon nanofibers were oriented along the direction of the applied electric field, the measured electrical properties were improved more efficiently.

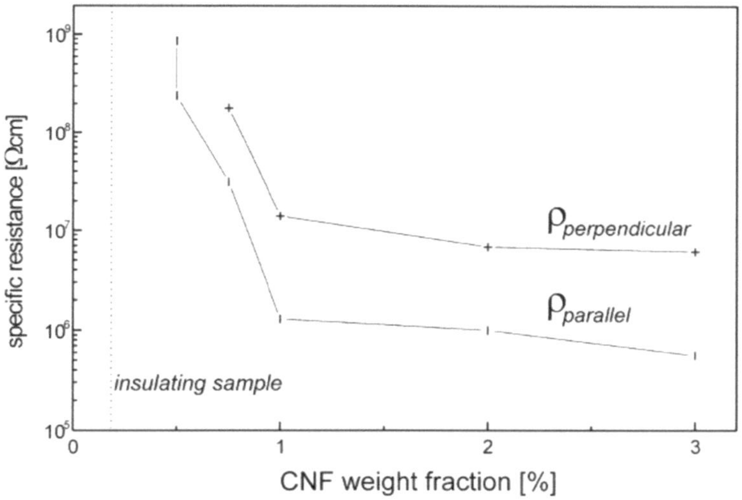

Fig. 12. Specific resistance of epoxy resin as a function of carbon nanofiber weight fraction with different orientations of carbon nanofibers (Torsten Prasse et al., 2003).

Metal/nanofiber complexes were also investigated to maximize the advantages of both metals and nanofibers. The self-assembly method of metal-complexed bolaamphiphiles was introduced by Masaki Kogiso's group (Masaki Kogiso et al., 2004). The illustration is shown in Fig. 13.
For example, a TiO_2 photocatalyst was embedded in electrospun fibers to optimize the photocatalytic ability shown in Fig. 14 (J.S. Im et al., 2008). Due to the low density of electrospun nanofibers, it was possible to make the photocatalysts floating on water for enhanced absorbance of UV light.

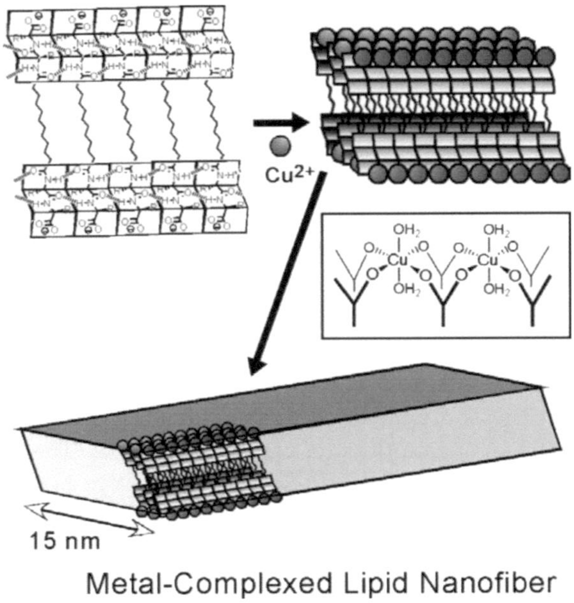

Fig. 13. A proposed structure of the copper(II) acetates complex and a schematic illustration of the molecular packing within the nanofibers (Masaki Kogiso et al., 2004).

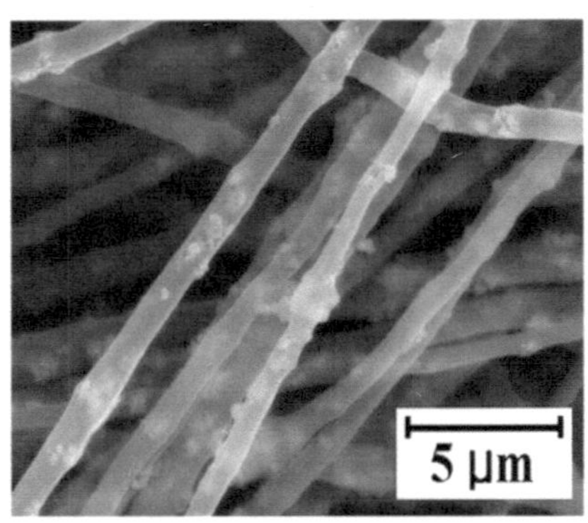

Fig. 14. SEM images of electrospun PAN-based nanofibers containing TiO_2 (J.S. Im et al., 2008).

4. Application of nanofibers

4.1 Energy storage materials

Nanofibers have been applied as a storage media for alternative energy sources such as hydrogen and natural gases. Porous carbon nanofibers have especially been investigated widely due to their large specific surface area and high pore volume. Hydrogen and natural gases can be stored by physical adsorption, indicating that the use of these gases is easy. The superior storage capacity of porous carbon nanofibers was presented through comparison with other porous carbon materials such as graphite, carbon nanotubes, and activated carbon, as shown in Fig. 15 (J.S. Im et al., 2009).

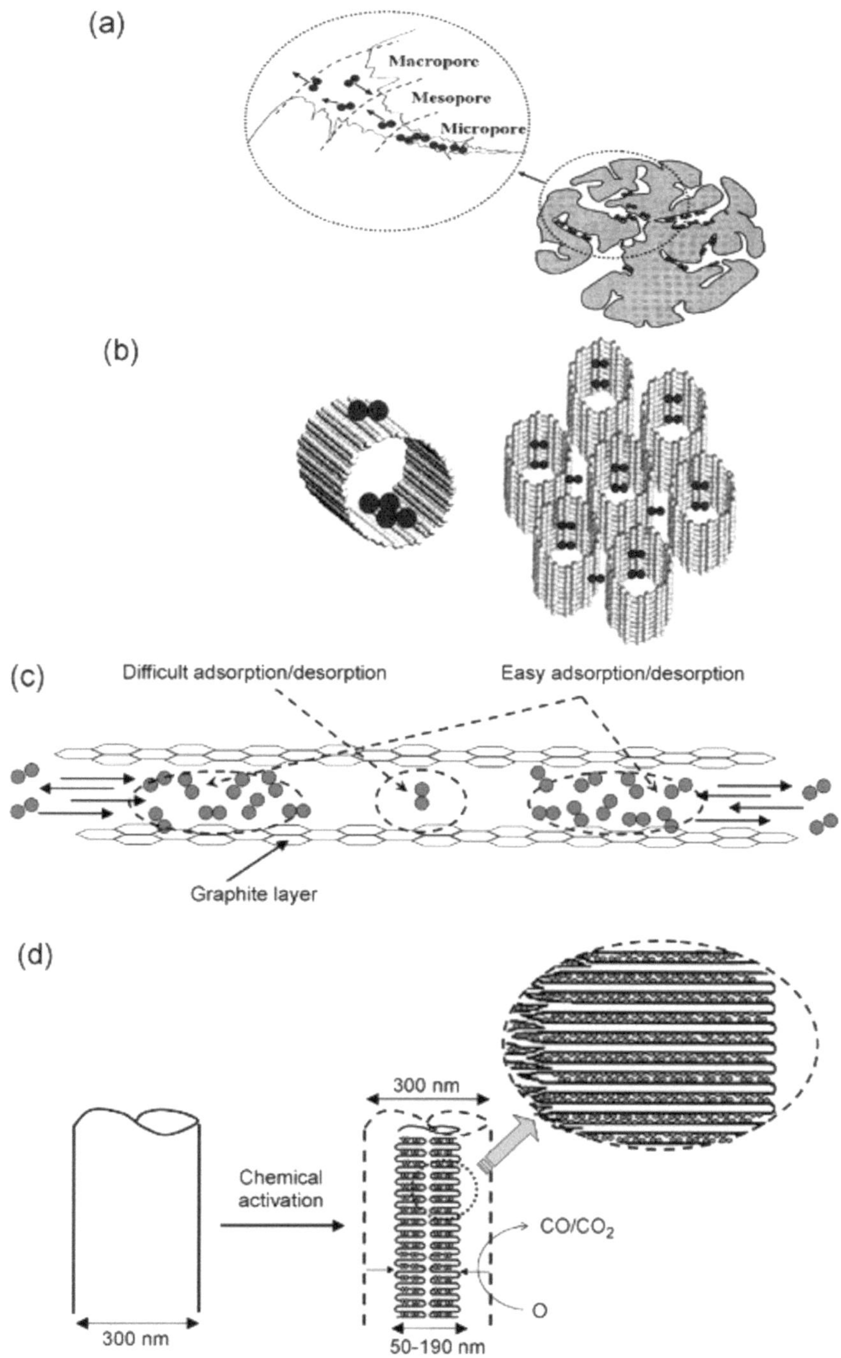

Fig. 15. The mechanism of hydrogen adsorption using various carbon materials; (a): activated carbon, (b): single walled carbon nanotube, (c): graphite, (d): electrospun activated carbon nanofibers (J.S. Im et al., 2009).

The main reason suggested for improved hydrogen adsorption was that electrospun activated carbon nanofibers might be expected to have an optimized pore structure with controlled pore size. This result may come from the fact that the diameters of electrospun fibers can be controlled easily, and optimized pore sizes can be obtained with a highly developed pore structure.
To find the optimized activation conditions, carbon nanofibers were activated based on varying the chemical activation agents, reaction time, reaction temperature, and the rate of inert gas flow. The role of the type of chemical activation and inert gas glow rate was explained by recent paper (A. Linares-Solano et al., 2009). When comparing potassium hydroxide and sodium hydroxide, a higher developed pore structure was observed with potassium hydroxide. A quicker flow rate of inert gas was also beneficial for an optimized pore structure for hydrogen adsorption of carbon nanofibers.

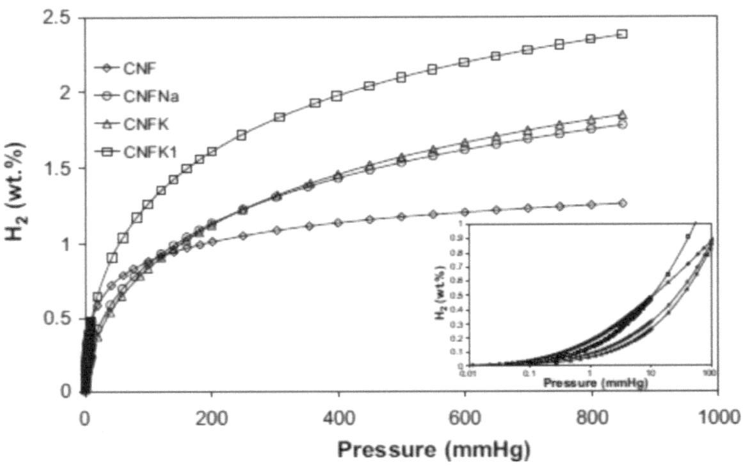

Fig. 16. Hydrogen adsorption isotherms at 77 K; CNF: carbon nanofiber, CNF Na: NaOH activated carbon nanofiber (N_2 flow of 500 ml/min), CNF K: KOH activated carbon nanofiber (N_2 flow of 500 ml/min), CNF K1: KOH activated carbon nanofiber (N_2 flow of 800 ml/min) (A. Linares-Solano et al., 2009).

The MCF (metal-carbon-fluorine) system was introduced by using electrospun carbon nanofibers (J.S. Im et al., 2009). The transition metal and fluorine were introduced on the activated carbon nanofibers to use the gap of electronegativity between the metal and fluorine. As a result of this gap, the hydrogen molecules were attracted into the carbon pores effectively.

4.2 Ecological materials

Photocatalysts (usually TiO_2) are used widely because of their high activity, chemical stability, robustness against photocorrosion, low toxicity and low cost. Nano-sized photocatalysts have been synthesized by many methods, such as the sol-gel process and thermal treatments, to enhance the photocatalytic ability with larger active sites. However, the recovery of nano-sized photocatalysts is not easy, making reuse difficult and secondary contamination likely. Therefore the immobilization method has been studied widely. Nanofibers have been investigated as one of the solutions using two methods: first,

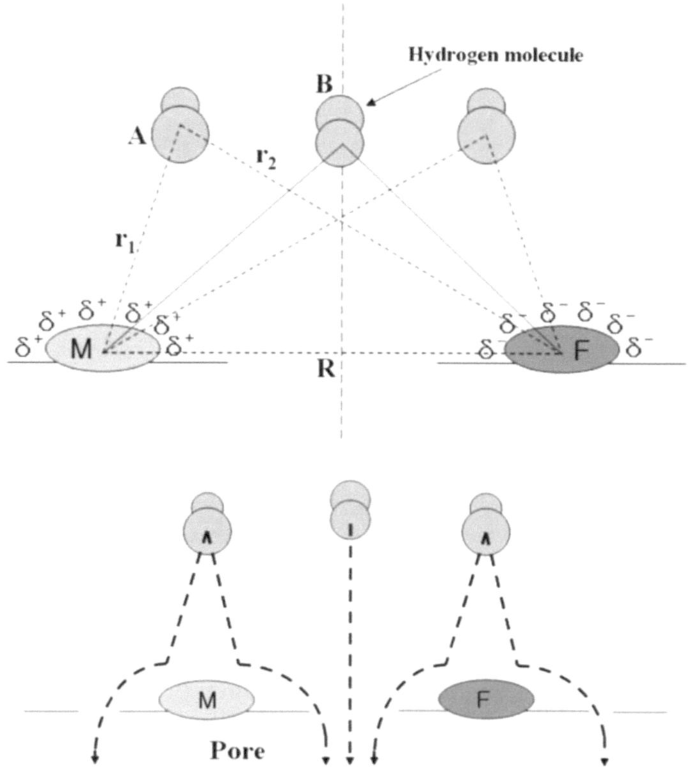

Fig. 17. Mechanism of hydrogen adsorption enhancement by the MCF system.

photocatalysts were fabricated as a non-woven mat by the electrospinning method; second, photocatalysts were embedded into a matrix of electrospun nanofibers. These materials can render large active sites for the photodegradation of pollutants. The TiO2 nanofibers were fabricated by C. Tekmen and A.K. Alves's groups (C. Tekmen et al., 2008, A.K. Alves et al., 2009). S.G. Lee's group synthesized SiO_2/TiO_2 nanofibers by electrospinning and presented their synergy effects (S.W. Lee et al., 2007). The TiO_2 embedded PAN-based carbon nanofibers were prepared and investigated based on the floating effect with a low density of nano-web (J.S. Im et al., 2008). The photocatalytic efficiency was maximized because photocatalysts can be activated by UV-light without any interruption such as UV absorption of dust water.

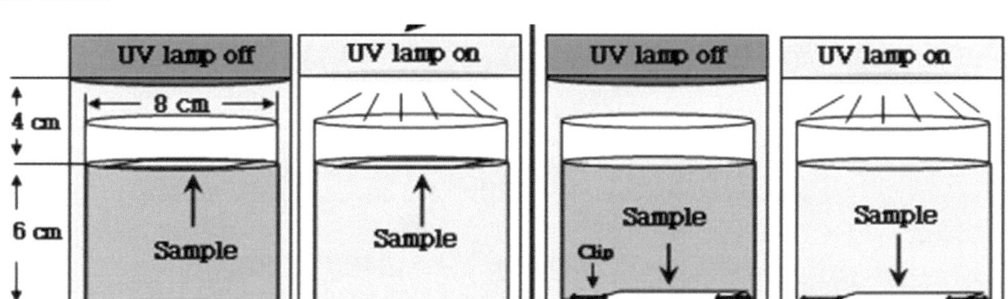

Fig. 18. Photocatalytic effects of nanofibers involving TiO_2 based on the floating property.

Nanofibers have also been studied as a membrane for the filtration of heavy metals in groundwater (Y. Sang et al., 2008). The experimental setup was shown in Fig. 19. From static adsorption experiments, an efficient removal of Cu^{2+}, Cd^{2+}, and Pb^{2+} was observed.

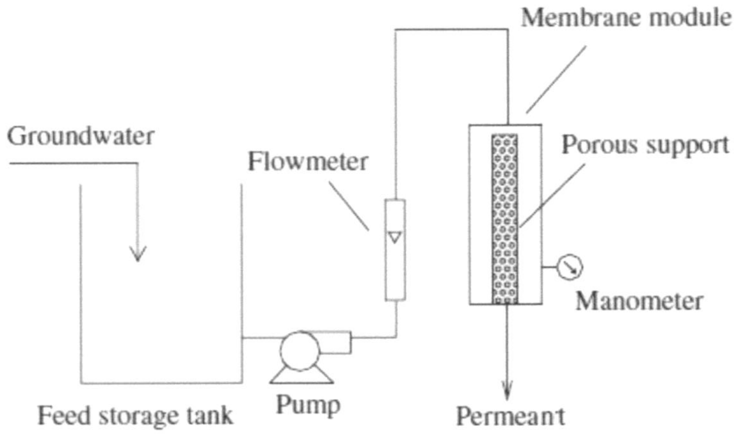

Fig. 19. Nanofiber membrane test as a groundwater treatment (Y. Sang et al., 2008).

4.3 Biomaterials

The biomedical applications of electrospun fibers remain the most interesting research area. Nanofibers can contribute in diverse emerging medical areas such as drug delivery, organogenesis, genomic medicine, rapid bedside clinical tests, and smart wound dressings. Among these applications, nanofiber-based three-dimensional scaffolds for tissue engineering and the design of nanofiber devices for sustained delivery systems have been attracted by the many advantages of electrospun fibers. The majority of nanofiber anti-

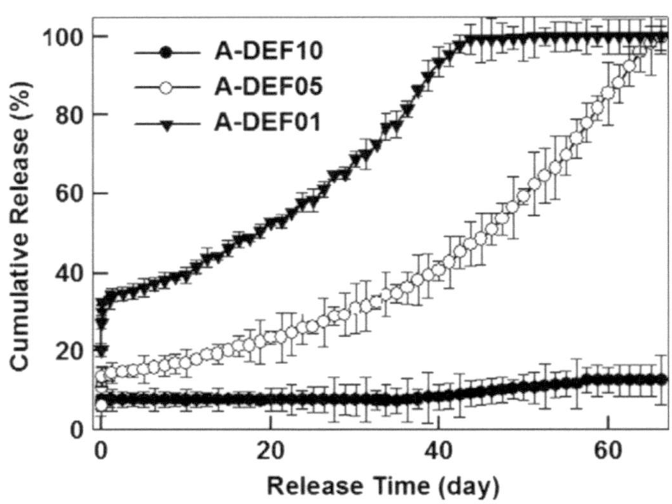

Fig. 20. Controlled drug release of electrospun hydrogel nanofibers by fluorination treatment.

neoplastic agent delivery systems have been envisioned for the treatment of malignant gliomas (a type of brain tumor). The current drug delivery system of choice is post tumor-resection implantation of a drug-eluting wafer. Thus, numerous studies have tried to elucidate the benefits of implanting a nanofiber delivery system over a wafer-based system. Recent results show that the released drug of electrospun nanofibers can be controlled by surface treatment (J.S. Im et al., 2009). Sustained drug release was observed by controlling the swelling rate of hydrogel nanofibers using a fluorination treatment.
The scaffolding application of electrospun nanofiber mats has already been applied in the industrial field because their size range approximates the structural features present in tissues surrounding animals' bodies. Based on the effects of various layered nanofiber matrices presented in Fig. 21, the ability of the scaffold was tested (S.H. Park et al., 2008).

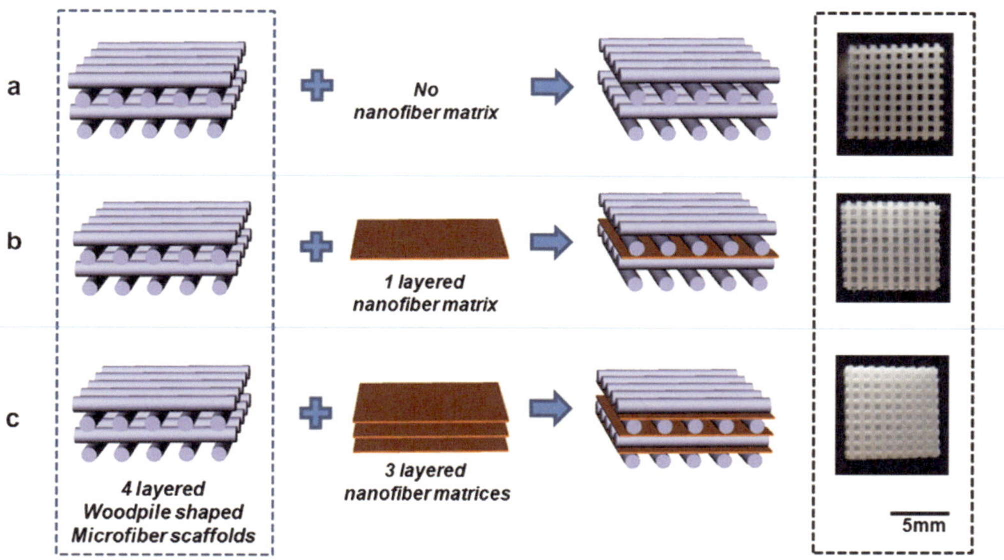

Fig. 21. Schematic diagrams and photographs of the three types of hybrid scaffolds used in the cell culture experiments. (a) Type I has no nanofiber matrix and only a PCL scaffold as a control specimen. PCL scaffolds combined with (b) a one-layer PCL/collagen nanofiber matrix (type II) and (c) a three-layer PCL/collagen nanofiber matrix (type III) (S.H. Park et al., 2008).

The M.P. Lutolf group presented the concept of the stem cell niche and the role of the extracellular matrix in regulating stem cell survival and signaling. They used electrospun nanofibers as a basement membrane between two phases.
By controlling the ratio of the polymer sources, the scaffold can easily have different properties. In Fig. 23, the lacunae in A1, B1 and C1 were enlarged and compared ((Hap/PLGA (%w/w); A1 (0/100), B1 (5/95) and C1 (10/90)) (H. Nie et al., 2009). Numerous osteoclast-like cells (identified by black circles) were observed to reabsorb the trabecular bone throughout the defects at this time-point. More isolated lacunae were observed in A1 and B1 than C1.

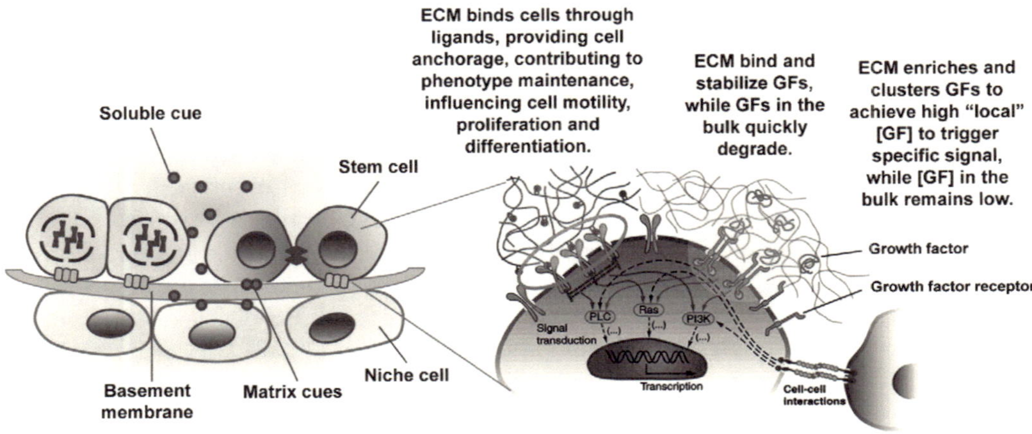

Fig. 22. The concept of the stem cell niche and role of the extracellular matrix in regulating stem cell survival and signaling. ECM: extracellular matrix; GF: growth factor; [GF]: growth factor concentration. Matrix (M.P. Lutolf and J.A. Hubbell, 2005).

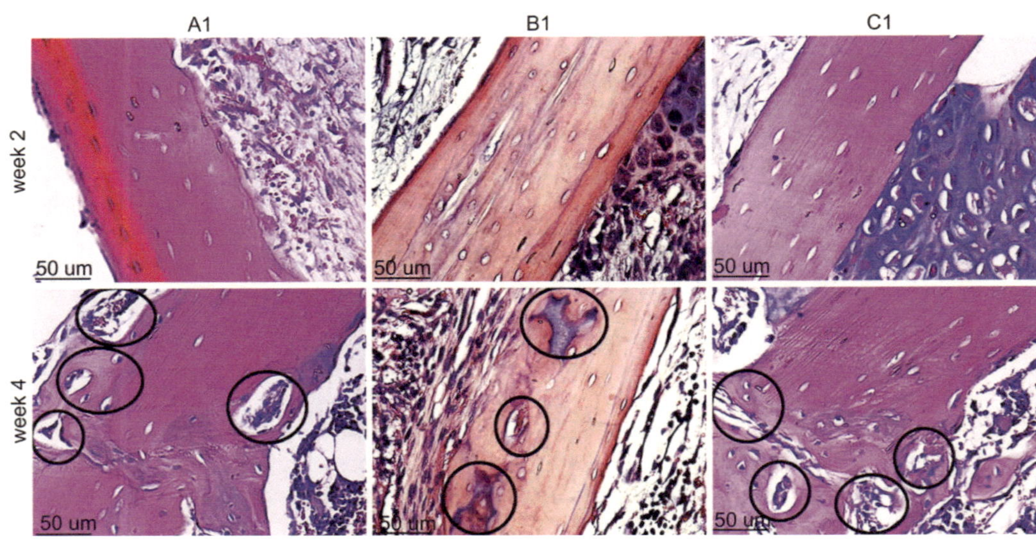

Fig. 23. Histological specimens from nude mice tibias after 2 and 4 weeks of implantation of A1, B1 and C1. Original magnification is 400 × for all (H. Nie et al., 2009).

The wound healing ability of nanofibers is shown in Fig. 24 (J.S. Choi et al., 2008). The wound of mouse was effectively healed by electrospun nanofibers within 14 days.

5. References

Alves, A.K., Berutti, F.A., Clemens, F.J., Graule, T., Bergmann, C.P. (2009). *Materials Research Bulletin*, 44., 312–317, 0025-5408

Dzenis, Y.A. (2004). Spinning continuous fibers for nanotechnology. *Science*, 304(5679), 1917-1919, 0036-8075

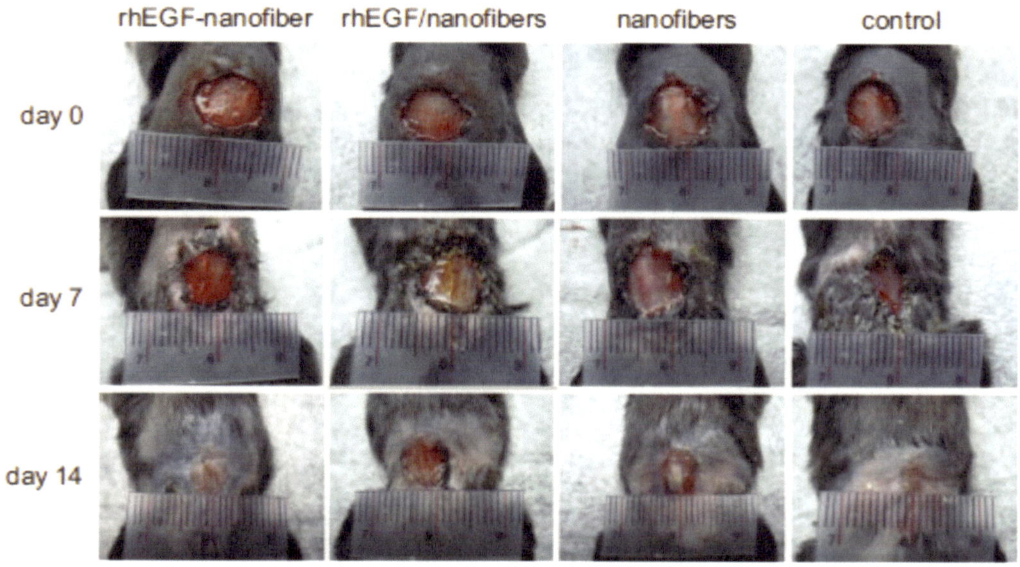

Fig. 24. Extent of wound healing in diabetic C57BL/6 mice treated with various formulations (J.S. Choi et al., 2008).

Garcia, S.F., Ortego E.V., Kunowsky, M., Kimura, M., Oya, A., Solano, A.L. (2009). *International Journal of Hydrogen Energy*, In Press, 0360-3199

Green, K.J., Dean, D.R., Vaidya, U.K. Nyairo, E. (2009). *Composites: Part A*, 40., 1470–1475, 1359-835X

Heintze, M., Bru¨ser, V., Brandl, W., Marginean, G., Bubert, H., Haiber, S. (2003). *Surface and Coatings Technology*, 174 –175., 831–834, 0257-8972

Hu, J.L., Huang, J.H., Chih, Y.K., Chuang, C.C., Chen, J.P., Cheng, S.H., Horng, J.L. (2009). *Diamond & Related Materials*, 18., 511–515, 0925-9635

Im, J.S., Jang, J.S., Lee, Y.S. (2009). Synthesis and characterization of mesoporous electrospun carbon fibers derived from silica template. *Journal of Industrial and Engineering Chemistry*, In Press, 1226-086X

Im, J.S., Kim, M.I., Lee, Y.S. (2008). Preparation of PAN-based electrospun nanofiber webs containing TiO_2 for photocatalyst degradation. Materials Letters, 62(21-22), 3652-3655, 0167-577X

Im, J.S., Kim, S.J., Kang, P.H., Lee, Y.S. (2009). The improved electrical conductivity of carbon nanofibers by fluorinated MWCNTs. *Journal of Industrial and Engineering Chemistry*, In Press, 1226-086X

Im, J.S., Park, S.J., Lee, Y.S., (2009). The metal–carbon–fluorine system for improving hydrogen storage by using metal and fluorine with different levels of electronegativity. *International Journal of Hydrogen Energy*, 34(3), 1423-1428, 0360-3199

Im, J.S., Park, S.J., Lee, Y.S., (2009). Superior prospect of chemically activated electrospun carbon fibers for hydrogen storage. *Materials Research Bulletin*, 44(9), 1871-1878, 0025-5408

Im, J.S., Yun, J., Lim, Y.M., Kim, H.I., Lee, Y.S. (2009). Fluorination of electrospun hydrogel fibers for a controlled release drug delivery system. *Acta Biomaterialia*, In Press, 1742-7061

Kam, J.S., Leong, K.W., Yoo, H.S. (2008). *Biomaterials*, 29., 587–596, 0966-0844

Lee, S.W., Kim, Y.U., Choi, S.S., Park, T.Y., Joo, Y.L., Lee, S.G. (2007). *Materials Letters*, 61., 889–893, 0167-577X

Lutolf, M.P., Hubbell, J.A. (2005). *Nature biotechnology*, 23., 47-55, 1087-0156

Nie, H., Ho, M.L., Wang, C.K., Wang, C.H., Fu, Y.C. (2009). *Biomaterials*, 30., 892–901, 0966-0844

Park, S.H., Kim, T.G., Kim, H.C., Yang, D.Y., Park, T.G. (2008). *Acta Biomaterialia*, 4., 1198–1207, 1742-7061

Park, S.J., Shin, J.S., Shin, J.W., Ryu, S.K. (2004). *Journal of Colloid and Interface Science*, 275., 342–344, 0021-9797

Prasse, T., Cavaillé, J.Y., Bauhofer, W. (2003). *Composites Science and Technology*, 63., 1835–1841, 0266-3538

Prolongo, S.G., Campo, M., Gude, M.R., Chaos-Moran, R., Urena, A. (2009). *Composites Science and Technology*, 69., 349–357, 0266-3538

Sang, Y. Li, F. Gu, Q. Liang, C. Chena, J. (2008). *Desalination*, 223., 349–360, 0011-9164

Tekmen, C., Suslu, A., Cocen, U. (2008). *Materials Letters*, 62., 4470–4472, 0167-577X

7

Keratin-based Nanofibres

Claudio Tonin, Annalisa Aluigi, Alessio Varesano and Claudia Vineis
ISMAC, Institute for Macromolecular Studies, CNR, National Research Council
Italy

1. Introduction

The interest in biopolymers from renewable resources as alternatives to polymers made from oil and other fossil resources has been increasing over the years. Biopolymers are also considered environmentally friendly over their entire live-cycle. There is much recent literature on carbohydrates and proteins derived from plants and animals and polyesters made from the fermentation of plant material. As to whether it is ethically justifiable to convert valuable foodstuffs into commodities is open to question. The focus here is on keratin, one of the most abundant and mostly unexploited non-food proteins, being the major component of hair, feathers, nails and horns of mammals and birds. In spite of their important and interesting characteristics keratin wastes represent a rather complicated disposal challenge because burning for fuel is inefficient and polluting due to the high sulphur content (3-4% wt). The total amount of keratin (including fibre by-products from the wool textile industry, poor quality raw wools from farms and butchery waste) has been estimated worldwide at more than 5 million tonnes per year (Barone et al., 2005). Ground horn and nail is used as a nitrogenous fertilizer for gardening and, more recently, has been processed by caustic hydrolysis to produce biodegradable surfactants for fire extinguisher foams. However, most keratin wastes made from unserviceable wools and feathers from poultry are not valorised and are simply disposed of (Martínez-Hernández et al., 2007; Schmidt, 1998). Pooling and processing into biopolymers might be a better way of exploiting such a large quantity of protein biomass. Keratin-based materials can be used in biotechnological and biomedical fields for tissue engineering and the production of affinity membranes, due to their biocompatibility, their ability to support fibroblast growth and absorb heavy metal ions and volatile organic compounds (VOCs). Transforming keratin into nanofibres by electrospinning combines the aforementioned properties of keratin with the high surface to volume ratio and the high porosity of nano-structured textiles. This may be an original and promising approach for the fabrication of scaffolds for tissue engineering and filtration devices.

Keratin distinguishes itself from other structural proteins by the quantity of cysteine residues in the protein molecules (7-20% of the total amino acid residues). In particular, cysteine amino acids form inter and intra molecular disulphide bonds (cysteine residues) giving rise to a compact three-dimensional structure that confers a high stability to the protein (Dowling et al., 1986).

Because of their low molecular weight (9–60 kDa), keratin-based materials have poor mechanical properties. Moreover, like most natural polymers, keratin is not thermoplastic. For electrospinning keratin should be blended with suitable polymers using common, volatile and easy to handle solvents. This is also recommended when mechanical performance is needed (Aluigi et al., 2008; Katoh et al., 2004; Zoccola et al., 2008). This chapter describes the extraction of keratin from wool and electrospinning of keratin-based blends with high molecular weight polymers. Blends of keratin with poly(ethylene-oxide) (PEO) and fibroin are suitable for biomedical application (tissue engineering and medical textiles), while keratin/polyamide 6 (PA6) nanofibres can be used for active filtration of air and water.

2. Keratin and regenerated keratin

Keratins represent a group of fibrous proteins with high sulphur content produced in some epithelial cells of vertebrate such as reptiles, birds and mammals. In particular, the cysteine amino acid residues form inter and intra molecular disulphide bonds (cystine residues) that give rise to a compact three-dimensional structure that confers to keratin proteins a high resistance to chemical and enzymatic attacks (Dowling et al., 1986). There are two kinds of keratins: the "hard-keratin" and the "soft-keratin" according to the physical and chemical properties, particularly the sulphur content. Soft keratins, with a sulphur content <3% wt, are found in the *stratum corneum* of the skin whereas the hard keratins found in hair, wool, feather, nails and horns and have a sulphur content > 3% wt (Fraser et al., 1972). A further classification is based on the X-ray diffraction pattern obtained from different keratin proteins. The α-helix appears to be the basic fibrillar element in all soft keratins and in the hard keratins from mammals (Crick, 1953). Studies carried out on the structure of feather keratin have shown that about 28% of the protein molecule has a β-conformation; the remainder does not possess a geometrically regular secondary structure (Fraser et al., 1971).

A key problem in blending keratin with other polymers is to find a suitable solvent capable of dissolving the complicated structure of the protein. Generally a mixture of solvent systems in which different components have different functions is used, one component breaks the hydrogen bonds while the reducing or oxidising agent breaks the disulphide bonds. Some multiple solvent systems that have been utilized are: urea/reducing agent (Yamauchi et al., 1996), carbamide/H_2O_2/H_2O and carbamide/2-mercaptoethanol (Garret & Grisham, 2002), or Cu-oxam metal complex system ($[Cu(NH_3)4(OH)_2]$) (Aluigi et al., 2004). In 2005, Xie and co-workers tested ionic liquids, particularly 1-butyl-3-methyllimidazolium-chloride ($BMIM^+Cl^-$), as a solvent for keratin proteins (Xie et al., 2005).

Recently, two chemical-free processes using just water and heat had been explored as methods of dissolving and converting keratinous materials (i.e. wool and feathers, respectively) : steam explosion (Tonin et al., 2006) and superheated water (Yin et al., 2007). Steam explosion is based on short time steam cooking of biomasses at a high temperature for several minutes, followed by explosive decompression. Using saturated steam at 220°C (~22 bar) for 10 min followed by a rapid decompression, wool fibres were disrupted into solid and liquid phases consisting of oligo-peptides, water-soluble peptides and free amino acids (Tonin et al., 2006). The second process consisted of treating feather barbs with superheated water at ~220°C and ~22 bar for 2 h followed by cooling. Most of the keratin was converted into oligo-peptides with a molecular weight of about 1.0 and 1.8 kDa (Yin et al., 2007).

However, for the electrospinning process, volatile and easy to handle solvents are needed and for this reason we have studied the solvation properties of water and formic acid for keratin.

2.1 Wool keratin

Wool fibre is a complex multi-cell system composed of inanimate cells which differ in composition, morphology and properties. The principal component of wool fibres is keratin and there is also a small amount of lipids (0.1%) and mineral salts (0.5%). Wool cleaning for scientific purposes is carried out by means of extraction with petroleum ether, ethanol and water (von Bergen, 1963).

The disulphide bonds of cystine form both inter and intra chain cross-links and are responsible for the greater stability and lower solubility of keratin compared with most proteins. Therefore, the presence of cystine plays the most important role in the chemical, thermal and mechanical properties of wool. Although disulphide bonds are responsible for keratin reticulation in wool, there are other different kinds of cross-links: hydrophobic bonds occur between non polar groups; hydrogen bonds have a very important influence on the mechanical properties of the wool fibre; ionic bonds occur between carboxylic anions and ammonium cations. These ionic bonds depend on the pH of the environment. In fact, at the isoelectric point (pH=4.9), the quantity of ionic bonds is greatest because the protein exists as zwitterions ($^{+}H3N-CHR-COO^{-}$). When the pH is very acid or very basic, there is a reduction in the number of ionic bonds. At low pH values the COO^{-} groups are protonated, while at high levels the $NH3^{+}$ groups are deprotonated. (von Bergen, 1963).

Proteins extracted from wool can be separated into four different groups: the low sulphur content keratins (LS) that have a molecular weight between 45 and 60 kDa; the high sulphur content keratins (HS) with a molecular weight of between 11 and 28 kDa and the high-glycine and tyrosine content proteins with a molecular weight of between 9 and 12 kDa (Jeffrey, P. D., 1972). The low sulphur content proteins have an α-helical crystalline structure (α-keratins), while the other proteins do not have a helicoidal structure.

X-ray studies showed that the secondary structure of α-keratins is dextrorotatory helicoidal. This α-helix is stabilized by the presence of hydrogen bonds between the N-H group of a peptide bond and the carbonyl group of a peptide bond placed over four residues. Although the strength of a single hydrogen bond (~5 kcal/mole) is only 5% of that of C-C and C-H covalent bonds the large number of hydrogen bonds confers a remarkable stability to the α-helix crystallites. The conformation of the α-helix is characteristic for the wool fibre in its natural state, therefore it is found in the fibre which is not stretched along its axis (Pauling & Corey, 1953). During stretching, the α-helix declines and β-keratin appears (Tonin et al., 2004).

2.2 Preparation and characterization of keratin solutions

Keratin extraction from wool may take place only after the reduction or oxidation of disulphide bonds. The reducing agents often used are thiols (thioglycolic acid, dithiothreitol and 2-mercaptoethanol) while oxidizing agents are peracetic or performic acids (Yamauchi et al., 1996; Thompson & O'Donnel, 1959). All these agents are always used in combination with protein denaturing agent (e.g. urea) that breaks hydrogen bonds and extraction yields range from 50 to 70% wt. However, for further scaled-up processes, we preferred to use sulphitolysis reaction, although the extraction yield is lower (37% wt), instead of harmful reducing and oxidizing agents. The sulphitolysis is a reaction of wool with sulphite ions that break disulphide bonds in thiols and S-sulphonate anions, known as Bunte salts (Cecil, 1963).

Five grams of cleaned fibres were cut into snippets and put in 100 ml of aqueous solution containing urea (8 M), m-bisulphite 0.5 M, adjusted to pH 6.5 with NaOH 5 N and treated by shaking for 2 h at 65°C. Successively, the solution was filtered through a 5 μm pore-size filter using a vacuum pump and then dialyzed against distilled water using cellulose tubes (molecular mass cut-off 12000-14000) for three days, changing the outer solution four times a day. The dialyzed part was first concentrated with a rotary vacuum evaporator in order to prepare solutions at different keratin concentrations (0.5, 1.5, 3.5 and 5% wt). The solution at 5%wt was cast onto a polyester plate and dried at 50°C overnight in order to prepare keratin films regenerated from water (samples labelled as KW). A part of keratin regenerated from water was dissolved in formic acid for 4h, so as to have solutions at different keratin concentrations (0.5, 1.5, 3.5 and 5% wt). The solution of keratin in formic acid at 5% wt was cast onto polyester plates and dried at 50°C overnight in order to prepare keratin films regenerated from water (samples labelled as KF).

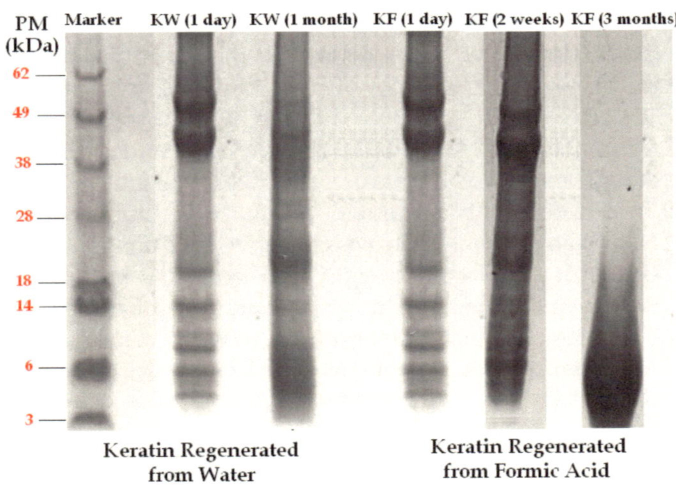

Fig. 1. SDS-PAGE of KW and KF.

The stability of keratin dissolved in water and in formic acid solutions was studied both following the protein degradation during time and the flocculation process by turbidity measurements. The molecular weight distribution analysis conducted by electrophoresis SDS-PAGE (reported in Fig. 1), revealed that the keratin regenerated from a one month old aqueous solution was degraded, in fact the high molecular weight proteins (at 45 and 60 kDa) disappear, while a series of proteins having molecular weights below 38 kDa appear. Regarding keratin in a formic acid solution, the keratin regenerated from a fresh solution (1 day) is not degraded while a slight degradation appears after two weeks and after three months there is a complete digestion of the protein.

Turbidity (τ) of solutions was calculated by measuring the transmittance T at 540 nm using the following Eq. 1:

$$\tau = -(\ln T)/c \tag{1}$$

where c is the cell length (1cm).

Fig. 2 shows that turbidity of the keratin aqueous solution increases with the increase in the keratin concentration, suggesting that higher concentrations promote protein chain aggregation (flocculation). On the other hand the turbidity of keratin dissolved in formic acid remains unchanged with increasing keratin concentration, indicating that flocculation does not occur, at least over a certain range of keratin concentration.

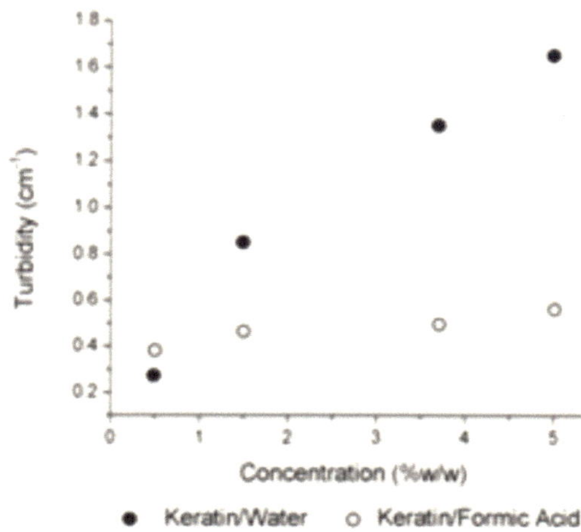

Fig. 2. Turbidity of keratin dissolved in water and in formic acid

Viscosities of the keratin solutions at 5% wt, were plotted against the shear rate (Fig. 3). In the whole shear range investigated, the viscosity of keratin dissolved in formic acid was higher than that of keratin in water. This behaviour indicates that formic acid has better solvation properties for keratin than water. This is because formic acid is more polar than water and forms strong interactions with the polar side chain groups of keratin such as CO, OH, COOH and NH_3^+. As a consequence, the molecular chains become closer in formic acid than in water and this is confirmed by the lower turbidity.

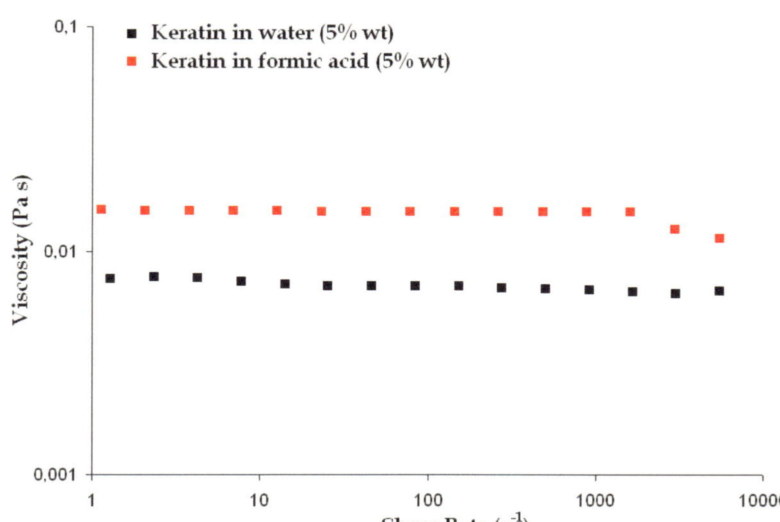

Fig. 3. Rheological behaviour of the keratin in water and keratin in formic acid solutions

3. Electrospinning

Several methods are dedicated to producing small diameter fibres for high-volume production such as fibrillation (Homonoff, 2008; Perez et al., 2000), island-in-sea (Pourdeyhimi et al., 2006), and a novel melt-blowing system (Bryner & Armantrout, 2006; Ellison et al., 2007); or highly accurate methods such as nanolithography (Tseng et al., 2005; Xie et al., 2006) and self-assembly (Zhang, 2002). However, their usefulness is restricted by combinations of narrow material ranges, high costs and low production rates. In comparison, electrospinning is a simple and low cost process, and has an intermediate production rate (Ramakrishna et al., 2006).

Electrospinning is a physical phenomenon classified as a branch of electro-fluid-dynamics. The process is based on the electrostatic repulsion of charges ruled by the well-known Coulomb law. By means of an electrostatic field, an electrospinning apparatus generates electrically-driven polymer jets that solidify into sub-micron scale polymer-based filaments (usually called nanofibres). The idea was patented in the early 20th century (Cooley, 1902; Morton, 1902; Formhals, 1934); however, the invention was disregarded until the 1990s (Huang et al., 2003; Li & Xia, 2004), probably due to lack of knowledge and interest of both industries and scientists about tiny fibres. Electrospinning works with polymer melts and polymer solutions, but few research papers deal with electrospinning of molten polymer (so-called "melt-electrospinning") compared with the huge literature regarding electrospinning of polymer solutions (usually simply referred to as "electrospinning").

Electrospinning can be usefully employed in a wide range of applications requiring sub-micron scale fibre diameter such as: filtration (Tsai et al., 2002; Qin & Wang, 2006; Li et al., 2006; Barhate & Ramakrishna, 2007), membrane separation (Shin et al., 2005; Gopal et al., 2006), protective clothing and breathable garments (Gibson et al., 2001), wound dressings and scaffolds for tissue engineering (Pham et al., 2006; Lee et al., 2008; Sill & von Recum, 2008), and drug delivery (Sill & von Recum, 2008).

A basic electrospinning setup consists of three elements: an electrical generator (high voltage supply), a capillary (jet source) and a metal collector (target). The solution is usually electrically charged by the generator and the collector is grounded, but it is also possible to invert the system by electrically charging the collector and grounding the solution (Kilic et al., 2008). A scheme of an electrospinning apparatus is shown in Fig. 4.

A pump pushes the polymer solution through the capillary with a fixed flow rate by means of a syringe. The generator, connected by a wire to the metal tip of the capillary, supplies the

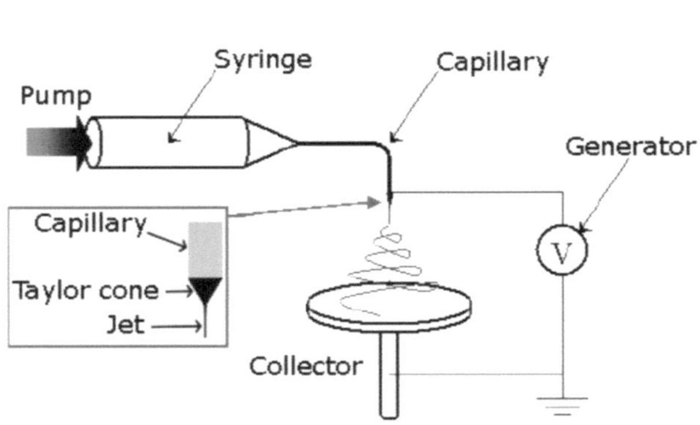

Fig. 4. Scheme of an electrospinning setup. In the box: enlargement of the capillary tip.

voltage to the solution. A stream of solution (jet) is produced from the tip toward the grounded metal collector on which electrospun nanofibres were collected. Usually, the production rate of a single jet is less than 10 g h^{-1} of nanofibres (Tsai et al., 2002), depending on polymer concentration and process conditions, mainly the flow rate of solution.

The electrospinning process starts when the voltage generator is turned on. The electrical potential of the droplet surface at the capillary tip is increased to a sufficiently high value that causes the droplet to assume a steady conical shape, known as the Taylor cone (see box in Fig. 4), instead of a spherical shape. The electrostatic forces act in opposition to the surface tension of the fluid. The Coulomb repulsion between the charges promotes the formation of the cone; on the opposite the surface tension of the fluid favours sphere-like shapes. As the electrostatic forces exceed the surface tension, a charged jet of fluid is ejected from the vertex of the cone (Yarin et al., 2001; Shin et al., 2001).

The trajectory of the jet solution begins with a straight segment. The diameter of the jet is one or two orders of magnitude smaller that the inner diameter of the capillary from which the jet is generated, and progressively decreases during its journey towards the collector. The diameter reduction occurs thank to two mechanisms: (*a*) the electrostatic repulsion forces of the charges in the jet tend to extend the jet itself, and (*b*) the solvent evaporation leaves the solid materials in the jet (e.g. polymer), this effect is self-induced because as the diameter decreases the evaporation further increases. Moreover, it worth noting that as the solvent evaporates the concentration and the viscoelastic properties of the liquid jet change.

After the straight segment, the jet is subjected to a bending instability and its trajectory becomes disordered; the oscillation eventually evolves into a spiral, similar to the movement of a whip (so-called whipping motion), as illustrated in Fig. 4. Finally, the jet completely solidifies into a nanofibre and deposits on the collector as a disordered continuous filament (non-woven structure). In spite of the ease of the electrospinning, the onset of bending instability is quite a complex problem; several theories and mechanisms have been recently proposed to explain this phenomenon (Shin et al., 2001; Deitzel et al., 2001; Reneker & Yarin, 2008).

The fibre-forming process depends on operating parameters (applied voltage, working distance between tip and collector, flow-rate of solution, tip diameter), properties of the polymer solution (polymer molecular weight, concentration, relaxation time, viscosity, electrical conductivity, surface tension, and vapour tension of the solvent), and environmental parameters such as temperature, humidity and ventilation (Theron et al., 2004). Depending on working conditions and polymer solution properties, the electrospun jet might be subjected to many phenomena such as branching, splitting, buckling, jet break-up (spraying), production of flat ribbon or beaded nanofibres that engenders different morphologies (Reneker & Yarin, 2008).

4. Keratin/PEO blend nanofibres

Recently in a leading opinion paper, Furth and co-workers (Furth et al., 2007) stated that they had been investigating the utility of keratin-based biomaterials for regenerative medicine applications. Keratin contains intrinsic sites of cellular recognition that mimic the extra-cellular matrix and keratin-based biomaterials demonstrated cell instructive capabilities. Certain keratin materials have been shown to be mitogenic and chemotactic for a variety of cell types, and to mediate changes in gene expression consistent with the promotion of wound healing. Moreover, other recent works (Tachinaba et al., 2002;

Tachinaba et al., 2005; Hamasaki et al., 2008) have highlighted the excellent cell adhesion onto wool keratin sponges produced by lyophilisation from aqueous solution. These results show that keratin is a useful biomaterial for scaffolds for cell cultivation.

The poor mechanical properties of regenerated keratin hinder its processability and restrict its practical applications to blends with appropriate polymers with better structural properties. Our attempts to obtain filaments of pure regenerated keratin from water were indeed unsuccessful; moreover, literature reports on the fabrication of regenerated keratin films (Tanabe et al., 2004) using cross linking agents, and the fabrication of composite nanofibres of regenerated silk fibroin blended with synthetic polymers such as poly(ethylene oxide) (PEO) (Jin et al., 2002). PEO is an amphiphilic water soluble and non-degradable polymer, with good biocompatibility (Desai & Hubbel, 1991) and low toxicity (Bergsma et al., 1995). This polymer is often used as an ideal model for the electrospinning process (Theron et al., 2004; Son et al., 2004) because it can be electrospun without defects from aqueous solutions in a rather narrow range of conditions.

We studied the production and the characterization of nanofibres produced by electrospinning of pure PEO and 50/50 wt/wt keratin/PEO blends, from aqueous solution of the polymers, in different operating conditions. The nanofibres were studied by Scanning Electron Microscopy (SEM) and Differential Scanning Calorimetry (DSC). The results were compared with those obtained by thin films produced by casting from the same solutions, with the aim of investigating the influence of the production processes on the structural arrangement of these materials.

4.1 Materials and methods

Keratin was obtained from wool by means of a sulphitolysis extraction method. PEO powder with an average molecular weight of 4×10^5 g mol^{-1} was dissolved in distilled water at ambient temperature for about 12 h. The concentrations used were 5, 7 and 10 % wt.

The keratin/PEO blend solutions were prepared at room temperature in about 12 h by simply adding PEO powder to the keratin aqueous solution. The solutions of keratin/PEO blend had total polymer concentrations of 5, 7 and 10 % wt with a keratin/PEO weight ratio of 50/50. These solutions were electrospun at 20 cm working distance to ensure that the nanofibres were dried. The applied voltages were between 10 and 30 kV. The process was stopped after about 10 minutes. During the electrospinning process, environmental conditions were kept in check; in particular, the temperature was from 20 to 25 °C and the relative humidity was in the range 55-65 %.

4.2 Solution characterization

Viscosity is an important factor for complete fibre formation in the electrospinning of polymer solutions. In particular, fibres without beads were produced when polymer chain entanglements are present at sufficient concentration (Shenoy et al., 2005). Moreover, fibre formation is promoted at low polymer concentrations, increasing solution conductivity (Son et al., 2004).

A great increase in conductivity was measured from solutions containing the keratin because of SDS that remains associated with keratin through ionic interaction conferring negative charges to the protein (Schrooyen et al., 2001). In fact the conductivity of pure PEO was about 0.120 mS cm^{-1} and the conductivity of keratin/PEO was 1.282 mS cm^{-1} (at polymer concentration 7% wt).

Fig. 5 shows the flow curves of PEO and keratin/PEO solutions at 5, 7 and 10% wt of polymers in water. All solutions behave like shear-thinning fluids. At low shear rates, disentanglement is balanced by formation of new entanglements, thus the fluid has a Newtonian behaviour which corresponds to a constant viscosity value (zero-shear viscosity). At higher values of shear rate, the disentanglement rate exceeds the rate of entanglement formation, therefore the viscosity decreases and shear-thinning behaviour is observed. The onset of shear thinning shifts to lower values of shear rate with the increase of polymer concentrations for both pure PEO and keratin/PEO solutions. The zero-shear viscosity increases when the polymer concentration increases. It is worth nothing that the keratin/PEO solutions at concentrations of 7 and 10% wt show flow curves Comparable with 5 and 7% PEO solutions, respectively. Thus, the keratin with a relatively low molecular weight slightly increases the viscosities of the keratin/PEO solutions, but its contribution is not negligible (Varesano et al., 2008).

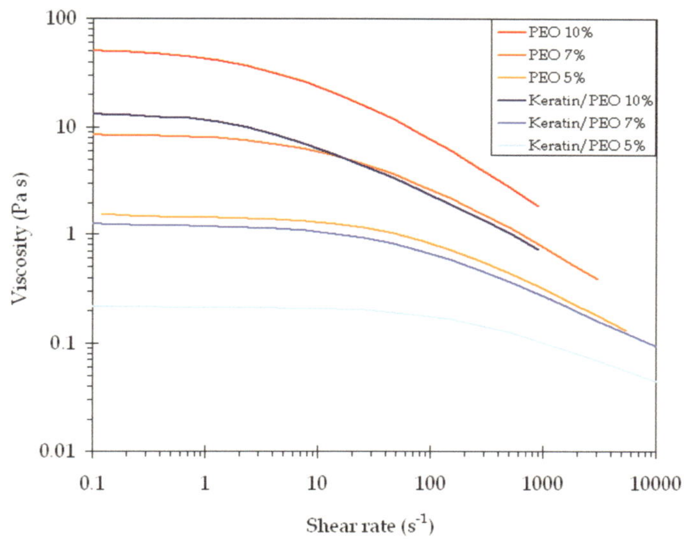

Fig. 5. Viscosity flow curves of PEO and keratin/PEO solutions

4.3 Electrospinning and morphological study

Morphologies of the electrospun materials from PEO and keratin/PEO solutions were investigated for comparison. The samples were produced by varying flow-rate, voltage and polymer concentrations.

PEO nanofibres without defects were produced from the 7% wt solution at 0.01 ml min^{-1} flow-rate from 20 to 30 kV of voltage. For higher flow-rate (0.05 ml min^{-1}) macroscopic drops fall from the capillary also at the highest voltage (30 kV). The electrospinning of 5% wt solution produced nanofibres without defects using a flow-rate of 0.01 ml min^{-1} with the voltage range from 13 to 30 kV. Increasing the flow-rate to 0.03 ml min^{-1} the nanofibres became more irregular with some beads due to the insufficient stretching of the jet at voltages below 25 kV. For higher flow-rate (0.05 ml min^{-1}), the process started only at 30 kV; whereas at lower voltages the solution dripped from the capillary.

Nanofibres of keratin/PEO blend with regular diameter distribution and few defects were electrospun from the solutions at 7 and 10 % wt of polymers concentration applying a voltage from 20 to 30 kV and a solution flow-rate of 0.01 ml min^{-1} (Fig. 6 a). The keratin/PEO solutions at high concentrations (7 and 10 % wt) produce fibres with few defects, like 5 and 7 % wt PEO solutions, probably because they have a similar flow behaviour. At low polymers concentration (5 % wt) the nanofibres were electrospun with many defects also at high voltage (25-30 kV), as Fig. 6 b shows, and the solution dripped from the capillary during the process at voltages below 20 kV because of the low viscosity.

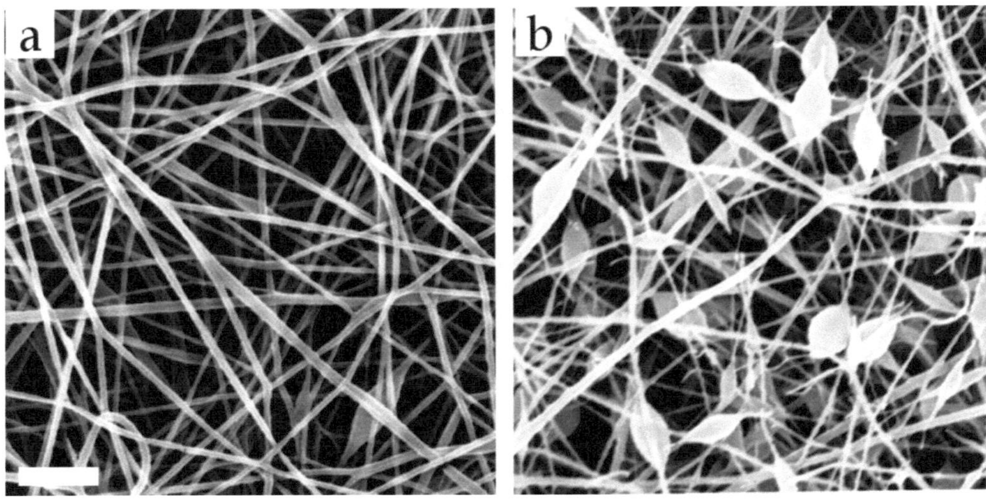

Fig. 6. Electrospun nanofibres of 50/50 keratin/PEO blend at 25 kV, 0.01 ml min^{-1} from (a) 7 % wt and (b) 5 % wt solutions. Scale bar = 2 μm.

The average diameters of the electrospun nanofibres, measured at 150 different points from SEM pictures for each sample produced, as a function of the applied voltage. The diameters of the nanofibres produced at the same voltage from 5 and 7 % wt of pure PEO solutions are generally comparable. The keratin/PEO solution produces nanofibres with small diameter at higher voltage (30 kV), whereas at 20 kV electrospun nanofibres have a diameter much higher than the pure PEO nanofibres at the same voltage. Moreover, the slope of the diameter/voltage trend-line for the keratin/PEO blend is the highest obtained in our experiment (Aluigi et al., 2007b). Thus, it seems that the presence of keratin strengthens the influence of the voltage on the size of the filaments. Since the keratin has many different functional groups, it is possible that inter- and intra-molecular bonds increase the jet rigidity when the solvent evaporates. Therefore, a higher voltage is required to stretch the solidifying keratin/PEO solution jet.

4.4 Thermal analysis

The DSC curves of electrospun keratin/PEO samples, compared with the films produced by casting from the same solution, are shown in Fig. 7 with the aim of highlighting the structural changes induced by the electrospinning process.

The endothermic overlapped peaks at around 60°C are due to the fusion of PEO crystalline phase and to the evaporation of water, especially absorbed by keratin. It is worth noting that in the keratin/PEO film, the water evaporation occurs at lower temperature with respect to keratin/PEO nanofibres (50°C in the nanofibres and 80°C in the film). This is probably due

to the high surface/volume ratio of the nanofibres which promotes water evaporation even at lower temperatures.
The DSC analysis the electrospun PEO exhibits a slight increase of the melting point. It is believed that the high stretching due to the electrospinning process promotes the orientation of the long polymer chains of PEO. This high degree of order shifts the melting point to a higher temperature.
The endothermic events observed in the range of 200-350°C are attributed to protein denaturation followed by protein degradation (Spei & Holzem, 1990). The thermograms show that the peaks related to the protein denaturation, which falls at 233°C in the film, shifts to lower temperature (213°C) in the electrospun sample. It could be presumed that the high draw, given by electrospinning process, and the quick water evaporation hinders the keratin self-assembly leading the protein chains to assume a less complex supermolecular organization which denatures at lower temperatures. This thermal behaviour is in agreement with the FT-IR observations (Aluigi et al., 2007b); in fact, the keratin in the electrospun fibres shows a molecular conformation characterized by weaker hydrogen bonds that make the protein less thermally stable.

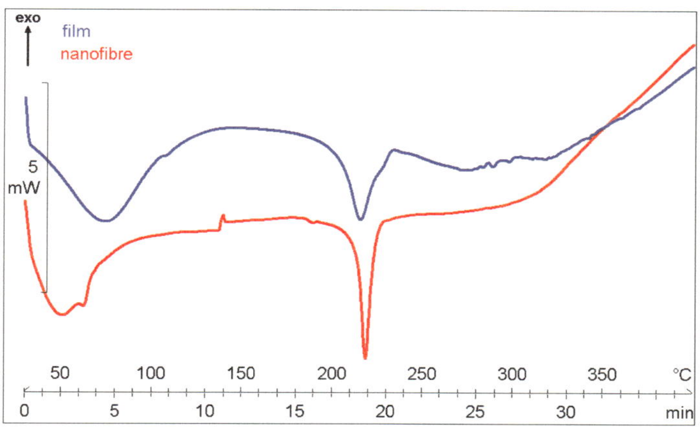

Fig. 7. DSC analysis of keratin/PEO film from casting and keratin/PEO nanofibres from electrospinning.

5. Keratin/PA6 blend nanofibres

Since mid-1990s, there has been an increasing interest in the application of nano-sized fibres for high efficiency filtration (Montefusco, 2005; Gopal et al., 2006; Dotti et al., 2007). Indeed, high surface area associated to great surface cohesion and small pore dimensions of nanofibrous mats allow the capture of submicron particles improving filtration efficiency, cleaning operations and filter life. As far as "active" filtration is concerned, keratin is a bio-polymer that can absorb and remove toxic, sensitizing and suspected carcinogenic agents substances such as heavy-metal ions (copper, chromium, lead, mercury, etc.) (Kar & Misra, 2004; Ki et al., 2007; Aluigi et al., 2009), formaldehyde (Wortmann et al., 1999) and other hazardous VOCs.
Keratin is rich in amino acids having polar and ionisable side chains able to bind charged species such as metal ions. In particular, having a pK_a about 4.5, free carboxyl groups of

aspartyl and glutamyl residues are considered the most likely binding sites over a wide pH range (Maclaren & Milligan, 1981; Taddei et al., 2003). Recent studies showed that formic acid is a good media to dissolve keratin having strong solvation properties for keratin regenerated from wool (Alemdar et al., 2005; Aluigi et al., 2007a). Formic acid is also a solvent for polyamides so that it can be used to prepare keratin-polyamide 6 (PA6) blends in different proportions.

5.1 Solution characterizations

Keratin/PA6 blend solutions with different blending ratios (100/0, 90/10, 70/30, 50/50, 30/70, 10/90 and 0/100), prepared by mixing solutions at 15% wt of pure keratin and pure PA6 separately dissolved in concentrated formic acid, can be processed into nanofibres by electrospinning, with different results according to viscosity and conductivity of the solutions. Viscosities measured at the shear rate 84 s^{-1} show deviation from linearity when compared with the theoretical values calculated by the additive rule (Eq. 2):

$$\ln \eta_T = \sum_i w_i \ln \eta_i \tag{2}$$

where: w_i is the weight fraction of the ith component, η_i is the solution viscosity of the ith component and η_T is the theoretical viscosity of the polymer blend. The fitting of the experimental measurements displays a negative deviation from the additive rule indicating a negligible interaction between the protein and the synthetic polymer. Moreover, viscosity of the blend solutions decreased with increasing the keratin content, while electrical conductivity (about of 2 mS cm^{-1}) did not change significantly.

5.2 Electrospinning and morphological study

Electrospinning is successful at room temperature, 25 kV applied voltage, 0.01 ml min^{-1} flow rate and 10 cm working distance, producing very thin nanofibres in the range from 70 to 300 nm, with mean diameter of about 150 nm.

SEM images of the nanofibres with different compositions randomly deposited on the collecting screen are shown in Fig. 8. Nanofibres obtained by electrospinning keratin/PA6 blends are thin and free from defects, but pure keratin nanofibres show many bead defects since the jet breaks up into droplets as a result of the lower viscosity of the solution. The mean diameter of the nanofibres does not change significantly with the blend composition; however, nanofibres rich in PA6 appear more homogeneous than keratin rich ones.

5.3 Testing and characterization

Chemical interactions between the polymers can be excluded, according to the viscosity measurements and by the study of FT-IR spectra in the region 4000–2600 cm^{-1} and 2000-900 cm^{-1}. Indeed, the nanofibre blend spectra appear as weighed overlapped spectra of pure keratin and pure PA6; a broadening of the Amide A and Amide I bands can be observed with increasing the keratin content and also the intensity of the 1025 cm^{-1} peak (Bunte salt) increases with increasing the keratin content. But the absence of both band shifts and new peaks exclude chemical interaction.

The nanofibres containing more than 50% wt of keratin, are totally destroyed due to swelling of regenerated keratin when it contacts with water. On the other hand, morphology doesn't change for pure PA6 and keratin/PA6 10/90 nanofibres while those of keratin/PA6

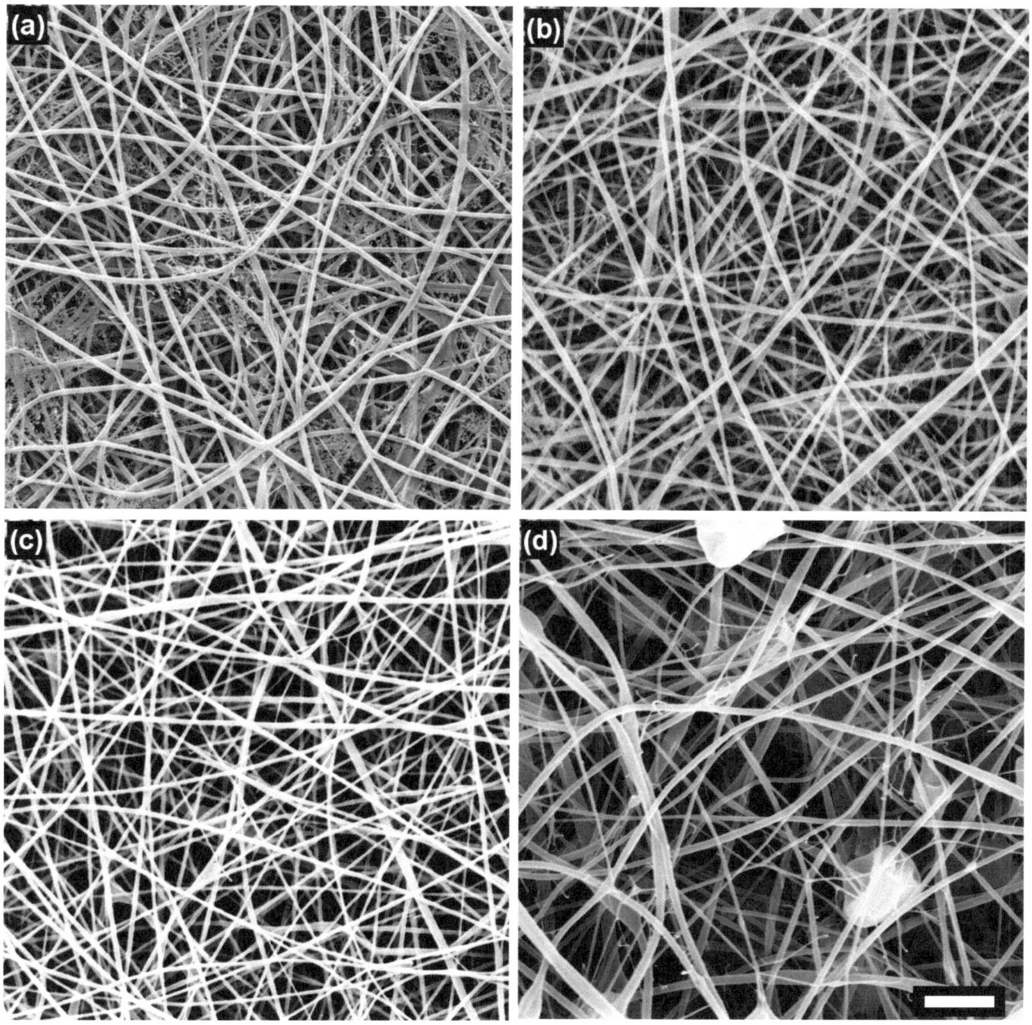

Fig. 8. Keratin/PA6 blend nanofibres: (a) 30% keratin, (b) 70% keratin, (c) 90% keratin and (d) pure keratin. Scale bar = 2 μm.

30/70 start swelling and become flat. The adsorption of Cr^{3+} ions by keratin can be studied only at low pH values in order to avoid the precipitation of chromium hydroxide above pH 4 (Maclaren & Milligan, 1981). Cr^{3+} uptake of pure PA6 and blend nanofibres was calculated using the following Eq. 3:

$$q\ (\mu g/mg) = \frac{q_0 - q_1}{m} \tag{3}$$

where q_0 and q_1 are the Cr^{3+} amount (μg) in the 10 ml of standard solution before and after adsorption, respectively, and m is the mass of the nanofibres. The Cr^{3+} adsorption of keratin/PA6 0/100, 10/90 and 30/70 nanofibres were compared with that of blend films obtained by casting from the same solutions. The chromium uptake of the nanofibres

slightly increases with the increasing the keratin content. Moreover, the adsorption capacity of the nanofibrous mats is much higher than that of films with the same composition, since the higher surface area of the nanofibres increases the number of available binding sites. Both films and nanofibrous membranes were cut into square shapes of 4 cm^2 and placed in a standard solution (10 ml) of chromium ions, with an initial concentration of 50 μg/L at pH 4, for 24 hours. The standard solution was prepared diluting with deionised water a stock solution of chromium (III) nitrate nonahydrate in nitric acid 1000 μg/ml. The Cr^{3+} concentration of the solution after immersion of the nanofibre mats was analysed by a Atomic Adsorption Spectrometer. Before the measurements, the calibration curve was made using standard solution of Cr^{3+} at 25, 50, 75 and 100 μg l^{-1}.
For the formaldehyde adsorption tests, multi-component filters of polypropylene (PP) and keratin/PA6 blend nanofibres were prepared by electrospinning the blend solutions directly onto the surface of the polypropylene filters. Adsorption tests were carried out using an apparatus composed as follows: a closed chamber having a volume of 3.3 L, a fan forcing air to pass through the filter, a filter holder and a Formaldemeter™ htV (PPM Technology Ltd.). About 0.6 ppm of formaldehyde was introduced in the chamber by using formaldehyde-releasing silica. When a stable concentration was reached in the chamber, the solid was removed, the multi-component filter was introduced in the chamber and the measurement started. The adsorption tests were performed at 20°C and 65% R.H. measuring the formaldehyde concentration versus time. The adsorption of gases on solid surfaces includes physiosorption and chemisorption. The physiosorption occurs when the interaction forces between adsorbent and adsorbate are intermolecular attractions (Van der Waals forces), while the chemisorption is a type of adsorption whereby the adsorbent adheres to the adsorbate through the formation of chemical bonds. Therefore, chemisorption is more selective than physiosoprtion.
It is known that formaldehyde has high reactivity towards proteins; in fact it is able to react with different side chains of amino acids such as the amino groups of lysine, arginine,

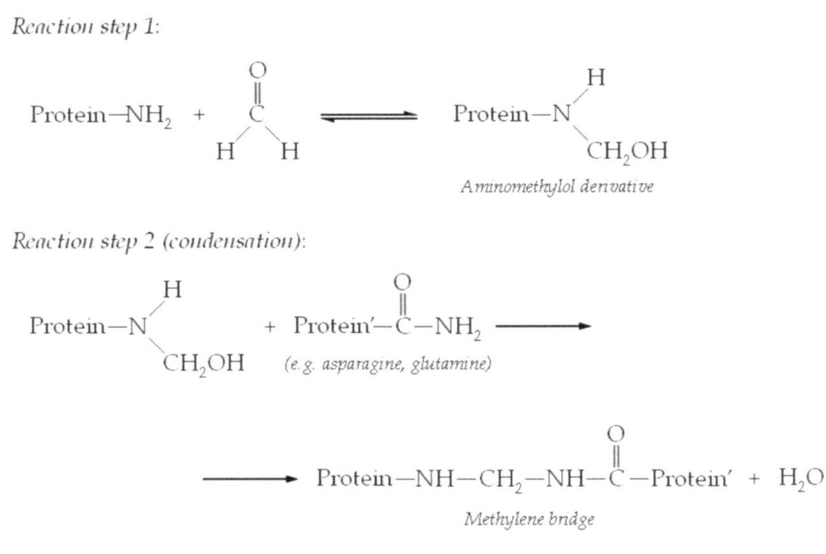

Fig. 9. Reaction scheme of formaldehyde with keratin proteins

glutamine and asparagine. The reaction scheme is reported in Fig. 9. The first reaction step involves amino groups and formaldehyde and produces an unstable product (aminomethylol derivative). The second reaction leads to a stable methylene bridge (Wortmann et al., 2005).

The adsorption performances of keratin-based nanofibres were compared with that of PP filters and pure PA6 nanofibres/PP filters. Because of the selectivity of keratin towards formaldehyde, only the nanofibres containing the 90% wt of keratin were studied. It was found that PP filters and multi-component filters of PP sheets and PA6 nanofibres can reduce the formaldehyde concentration of about 30% and 40% respectively, but the reduction reaches 70% in the presence of keratin based nanofibres. The main reason is that PP and PA6 nanofibres/PP filters adsorb formaldehyde only by physiosorption. On the other hand keratin based nanofibres have stronger capability of adsorbing formaldehyde because in the adsorbing process both physiosorption and chemisorption occur simultaneously.

6. Research trends

In the biomedical field great efforts have recently been made to mimic nature by producing nanofibrous materials solely composed of natural polymers in the assumption that such materials would be less subject to rejection by the host. Moreover, electrospinning of natural proteins offers a promising method to produce nanofibres with a similar nanofibrillar structure to that of a native extra-cellular matrix. Unfortunately, the electrospinning of natural macromolecules has demonstrated to be a challenge because they frequently do not possess the viscoelastic properties needed for good electrospinning. One exception is silk fibroin, a natural high molecular weight polypeptide. Silk fibroin is spun by *Bombix mori* and can be extracted in two molecular weights: 25 kDa (light chains) and 320-390 kDa (heavy chains). Silk fibroin is rich in glycine, alanine and serine amino acids. In the last years, regenerated silk fibroin has been proposed for the fabrication of a variety of biomaterials such as films, sponges, gels and fibres for biomedical applications.

Recently, keratin/fibroin blend has received much attention due to their potential applications in tissue engineering (Vasconcelos et al., 2008), and also for removal of metal ions and water purification (Ki et al., 2007; Baek et al., 2007).

Keratin/fibroin blended materials are a totally protein material (biodegradable and natural) that could combine the interesting properties of keratin (VOCs and metal ion adsorption) with the mechanical properties and electrospinnability of silk fibroin. Moreover, synergic effects could enhance some properties: for instance, it was found that films cast from silk fibroin and keratin decreased blood coagulation compared with silk fibroin or keratin alone (Lee et al., 1998).

Concerning the production of keratin/fibroin nanofibres by electrospinning, both silk fibroin and wool keratin can be dissolved in formic acid and mixed in several blending ratios (Ki et al., 2007; Baek et al., 2007; Zoccola et al., 2008). Blend nanofibers rich in fibroin, show branched and flatten morphology characterized by a bimodal diameter distribution. In spite of the lower viscosities of the solutions, nanofibers rich in keratin have more homogeneous diameter distribution, with a round cross section. This may be due to the higher conductivity that promotes nanofiber formation. Keratin rich nanofibers are also thinner than those rich in fibroin.

Particular behaviour was observed for the 50/50 blend solutions that showed the highest zero-shear viscosity and the longest relaxation time (i.e. lowest shear thinning onset). The resulting nanofibers were thin with respect to those produced with the other blend ratios, and the trend was repeated at different operating conditions (Zoccola et al., 2008). It had been supposed some interactions between keratin and fibroin promote the formation of entangled chain networks. It has also been pointed out that keratin/fibroin 50/50 blend nanofibers have excellent mechanical properties (Baek et al., 2007). These studies encourage further investigations into these types of natural matrices for use in the biomedical field, biotechnology and water purification.

7. References

Alemdar, A.; Iridag, Y. & Kazanci, M. (2005). Flow behavior of regenerated wool-keratin proteins in different mediums. *International Journal of Biological Macromolecules.*, 35, 151-153.

Aluigi, A.; Innocenti, R.; Tonin, C. & Vineis, C. (2004). Studies on keratin/cellulose films from cuprammonium solution. *Autex Research Journal*, 4, 175-181.

Aluigi, A.; Zoccola, M.; Vineis, C.; Tonin, C.; Ferrero, F. & Canetti, M. (2007a). Study on the structure and properties of wool keratin regenerated from formic acid. *International Journal of Biological Macromolecules.*, 41, 266-273.

Aluigi, A.; Varesano, A.; Montarsolo, A.; Vineis, C.; Ferrero, F; Mazzuchetti, G. & Tonin, C. (2007b). Electrospinning of keratin/PEO blend nanofibres. *Journal of Applied Polymer Science*, 104, 863-870.

Aluigi, A.; Vineis, C.; Varesano, A.; Mazzuchetti, G.; Ferrero, F. & Tonin, C. (2008). Structure and properties of keratin/PEO blend nanofibres. *European Polymer Journal*, 44, 2465-2475.

Aluigi, A.; Vineis, C.; Tonin, C.; Tonetti, C.; Varesano, A. & Mazzuchetti, G. (2009). Wool keratin-based nanofibres for active filtration of air and water. *Journal of Biobased Materials and Bioenergy*, 3, 311–319.

Baek, D.H.; Ki, C.S.; Um, I.C. & Park, Y.H. (2007). Metal ion adsorbability of electrospun wool keratose/silk fibroin blend nanofiber mats. *Fibers and Polymers*, 8, 271-277.

Barhate, R.S. & Ramakrishna, S. (2007). Nanofibrous filtering media: filtration problems and solutions from tiny materials. *Journal of Membrane Science*, 296, 1–8.

Barone, J.R.; Schmidt, W.F. & Liebner, C.F (2005). Thermally processed keratin films. *Journal of Applied Polymer Science*, 97, 1644-1651.

Bergsma, J.E.; Rozema, F.R.; Bos, R.R.M.; de Bruijn, W.C.; Boering, G. & Pennings, A.J. (1995). In vivo degradation and biocompatibility study of in vitro pre-degraded as-polymerized polylactide particles. *Biomaterials*, 16, 267–274.

Bryner M. & Armantrout, J. (2006). *U.S. Patent* 2006138710.

Cecil, R. (1963). In: *The proteins*, Vol. 1, Neurath H. (Ed.), Academic Press, New York, USA.

Cooley, J.F. (1902). *U.S. Patent* 692631.

Crick, F.H.C. (1953). The Fourier transform of a coiled-coil. *Acta Crystographica*, 6, 685-689.

Deitzel, J.M.; Kleinmeyer, J.; Hirvonen, J.K. & Beck Tan, N.C. (2001). Controlled deposition of electrospun poly(ethylene oxide) fibers. *Polymer*, 42, 8163–8170.

Desai, N.P. & Hubbel, J.A. (1991). Solution technique to incorporate polyethylene oxide and other water-soluble polymers into surfaces of polymeric biomaterials. *Biomaterials*, 12, 144–153.

Dotti, F.; Varesano, A.; Montarsolo, A.; Aluigi, A; Tonin, C. & Mazzuchetti, G. (2007). Electrospun porous mats for high efficiency filtration. *Journal of Industrial Textiles*, 37, 151-162.

Dowling, L.M.; Crewther, W.G. & Parry, D.A.D. (1986). Secondary structure of component 8c-1 of α-keratin. *Biochemical Journal*, 236, 705-712.

Ellison, C.J.; Phatak, A.; Giles, D.W.; Macosko, C.W. & Bates, F.S. (2007). Melt blown nanofibers: fiber diameter distributions and onset of fiber break-up. *Polymer*, 48, 3306–3316.

Formhals, A. (1934). *U.S. Patent* 1975504.

Fraser, R.D.B.; MacRae, T.P.; Parry, D.A.D. & Suzuki, E. (1971). The structure of feather keratin. *Polymer*, 12, 35-56.

Fraser, R.D.B.; MacRae, T.P. & Rogers, G.E. (1972). In: *Keratins: Their composition, structure and biosynthesis*, Charles C. Thomas Publisher, Springfield, Illinois, USA.

Furth, M.E.; Atala, A. & Van Dyke, M.E. (2007). Smart biomaterials design for tissue engineering and regenerative medicine. *Biomaterials*, 28, 5068-5073.

Garret, R.H. & Grisham, C.M. (2002). In: *Biochemistry 2nd Edition*, Higher Education Press, Beijing, China.

Gibson, P.; Schreuder-Gibson, H. & Rivin, D. (2001). Transport properties of porous membranes based on electrospun nanofibers. *Colloid Surface A*, 187–188, 469–481.

Gopal, R.; Kaur, S.; Ma, Z.; Chan, C.; Ramakrishna, S. & Matsuura, T. (2006). Electrospun nanofibrous filtration membrane. *Journal of Membrane Science*, 281, 581–586.

Hamasaki, S.; Tachinaba, A.; Tada, D.; Yamauchi, K. & Tanabe, T. (2008). Fabrication of highly porous keratin sponges by freeze-drying in the presence of calcium alginate beads. *Material Science and Engineering: C*, 28, 1250-1254.

Homonoff, E. (2008). Nanofibrillated fibers: opening newmarkets to nanofibre usage. *International Fibers Journal* 23, 22–24.

Huang, Z.M.; Zhang, Y.Z.; Kotaki, M. & Ramakrishna, S. (2003). A review on polymer nanofibers by electrospinning and their applications in nanocomposites. *Composites Science Technology*, 63, 2223–2253.

Jeffrey, P.D. (1972). The molecular weights of the low-sulphur proteins from wool: A review. *Journal of the Textile Institute*, 63, 91-113.

Jin, H.J.; Fridrikh, S.V.; Rutledge, G.C. & Kaplan, D.L. (2002). Electrospinning Bombyx mori silk with poly(ethylene oxide). *Biomacromolecules*, 3, 1233–1239.

Kar, P. & Misra, M. (2004). Use of keratin fiber for separation of heavy metals from water. *Journal of Chemical Technology & Biotechnology*, 79, 1313-1319.

Katoh, K.; Shibayama, M.; Tanabe, T. & Yamauchi, K. (2004). Preparation and properties of keratin-poly(vinyl alcohol) blend fiber. *Journal of Applied Polymer Science*, 91, 756-762.

Ki, C.S.; Gang, E.H.; Um, I.C. & Park, Y.H. (2007). Nanofibous membrane of wool keratose/silk fibroin blend for heavy metal ion adsorption. *Journal of Membrane Science*, 302, 20-26.

Kilic, A.; Oruc, F. & Demir, A. (2008). Effects of polarity on electrospinning process. *Textile Research Journal*, 78, 532–539.

Lee, K.; Kong, S.; Park, W.; Ha, W. & Kwon, I. (1998). Effect of surface properties on the antithrombogenicity of silk fibroin/S-carboxymethyl kerateine blend films. *Journal of Biomaterial Science Polymer Edition*, 9, 905-914

Lee, J.; Cuddihy, M.J. & Kotov, N.A. (2008). Three-dimensional cell culture matrices: State of the art. *Tissue Engineering B*, 14, 61–86.

Li, D. & Xia, Y. (2004). Electrospinning of nanofibers: reinventing the wheel. *Advanced Materials*, 16, 1151–1170.

Li, L.; Frey, M.W. & Green, T.B. (2006). Modification of air filter media with nylon-6 nanofibers. *Journal of Engineered Fibers and Fabrics*, 1, 1–24.

Maclaren, J.A. & Milligan, B. (1981). The chemical reactivity of the wool fibre, In: *Wool science*, pp. 109-127, Science Press, Marrickville, Australia.

Martínez-Hernández, A.L.; Velasco-Santos, C.; De-Icaz, M. & Castaño, V.M. (2007). Dynamical-mechanical and thermal analysis of polymeric composites reinforced with keratin biofibers from chicken feathers. *Composites Part B: Engineering*, 38, 405-410.

Montefusco F. (2005). The use of nonwovens in air filtration. *Filtration and Separation*, 42, 30-31.

Morton, W.J. (1902). *U.S. Patent* 705691.

Pauling, L. & Corey, R.B. (1953). Compound helical configurations of polypeptide chains: Structure of proteins of the α-keratin type. *Nature*, 171, 59-61.

Perez, M.; Swan, M. & Louks, J. (2000). *U.S. Patent* 6110588.

Pham, Q.P.; Sharma, U. & Mikos, A.G. (2006). Electrospinning of polymeric nanofibers for tissue engineering applications: A review. *Tissue Engineering*, 12, 1197–1211.

Pourdeyhimi, B.; Fedorova, N. & Sharp, S. (2006). *U.S. Patent* 2006292355.

Qin, X.H. & Wang, S.Y. (2006). Filtration properties of electrospinning nanofibers. *Journal of Applied Polymer Science*, 102, 1285–1290.

Ramakrishna, S.; Fujihara, K.; Teo, W.E.; Yong, T.; Ma, Z. & Ramaseshan, R. (2006). Electrospun nanofibers: Solving global issues. *Materials Today*, 9, 40–50.

Reneker, D.H. & Yarin, A.L. (2008). Electrospinning jets and polymer nanofibers. *Polymer*, 49, 2387-2425.

Schmidt, W.F. (1998). Innovative feather utilization strategies, *Proceedings of the National Poultry Waste Management Symposium*, pp. 276-282, Springdale, AR, October 19-22, 1998.

Schrooyen, P.M.M.; Dijkstra, P.J.; Oberthür, R.C.; Bantjes, A. & Feijen J. (2001). Stabilization of solutions of feather keratins by sodium dodecyl sulfate. *Journal of Colloid and Interface Science*, 240, 30-39.

Shenoy, S.L.; Bates, W.D.; Frisch, H.L. & Wnek, G.E. (2005). Role of chain entanglements on fiber formation during electrospinning of polymer solutions: good solvent, non-specific polymer–polymer interaction limit. *Polymer*, 46, 3372-3384.

Shin, Y.M.; Hohman, M.M.; Brenner, M.P. & Rutledge, G.C. (2001). Experimental characterization of electrospinning: the electrically forced jet and instabilities. *Polymer*, 42, 9955–9967.

Shin, C.; Chase, G.G. & Reneker, D.H. (2005). The effect of nanofibers on liquid-liquid coalescence filter performance. *AIChE Journal,* 51, 3109-3113.

Sill, T.J. & von Recum, H.A. (2008). Electrospinning: Applications in drug delivery and tissue engineering. *Biomaterials,* 29, 1989-2006.

Son, W.K.; Youk, J.H.; Lee, T.S. & Park W.H. (2004). The effects of solution properties and polyelectrolyte on electrospinning of ultrafine poly(ethylene oxide) fibers. *Polymer.* 45, 2959-2966.

Spei, M. & Holzem, R. (1990). Further thermoanalytical investigations of annealed keratins: The time and temperature dependence. *Colloid and Polymer Science,* 268, 630-635.

Tachinaba, A.; Furuta, Y.; Takeshima, H.; Tanabe, T. & Yamauchi, K. (2002). Fabrication of wool keratin sponge scaffolds for long-term cell cultivation. *Journal of Biotechnology,* 93, 165-170.

Tachinaba, A.; Kaneko, S.; Tanabe, T. & Yamauchi, K. (2005). Rapid fabrication of keratin-hydroxyapatite hybrid sponges toward osteoblast cultivation and differentiation. *Biomaterials,* 26, 297-302.

Taddei, P.; Monti, P.; Freddi, G.; Arai, T. & Tsukada, M. (2003). Binding of Co(II) and Cu(II) cations to chemically modified wool fibres: An IR investigation. *Journal of Molecular Structure,* 650, 105-113.

Tanabe, T.; Okitsu, N. & Yamauchi, K. (2004). Fabrication and characterization of chemically crosslinked keratin films. *Materials Science and Engineering: C,* 24, 441-446.

Theron, S.A.; Zussman, E. & Yarin, A.L. (2004). Experimental investigation of the governing parameters in the electrospinning of polymer solutions. *Polymer,* 45, 2017-2030.

Thompson, E.O.P. & O'Donnel I.J. (1959). Studies on oxidized wool: A comparison of the completeness of oxidation with peracetic and performic acids. *Australian Journal of Biological Sciences,* 12 282-292.

Tonin, C.; Aluigi, A.; Bianchetto Songia, M.; D'Arrigo, C.; Mormino, M. & Vineis, C. (2004). Thermoanalytical characterization of modified keratin fibres. *Journal of Thermal Analalysi and.Calorimetry,* 77, 987-996.

Tonin, C.; Zoccola, M.; Aluigi, A.; Varesano, A.; Montarsolo, A.; Vineis, C. & Zimbardi, F. (2006). Study on the conversion of wool keratin by steam explosion. *Biomacromolecules,* 7, 3499-3504.

Tsai, P.P.; Schreuder-Gibson, H. & Gibson, P. (2002). Different electrostatic methods for making electret filters. *Journal of Electrostatics,* 54, 333-341.

Tseng, A.A.; Notargiacomo, A. & Chen, T.P. (2005). Nanofabrication by scanning probe microscope lithography: A review. *Journal of Vacuum Science and Technology B,* 23, 877-894.

Varesano, A.; Aluigi, A.; Vineis, C. & Tonin, C. (2008). Study on the shear viscosity behaviour of keratin/PEO blends for nanofibre electrospinning. *Journal of Polymer Science Part B: Polymer Physics,* 46, 1193-1201.

Vasconcelos, A.; Freddi, G. & Cavaco-Paulo, A. (2008). Biodegradable materials based on silk fibroin and keratin. *Biomacromolecules,* 9, 1299-1305.

von Bergen, W. (1963). In: *Wool Handbook 3rd Edition,* Vol. 1, John Wiley & Sons, Inc., New York, USA.

Wortmann, G.; Zwiener, G.; Sweredjiuk, R.; Doppelmayer, F. & Wortmann, F.J. (1999). Sorption of indoor air pollutants by sheep's wool: Formaldehyde as an example. *Proceedings of the International Wool Textile Organization*, Report CTF 3, Florence, Italy, June 1999.

Wortmann, G.; Thome, S.; Sweredjuk, R. & Zwiener, G. (2005). Chemisorption of protein reactive indoor air pollutants by wool, *Proceedings of the International Wool Textile Organization Conference*, Report CTF03, pp. 1-7, Hobart, Australia, April 2005.

Xie, H.; Li, S. & Zhang, S. (2001). Ionic liquids as novel solvents for the dissolution and blending of wool keratin fibres. *Green Chemistry*, 7, 606-608.

Xie, X.N.; Chung, H.J.; Sow, C.H. & Wee, A.T.S. (2006). Nanoscale materials patterning and engineering by atomic force microscopy nanolithography. *Materials Science and Engineering R: Reports*, 54, 1–48.

Yamauchi K.; Yamauchi A.; Kusunoki T.; Kohda A. & Konishi Y. (1996) Preparation of stable aqueous solution of keratins and physiochemical and biodegradational properties of films. *Journal of Biomedical Material Research*, 31, 439-444.

Yarin, A.L.; Koombhongse, S. & Reneker, D.H. (2001). Taylor cone and jetting from liquid droplets in electrospinning of nanofibers. *Journal of Applied Physics*, 90, 4836–4846.

Yin, J.; Rastogi, S.; Terry, A.E. & Popescu, C. (2007). Self-organization of oligopeptides obtained on dissolution of feather keratins in superheated water. *Biomacromolecules*, 8, 800-806.

Zhang, S. (2002). Emerging biological materials through molecular self-assembly. *Biotechnology Advances*, 20, 321–339.

Zoccola, M.; Aluigi, A.; Vineis, C.; Ferrero, F.; Piacentino, M. & Tonin, C. (2008). Study on cast membranes and electrospun nanofibers made from keratin/fibroin blends. *Biomacromolecules*, 9, 2819-2825.

8

In Situ Probing of Oxygen-containing Groups on Acid-treated Carbon Nanofibers using Aromatic Molecules

Hiromasa Nishikiori, Satoshi Kubota, Nobuaki Tanaka,
Morinobu Endo, and Tsuneo Fujii
Shinshu University
Japan

1. Introduction

Development of chemical nanotechnology to control the structure of materials on a nanosize scale is necessary in order to obtain certain physical and chemical properties of the nanomaterials. Carbon nanofibers (CNFs) (Oberlin et al., 1976; Endo, 1988; Endo et al., 2001) are very large multi-walled carbon nanotubes and are technologically easier and economically more favorable to produce than individual single- or double-walled carbon nanotubes (Iijima, 1991; Iijima & Ichihashi, 1993). The CNFs are valuable materials for electronic, mechanical, and optical devices because of their unique structural and quantum characteristics that are similar to small-sized carbon nanotubes (Oberlin et al., 1976; Endo, 1988; Endo et al., 2001; Endo et al., 2002; Yang et al., 2003; Wang et al., 2005; Tan et al., 2006). For practical use, such carbon nanomaterials need to be well dispersed throughout other raw materials. An example of this is the incorporation of carbon nanomaterials into plastics or ceramics, which provide practical materials with well-defined shape and increased strength. Composites of matrices with dispersed carbon nanotubes have been prepared by the polymerization of a polyimide under sonification (Park et al., 2002) and by the sol−gel reaction of a system containing a relatively large amount of N,N′-dimethylformamide as the starting material (Hongbinget al., 2004). However, carbon nanomaterials have a high specific surface area and easily aggregate. Surface functionalization of the carbon nanomaterials is an effective method to disperse them throughout various media for producing new functional materials, which utilize their unique characteristics (Zhu et al., 2003; Gao et al., 2005; Singh et al., 2005). In order to functionalize these carbon nanomaterials one must treat their surface with acids or other chemicals. Treatment of the carbon nanomaterials with nitric acid and sulfuric acid leads to the oxidation of their surface that forms oxidized groups such as −COOH and −C=O within the graphene sheet (Liu et al., 1998; Hamon et al., 2001; Hamon et al., 2002). Generally, the surface functional groups of the modified CNFs are characterized by IR or Raman spectroscopy. It is, however, difficult to obtain quantitative information of the chemical species existing in a monolayer or only a few layers of the oxidized surface of the CNFs using these analyses.

We have previously shown that observing the fluorescence spectra of 1-naphthol (1-NP) is a useful probe on a molecular level for studying the physicochemical properties of the surrounding environment around the 1-NP (Suzuki et al., 1977; Fujii et al., 1995). Based on these investigations, significant physicochemical information of the CNF surface was obtained by in situ spectrometry using 1-NP as a fluorescent probe (Nishikiori et al., 2004; Kubota et al., 2005a; Kubota et al., 2005b). Our unique procedure to create a highly disperse system of CNFs throughout solvents allowed these observations even though the fluorescence of aromatic molecules adsorbed on carbon materials has scarcely been observed due to strong quenching. In this procedure, 1-NP was adsorbed on the untreated CNFs and acid-treated CNFs and they were then dispersed in solvents (Nishikiori et al., 2004; Kubota et al., 2005a). Two types of 1-NP fluorescence, the ^1L_b fluorescence (Suzuki et al., 1977; Fujii et al., 1992) and the ion-pair fluorescence (Mishra et al., 1991; Fujii et al., 1992), were observed from the following two adsorbed forms. These are generated by the π−π interaction between 1-NP and the graphene sheet (Chen et al., 2001b; Long & Yang, 2001) and the hydrogen-bonding interaction between 1-NP and proton-accepting groups the oxidized groups, such as -COOH ($-COO^-$) and -C=O (Nishikiori et al., 2004; Kubota et al., 2005a; Kubota et al., 2005b), generated at the acid-treated CNF surface (Hammon et al., 2002; Lakshminarayanan et al., 2004).

1-Aminopyrene (1-AP) is a Brönsted base and is expected to interact with acidic groups on a solid surface (Hite et al., 1986; Miller et al., 2005). Therefore, oxygen-containing functional groups produced on the surface of the acid-treated CNFs, especially acidic groups such as -COOH (Chen et al., 2001a; Kahn et al., 2002), are characterized by the fluorescence measurements using 1-AP as a molecular probe (Nishikiori et al., 2008). 1-AP is suitable as a fluorescence probe since its spectrum drastically changes with the acid-base equilibrium compared with those of aminonaphthalene or aminoanthracene.

However, unlike the 1-NP species, the low-polar species fluorescence is not observed from the 1-AP species on the graphene sheet of the CNFs. This is because the polar amino group prevents the π−π interaction between the pyrene ring and graphene sheet (Nishikiori et al., 2008) even though pyrene derivatives without amino group are adsorbed onto the CNTs by this interaction (Chen et al., 2001; Tomonari et al., 2006). Therefore, the fluorescence observation of the pyrene adsorbed on the surface of CNFs presents the information of the π−π interaction between the pyrene ring and graphene sheet (Tanigaki et al., 2007).

In this chapter, we will report that in situ fluorescence measurements using aromatic probe molecules are useful for studying the physicochemical properties on the CNF surface. The relationship between the CNF dispersion throughout the solution of the probe molecules and their adsorption onto the CNFs will be discussed by analyzing the UV-visible absorption and fluorescence spectra of the suspension containing the probe molecules and the untreated or acid-treated CNFs.

2. Acid treatment of carbon nanofibers

The CNFs (VGCF, vapor grown carbon fiber) having a diameter ca. 200 nm, a length of ca. 10−20 μm and a surface area of ca. 15 m^2g^{-1}, were provided by Showa Denko Co, Ltd. (Endo et al., 2001). Functionalization of the CNF surfaces was carried out in two ways with liquid acid, as reported in the literature (Liu et al., 1998; Hamon et al., 2001; Hamon et al., 2002, Nishikiori et al., 2004; Kubota et al., 2005b). The first was by refluxing the CNFs in concentrated nitric acid at 393 K for 1 or 24 h followed by rinsing them with copious amounts of water and allowing them to dry at room temperature under vacuum. For a stronger treatment procedure, the CNFs were sonicated in a concentrated H_2SO_4/HNO_3

mixture (3/1 in volume) at 313 K for 24 h. They were then refluxed in a mixture of concentrated sulfuric acid and 30% aqueous hydrogen peroxide (4/1 in volume) at 343 K for 24 h and were refluxed in concentrated nitric acid at 393 K for 24 h. They were then rinsed with copious amounts of water until the washings were confirmed neutral using a pH test paper and allowed to dry at room temperature under vacuum. Designation is held throughout this chapter as the following: Untreated CNFs (N-CNF); CNFs treated solely with nitric acid for 1 and 24 h (A1-CNF and A24-CNF, respectively); CNFs treated with both sulfuric and nitric acids (AA-CNF).

The SEM images and IR spectra of these samples have been described elsewhere (Kubota et al., 2005b). The surface structures of the three CNF samples were hardly distinguished even by their TEM images as previously reported (Toebes et al., 2004; Lakshminarayanan et al., 2004).

3. Quantitative characterization of surface adsorption sites of carbon nanofibers by in-situ fluorescence measurement using 1-naphthol

3.1 Adsorption characteristics of 1-NP on N-CNF

The N-CNF (1.0 mg) was dispersed in aqueous solutions of 1-NP (10 cm^3) by ultrasonic irradiation for 120 h. The 1-NP concentrations ranged from 5×10^{-5} to 5×10^{-4} mol dm^{-3}. The resulting suspensions were centrifuged to remove any precipitates. The absorption and fluorescence spectra of the resulting transparent suspensions were measured.

Figure 1 shows the fluorescence spectra of 1-NP observed in aqueous suspensions including the N-CNF prepared at different initial concentrations of 1-NP, and the resulting Langmuir isotherm for the adsorption of 1-NP on the N-CNF. The fluorescence originating from the ^1L_b state and anion of 1-NP were observed at around 330–350 and 460 nm, respectively (Fujii et al., 1992; Nishikiori et al., 2004). The ^1L_b fluorescence indicates that 1-NP exists in a nonpolar environment as a result of the π–π interaction with the CNF surface (Chen et al., 2001a; Long & Yang, 2001; Nishikiori et al., 2004). Almost all the 1-NP molecules in water exist as neutral species in the ground and the anion species in the excited states because the values of pK_a are 9.2 and 0.4 in their states, respectively (Fujii et al., 1992; Nishikiori et al., 2004). The intensity of the ^1L_b fluorescence relative to that of the anion fluorescence decreases with an increase in the 1-NP concentration, i.e., the ratio of 1-NP in the liquid phase becomes higher, indicating that the adsorption approaches saturation. The total concentration of 1-NP, C, is written as

$$C = [N] + [NF], \tag{1}$$

where N and NF are 1-NP existing in the liquid phase and on the CNF surface by a π–π interaction, respectively. The ratio of the fluorescence intensity of N, I_N, to that of NF, I_{NF}, is defined as R_1. The desorption in the excited states can be ignored because the diffusion rate of molecules is much lower than the fluorescence life time.

$$R_1 = I_N/I_{NF} = A\,[N]/[NF], \tag{2}$$

where A indicates the ratio of the spectroscopic constant of N to that of NF defined by the molar extinction coefficient, ε, and fluorescence quantum yield, Φ.

$$A = \varepsilon_N\Phi_N/\varepsilon_{NF}\Phi_{NF}\,. \tag{3}$$

These equations can be applied to the Langmuir isotherm as follows:

$$\theta = [NF]/[F_1] = K_1[N]/(1+ K_1[N]), \tag{4}$$

where θ is the fractional surface coverage and F_1 is the adsorption site due to the π–π interaction on the N-CNF surface, and K_1 is the adsorption equilibrium constant on the F_1 site. The value $[F_1]$ is the molar amount of the F_1 sites existing on the CNFs dispersed in 1 dm^3 of water. The weight of the dispersed CNFs is about 100 times lower than the initial sampling weight, 1.0 mg, in 10 cm^3 of water.

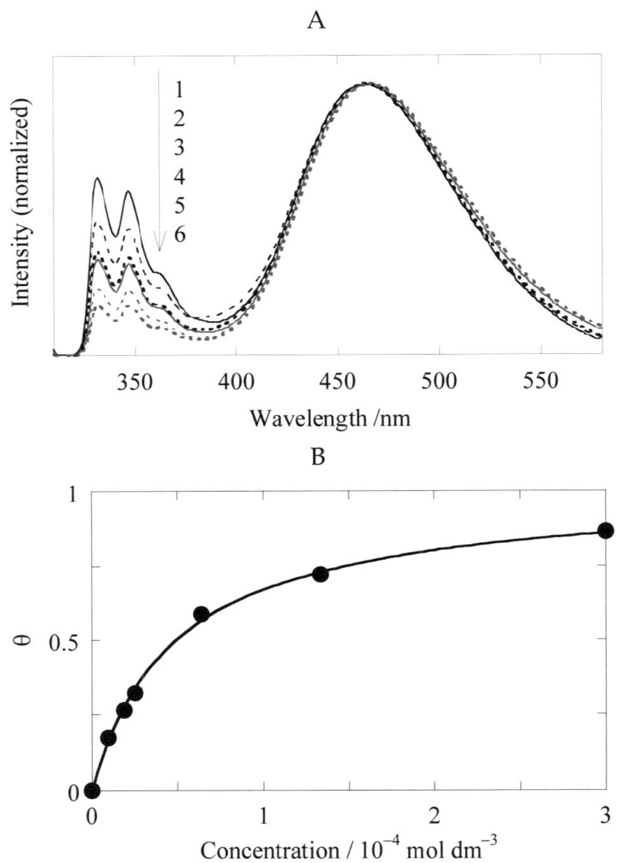

Fig. 1. (A) fluorescence spectra of 1-NP observed in aqueous suspension including N-CNF and (B) resulting Langmuir isotherm for F_1 site. The initial concentrations of 1-NP in water are (1) 5.0×10^{-5}, (2) 8.0×10^{-5}, (3) 1.0×10^{-4}, (4) 2.0×10^{-4}, (5) 3.0×10^{-4}, and (6) 5.0×10^{-4} mol dm^{-3}. The excitation wavelength is 297 nm.

The Langmuir isotherm of 1-NP on the surface of N-CNF was obtained by the curve fitting calculation using the known variables, C and R_1, and the parameters, A, K_1, and $[F_1]$. The experimental and simulated relationships between θ and [N] are plotted in Figure 1B. The experimental values are well-fitted to the Langmuir isotherm. The resulting constants of A, K_1, and $[F_1]$ are summarized in Table 1.

	Spectroscopic constant	Adsorption equilibrium constant / mol^{-1} dm^3
Characters of adsorption sites		
F_1 site	6.5 (*A*)	2.0×10^4
F_2 site (A1-CNF)	1.0 (*B*)	1.7×10^5
F_2 site (A24-CNF)	1.1 (*B*)	2.2×10^5

	F_1 site / mol dm^{-3}	F_2 site / mol dm^{-3}	Total / mol dm^{-3}
Concentration of adsorption sites			
N-CNF	2.3×10^{-4}	0	2.3×10^{-4}
A1-CNF	1.7×10^{-4}	7.4×10^{-5}	2.4×10^{-4}
A24-CNF	1.4×10^{-4}	1.9×10^{-4}	3.3×10^{-4}

Table 1. Constants for CNFs estimated by fluorescence measurements using 1-NP

3.2 Adsorption characteristics of 1-NP on A-CNF

The HNO_3-treated samples, A1-CNF and A24-CNF (1.0 mg) were individually dispersed in aqueous solutions of 1-NP (10 cm^3) by ultrasonic irradiation for 120 h. The 1-NP concentrations ranged from 5×10^{-5} to 5×10^{-4} mol dm^{-3}. The absorption and fluorescence spectra of the resulting transparent supernatants were measured after centrifugation of the suspensions.

Figure 2 shows the fluorescence spectra of 1-NP observed in the A24-CNF suspensions, and resulting Langmuir isotherm for the adsorption of 1-NP on the A24-CNF. The fluorescence band at around 380–450 nm was observed as well as the ^1L_b band. This new band is assigned to the ion pair fluorescence of 1-NP located at the shoulder of the anion band (Mishra et al., 1991; Fujii et al., 1992; Nishikiori et al., 2004). The ion-pair fluorescence indicates that a chemical modification occurred on the CNF surface by the HNO_3 treatment. The ion pair of 1-NP was produced by the hydrogen-bonding interaction between the neutral 1-NP and the oxidized surface of the CNFs (Hamon et al., 2002). The spectral shift in the band from 400 to 440 nm can be observed with an increase in the initial concentration of 1-NP, caused by a decrease in the relative intensity of the ion-pair band and increase in that of the anion band. This phenomenon results from the fact that the adsorption is approaching saturation. On the other hand, the relative intensity of the ^1L_b band increases with an increase in the initial concentration of 1-NP. This result is somewhat strange considering the adsorption equilibrium.

The degree of the CNF dispersions in the water phase has to be considered for each system. The degree corresponds to the absorbance at 500 nm of a CNF suspension caused by the electronic transition of the CNFs (Bahr et al., 2001; Haiber et al., 2003). Figure 3 shows the 1-NP concentration dependence of the absorbance at 500 nm observed in the water suspension including 1-NP and the CNF samples. The degree of dispersion of the A1-CNF and A24-CNF increases with an increase in the concentration of 1-NP in contrast to the N-CNF. The A1-CNF and A24-CNF easily aggregate because of the hydrogen-bonding between the oxidized groups, whereas the results show that the adsorption of 1-NP on the groups effectively prevents the aggregation of the CNFs (Shaffer et al., 1998). Our study suggests that the oxidized groups promote the aggregation of the CNFs since the A1-CNF and A24-CNF are highly dispersed in basic solvents in which the oxidized groups are deprotonated.

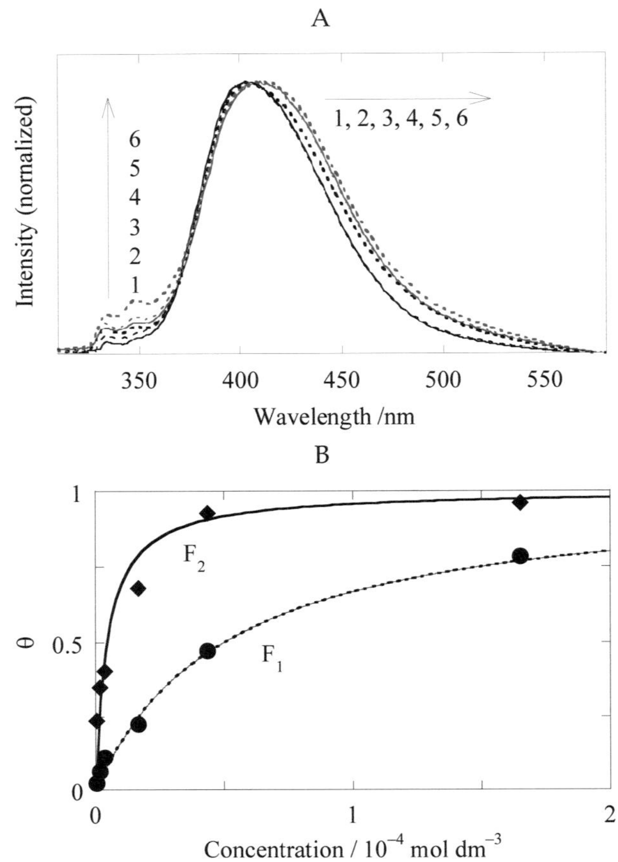

Fig. 2. (A) fluorescence spectra of 1-NP observed in aqueous suspension including A24-CNF and (B) resulting Langmuir isotherm for F_1 and F_2 sites. The initial concentrations of 1-NP in water are (1) 5.0×10^{-5}, (2) 8.0×10^{-5}, (3) 1.0×10^{-4}, (4) 2.0×10^{-4}, (5) 3.0×10^{-4}, and (6) 5.0×10^{-4} mol dm^{-3}. The excitation wavelength is 297 nm.

The average pK_a value of the groups, such as −COOH and −OH, on the CNFs was around 4.0, similar to that in a previous report (Toebes et al., 2004). The spectral data observed in the A1-CNF and A24-CNF systems shown in Figure 2 must be corrected with respect to the apparent amount of the dispersed CNFs. The intensities of the ^1L_b and the ion-pair fluorescences relative to that of the anion's were, therefore, reduced to the values per a certain amount of the CNFs corresponding to that of the dispersed N-CNF using the results shown in Figure 3.

Considering the ion-pair species of 1-NP interacting with the oxidized groups, IF, the total concentration of 1-NP, C, is given by

$$C = X[\mathrm{N}] + [\mathrm{IF}], \tag{5}$$

where

$$X = 1+[\mathrm{NF}]/[\mathrm{N}] = 1+AI_{\mathrm{NF}}/I_{\mathrm{N}}. \tag{6}$$

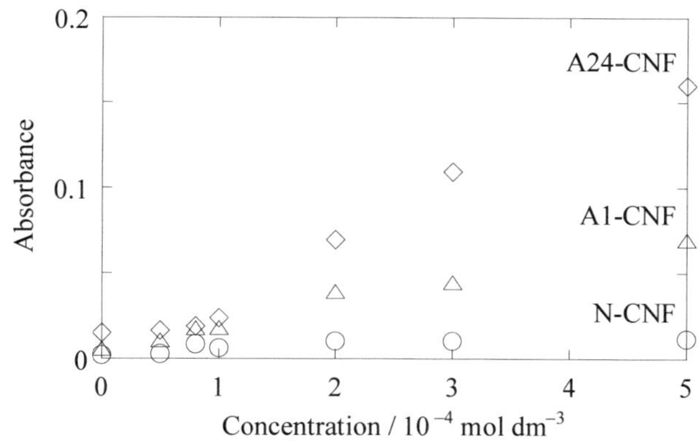

Fig. 3. 1-NP concentration dependence of absorbance at 500 nm observed in aqueous suspension including 1-NP and the N-CNF, A1-CNF, and A24-CNF.

The ratio of the fluorescence intensity of N, I_N, to that of IF, I_{IF}, is defined as R_2.

$$R_2 = I_N / I_{IF} = B[N]/[IF], \tag{7}$$

where B indicates the constant ratio of the fluorescence efficiency of N to that of IF defined similar to A.

$$B = \varepsilon_N \Phi_N / \varepsilon_{IF} \Phi_{IF}. \tag{8}$$

These equations can be adopted to the Langmuir isotherm as follows:

$$\theta = [IF]/[F_2] = K_2[N]/(1+ K_2[N]), \tag{9}$$

where F_2 is the adsorption site of the oxidized groups on the A1- or A24-CNF surface, and K_2 is the adsorption equilibrium constants on the F_2 sites. The $[F_2]$ value is the molar amount of the F_2 sites existing on the CNFs dispersed in 1 dm^3 of water similar to $[F_1]$. The equilibrium equations for the site F_1 are also adopted for the A1-CNF and A24-CNF in order to estimate the values of $[F_1]$ and K_1. The values, $[F_1]$ and $[F_2]$, on the A1-CNF and A24-CNF are the corrected ones with respect to the amount of the dispersed CNFs.

The Langmuir isotherms of 1-NP on the surface of the A24-CNF for the F_2 sites were obtained by the curve fitting calculation using the variables, C, R_2, and X, and the parameters, B, K_2, and $[F_2]$. The isotherms for the F_1 site were also obtained in the same way for the N-CNF. The experimental and simulated relationships between the θ and [N] for the F_1 and F_2 sites are plotted in Figure 2B. The experimental values are well-fitted to the Langmuir isotherms. The constants for A1-CNF and A24-CNF are summarized in Table 1 in addition to those of N-CNF. The spectroscopic constants, A and B, are the magnitudes of the fluorescent quenching, so that a higher value means that the fluorescence is harder to be emitted. Our experimental results indicate that the F_1 site easily quenches the 1-NP fluorescence by about six times faster than the F_2 site. The value of K_2 is nearly 10 times higher than that of K_1, so that the interaction of the 1-NP molecules with the F_2 site is much larger than with the F_1 site. The relative amount of the F_2 sites to all the sites is 30% on the A1-CNF and 58% on the A24-CNF, indicating that the amount of oxidized groups increases with an increase in the acid treatment time. Considering that the N-CNF adsorbs a naphthalene ring unit per 0.70 nm^2 based on a simulation of benzene molecules adsorbed on graphite (Vernov & Steele, 1991), we estimated the number of F_2 sites per nm^2 on the A1-CNF and A24-CNF as 0.46 and 1.18, respectively.

4. In situ probing of acidic groups on acid-treated carbon nanofibers using 1-aminopyrene

4.1 Changes in UV-visible absorption spectra of CNF suspensions

The N-, A24-, and AA-CNFs (0.50 mg) were individually dispersed in the 1-AP solutions (10 cm^3) containing water and ethanol (4/1 in volume) at 1.0×10^{-4} mol dm^{-3} by ultrasonic irradiation for 1–18 days. The resulting suspensions were centrifuged to remove any precipitates. The UV-visible absorption spectra of the resulting supernatant suspensions were measured in order to examine the adsorption of 1-AP on the CNFs and the dispersion of the CNFs into the liquid phase. The suspensions contained slightly amounts of acids and were almost neutral.

In the N-CNF suspension, only slight temporal changes in the spectrum are observed. This suggests that 1-AP was not adsorbed on the N-CNF surface as readily as 1-NP (Nishikiori et al., 2004; Kubota et al., 2005a). Our experiments revealed that the N-CNF clearly adsorbed pyrene better than 1-AP as shown in Figures 9A and 10A. This adsorption property indicates that the polar amino group prevents the π–π interaction between the pyrene ring and graphene sheet even though pyrene molecules are adsorbed onto the carbon nanotubes by this interaction (Chen et al., 2001a; Long & Yang, 2001). Figure 4 shows the absorption spectra of the AA-CNF suspension containing 1-AP observed immediately after the preparation, and after the ultrasonic irradiation for 4–11 days. In the AA-CNF and A24-CNF suspensions, the absorbance gradually decreases with time. We conclude from these data that there is an interaction between 1-AP and the acid-treated CNFs.

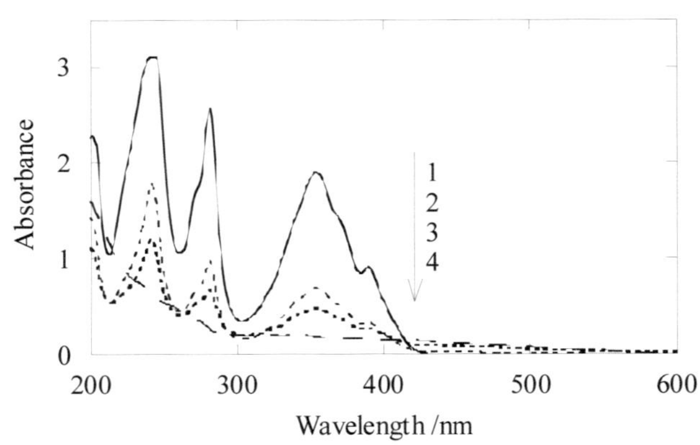

Fig. 4. Absorption spectra of AA-CNF suspensions containing 1-AP observed (1) just after the preparation and after ultrasonic irradiation of (2) 4, (3) 7, and (4) 11 days.

Figure 5 displays the changes in the absorbance of each CNF suspension at 350 and 500 nm versus the ultrasonic irradiation time. The absorbance at 350 nm indicates the amount of 1-AP existing in the liquid phase. 1-AP was hardly adsorbed on the N-CNF, but was adsorbed on both the A24-CNF and the AA-CNF. The AA-CNF was seen to adsorb the 1-AP at a faster rate than the A24-CNF. The absorbance at 500 nm corresponds to the degree of CNF dispersion according to a good correlation between the concentration and the absorbance of the carbon nanotubes in a solvent (Bahr et al., 2001). Such nanocarbon materials exhibit the broad absorption spectra over a wide range of UV-visible-IR due to the superposition of various electric structures originating from many species (Saito et al., 1992; Saito et al., 2000). The wavelength of 500 nm was selected to observe the dispersion because 1-AP have no absorption at longer wavelength than around 450 nm and the CNFs have higher absorption at shorter wavelength. The absorbance increased with the ultrasonic irradiation time until reaching saturation. These results indicate that the degree of CNF dispersion of each sample is ordered in this way; AA-CNF, A24-CNF, N-CNF, with AA-CNF being most highly dispersed. This order is closely correlated to the amount of adsorbed 1-AP. In addition to the surface modification by the acid treatment the adsorption of aromatic molecules should also play an important role in the CNF dispersion (Kubota et al., 2005a).

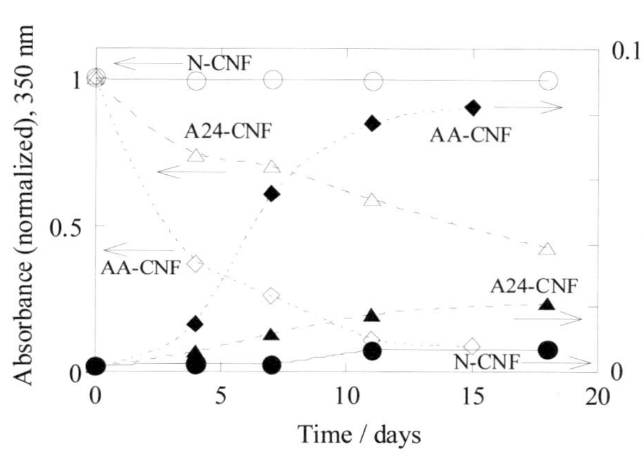

Fig. 5. Changes in absorbance vs. ultrasonic irradiation time monitored at 350 (open dots) and 500 nm (closed dots) in N-CNF, A24-CNF, and AA-CNF suspensions.

4.2 Changes in fluorescence and fluorescence excitation spectra of CNF suspensions

1-AP exhibits a protonation equilibrium; the original species is called AP, and the protonated species is called APH^+. The proton dissociation equilibrium constant of the ground state (pK_a) and the excited state (pK_a^*) are 2.8 and −1.2, respectively (Shizuka et al., 1979). Therefore, in moderately low-pH solutions, 1-AP exists as APH^+ in the ground state, deprotonates to form AP in the excited state, and then emits fluorescence.

Figure 6 shows the fluorescence and fluorescence excitation spectra of 1-AP in the AA-CNF suspension immediately after preparation and after ultrasonic irradiation for 4−11 days. The excitation wavelength for the fluorescence spectra was 350 nm, and the emission wavelength for the excitation spectra was 420 nm. The broad bands of fluorescence around 440 nm and fluorescence excitation around 350−400 nm are assigned to AP in the liquid phase of the suspension since they coincide with the spectra of 1-AP in neutral polar solvents (Hite et al., 1986; Miller et al., 2005). The shape of this excitation spectral band is also similar to that of the absorption spectra of 1-AP shown in Figure 4. The fluorescence spectral bands around 360−400 nm and fluorescence excitation spectra around 300−360 nm are similar to those of APH^+ in the acidic solution. Their relative intensities increased with ultrasonic irradiation time. The spectra of APH^+ are structurally similar to those of pyrene because the interaction between the free electron pair of nitrogen and the π-electrons in the pyrene ring is blocked by protonation of the amino group (Hite et al., 1986; Miller et al., 2005). These results of the fluorescence indicate that 1-AP is adsorbed on the CNF surface to form APH^+-like species.

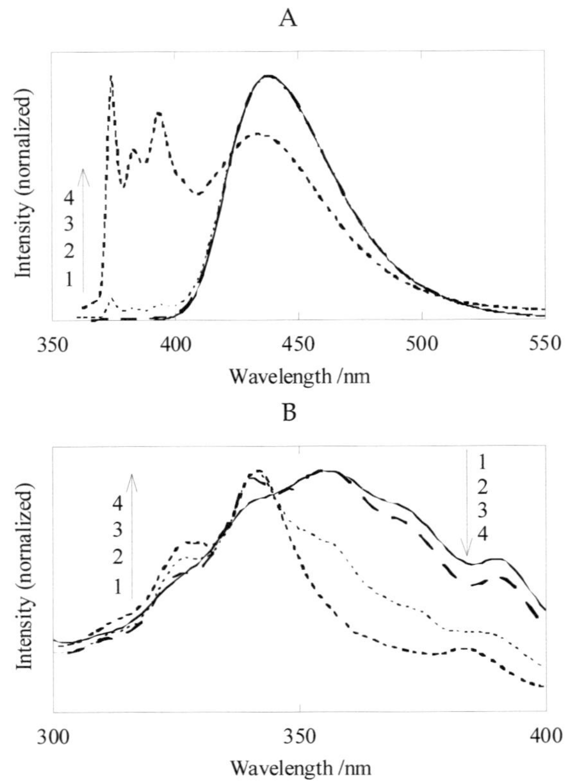

Fig. 6. (A) Fluorescence and (B) excitation spectra of 1-AP in AA-CNF suspension observed (1) just after the preparation and after ultrasonic irradiation of (2) 4, (3) 7, and (4) 11 days.

The ^1L_b fluorescence of 1-NP, which is observed in nonpolar environments, was seen on the CNFs due to π–π interaction between 1-NP and the graphene sheet (Nishikiori et al., 2004; Kubota et al., 2005a; Kubota et al., 2005b). The N-CNF cannot readily adsorb 1-AP molecules due to its low dispersibility throughout the solvent (Figure 6B). Some amount of 1-AP should be adsorbed onto the graphene sheet of the acid-treated CNFs because they are better dispersed throughout the liquid phase (Toebes et al., 2004; Lakshminarayanan et al., 2004). Unlike the 1-NP species, fluorescence was not observed, however, from the 1-AP species adsorbed through π–π interaction onto the CNFs. This is due to quenching that occurs in the more strongly interacting 1-AP/CNF systems.

Figure 7 shows the changes in the fluorescence intensities of 1-AP in the CNF suspensions as a function of the ultrasonic irradiation time. The fluorescence intensities of each CNF suspension at 375 nm per unit dispersed-CNF amount are plotted versus time. These values were obtained by dividing the original fluorescence intensities by the absorbance at 500 nm for each sample. The N-CNF did not adsorb 1-AP whereas the AA-CNF readily adsorbed 1-AP as the APH^+-like species. The A24-CNF adsorbed 1-AP at an intermediate rate. The order of the adsorption ability of the CNF samples agreed with that estimated from the UV-visible absorption spectra depending on the CNF dispersion into the solvent.

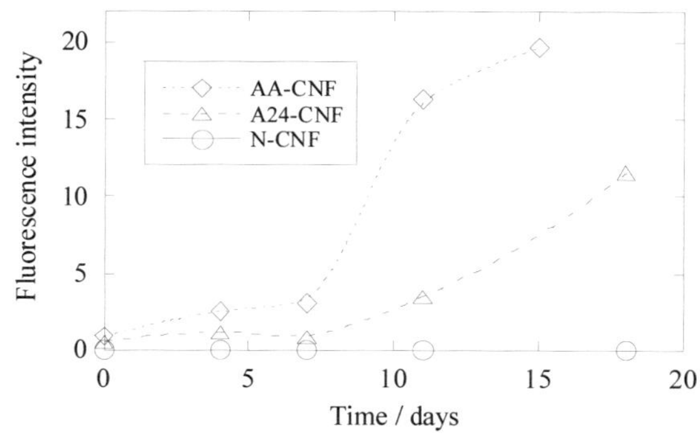

Fig. 7. Changes in fluorescence intensities of 1-AP per unit dispersed-CNF amount vs. ultrasonic irradiation time monitored at 375 nm in N-CNF (○), A24-CNF (Δ), and AA-CNF (◊) suspensions. The values were obtained by dividing the original fluorescence intensities by the absorbance at 500 nm for each sample.

The ion-pair fluorescence of 1-NP was generated by the relatively strong hydrogen-bonding between 1-NP and the oxidized groups, such as −COOH, −C=O, and −OH (Nishikiori et al., 2004; Kubota et al., 2005a). As a Brönsted base, 1-AP is expected to interact with the acidic oxygen-containing groups and selectively detected the acidic groups such as −COOH. It is suggested that 1-AP was immobilized by the hydrogen bonding between its amino group and the Brönsted-acidic groups on the CNF surface, leading to the formation of APH^+-like species. This behavior agrees with previously reported results that the carboxyl group on the carbon nanotube surface is modified by amines forming the ionic bond of $-COO^- \, ^+H_3N-$ (Chen et al., 2001b; Kahn et al., 2002). The fluorescence intensities per unit dispersed-CNF amount for each CNF sample indicate that its adsorption ability depends on not only the CNF dispersion into the solvent but also the amounts of the acidic groups on the CNFs.

4.3 Confirmation of formation of APH^+-like species

The desorption of 1-AP from the AA-CNF surface was examined by fluorescence measurements in order to confirm the adsorption of 1-AP on the CNFs as shown in Figure 8. Figure 8A shows the fluorescence spectra of 1-AP in the AA-CNF suspension observed before and after adding sodium hydroxide. The fluorescence of the APH^+-like species was seen in the original suspension even though the suspension is nearly neutral. After adding the basic NaOH, the structural band of the fluorescence spectra disappeared. This result indicates that the 1-AP molecules that were adsorbed onto the CNF surface as the APH^+-like species were desorbed from the surface into the liquid phase. The APH^+-like species and the acidic groups should be deprotonated and would then hardly interact with one another since the pH value in the suspension was greater than 13. The adsorption and desorption equilibrium was shifted to the desorption process under this condition.
We examined the changes in the fluorescence spectra of the suspension re-dispersing the 1-AP-adsorbing AA-CNF as shown in Figure 8B. The suspension of the 1-AP-adsorbing AA-CNF was filtered, and then re-dispersed into pure water. The APH^+-like band intensity decreased with time while the AP band intensity increased with time. This result indicates

that a portion of the 1-AP molecules that were adsorbed onto the CNF surface as an APH^+-like species were desorbed from the surface and then diffused into the liquid phase. The results shown in Figure 8 support the adsorption of 1-AP onto the CNF surfaces as an APH^+-like species.

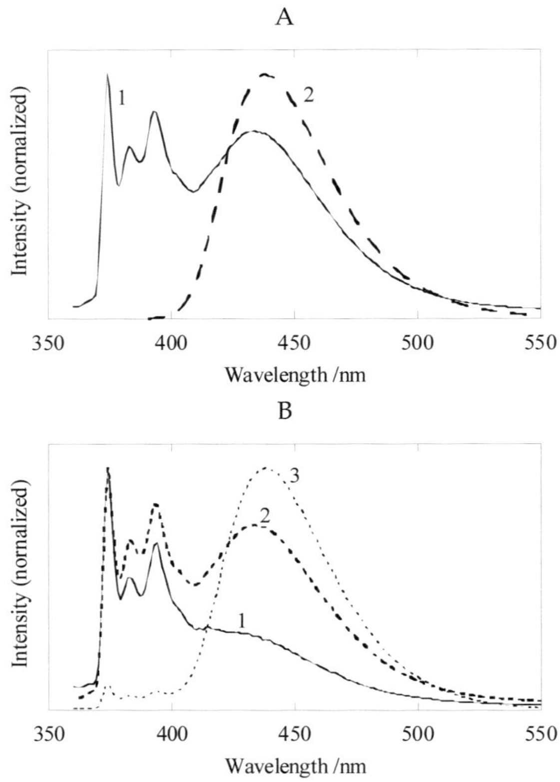

Fig. 8. Fluorescence spectra of 1-AP (A) in AA-CNF suspension observed before (1) and after (2) adding sodium hydroxide and (B) in suspension re-dispersing the 1-AP-adsorbing AA-CNF observed after ultrasonic irradiation of (1) 1, (2) 2, and (3) 5 h.

A stronger acid treatment caused the chemical modification to generate higher amounts of the acidic functional groups on the CNF surface. For this reason the more strongly treated CNFs (AA-CNF) was better dispersed in the 1-AP solution and adsorbed a higher amount of 1-AP than the weakly treated CNFs (A24-CNF). The quantitative analysis of the adsorption sites is now in progress and will be completed in the near future.

5. Fluorescence observation of pyrene adsorbed on carbon nanofibers

5.1 Changes in UV-visible absorption spectra of CNF suspensions

The N- or AA-CNF (0.50 mg) was individually dispersed in the pyrene solutions (10 cm^3) containing water and ethanol (4/1 in volume) at 1.0×10^{-5} mol dm^{-3} by ultrasonic irradiation for 1–8 days. The resulting suspensions were centrifuged to remove any precipitates.

The adsorption of pyrene on the CNF surface and the dispersion of the CNFs into the liquid phase were examined by UV-visible absorption of the CNF suspensions. Figure 9 shows the absorption spectra of the N-CNF and AA-CNF suspensions containing pyrene observed just after the preparation (day 0) and after ultrasonic irradiation for several days. The absorption spectrum of the N-CNF suspension as prepared, which ranges around 220–350 nm, is almost same as that of pyrene in the water/ethanol solution. The absorbance of pyrene gradually decreased with the ultrasonic irradiation time due to its adsorption onto the N-CNF. The N-CNF was obviously found to adsorb pyrene (Chen et al., 2001; Tomonari et al., 2006) similar to 1-NP (Nishikiori et al., 2004; Kubota et al., 2005a). In the AA-CNF suspension, even the absorbance observed just after the preparation was lower than that in the solution, and the absorbance progressively decreases at a faster rate than in the N-CNF suspension. The ultrasonic irradiation makes the CNF bundles disentangled and dispersed into the liquid phase, and then the number of collision of the CNFs with pyrene molecules increases. Oxygen-containing functional groups on the surface of the acid-treated CNFs prevent the π–π interaction between the CNF graphene sheets and interact with solvents (Nishikiori et al., 2004; Kubota et al., 2005a; Kubota et al., 2005b). Therefore, the more highly dispersed AA-CNF was suggested to adsorb a higher amount of pyrene than the low dispersed N-CNF.

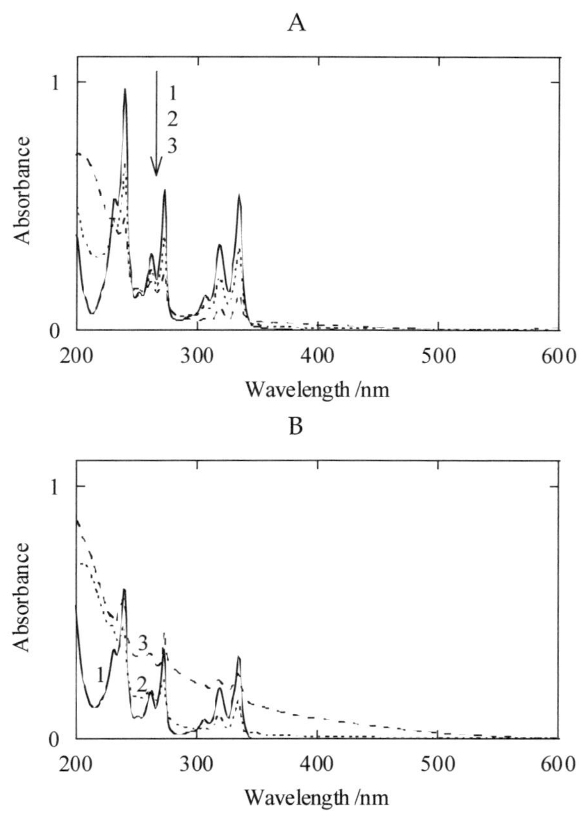

Fig. 9. Absorption spectra of (A) N-CNF and (B) AA-CNF suspensions containing pyrene; (A) (1) just after the preparation and after ultrasonic irradiation of (2) 2 and (3) 5 days, (B) (1) just after the preparation and after ultrasonic irradiation of (2) 1 and (3) 3 days.

Figure 10 shows the changes in the absorbance of each CNF suspension at 334 and 500 nm versus the ultrasonic irradiation time. The absorbance at 334 nm reflects the amount of pyrene existing in the liquid phase. The absorbance of pyrene in the solution without CNFs was also denoted by the closed circles located at day 0. However, unlike the N-CNF suspension, the spectra of the AA-CNF suspension include the strong absorption of the dispersed CNFs in the wavelength region of pyrene absorption. The net pyrene absorbance was obtained by subtracting the CNF absorption spectra from the original ones. The absorbance of pyrene in the solution before adding the CNFs was 0.52. The absorbance of pyrene in the N-CNF suspension, which was 0.52 at day 0, decreased with the irradiation time due to its adsorption. The adsorption equilibrium on the N-CNF was almost reached after the 5-days irradiation, when the absorbance was 0.12. This result indicates that the concentration of pyrene in the liquid phase decreased until reaching saturation. The AA-CNF was seen to adsorb the pyrene at a faster rate than the N-CNF. The absorbance of pyrene in the AA-CNF suspension, which was 0.33 at day 0, decreased with the irradiation time until reaching saturation after the 3-days irradiation, when the absorbance was 0.043.

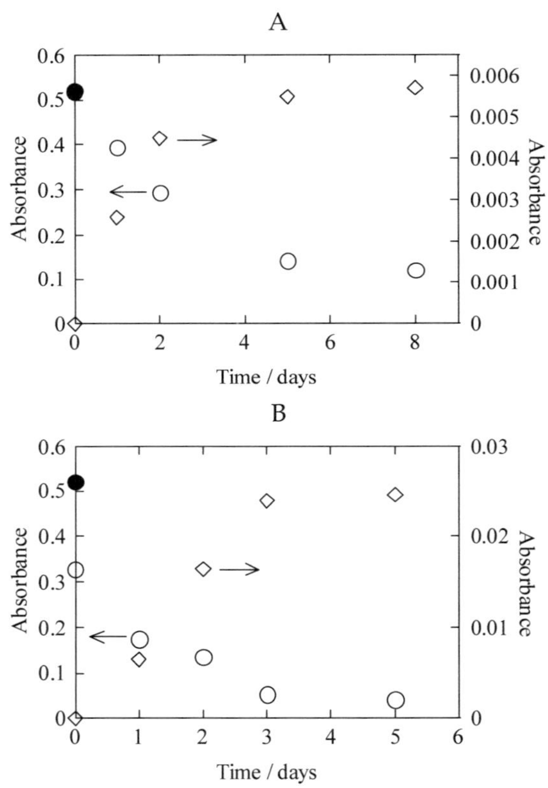

Fig. 10. Changes in absorbance vs. ultrasonic irradiation time monitored at 334 (○) and 500 (◊) nm in (A) N-CNF and (B) AA-CNF suspensions. The closed circles at day 0 denote the absorbance of pyrene in the solution.

The absorbance at 500 nm corresponds to the degree of CNF dispersion according to a good correlation between the concentration and the absorbance of the CNTs in a solvent (Bahr et al., 2001). Such nanocarbon materials exhibit the broad absorption spectra over a wide range of UV-visible-IR due to the superposition of various electric structures originating from many species, regardless of whether they are acid-treated or not. The each absorbance of the N-CNF and AA-CNF suspensions, which was almost zero at day 0, increased with the ultrasonic irradiation time and reached saturation after the 5- and 3-days irradiation, respectively. The absorbance values of the N-CNF and AA-CNF suspensions at the dispersion saturation were 0.0057 and 0.025, respectively. These results indicate that the dispersion of the AA-CNF is higher than that of N-CNF due to the functional groups produced on the CNF surface by the acid treatment and it is closely correlated to the amount of the adsorbed pyrene.

5.2 Changes in fluorescence spectra of CNF suspension

It is well known that the ratio of vibronic peak intensities in fluorescence spectra of pyrene changes with the surrounding polarity (Nakajima, 1971; Kalyanasundaram & Thomas, 1977). Therefore, pyrene is available to use as an in-situ probe of polarity in the surrounding media.

Figure 11 shows the fluorescence spectra of pyrene in the N-CNF and AA-CNF suspensions observed just after the preparation (day 0) and after ultrasonic irradiation for 1−5 days compared with that in cyclohexane. The excitation wavelength for the fluorescence spectra was 350 nm in the weak ^1L_b (S_1) band of pyrene. The emission wavelength for the excitation spectra was 392 nm. The fluorescence peaks at 372, 382, and 392 nm observed at day 0 are assigned to pyrene present in liquid phase of the suspension. The intensity of the 0−0 vibronic bands is significantly enhanced at the expense of other bands in the polar solvents due to solute−solvent dipole−dipole interaction (Nakajima, 1971; Kalyanasundaram & Thomas, 1977). The relative peak intensity at 372 nm to that at 382 nm decreased with the ultrasonic irradiation time in both the N-CNF and AA-CNF systems, and the spectrum became a feature similar to that observed in non-polar solvent such as cyclohexane. These results indicate that pyrene is absorbed on graphene sheet of the CNF surface by the π−π interaction similar to 1-NP (Nishikiori et al., 2004; Kubota et al., 2005a; Kubota et al., 2005b). Although such fluorescence was easily quenched by a much stronger π−π interaction, the spectrum was clearly observed in our systems due to the highly dispersion system of the CNFs throughout the solution. The AA-CNF is oxidized on only a small part of the surface and has a relatively large area of the unoxidized low-polar surface. The fluorescence spectra indicate that pyrene molecules are adsorbed on the low-polar area of the graphene sheet of both the N-CNF and AA-CNF.

The excitation spectra of the samples were also measured and obtained the peaks at 320 and 335 nm corresponding to the ^1L_a absorption and the weak band at 350−380 nm corresponding to the ^1L_b absorption. As shown in Figure 12, it was barely found that the relative peak intensity at 372 nm (0−0 band) to that at 361 nm for the AA-CNF system decreased with the ultrasonic irradiation time (Nakajima, 1971).

Figure 13 shows the changes in ratio of the fluorescence intensity of the pyrene in the CNF suspensions at 372 nm to that at 382 nm as a function of the ultrasonic irradiation time. The fluorescence intensity ratio for the CNF suspension containing pyrene before ultrasonic irradiation (day 0) is 1.6, which is confirmed to be same as that of pyrene in the water/ethanol solution (4/1 in volume) although the results are not shown here. Both the ratios for the N-CNF and AA-CNF suspensions decreased with the ultrasonic irradiation time and then were 0.77 after 5-days irradiation and 0.76 after 3-days irradiation,

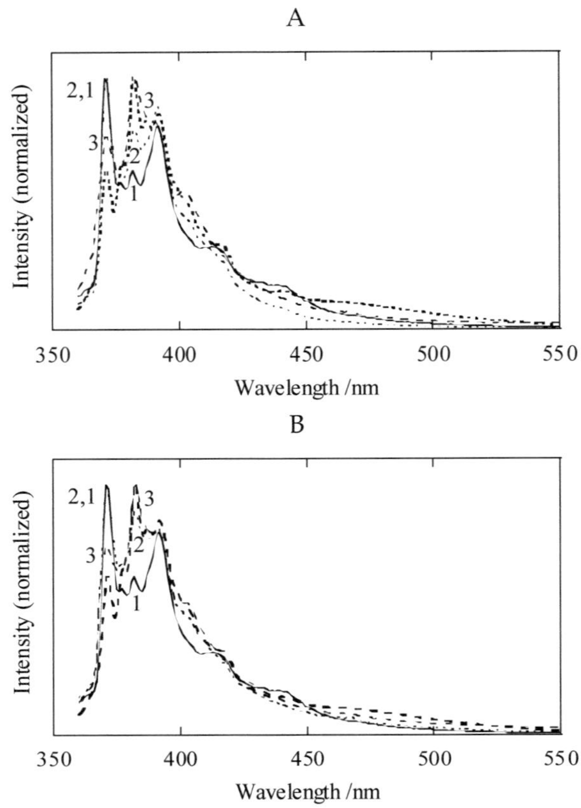

Fig. 11. Fluorescence spectra of pyrene in (A) N-CNF and (B) AA-CNF suspensions upon 350 nm excitation; (a) (1) just after the preparation and after ultrasonic irradiation of (2) 2 and (3) 5 days, (b) (1) just after the preparation and after ultrasonic irradiation of (2) 1 and (3) 3 days. Bold broken lines indicate fluorescence spectra of pyrene in cyclohexane.

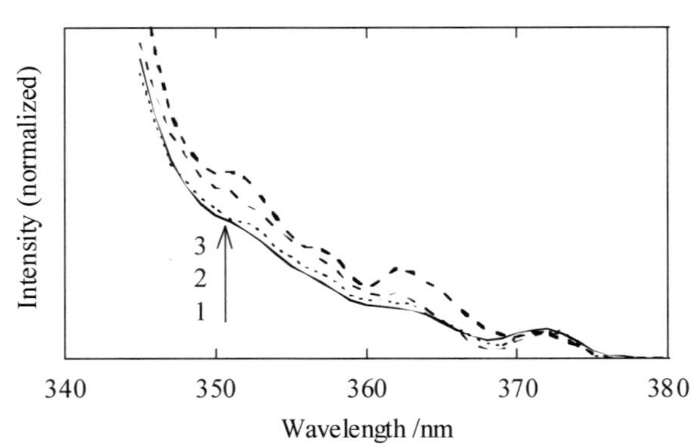

Fig. 12. Fluorescence excitation spectra of pyrene in AA-CNF suspension monitored at 392 nm; (1) just after the preparation and after ultrasonic irradiation of (2) 1 and (3) 3 days. Bold broken line indicates spectrum of pyrene in cyclohexane.

respectively, when they reached the constant values. They became close to the value in cyclohexane as a non-polar solvent, 0.63. Therefore, pyrene molecules were adsorbed onto the low-polar surface of the CNF graphene sheet by π−π interaction. Obviously, the adsorption of pyrene onto the AA-CNF reached its equilibrium earlier than that onto the N-CNF due to the higher dispersion of CNFs throughout the solution. The changes in the fluorescence intensity during ultrasonic irradiation well-corresponded to those in the CNF dispersion throughout the solutions and the pyrene adsorption onto the CNFs.

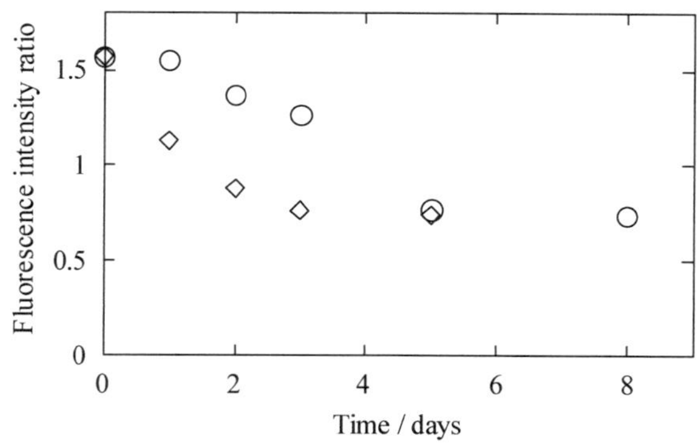

Fig. 13. Changes in ratio of fluorescence intensity of pyrene at 372 nm to that at 382 nm in N-CNF (○) and AA-CNF (◊) suspensions vs. ultrasonic irradiation time.

6. Conclusions

A unique procedure to create a highly disperse system of CNFs throughout solvents allows in situ fluorescence measurements using aromatic probe molecules even though the fluorescence of those adsorbed on carbon materials is scarcely observed due to strong quenching.

Oxidized groups on the outer surface of acid-treated CNFs are quantified using 1-naphthol (1-NP), whereas it is difficult to obtain quantitative information of the chemical species existing in a monolayer or only a few layers of the oxidized surface of the CNFs by IR spectroscopy. In situ fluorescence measurements of CNFs using 1-NP revealed two types of adsorption onto the surface of the CNFs when they were dispersed in solvents. One is generated by the π−π interaction between 1-NP and the graphene sheet and the other is the hydrogen-bonding interaction between 1-NP and proton-accepting groups such as −COOH ($-COO^-$) and −C=O.

Furthermore, acidic functional groups such as −COOH produced on the surface of acid-treated CNFs are characterized using 1-aminopyrene (1-AP) as a Brönsted base. The 1-AP molecules only slightly interact with the untreated CNF surface, whereas the 1-AP cation-like bands are observed on the acid-treated CNF surfaces. These results indicate that 1-AP is tightly immobilized by the hydrogen bonding interaction between its amino group and the Brönsted-acidic groups on the CNF surface.

However, unlike the 1-NP species, the low-polar species fluorescence is not observed from the 1-AP species on the graphene sheet of the CNFs. This is because the polar amino group prevents the π−π interaction between the pyrene ring and graphene sheet. The fluorescence spectra of pyrene as observed in low-polar solvents are clearly found in the suspensions

containing pyrene and the CNFs due to its adsorption onto the CNF graphene sheet by π–π interaction. The CNF dispersion well-corresponds to the adsorption of pyrene onto the CNF surface.
In situ fluorescence measurements using aromatic probe molecules are useful for studying the physicochemical properties on the CNF surface.

7. Acknowledgements

This work was supported by the CLUSTER of the Ministry of Education, Culture, Sports, Science and Technology.

8. References

Bahr, J. L.; Mickelson, E. T.; Bronikowski, M. J.; Smalley, R. E. & Tour, J. M. (2001). Dissolution of small diameter single-wall carbon nanotubes in organic solvents? *Chem. Commun.*, 2001, 2, 193–194.

Chen, R. J.; Zhang, Y.; Wang, D. & Dai, H. (2001a). Noncovalent sidewall functionalization of single-walled carbon nanotubes for protein immobilization. *J. Am. Chem. Soc.*, 123, 16, 3838–3839.

Chen, J.; Rao, A. M.; Lyuksyutov, S.; Itkis, M. E.; Hamon, M. A.; Hu, H.; Cohn, R. W.; Eklund, P. C.; Colbert, D. T.; Smalley, R. E. & Haddon, R. C. (2001b). Dissolution of full-length single-walled carbon nanotubes. *J. Phys. Chem. B*, 105, 13, 2525–2528.

Endo, M. (1988). Grow carbon fibers in the vapor phase. *CHEMTECH*, 18, 9, 568–576.

Endo, M.; Kim, Y. A.; Hayashi, T.; Nishimura, K.; Matsushita, T. & Dresselhaus, M. S. (2001). Vapor-grown carbon fibers (VGCFs): Basic properties and their battery applications. *Carbon*, 39, 9, 1287–1297.

Endo, M.; Kim, Y.A.; Fukai, T.; Hayashi, T.; Oshida, K.; Terrones, M.; Yanagisawa, T.; Higaki, S. & Dresselhaus, M. S. (2002). Structural characterization of cup-stacked-type nanofibers with an entirely hollow core. *Appl. Phys. Lett.*, 80, 7, 1267–1269.

Fujii, T.; Mabuchi, T.; Kitamura, H.; Kawauchi, O.; Negishi, N. & Anpo, M. (1992). Fluorescence spectra of 1-naphthol during the sol-gel process of a mixed aluminum-silicon alkoxide (Si : Al = 94 : 6). *Bull. Chem. Soc. Jpn.*, 65, 3, 720–727.

Fujii, T.; Sugawara, Y.; Kodaira, K.; Mabuchi, T. & Anpo, M. (1995). Characteristic changes in the fluorescence spectra of 1-naphthol at the gelation point during the sol-gel-xerogel transitions. *Res. Chem. Intermed.*, 21, 6, 643–652.

Gao, J.; Itkis, M. E.; Yu, A.; Bekyarova, E.; Zhao, B. & Haddon, R. C. (2005). Continuous spinning of a single-walled carbon nanotube–nylon composite fiber. *J. Am. Chem. Soc.*, 127, 11, 3847–3854.

Hamon, M. A.; Hu, H.; Bhowmik, P.; Niyogi, S.; Ahao, B.; Itkis, M. E. & Haddon, R. C. (2001). End-group and defect analysis of soluble single-walled carbon nanotubes. *Chem. Phys. Lett.*, 347, 1–3, 8–12.

Hamon, M. A.; Hui, H.; Bhowmik, P.; Itkis, H. M. E. & Haddon, R. C. (2002). Ester-functionalized soluble single-walled carbon nanotubes. *Appl. Phys. A*, 74, 3, 333–338.

Hite, P.; Krasnansky, R. & Thomas, J. K. (1986). Spectroscopic investigations of surfaces by using aminopyrene. *J. Phys. Chem.*, 90, 22, 5795–5799.

Hongbing, Z.; Wenzhe, C.; Minquan, W.; Zhengchan & Chunlin, Z. (2004). Optical limiting effects of multi-walled carbon nanotubes suspension and silica xerogel composite. *Chem. Phys. Lett.*, 382 313–317.

Iijima, S. (1991). Helical microtubules of graphitic carbon. *Nature*, 354, 6348, 56–58.

Iijima, S. & Ichihashi, T. (1993). Single-shell carbon nanotubes of 1-nm diameter. *Nature*, 363, 6430, 603–605.

Kahn, M. G.; Banerjee, S. & Wong, S. S. (2002). Solubilization of oxidized single-walled carbon nanotubes in organic and aqueous solvents through organic derivatization. *Nano Lett.*, 2, 11, 1215–1218.

Kalyanasundaram, K. & Thomas, J. K. (1977). Environmental effects on vibronic band intensities in pyrene monomer fluorescence and their application in studies of micellar systems. *J. Am. Chem. Soc.*, 99, 7, 2039–204.

Kubota, S.; Nishikiori, H.; Tanaka, N.; Endo, M. & Fujii, T. (2005a). Quantitative characterization of surface adsorption sites of carbon nanofibers by in-situ fluorescence measurement using 1-naphthol. *Chem. Phys. Lett.*, 412, 1–3, 223–227.

Kubota, S.; Nishikiori, H.; Tanaka, N.; Endo, M. & Fujii, T. (2005b). Dispersion of acid-treated carbon nanofibers into gel matrices prepared by the sol–gel method. *J. Phys. Chem. B*, 109, 43, 23170–23174.

Lakshminarayanan, P. V.; Toghiani, H. & Pittman Jr., C. U. (2004). Nitric acid oxidation of vapor grown carbon nanofibers. *Carbon*, 42, 12–13, 2433–2442.

Liu, J.; Rinzer, G.; Dai, H.; Hafner, J. H.; Bradley, R. K.; Boul, P. J.; Lu, A.; Iverson, T.; Shelimov, K.; Huffman, C. B.; Rodriguez-Macias, F.; Shon, Y. S.; Lee, T. R.; Colbert, D. T. & Smalley, R. E. (1998). Fullerene Pipes. *Science*, 280, 5367, 1253–1256.

Long, R. Q. & Yang, R. T. (2001). Carbon nanotubes as superior sorbent for dioxin removal. *J. Am. Chem. Soc.*, 123, 9, 2058–2059.

Miller, E.; Wysocki, S. & Jóźwik, D. (2005). Fluorescence studies of the sol–gel transition using aminopyrene. *J. Photochem. Photobiol. A*, 169, 3, 221–228.

Nakajima, A. (1971). Solvent effect on the vibrational structures of the fluorescence and absorption spectra of pyrene. *Bull. Chem. Soc. Jpn.*, 44, 12, 3272–3277.

Nishikiori, H.; Tanaka, N.; Kubota, S.; Endo, M. & Fujii, T. (2004). In situ characterization of surface physicochemical properties of carbon nanofibers using 1-naphthol as a fluorescent probe. *Chem. Phys. Lett.*, 390, 4–6, 389–393.

Nishikiori, H.; Tanaka, N.; Tanigaki, T.; Endo, M. & Fujii, T. (2008). In situ probing of acidic groups on acid-treated carbon nanofibers using 1-aminopyrene. *J. Photochem. Photobiol. A*, 193 2–3, 161–165.

Oberlin, A.; Endo, M. & Koyama, T. (1976). Filamentous growth of carbon through benzene decomposition. *J. Cryst. Growth.*, 32, 3, 335–249.

Park, C.; Ounaies, Z.; Watson, K. A.; Crooks, R. E.; Smith, J.; Lowther, S. E.; Connell, J. W.; Siochi, E. J.; Harrison, J. S. & St. Clair, T. L. (2002). Dispersion of single wall carbon nanotubes by in situ polymerization under sonication. *Chem. Phys. Lett.*, 364, 3–4, 303–308.

Saito, R.; Fujita, M.; Dresselhaus, G. & Dresselhaus, M. S. (1992). Electronic structure of chiral graphene tubules. *Appl. Phys. Lett.*, 60, 18, 2204–2207.

Saito, R.; Dresselhaus, G. & Dresselhaus, M. S. (2000). Trigonal warping effect of carbon nanotubes. *Phys. Rev. B*, 61, 4, 2981–2990.

Shizuka, H.; Tsutsumi, K.; Takeuchi, H. & Tanaka, I. (1979). Direct measurement of proton dissociation in thhe excited state of protonated 1-aminopyrene with picosecond pulses. Chem. Phys. Lett., 62, 2, 408–411.

Singh, R.; Pantarotto, D.; McCarthy, D.; Chaloin, O.; Hoebeke, J.; Partidos, C. D.; Briand, J. P.; Prato, M.; Bianco, A. & Kostarelos, K. (2005). Binding and condensation of plasmid dna onto functionalized carbon nanotubes: Toward the construction of nanotube-based gene delivery vectors. *J. Am. Chem. Soc.*, 127, 12, 4388–4396.

Suzuki, S.; Fujii, T.; Imai, A. & Akahori, H. (1977). The fluorescent level inversion of dual fluorescences and the motional relaxation of excited state molecules in solutions. *J. Phys. Chem.*, 81, 16, 1592–1598.

Tan, T. T.; Sim, H. S.; Lau, S. P.; Yang, H. Y.; Tanemura, M. & Tanaka, J. (2006). X-ray generation using carbon-nanofiber-based flexible field emitters. *Appl. Phys. Lett.*, 88, 10, 103105.

Tanigaki, T.; Nishikiori, H.; Kubota, S.; Tanaka, N.; Endo, M. & Fujii, T. (2007). Fluorescence observation of pyrene adsorbed on carbon nanofibers. *Chem. Phys. Lett.*, 448, 4–6, 218–222.

Toebes, M. L.; van Heeswijk, J. M. P.; Bitter, J. H.; Jos van Dillen, A. & de Jong, K. P. (2004). The influence of oxidation on the texture and the number of oxygen-containing surface groups of carbon nanofibers. *Carbon*, 42, 2, 307–315.

Tomonari, Y.; Murakami, H. & Nakashima, N. (2006). Solubilization of single-walled carbon nanotubes by using polycyclic aromatic ammonium amphiphiles in water—strategy for the design of high-performance solubilizers. *Chem. Eur. J.*, 12, 15, 4027–4034.

Wang, F.; Arai, S. & Endo,M. (2005). Preparation of nickel–carbon nanofiber composites by a pulse-reverse electrodeposition process. *Electrochem. Commun.*, 7, 7, 674–678.

Yang, X.; Guillorn, M. A.; Austin, D.; Melechko, A. V.; Cui, H.; Meyer III, H. M.; Merkulov, V. I.; Caughman, J. B. O.; Lowndes, D. H. & Simpson, M. L. (2003). Fabrication and characterization of carbon nanofiber-based vertically integrated schottky barrier junction diodes. *Nano Lett.*, 3, 12, 1751–1755.

Zhu,J.; Kim, J. D.; Peng, H.; Margrave, J. L.; Khabashesku, V. N. & Barrera, E. V. (2003). Improving the dispersion and integration of single-walled carbon nanotubes in epoxy composites through functionalization. *Nano Lett.*, 3, 8, 1107–1113.

9

Preparation of Cellulose-based Nanofibers Using Electrospinning

Youn-Mook Lim, Hui-Jeong Gwon, Joon Pyo Jeun and Young-Chang Nho
Advanced Radiation Technology Institute, Korea Atomic Energy Research Institute
Republic of Korea

1. Introduction

Eelctrospinning is a straightforward method to prepare fibers with diameters as small as several tens of nanometers (Doshi & Reneker, 1995). In eclectrospinning, a high electrostatic voltage is imposed on a drop of polymer solution held by its surface tension at the end of a capillary. The surface of the liquid is distorted into a conical shape known as the Taylor cone. Once the voltage exceeds a critical value, the electrostatic force overcomes the solution surface tension and a stable liquid jet is ejected from the cone tip. Solvent evaporates as the jet travels through the air, leaving behind ultrafine polymeric fibers collected on an electrically grounded target (Fong et al., 1999, 2002; Shin et al., 2001). Electrospun mats have a larger specific surface area and small pore size compared to commercial non-woven fabrics. They are of interest in a wide variety of applications including semi-permeable membranes, tissue engineering scaffolds and drug delivery systems (Tsai et al., 2002; Gibson et al., 2001; Kenawy et al., 2002; Luu et al., 2003).

Recently, electrospun nanofibers (NFs) based on cellulose and its derivatives have been studied as potential candidates for applications within the field of pharmaceuticals. For instance, several reports deal with the investigation of electrospun fiber mats as delivery vehicles, showing dosage forms with useful and controllable dissolution properties. This interest in cellulose-based NFs is primarily driven by its environmental value as a biomaterial. The cellulose is an abundant and renewable resource found in most parts of the world, which makes it a cheap raw material for various applications (Zeng et al., 2003; Jiang et al., 2004; Verreck et al., 2003; Liu & Hsieh, 2002). However, little research has been done on the use of cellulose and cellulose derivatives as a raw material within electrospinning. The complications involved in electrospinning of cellulose are mainly due to the many difficulties ascribed to the material, one being its reluctance to interact with conventional solvents. Therefore, the choice of solvent systems is very important.

Ethyl-cellulose (EC) is a kind of cellulose ether, and it shows a non-biodegradable and biocompatible polymer. EC is one of the extensively studied encapsulating materials for the controlled release of pharmaceuticals (Prasertmanakit et al., 2009). The film made from EC has quite good permeability, it has been widely used industrial air filter (Park et al., 2007).

Hydroxypropyl methylcellulose (HPMC) is frequently used as the basis for sustained release hydrophilic matrix tablets (Ford, 1999). HPMC backbone is composed of glucose

units linked by β-1,4-glycosidic bonds (Zhao et al., 2009; Péreza et al., 2008). The hydrophobic (methyl) and hydrophilic (hydropropyl) groups are distributed along the cellulose backbone. It can be lower the surface tension when adsorbed at fluid interface (Zhao et al., 2009; Kulicke et al., 1998). The HPMC molecule is rod-like and has many hydroxyl groups to participate in intermolecular hydrogen bonds, which allows the cellulose chains to form bundles or fibrils (Zhao et al., 2009; Péreza et al., 2008).
However, this chapter deals with the electrospinning of cellulose derivatives such as EC and HPMC. The fiber morphology and influence of experimental parameters including the solution concentration, flow rate and electric voltage and pattern of the derivatives on the eclectrospinning process were discussed.

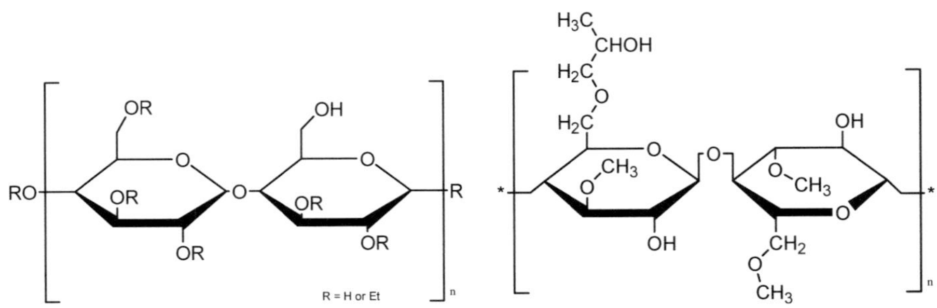

(a) (b)

Fig. 1. The structure of (a) ethylcellulose (EC) and (b) hydroxypropyl methylcellulose (HPMC).

2. Experimental

2.1 Materials

EC (ethoxy content, 48%; viscosity, 46 cps) and HPMC (M_n, 120000; D.S. (methoxy), 1.10-1.60; M.S. (propylene oxide), 0.10-0.30; viscosity, 100000cps 2wt% in H_2O at 20°C) were obtained from Aldrich Chemical and used without further purification. 2,2,2-Trifluoroethanol (Aldrich, USA) was used as a solvent and used as received from commercial supplier.

2.2 Preparation and characterization of cellulose derivatives solutions

EC and HPMC solutions were prepared by dissolving a measured amount of EC and HPMC at room temperature. Four concentrations were prepared about EC: 6, 8, 10 and 12 wt%. Two concentrations were prepared about HPMC: 0.5, 1.0 wt%. These solutions were characterized for their viscosity using a Brookfield DV-II viscometer.

2.3 Electrospinning of cellulose derivatives solutions

In the electrospinning process, a high electric potential was applied to a droplet of EC solution at the tip (ID 0.15 mm) of a syringe needle. The electrospun EC NFs were collected on a target drum which was placed at a distance of 5 cm from the syringe tip. A voltage of 5-28 keV was applied to the collecting target by a high voltage power supply. The flow rate of polymer solution was 0.001-0.01 ml/min. The HPMC NFs were collected on a target plate which was placed at a distance of 5-15 cm from the syringe tip. A voltage of 10, 15 keV was applied to the collecting target by a high voltage power supply. The flow rate of polymer solution was 0.05 ml/min. Fig. 2 shows the photograph of electrospinning apparatus.

Fig. 2. The photograph of electrospinning apparatus.

2.4 Characterization of cellulose NFs

The morphological changes of the samples were observed using a scanning electron microscope (SEM, JMS6390 JEOL Company). Specimen surfaces were coated with a thin layer of gold palladium alloy by sputtering to provide a conductive surface. The fiber diameter of the electrospun fibers was measured by Adobe Photoshop 5.0 software from the SEM pictures in original magnification of 10k X.

Polymers	Concentration [wt%]	Electric voltage [keV]	Flow rate [ml/min]	Tip-target distance [cm]	Note
EC	6	10	0.001	5	
	8	15	0.003	7	
	10	20	0.005	9	
	12	25	0.01	11	
HPMC	0.5	10	0.05	5	
	1.0	15		10	
	-	-		15	

Table 1. Electrospinning condition of the cellulose solutions

3. Result and discussion

3.1 Effect of the solution concentration

The viscosity of the polymer solutions are summarized in Table 1. Increasing the concentration of the cellulose solution increased. Viscosity plays an important role in determining the rage of concentrations from which continuous fibers can be obtained by electrospinning. At low viscosity, surface tension is the dominant effect on fiber morphology. Below a certain concentration, drops will form instead of fibers. At high concentrations, processing will be prohibited by an inability to control and maintain the flow of a polymer solution to the tip of the needle and by the cohesive nature of the high viscosity. As shown in Fig. 3, with increasing concentration, the morphology was changed from beaded fiber to uniform fiber structure and the fiber diameter was also increased gradually. At 6wt%, a large amount of smooth fibers and beaded fibers were obtained. (With increasing the concentration of the solution to 8 wt%, very small amount of beads was present.) With further increasing concentration of the solution to 10 wt%, only smooth, bead-less fibers were obtained. At 12 wt%, the average fiber diameter was much larger than those of fibers spun at lower concentrations.

Polymers	Concentration [wt%]	Viscosity [cP]	note
EC	6	461	
	8	1245	
	10	2918	
	12	6347	
HPMC	0.5	1284	
	1.0	8614	

Table 2. Viscosity of polymer solutions as a function of concentration.

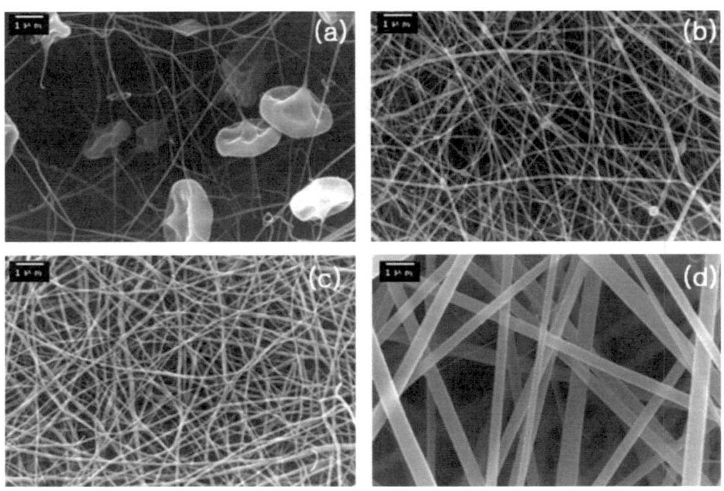

Fig. 3. SEM micrographs of electrospun fibers from EC solution with different solution concentration (voltage=10 keV, flow rate=0.005 ml/min). EC concentration; (a) 6 wt%; (b) 8 wt%; (c) 10 wt%; (d) 12 wt%.

The morphologies of HPMC NFs in Fig. 4. (??) No big change in the morphology of fibers as a function of concentration. At 0.5wt%, a large amount of smooth fibers and thick fibers was obtained. Increasing the concentration of the solution to 1.0 wt%, only smooth, bead-less fibers were obtained. The average fiber diameter was investigated 155 and 162nm, respectively.

In electrospinning, the coiled macromolecules in solution were transformed by the elongational flow of the jet into oriented entangled networks that persist with fiber solidification. Below this concentration, chain entanglements were insufficient to stabilize the jet and the contraction of the diameter of the jet driven by surface tension caused the solution to form beads or beaded fibers. At high concentration, viscoelastic force which resisted rapid changes in fiber shape resulted in uniform fiber formation. However, it was impossible to electrospin if the solution concentration or the corresponding viscosity was too high due to the difficulty in liquid jet formation.

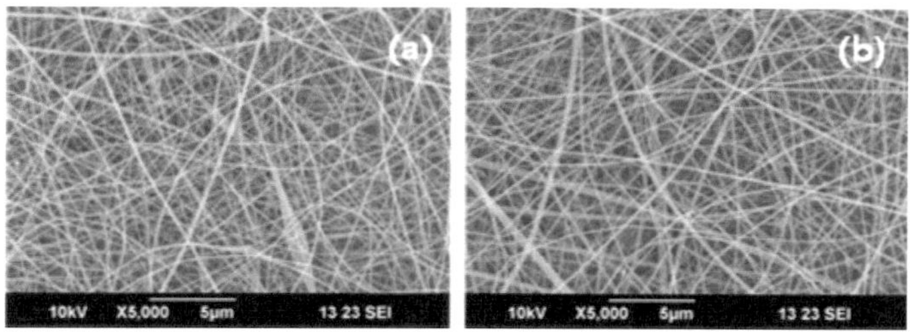

Fig. 4. SEM micrographs of electrospun fibers from HPMC solution with different solution concentration (voltage=15 keV, flow rate=0.05 ml/min). HPMC concentration; (a) 0.5 wt%; (b) 1.0 wt%.

3.2 Effect of the electrospinning flow rate

The morphological structure can be changed by changing the solution flow rate as shown in Fig. 5. At the flow rate of 0.001 ml/min, uniform fiber structure was observed. But increasing flow rate, the average fiber diameter was much larger than those of fibers spun at lower flow rate. At the flow rate of 0.01 ml/min, a considerable amount of thick fibers with diameters above 1mm were found. When the flow rate exceeded a critical value, the delivery rate of the solution jet to the capillary tip exceeded the rate at which the solution was removed from the tip by the electric forces. This shift in the mass-balance resulted in sustained but unstable jet and fibers with broad distribution in the fiber diameter were formed.

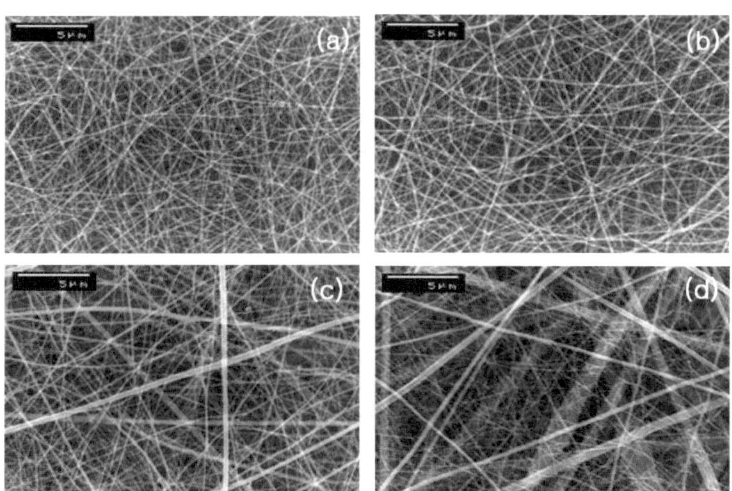

Fig. 5. Effect of flow rate of 10 wt% EC solution on fiber morphology (voltage=15 kV). Flow rate: (a) 0.001 ml/min; (b) 0.003 ml/min; (c) 0.005 ml/min; (d) 0.01 ml/min.

3.3 Effect of the electric voltage

A series of experiments were carried out when the applied voltage was varied from 5 to 28 keV. The results are shown in Fig. 6 and Fig. 7. Weaker voltage, such as 5 keV, was not strong enough to overcome the surface tension and viscoelastic forces of the polymer solution, but at higher voltages at this distance, electrical discharge would occur. There was a slight decrease in average fiber diameter with increasing applied electric field. A considerable amount of thin fibers with diameters below 100 nm were found when the applied voltage above 20 keV. Increasing the applied voltage, i.e., increasing field strength will increase the electrostatic repulsion force on the fluid jet which favors the thinner fiber formation. On the other hand, the solution will be removed from the capillary tip more quickly as jet is ejected from Taylor cone. This results in the increase of the fiber diameter. Corona discharge was observed at voltage above 28 keV, making electrospinning impossible.

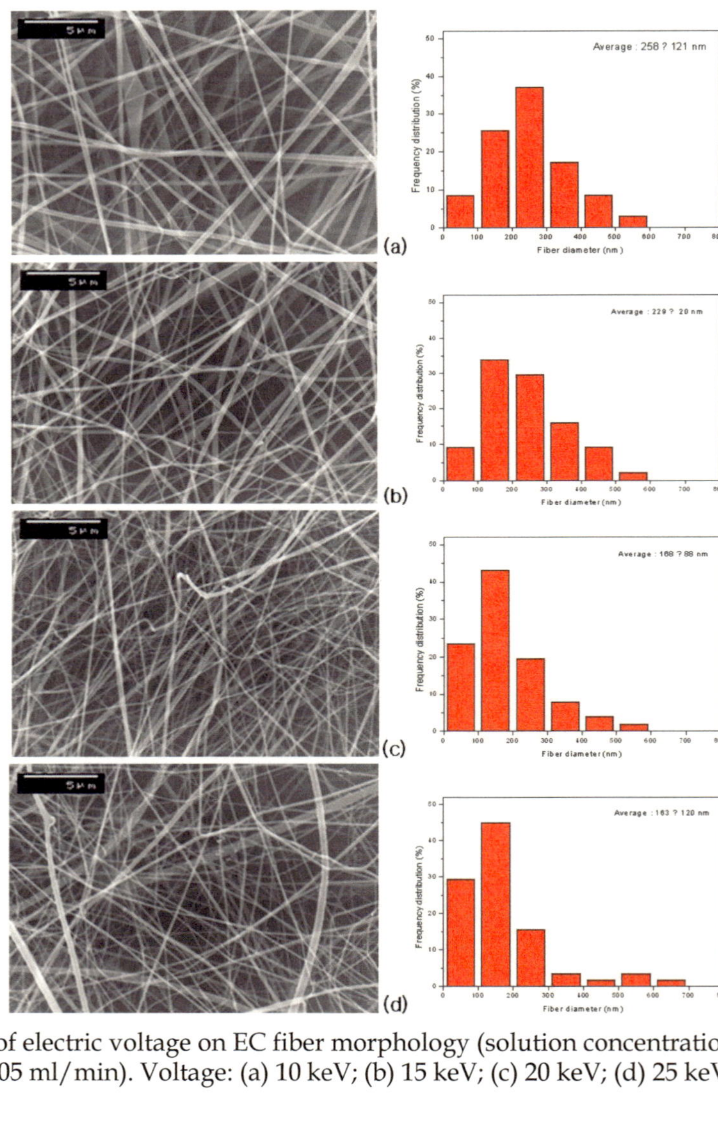

Fig. 6. Effect of electric voltage on EC fiber morphology (solution concentration=10 wt%, flow rate=0.005 ml/min). Voltage: (a) 10 keV; (b) 15 keV; (c) 20 keV; (d) 25 keV.

The diameter of HPMC NFs was measured (??) a narrow distribution when the electric voltage decreased. It was found that the fibers had an average diameter of 129nm at a voltage of 10keV and an average diameter of 140nm at a voltage of 15keV. Electrospun NFs of HPMC showed similar results to EC NFs

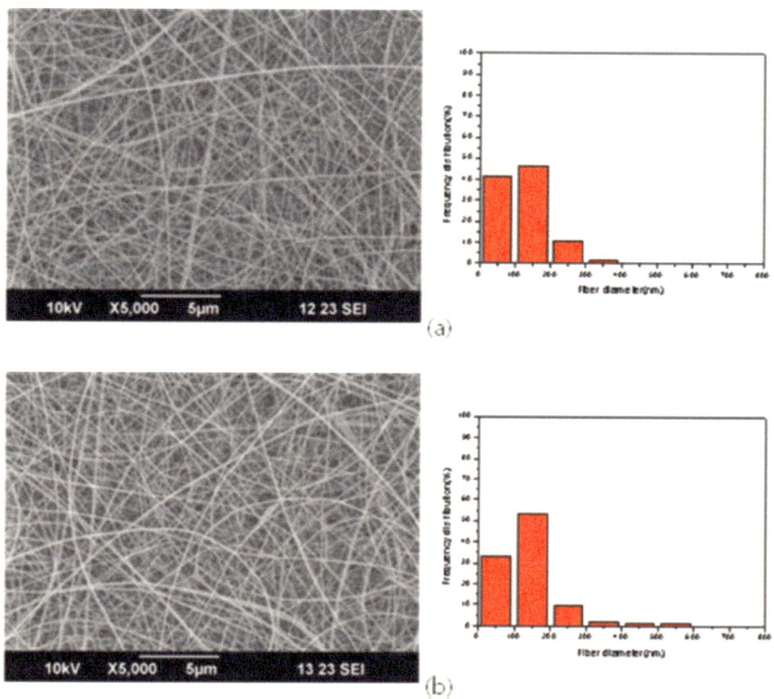

Fig. 7. Effect of electric voltage on HPMC fiber morphology (solution concentration=1.0 wt%, flow rate=0.05 ml/min). Voltage: (a) 10 keV; (b) 15 keV.

3.4 Effect of the needle tip-target distance

Fig. 8 shows that NFs formed with decreasing tip-target distance of EC. As tip-target distance of NFs was increased, the diameter of NFs. The NFs had non-uniform diameters. It may be assumed that the lower electric strength solvent from evaporating. However, (??we were observed that formation of HPMC NFs regardless of tip-target distance. ??) The most uniform NFs derived with our experimental conditions were afforded at 5cm of tip-target distance. Increasing diameter of NFs at 10cm and decreasing diameter of NFs at 15cm (??).

4. Conclusions

Submicron EC and HPMC fibers have been successfully prepared by electrospinning of polymer solutions. The morphology of the fibers was strongly affected by the parameters such as polymer concentration, tip-target distance, solution flow rate, and applied voltage. At below 6 wt% EC solutions, electrospinning was not enhanced due to the low viscosity. (the concentration of EC solution,) the morphology was changed from beaded fiber

Fig. 8. Effect of needle tip-target distance of EC NFs (solution concentration, 10wt%; electric voltage, 15 keV; flow rate, 0.005 ml/min). (a) 5 cm, (b) 7 cm, (c) 9 cm, (d) 11 cm.

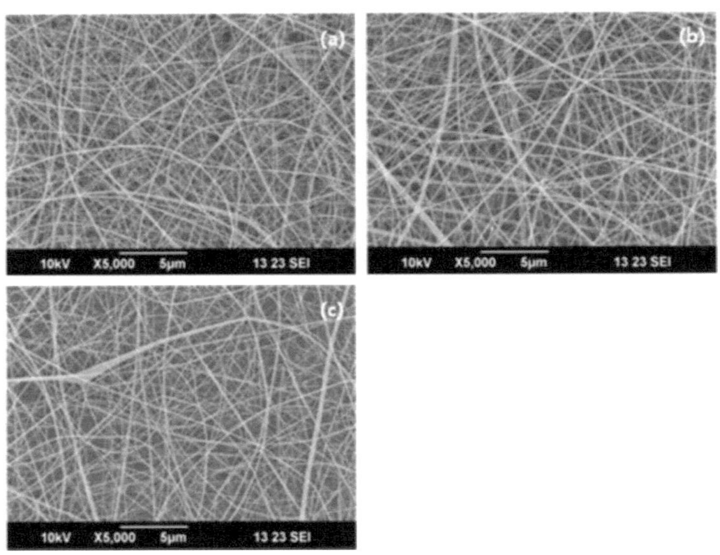

Fig. 9. Effect of needle tip-target distance of HPMC NFs (solution concentration, 1.0wt%; electric voltage, 15 keV; flow rate, 0.05 ml/min). (a) 5 cm, (b) 10 cm, (c) 15 cm.

to uniform fiber structure and the fiber diameter was also increased. There was an increase in fiber diameter and fiber distribution with increasing solution flow rate, while there was a slightly decrease in average fiber diameter with increasing applied electric field. The HPMC NFs are feasible for use as wound healing and drug delivery systems for biomedical applications.

5. References

Doshi, J. & Reneker, D.H. (1995). Electrospinning Process and applications of Electrospun Fibers. *J. Electrostat.*, 35, 151-160, 0304-3886

Fong, H; Chun, I. & Reneker, D.H. (1999). Beaded nanofibers formed during electrospinning. *Polym.*, 40, 4585-4592, 0032-3861

Fong, H.; Liu, W.; Wang, C.S. & Vaia, R.A. (2002). Generation of electrospun fibers of nylon 6 and nylon 6-montmorillonite nanocomposite. *Polym.*, 43, 775-780, 0032-3861

Shin, Y.M.; Hohman, M.M.; Brenner, M.P. & Rutledge, G.C. (2001). Experimental characterization of electrospinning: the electrically forced jet and instabilities. *Polym.*, 42, 9955-9967, 0032-3861

Tsai, P.P.; S-Gibson, H. & Gibson, P. (2002). Different electroststic methods for making electrets filters. *J. Electrostat.*, 54, 333-341, 0304-3886

Gibson, P.; S-Gibson, H. & Rivin, D. (2001). Transport properties of porous membranes based on electrospun nanofibers. *Colloid. Surface A.*, 187-188, 469-481, 0927-7757

Kenawy, El-R.; Bowlin, G.L.; Mansfield, K.; Layman, J.; Simpson, D.G.; Sanders, E.H. & Wnek, G. E. (2002). Release of tetracycline hydrochloride from electrospun poly(ethylene-co-vinylacetate), poly(lactic acid), and a blend. *J. Control. Release*, 81, 57-64, 0168-3659

Luu, Y.K.; Kim, K.; Hsiao, B.S.; Chu, B. & Hadjiargyou, M. (2003). Development of a nanostructured DNA delivery scaffold via electrospinning of PLGA and PLA-PEG block copolymers. *J. Control. Release*, 89, 341-353, 016-3659

Zeng, J.; Xu, X.; Chen, X.; Liang, Q.; Bian, X.; Yang, L. & Jing, X. (2003). Biodegradable electrospun fibers for drug delivery. *J. Control. Release*, 92, 227-231, 016-3659

Jiang H.; Fang, D.; Hsiao, B.; Chu, B. & Chen, W. (2004). Preparation and characterization of ibuprofen-loaded poly(lactide-co-glycolide)/poly(ethylene glycol)-g-chitosan electrospun membranes. *J. Biomater. Sci. Polym. Ed.* 15, 279-, 0920-5063

Verreck, G.; Chun, I.; Peeters, J.; Rosenblatt, J. & Brewster, M.E. (2003). Preparation and Characterization of Nanofibers containing Amorphous Drug Dispersions Generated by Electrotatic Spinning. *Pharmcol. Res.*, 20, 810-817, 1043-6618

Liu, H.Q. & Hsieh, Y.L. (2002). Ultrafine Fibrous Cellulose Membranes from Electrospinning of Cellulose Acetate. *J. Polym. Sci. Part B: Polym. Phys.*, 40, 2119-2129, 0887-6266

Prasertmanakit, S.; Praphairaksit, N.; Chiangthong, W. & Muangsin, N. (2009). Ethyl Cellulose Microcapsules for Protecting and Controlled Release of Folic Acid. *AAPS PharmSciTech.*, article in press, 1530-9932

Park, J.Y.; Han, S.W. & Lee, I.H. (2007). Preparation of Electrospun Porous Ethyl Celluolose Fiber by THF/DMAc Binary Solvent System. *J. Ind. Eng. Chem.*, 13, 6, 1002-1008, 1226-086X

Ford, J.L. (1999). Thermal analysis of hydoxypropylmethylcellulose and methylcellulose: powders, gels and matrix tablets. *Int. J. Pharm.*, 179, 209-228, 0378-5173

Zhao, Q.; Zhao, M.; Li, J.; Yang, B.; Su, G.; Cui, C. & Jiang Y. (2009). Effect of hydroxypropyl methylcellulose on the textural and whipping properties of whipped cream. *Food Hydrocolloids*, 23, 2168-2173, 0268-005X

Péreza, O.E.; Sánchezb, C.C.; Pilosofa, A.M.R. & Patino, J.M.R. (2008). Dynamics of adsorption of hydroxypropyl methylcellulose at the air-water interface. *Food Hydrocolloids*, 22, 397-402, 0268-005X

Kulicke, M.W.; Aredt, O. & Berger, M. (1998). Characterization of hydroxypropyl methylcellulose-stabilized emulsion. Part I. Particle size distribution. *Colloid polym. Sci.*, 276, 1019-1023, 0303-402X

10

Nanofibrous Scaffolds of Bio-polyesters: *In Vitro* and *In Vivo* Characterizations and Tissue Response

Hui Ying Tang[1,2], Daisuke Ishii[3], Kumar Sudesh[1], Tetsuji Yamaoka[4] and Tadahisa Iwata[2,5]
[1]School of Biological Sciences, Universiti Sains Malaysia, 11800 Penang,
[2]Bioengineering Laboratory, RIKEN Institute / Hirosawa 2-1, Wako-shi, Saitama 351-0198,
[3]Department of Materials Chemistry, Faculty of Science and Engineering, Ryukoku University / 1-5 Yokotani, Seta Oe-cho, Otsu-shi, Shiga 520-2194,
[4]Department of Biomedical Engineering, Advanced Medical Engineering Center, National Cardiovascular Center Research Institute, 5-7-1 Fujishirodai, Suita-shi, Osaka 565-8565,
[5]Department of Biomaterial Sciences, Graduate School of Agricultural and Life Sciences, The University of Tokyo / 1-1-1 Yayoi, Bunkyo-ku, Tokyo 113-8657,
[1]Malaysia
[2,3,4,5]Japan

1. Introduction

1.1 Nanofibers as tissue-engineering scaffolds

Tissue engineering can be broadly defined as the process of creating living, physiological, three-dimensional tissues and organs utilizing specific combination of cells, cell scaffolds, and cell signals, both chemical and mechanical (Griffith, 2000). It offers a promising alternative to donor tissues which are often in short supply. In tissue-engineering, biomaterials are fashioned into scaffolds that can replace the natural extracellular matrix (ECM) until host cells can repopulate and re-synthesize a new natural matrix.

For tissue engineering, apart from the supercellular and cellular scale structures, the subcellular and nanoscale structures are also essential for the control of cellular environment, cell-molecular interactions and cell-cell interactions. It is not surprising that more tissue engineering studies are gearing towards miniaturization at these scales in efforts to mimic the fibrillar structure of the ECM. One such effort is the use of electrospinning, a simple yet versatile method, to produce nanofibers with morphological similarities to the ECM. Nanofibers generally refer to fibers with diameter less than 1 micron. Nanofibers are attractive for tissue engineering in several ways. First, very high surface area-to-volume ratio and high porosity can be achieved for better cell adhesion and growth. Second, the nanofibers possess enhanced mechanical properties compared to the solid-walled equivalents. Third, the ECM-mimicking structures may improve cellular response and biocompatibility.

Polyhydroxyalkanoates (PHAs), a class of naturally derived polyesters, are popular biomaterials for biomedical applications because they are bioabsorbable and biocompatible.

They are produced by various bacteria as intracellular carbon and energy compound under unfavorable growth conditions such as limitation of nitrogen, phosphorus, oxygen or magnesium in the presence of excess supply of carbon source (Anderson & Dawes, 1990).
Of the PHAs, the poly(4-hydroxybutyrate) [P(4HB)] is rapidly gaining recognition as a new absorbable material for implantable medical applications because it offers new set of properties that are not currently available (Martin & Williams, 2003). A recent major breakthrough for P(4HB) is the clearance obtained from the Food and Drug Administration of the United States of America for the use of P(4HB)-derived TephaFLEX® absorbable suture (http://www.fda.gov/bbs/topics/NEWS/2007/NEW01560.html). Hence, this book chapter focuses on the potential of the copolymer of P(4HB), that is the poly[(*R*)-3-hydroxybutyrate-*co*-97mol%-4-hydroxybutyrate] [P(3HB-*co*-97mol%-4HB)] as a biomaterial for electrospun nanofibrous scaffold. Its performance as a tissue-engineering scaffold was compared with scaffolds derived from poly[(*R*)-3-hydroxybutyrate] [P(3HB)] and its copolymers; poly[(*R*)-3-hydroxybutyrate-*co*-5mol%-(*R*)-3-hydroxyhexanoate] [P(3HB-*co*-5mol%-3HHx)] and poly[(*R*)-3-hydroxybutyrate-*co*-7mol%-4-hydroxybutyrate] [P(3HB-*co*-7mol%-4HB)]. Since bio-polyesters based on lactide monomers are the workhorse synthetic polymers for scaffold fabrication, the performance aspects of nanofibrous scaffolds made from PHAs were compared with those derived from synthetic polyesters; poly(L-lactide) (PLLA) and stereocomplexed PLA. The chemical structures of these biopolyesters are shown in Figure 1. Subcutaneous implantation of the nanofibrous scaffolds in rats was performed to investigate their bioabsorption behavior and tissue response. The nanofibrous scaffolds before and after the *in vitro* and *in vivo* experiments were characterized using scanning electron microscopy (SEM), X-ray diffraction, gel permeation chromatography and tensile test. Tissue response was also determined by histological evaluation. Our findings reported here had been published in our earlier works (Ishii et. al., 2007; Tang et al., 2009; Ishii et al., 2009).

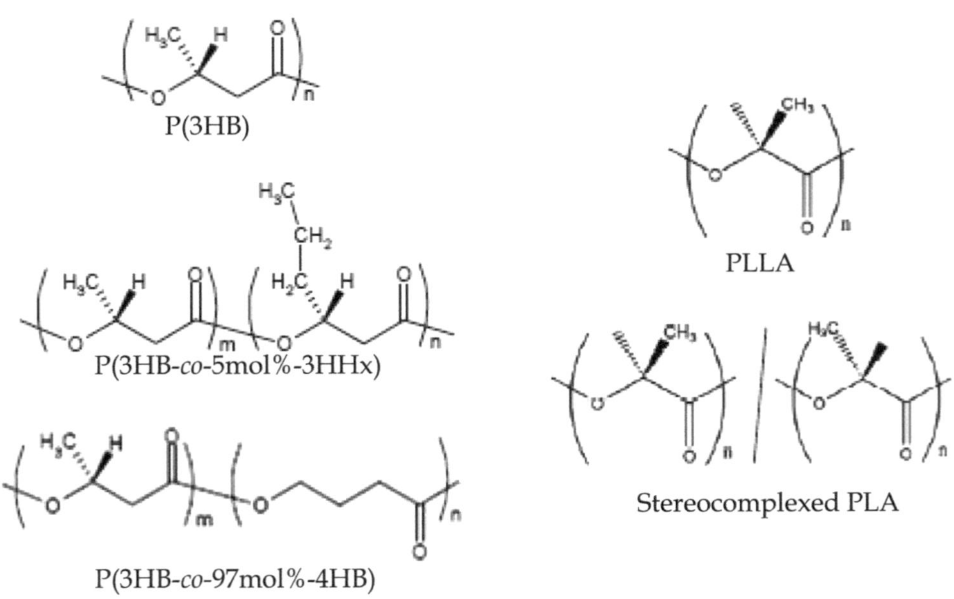

Fig. 1. Chemical structures of PHAs, PLLA and stereocomplexed PLA.

2. Preparation of PHA and PLA-based nanofibrous scaffolds by electrospinning

The electrospinning apparatus is quite simple, consisting only three major components: a high voltage power supply, a polymer solution reservoir (usually a syringe with a small diameter needle) with or without a flow control pump and a metal collecting plate. To initiate the electrospinning process, the polymers are dissolved in the appropriate solvent and loaded into the polymer reservoir. Once a source electrode is attached to the syringe, an electric field between the polymer solution and a grounded target is created by a high-voltage power source. The electrostatic field at the syringe-air interface competes with the surface tension of the liquid solution. A semispherical polymer solution droplet is formed at the tip of the needle when the polymer solution is extruded slowly by a syringe pump and/or gravity. With increasing voltage, the polymer droplet elongates to form a conical shape known as the Taylor cone (Subbiah et al., 2005). When the critical voltage is reached, the surface tension is overcome by the electrostatic force and a polymer jet is initiated (Taylor, 1964). The surface charge on the jet is further increased as the solvent in the polymer jet evaporates when drawn to the collecting plate. This increase in surface charge induces instability in the polymer jet as it passes through the electric field (Demir et al., 2004). Hence, the polymer jet divides geometrically, first into two jets, and then into many more as the process repeats itself to compensate for the instability. The formation of fibers results from the action of the spinning force provided by the electrostatic force on the continuously splitting polymer droplets. A non-woven nanofibrous scaffold is eventually formed when the nanofibers are deposited layer-by-layer on the metal collecting plate.

3. *In Vitro* and *In Vivo* characterizations of PHA and PLA-based nanofibrous scaffolds

3.1 Morphology of nanofibrous scaffolds

As revealed by SEM, all electrospun nanofibrous scaffolds spun from 1,1,1,3,3,3- hexafluoro-2-propanol (HFIP) consist of randomly oriented nanofibers (Figs. 2, 3 and 4). The width of the nanofibers between the junctions was quite uniform. The width decreased in the order of P(3HB) ≈ P(3HB-*co*-5mol%-3HHx) (520 nm) > PLLA ≈ stereocomplexed PLA (300 nm) > P(3HB-*co*-97mol%-4HB) (220 nm) > P(3HB-*co*-7mol%-4HB) (190 nm). The increment in width is proportional to the molecular weight of the polymer (Dong, 2004) because higher degree of chain entanglement due to high molecular weight is assumed to make it harder for the electrostatic forces to pull, or extend individual chains (Lyons et al., 2004). As such, the matrices of the electrospun P(3HB) and P(3HB-*co*-5mol%-3HHx) consisted of larger nanofibers compared to the electrospun P(3HB-*co*-97mol%-4HB) and PLA-based nanofibrous scaffolds because of their high molecular weight (Table 1). Interestingly, only the electrospun P(3HB-*co*-7mol%-4HB) formed nanofibers with irregular shapes with intermittent spindle-like beads on string (Fig. 3A). Possibly the formation of the beaded P(3HB-*co*-7mol%-4HB) nanofibers is the result of low net charge density as shown by previous studies (Fong et al. 1999; Zuo et al., 2005). According to Dong et al., the net charge density is inversely proportional to the mass of dry polymer (i.e. mass of scaffolds collected from electrospinning), if the other experimental conditions such as jet current, collecting time and polymer concentration are the same. The net charge density decreases in the order of P(3HB-*co*-5mol%-3HHx) (1058 Coulomb/liter) > P(3HB) (1002 Coulomb/liter) > P(3HB-*co*-97mol%-4HB) (778 Coulomb/liter) > P(3HB-*co*-7mol%-4HB) (484 Coulomb/liter).

After sterilization, the morphologies observed for the electrospun P(3HB), P(3HB-*co*-5mol%-3HHx), P(3HB-*co*-7ml%-4HB) and PLA-based nanofibrous scaffolds remained unchanged (Figs. 2, 3 and 4). The matrix of P(3HB-*co*-97mol%-4HB), however, became less porous (Fig. 3 B2). The temperature (40 °C) of the ethylene oxide (EtO) sterilization is close to the melting temperature of P(3HB-*co*-97mol%-4HB) (T_m = 47 °C), and thus led to the fusing of some nanofibers to each other.

3.1 (a) *In Vitro* study

After 4 weeks of immersion in phosphate buffered saline (PBS), there was no evidence of degradation on the surface of all the electrospun PHA scaffolds. The structural integrities of the PHA nanofibrous scaffolds were maintained even after 12 weeks. This is because PHA hardly undergoes hydrolysis at pH value around neutrality. Interestingly, the electrospun P(3HB-*co*-97mol%-4HB) and PLA-based nanofibrous scaffolds appeared to have swollen. Possibly, the swelling of these nanofibers was caused by the penetration of water into their amorphous regions as reported in the literature concerning electrospun Poly(D,L-lactic-*co*-glycolic acid), poly(D,L-lactic acid) and poly(butylene succinate) fibers after immersion in PBS (Zong et al., 2003; Jeong et al., 2005; Li et al., 2006). The PLLA nanofibers showed the most pronounced swelling (width of nanofibers increased from 300 nm to 1200 nm) compared to the stereocomplexed PLA nanofibers (width of nanofibers increased from 300 nm to 600 nm). It is believed that the strong interaction between the molecular chains of PLLA and PDLA in the stereocomplexed PLA fibers might limit their swelling in PBS. In accordance to this observation, only the PLLA fibers showed fragmentation after 4 weeks in PBS. However, after 12 weeks in PBS, both the PLLA and stereocomplexed PLA fibers showed cleavage.

3.1 (b) *In Vivo* study

In SEM, only the remaining scaffolds that could be retrieved from rat after the *in vivo* experiments were observed. Nevertheless, such observations provided the information on the morphological changes by implantation. No remarkable change was observed for the P(3HB) and P(3HB-*co*-5mol%-3HHx) nanofibrous scaffolds during the period of implantation (Fig. 2). In contrast, the nanofibers of the copolymers with 4HB unit were affected by the implantation. At 4 weeks of implantation, the nanofibers of P(3HB-*co*-97mol%-4HB) showed fragmentation (Fig. 3 B3). After 12 weeks, surface erosion became more evident as the density of the nanofibers decreased remarkably due to fragmentation of the fibers to shorter segments. The progression of bioabsorption was also evidenced by the formation of pores on the surface of these fibers as indicated by the arrow in Fig. 3 B4. These evidences indicate that the existence of 4HB monomer units enhances the degradability, or the bioabsorption of PHA. As for the PLLA nanofibrous scaffold, cleavage of the nanofiber strands occurred after 4 weeks of implantation (Fig. 4 A3). A decrease in the density of the nanofibers in the scaffold was observed after 12 weeks. The stereocomplexed PLA nanofibrous scaffold remained in its original state even after 12 weeks of implantation (Fig. 4 B4).

The *in vitro* and *in vivo* observations using SEM revealed significant difference in the bioabsorption behavior of the electrospun PHA and PLA-based scaffolds. For the electrospun PHA scaffolds, fibers with smaller diameter were more prone to fragmentation because of increased water contact due to large surface area. However, for the electrospun

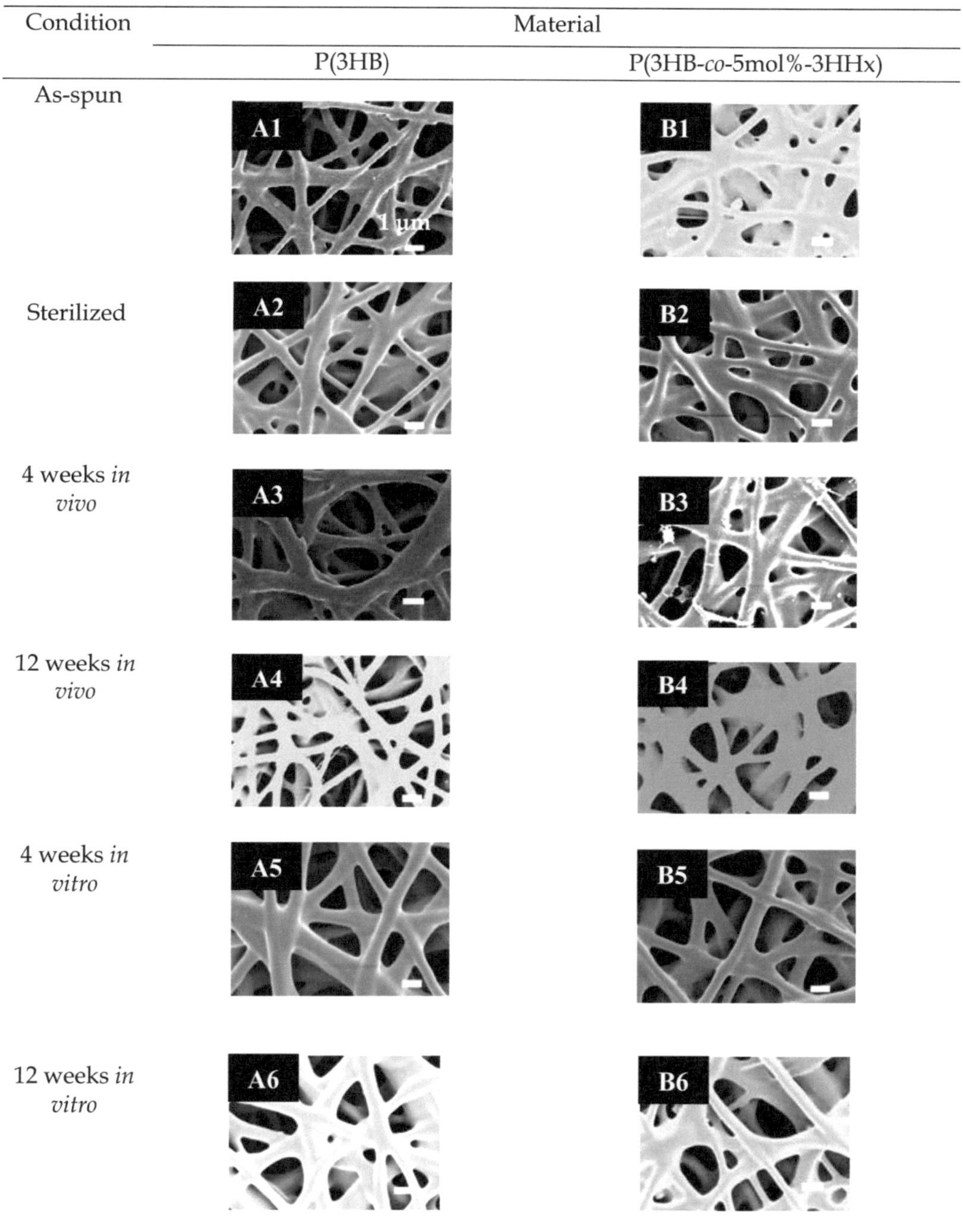

Fig. 2. SEM micrographs of the electrospun P(3HB) and P(3HB-*co*-5mol%-3HHx) in various conditions.

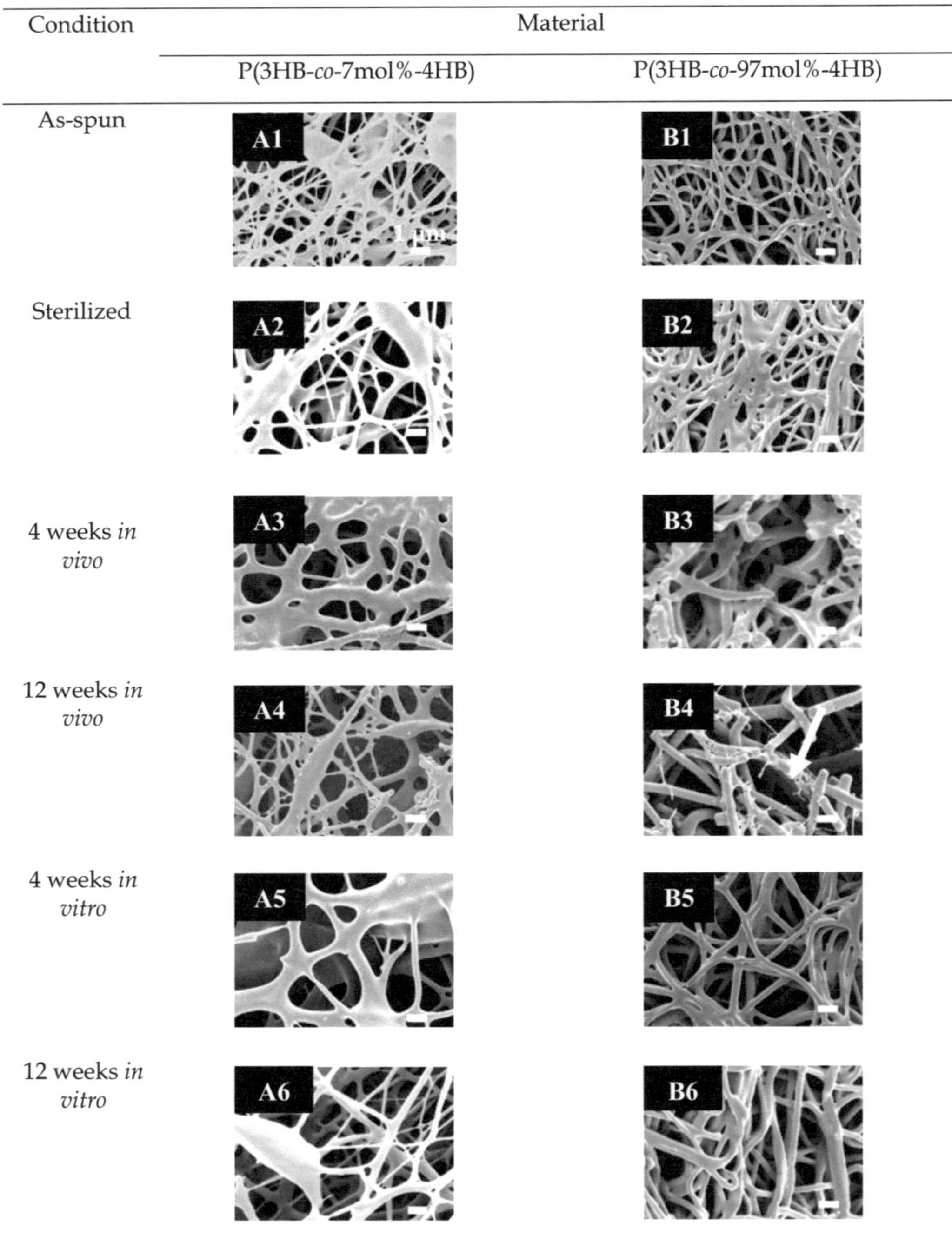

Fig. 3. SEM micrographs of the electrospun P(3HB-*co*-7mol%-4HB) and P(3HB-*co*-97mol%-4HB) in various conditions. The arrow shows the pores on the nanofiber.

Condition	Material	
	PLLA	Stereocomplexed PLA
As-spun	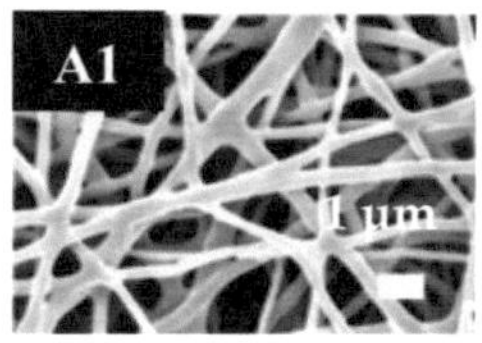	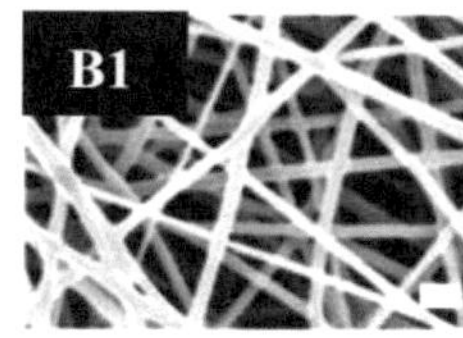
Sterilized	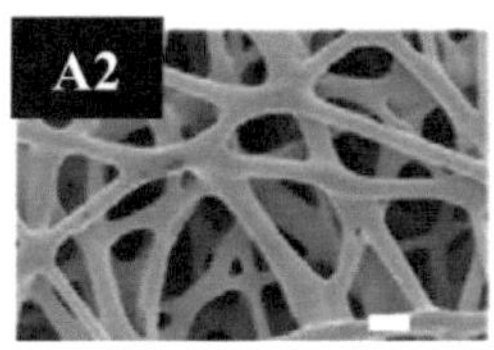	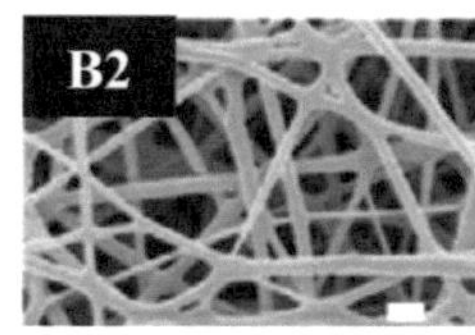
4 weeks *in vivo*	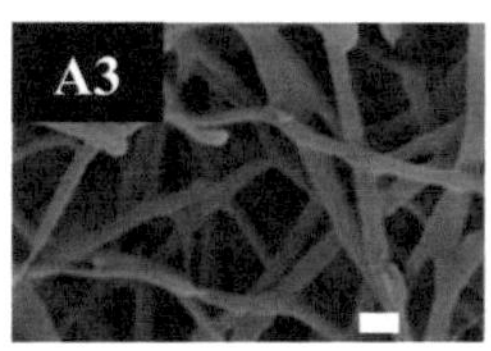	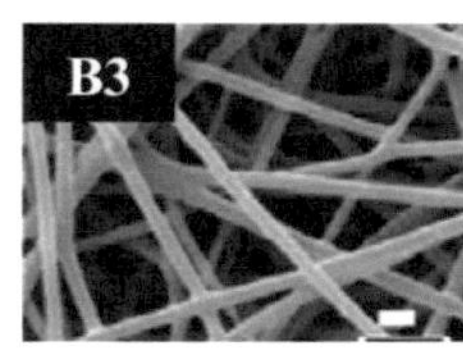
12 weeks *in vivo*	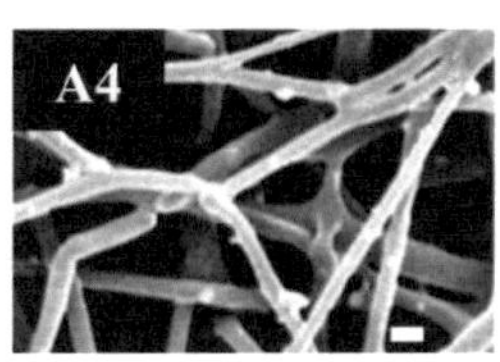	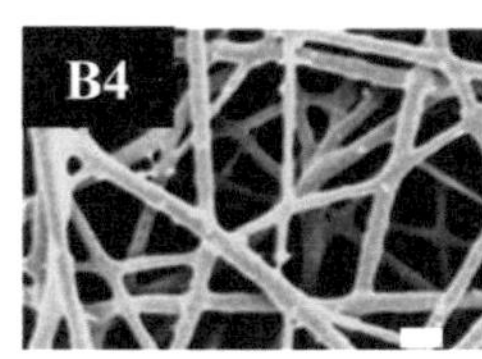
4 weeks *in vitro*		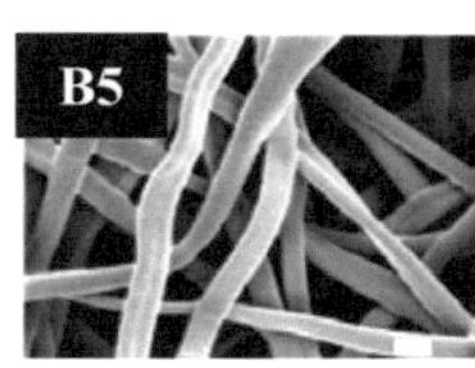
12 weeks *in vitro*		

Fig. 4. SEM micrographs of the electrospun PLLA and stereocomplexed PLA in various conditions.

PLA-based scaffolds consisting of the same average fiber diameter, the water uptake ability of the fibers becomes the determinating factor of their bioabsorption rate. Unlike the electrospun PHA scaffolds, the exposure to PBS brings more detrimental effects to the fiber structure of the PLA-based scaffolds which are known to be susceptible to hydrolysis. Thus, it can be concluded that while the surface erosion of the electrospun PHA scaffolds is dependent on the individual fiber dimensions and monomeric content, the bioabsorption of the electrospun PLA-based scaffolds is dependent on the water uptake ability.

3.2 Crystallinity of nanofibrous scaffolds

The WAXD profiles of the as-spun PHA nanofibrous scaffolds are displayed in Fig. 5. The profiles are the ones after subtraction of background. The crystalline reflections for the P(3HB) and the 3HB-rich copolymers could be indexed on the basis of P(3HB) α-form structure (Yokouchi et al., 1973) while the crystalline phase of P(3HB-*co*-97mol%-4HB) fibers adopted the P(4HB) crystal structure (Su et al., 2003).

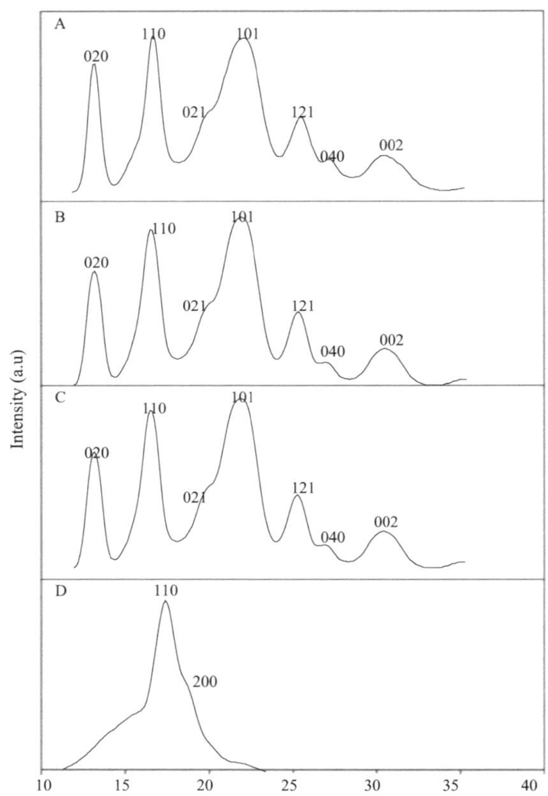

Fig. 5. Integrated 1D profiles from 2D WAXD patterns of as-spun: A. P(3HB) , B. P(3HB-*co*-5mol%-3HHx), C. P(3HB-*co*-7mol%-4HB) and D. P(3HB-*co*-97mol%-4HB).

The WAXD profiles of the as-spun PLLA and stereocomplexed PLA are shown in Fig. 6. The PLLA nanofibers show diffraction peaks at 2θ = 15.1°, 16.5° (assigned to (100)/ (200)), and 18.1° that are assigned to the α-form homocrystal of PLLA. The stereocomplexed PLA nanofibers showed diffraction peaks at 2θ = 12.0° (assigned to (100)), 20.8° and 24.1° that are assigned only to the stereocomplex crystal of PLA (Hasirci et al., 2001). This result confirms that the stereocomplexed PLA nanofibers solely consist of the stereocomplex crystal and not the homocrystal of PLLA and PDLA (Ishii et al., 2007).

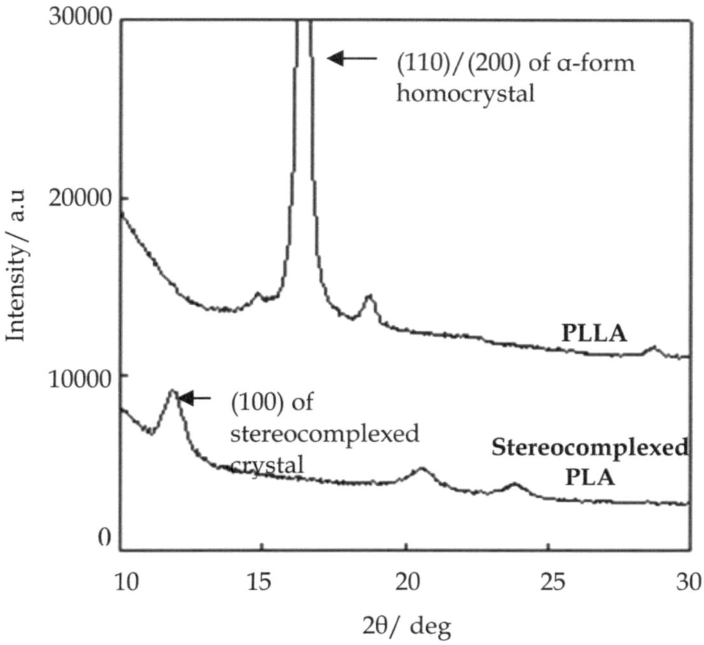

Fig. 6. Wide-angle X-day diffraction patterns of the electrospun PLLA and stereocomplexed PLA.

From the 1D profiles, the crystallinities of as-spun, sterilized and nanofibrous scaffolds after *in vivo* and *in vitro* studies are estimated, according to the method described above. As shown in Figs. 7 and 8, the crystallinities of the as-spun scaffolds increased in the order of P(3HB-*co*-97mol%-4HB) << P(3HB-*co*-5mol%-3HHx) ≈ P(3HB-*co*-7mol%-4HB) < Stereocomplexed PLA < P(3HB) < PLLA. The bulk material equivalents of PHA show the same tendency in crystallinity content (Doi et al., 1992). It has been reported that the crystallinity of P(4HB) homopolymer is much lower than that of P(3HB) homopolymer (Saito et al., 1996). The slight lowering of the crystallinities of the 3HB-rich copolymers is due to the exclusion of second monomer unit from the crystalline lattice (Di Lorenzo et al., 2001). It is evident that the degradability of the nanofibrous scaffolds, as shown in Figs. 2 and 3, strongly depends on the crystallinity.

It was confirmed that the EtO sterilization did not appear to have any effect on the crystallinities of all the PHA and PLA-based nanofibrous scaffolds. It was described earlier that partial melting might occur during the sterilization of P(3HB-*co*-97mol%-4HB). Even if so, the crystallinity will surely recover after the sterilization. The crystallinities of the PHA nanofibrous scaffolds after the *in vivo* and *in vitro* studies remained unchanged. But, it should be noted that the scaffolds for X-ray measurements are the retrieved or remained ones after *in vivo* and *in vitro* experiments. The crystallinity of the P(3HB-*co*-97mol%-4HB), which shows obvious bioabsorption or degradation in SEM observation , also little changed even after implantation in rat. This means that the degradation of nanofibrous scaffolds progresses preferentially from the surface of the nanofibrous scaffolds or interface which contacts with the tissues of rat. It is deduced that some substance such as tissue enzymes facilitate the degradation (Gogolewski, 1992).

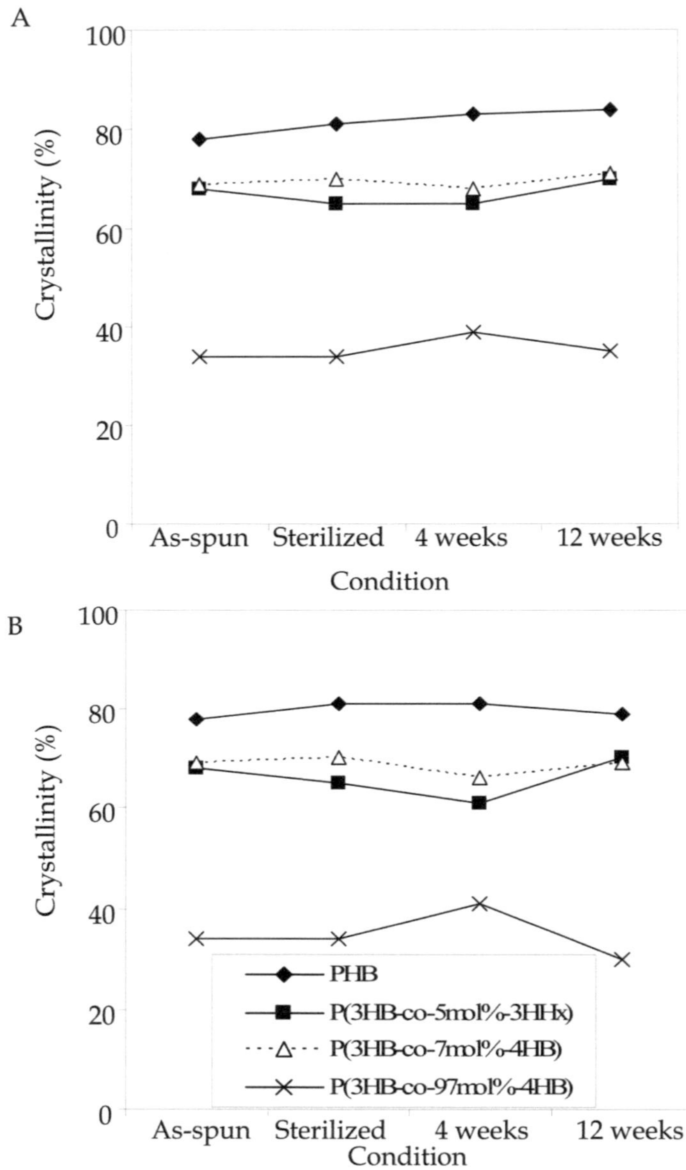

Fig. 7. Crystallinities of the electrospun PHA in different conditions: A. *in vivo* and B. *in vitro*.

After 4 weeks of implantation, the PLLA nanofibrous scaffold showed 28% decrease in crystallinity (86% to 58%) while the electrospun stereocomplexed PLA scaffold showed 12%

decrease in crystallinity (61% to 49%) (Fig. 8). These results support the higher stability of the stereocomplexed PLA nanofibrous scaffold than the PLLA nanofibrous scaffold, as seen from the visual inspection of the explanted nanofibrous scaffolds.

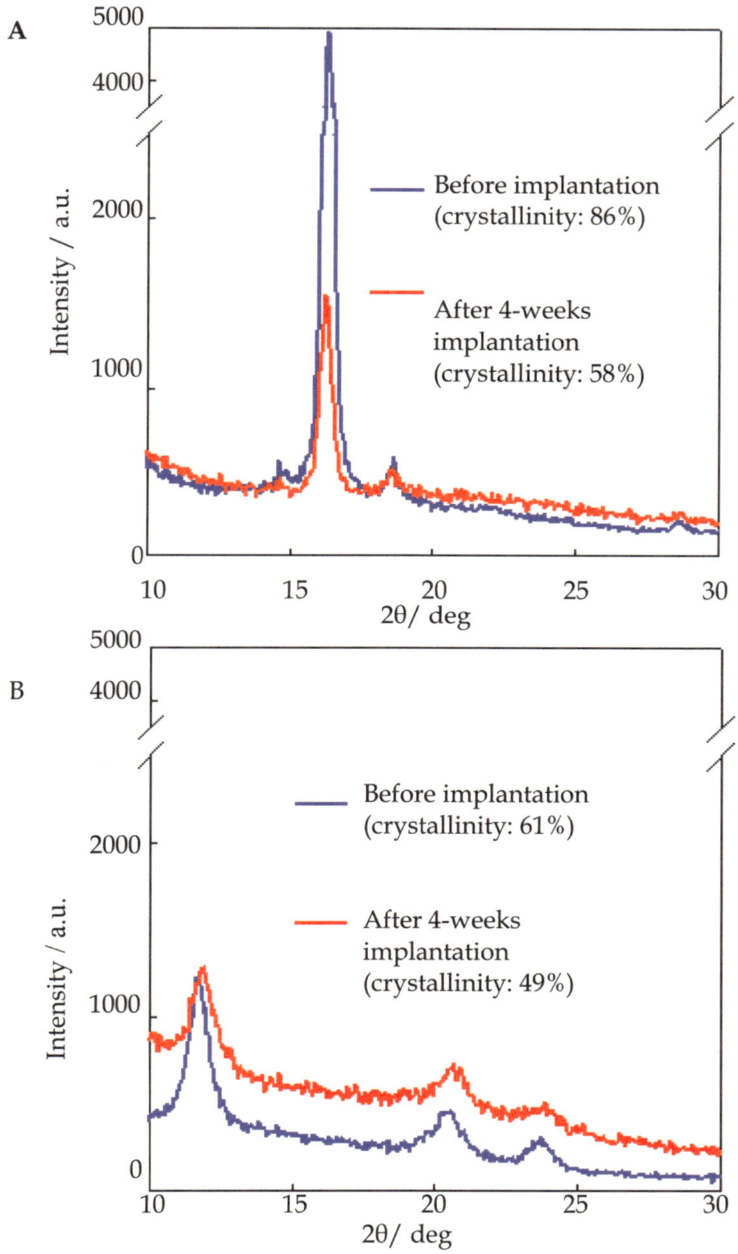

Fig. 8. WAXD patterns of A. PLLA and B. stereocomplexed PLA nanofibrous scaffolds before and after 4 weeks of implantation.

3.3 Molecular weight changes of nanofibrous scaffolds

GPC analysis was carried out to investigate the possible occurrence of molecular chain cleavage (as suggested by SEM and WAXD observations) *in vivo*. Table 1 summarizes the change in M_w and polydispersity index (M_w / M_n) for the as-spun scaffolds and scaffolds following sterilization, 4 and 12 weeks of *in vivo* and *in vitro* studies.

Material	Condition	M_w x 10^5	M_w / M_n
P(3HB)	As-spun	11	3.1
	Sterilized	9.1	3.3
	4 weeks *in vivo*	12	2.6
	in vitro	11	2.6
	12 weeks *in vivo*	8.5	2.9
	in vitro	17	3.3
P(3HB-*co*-5mol%-3HHx)	As-spun	13	3.6
	Sterilized	12	3.6
	4 weeks *in vivo*	12	3.1
	in vitro	11	4.3
	12 weeks *in vivo*	11	3.3
	in vitro	13	4.3
P(3HB-*co*-7mol%-4HB)	As-spun	7.0	3.0
	Sterilized	6.8	3.0
	4 weeks *in vivo*	6.4	2.6
	in vitro	6.1	2.6
	12 weeks *in vivo*	3.9	2.7
	in vitro	5.0	2.6
P(3HB-*co*-97mol%-4HB)	As-spun	1.7	1.5
	Sterilized	1.9	1.8
	4 weeks *in vivo*	1.0	1.6
	in vitro	2.5	2.0
	12 weeks *in vivo*	1.2	2.0
	in vitro	2.2	1.8
PLLA	As-spun	8.7	2.3
	Sterilized	8.7	2.3
	4 weeks *in vivo*	7.2	2.4
	in vitro	4.9	3.5
	12 weeks *in vivo*	3.9	2.3
	in vitro	5.4	3.0
Stereocomplexed PLA	As-spun	2.8	3.3
	Sterilized	2.9	3.3
	4 weeks *in vivo*	2.6	2.3
	in vitro	2.5	3.0
	12 weeks *in vivo*	ND[a]	ND[a]
	in vitro	2.5	2.2

Table 1. Molecular weight properties of the PHA and PLA-based nanofibrous scaffolds.
[a]Not determined due to poor solubility in chloroform.

After sterilization, all of the nanofibrous scaffolds showed no significant differences in their molecular weight. Despite the large surface area of the fibers, the PHA nanofibrous scaffolds remained intact in the *in vitro* study because they have higher resistance to hydrolysis in non-biological environment where specific enzymes are absent (Doi et al, 1992; Marois et al., 2000). Furthermore, the immersion in PBS (pH 7.4) under sterile conditions up to only 12 weeks is short and the temperature is too low for any significant hydrolysis to occur. The subcutaneous implantation, however, seems to cause decrease in the M_w of PHA copolymers with 4HB unit. At 4 weeks, bioabsorption was the most pronounced for the P(3HB-*co*-97mol%-4HB) nanofibrous scaffold with 47% loss M_w, while the M_w of P(3HB-*co*-7mol%-4HB) showed no decrease. Following longer implantation period, the P(3HB-*co*-7mol%-4HB) lost 43% of M_w. Unexpectedly, the P(3HB-*co*-97mol%- 4HB) recorded only 37% of M_w loss after 12 weeks. A possible reason for this observation is that the remaining P(3HB-*co*-97mol%-4HB) nanofibrous scaffold that was retrieved could consist of only higher molecular weight fraction as the biodegraded products of lower molecular weight diffused away more easily. The GPC data confirmed that the number of main-chain carbon atom strongly influences the rate of hydrolysis for the PHA nanofibrous scaffolds.

In the case of the PLA-based nanofibrous scaffolds, bioabsorption was obvious after 4 weeks implantation in rat. The PLLA nanofibers recorded a higher bioabsorption rate (17%) than the stereocomplexed PLA nanofibers (10%) and continued to decrease until 12 weeks of implantation in rat. The GPC data of the stereocomplexed PLA nanofibrous scaffold after 12 weeks implantation were not obtained due to its poor solubility in chloroform. After 4 weeks of exposure to PBS, the stereocomplexed PLA nanofibrous scaffold only lost 14% of its M_w but the PLLA nanofibrous scaffold lost 44% of its M_w. However, the degradation rate of the PLLA nanofibrous scaffold slowed down after 12 weeks in PBS while the stereocomplexed PLA nanofibrous scaffold continued to show increase of up to 31% M_w loss. The GPC data of the *in vitro* experiment suggest that the stereocomplexed PLA nanofibrous scaffold has higher resistance to hydrolysis compared to the PLLA nanofibrous scaffold and are in accordance to the SEM observations. Literature have reported that the strong interaction (even when they are in an amorphous state) between the PLLA chains and the PDLA chains in the stereocomplexed PLA has made it more hydrolysis-resistant than non-blended PLLA and PDLA specimens (Tsuji, 2005). Tsuji et al., confirmed the above by preparing various types of equimolarly blended specimens from PLLA and PDLA and performed their hydrolytic degradation in PBS (pH 7.4) at 37°C together with non-blended PLLA and PDLA specimens (Tsuji, 2002; Tsuji et al.,2003). They observed retarded hydrolytic degradation of the blended specimens compared with the non-blended specimens, irrespective of their state, amorphous or homo-crystallized.

Overall, the rate of bioabsorption of the PLA-based nanofibrous scaffolds was confirmed to be slower than those of the 4HB-rich scaffold but faster than the 3HB-rich nanofibrous scaffolds. Thus, the GPC data are in good agreement with the results obtained in SEM and WAXD analyses.

3.4 Mechanical properties of nanofibrous scaffolds

Table 2 summarizes the mechanical properties of the obtained PHA and PLA-based nanofibrous scaffolds.

Material	Condition	Mechanical properties	
		Tensile strength (MPa)	Young's Modulus (MPa)
P(3HB)	As-spun	17	223
	Sterilized	15	234
	4 weeks *in vivo*	12	182
	in vitro	14	220
	12 weeks *in vivo*	15	152
	in vitro	13	194
P(3HB-*co*-5mol%- 3HHx)	As-spun	15	277
	Sterilized	12	272
	4 weeks *in vivo*	12	268
	in vitro	13	208
	12 weeks *in vivo*	ND [b]	ND [b]
	in vitro	15	230
P(3HB-*co*-7mol%-4HB)	As-spun	8	184
	Sterilized	8	139
	4 weeks *in vivo*	ND [b]	ND [b]
	in vitro	8	163
	12 weeks *in vivo*	ND [b]	ND [b]
	in vitro	9	110
P(3HB-*co*-97mol%- 4HB)	As-spun	13	9
	Sterilized	15	16
	4 weeks *in vivo*	4	12
	in vitro	11	14
	12 weeks *in vivo*	ND [b]	ND [b]
	in vitro	14	16
PLLA	As-spun	6	143
	Sterilized	6	143
	4 weeks *in vivo*	ND [b]	ND [b]
	in vitro	11	130
	12 weeks *in vivo*	ND [b]	ND [b]
	in vitro	12	123
Stereocomplexed PLA	As-spun	6	94
	Sterilized	6	94
	4 weeks *in vivo*	9	80
	in vitro	13	78
	12 weeks *in vivo*	ND [b]	ND [b]
	in vitro	12	83
Skin [a]		5-30	15-150

Table 2. Mechanical properties of the nanofibrous scaffolds.
[a] Data obtained from Zong et al., 2005.
[b] Not determined as the retrieved scaffolds from rat had cracks on the surface that prevented tensile test.

The mechanical properties of all the as-spun scaffolds were comparable to those of human skin, and hence suggest they are mechanically stable in supporting regenerated tissues. The

Young's modulus of the as-spun nanofibrous scaffolds increased in the order of P(3HB-*co*-97mol%-4HB) << Stereocomplexed PLA < P(3HB-*co*-7mol%-4HB) < PLLA < P(3HB) < P(3HB-*co*-5mol%-3HHx). Low Young's modulus, that is, high elasticity is a characteristic property in rubber-state amorphous polymers. Accordingly, this indicates that the P(3HB-*co*-97mol%-4HB) fibers are more amorphous than the other nanofibrous scaffolds, and this is consistent with the WAXD results. The distinct mechanical properties of the nanofibrous scaffolds could find different use in tissue engineering. For example, the 3H-Brich and PLA-based nanofibrous scaffolds which are more rigid could serve as preferential substrates for directional cell migration (Lo et al., 2000) while the compliant 4HB-rich nanofibrous scaffold could be used to promote cell motility (Pelham et al., 1997) or for soft-tissue engineering. The EtO sterilization and the immersion in PBS buffer little affected the mechanical properties of all PHA nanofibrous scaffolds. Interestingly, the bioabsorption in rat and degradation in PBS resulted in greater plasticization of the PLA-based nanofibrous scaffolds than those of the PHA nanofibrous scaffolds, as evidenced by the increase in their tensile strength.

4. Tissue response and mechanism of bioabsorption of PHA and PLA-based nanofibrous scaffolds

4.1 Tissue response of nanofibrous scaffolds

Histological observations give important information about the degree of inflammatory reactions and the penetration of the surrounding tissues into the implanted scaffolds. The histological sections of the nanofibrous scaffolds at different period of subcutaneous implantation are shown in Figs. 9 and 10. The nuclei of the inflammatory cells or macrophages were stained blue by the hematoxylin dye and their presence is an indication of tissue response towards the implanted scaffolds.

Histological observations indicate that all the three PHA copolymers nanofibrous scaffolds elicited fairly mild tissue response relative to that of the P(3HB) nanofibrous scaffold throughout the course of study. After 4 weeks of implantation, some parts of the P(3HB-*co*-97mol%-4HB) nanofibrous scaffold bordering the interface were degraded as evidenced by the small fragments broken off from the main scaffold (Fig. 9 D4). More macrophages were found to be present along the interface connected to this copolymer in comparison to the P(3HB-*co*-7mol%-4HB) and P(3HB-*co*-5mol%-3HHx) nanofibrous scaffolds (Figs. 9 C4 and 9 D4). This phenomenon is desirable during wound healing because the presence of macrophages is necessary for the regeneration of many cell types (Rappolee et al., 1988). Macrophages clear the way for tissue ingrowths by phagocytosis of damaged tissue, necrotic and apoptotic cells and environment particles. They also secrete a spectrum of cytokines and growth factors to regulate cell recruitment, proliferation and differentiation which lead to effective tissue regeneration and angiogenesis. In fact, studies of wound healing have shown that if macrophage infiltration is prevented, then healing is severely impaired (Martin, 1997; Leiboyich et al., 1975).

The presence of thin connective tissue surrounding the P(3HB-*co*-97mol%-4HB) nanofibrous scaffold was also observed. The most promising finding was the tissue response after 12 weeks of implantation for the P(3HB-*co*-97mol%-4HB) nanofibrous scaffold. No fibrous encapsulation was observed around the degraded copolymer and there was also a substantial drop in the number of inflammatory cells (Fig. 9 B12). This observation is similar to a study done on the biocompatibility of P(4HB) implanted subcutaneously in rats by Martin and co-workers (Martin et al.,1999) that reported minimal inflammatory responses.

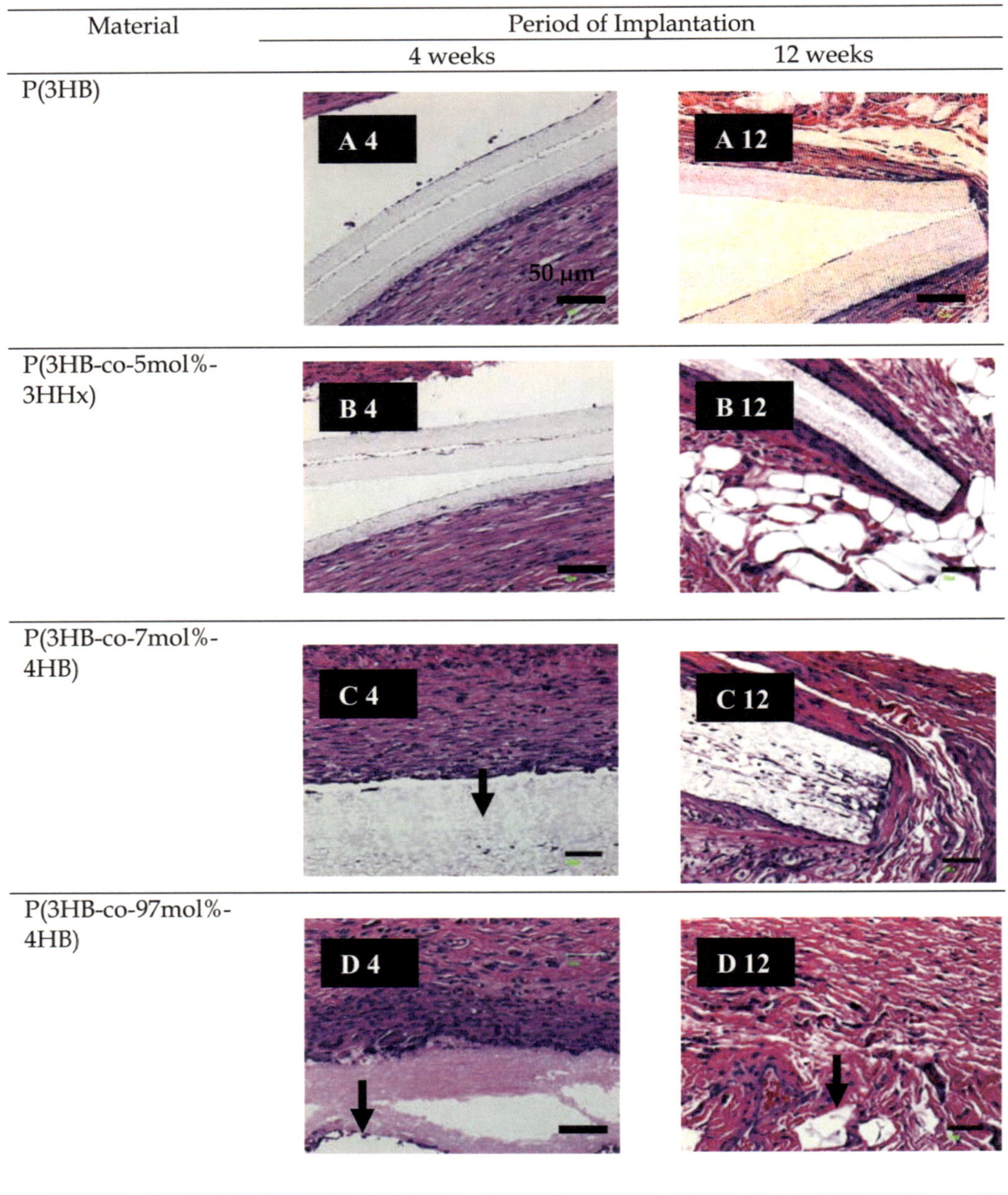

Fig. 9. Histological sections of the PHA nanofibrous scaffolds at different period of subcutaneous implantation. Arrows indicate the polymer surface.

In this study, the number of inflammatory cells surrounding the P(3HB-*co*-7mol%-4HB) and P(3HB-*co*-5mol%-3HHx) nanofibrous scaffolds did not appear to have lessen. The muscle cells surrounding these two scaffolds appeared compact as a result of inflammatory reaction (Figs. 9B and 9C). After 12 weeks of implantation, the number of macrophages bordering the

P(3HB) increased. Inflammation was obvious due to the compacted muscle cells surrounding the scaffold. The difference in tissue response to the P(3HB-*co*-97mol%-4HB) and the nanofibrous scaffolds with higher molar fraction of 3HB reflected their distinct physical properties. It has been reported that rigid polymer, such as P(3HB), elicit acute inflammatory reaction because it exerts a continuous mechanical stimulus to the surrounding tissues of the implants (Qu et al., 2006). Although the tissue response to the P(3HB-*co*-7mol%-4HB) and electrospun P(3HB-*co*-5mol%-3HHx) nanofibrous scaffolds was slightly more pronounced than that of the P(3HB-*co*-97mol%-4HB), the overall local tissue response to all three copolymers were found to be mild. The results have confirmed the biocompatibility of all three types of PHA copolymers.

For the PLA-based nanofibrous scaffolds, their presence *in vivo* brought stronger tissue response in comparison to the PHA nanofibrous scaffolds. As indicated in Fig. 10 A4, a thick layer of inflammatory cells was accumulated at the interface between the PLLA nanofibrous scaffold and the surrounding tissues. In contrast, the layer of accumulated inflammatory cells was thinner for the stereocomplexed PLA nanofibrous scaffold (Fig. 10 B4). This indicates that the stereocomplexed PLA nanofibrous scaffold induces smaller degree of tissue response than the PLLA nanofibrous scaffold. Interestingly, delamination (indicated by the ellipsoid in Fig. 10 A4) occurred on the surface of the PLLA nanofibrous scaffold and hence the infiltration of the surrounding tissues was observed. After 12 weeks of implantation, the PLLA nanofibrous scaffold was significantly fragmented as indicated by

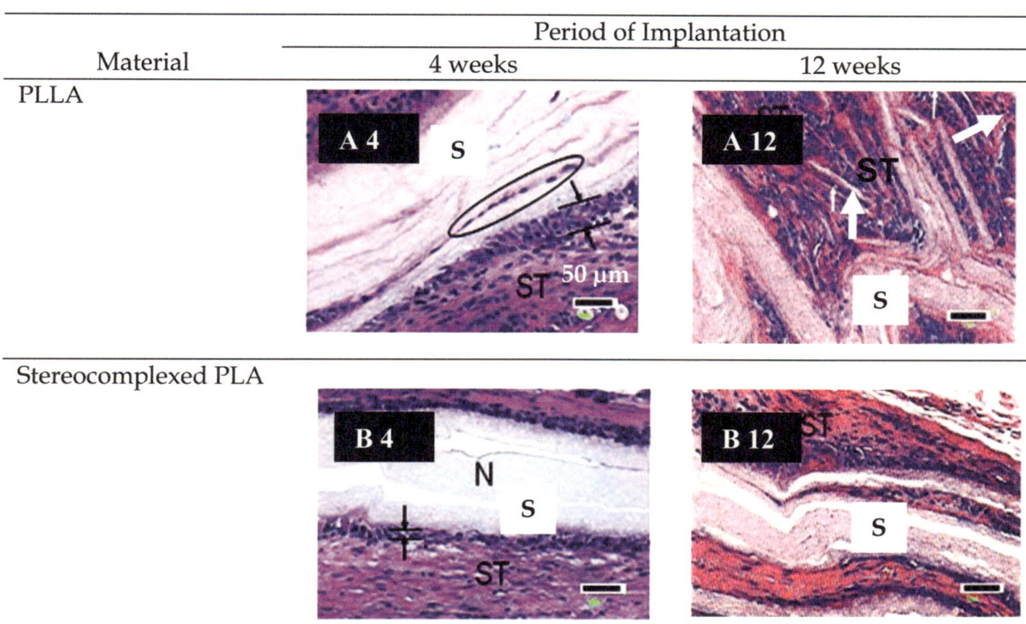

Fig. 10. Histological sections of the PLA-based nanofibrous scaffolds at different period of subcutaneous implantation. The width of inflammatory cells is indicated by the arrows and lines in A4 and B4. Ellipsoid region in A4 and white arrows in A12 indicate the infiltration of surrounding tissues and fragmented fibers, respectively. ST: surrounding tissues; S: scaffold.

the white arrows in Fig. 10 A12, but the stereocomplexed PLA nanofibrous scaffold retained its original morphology. These trends are well correlated with the bulk appearances of the scaffolds and further confirm that the bioabsorption of the stereocomplexed PLA nanofibrous scaffold proceeds slower than that of the PLLA nanofibrous scaffold.

From the histological observations, the biocompatibility of the P(3HB-*co*-97mol%-4HB) nanofibrous scaffold is by far the most superior among all of the electrospun scaffolds investigated in this current work. In addition, the *in vitro* toxicity of the P(3HB-*co*-4HB) biosynthesized from *D. acidovorans* which was determined by Siew and co-workers also demonstrated that this copolymer has good biocompatibility (Siew et al., 2007). Their recent genotoxicity study also revealed encouraging results for this PHA copolymer (Siew et al., 2008). The acidic products from the degradation of PLA-based scaffolds act to catalyze further degradation which brings on accumulation of acidic products at the implant site and elicit stronger inflammatory response. Hence, it is not surprising that all PHA nanofibrous scaffolds outperformed the PLA-based nanofibrous scaffolds in terms of tissue response.

4.2 Mechanism of bioabsorption of nanofibrous scaffolds

4.2 (a) PHA nanofibrous scaffolds

The results from various analyses clearly demonstrated that the bioabsorption rate of the P(3HB-*co*-97mol%-4HB) was the fastest relative to the other two PHA copolymers. Three possible reasons for this observation are as follows: Firstly, the P(3HB-*co*-97mol%- 4HB) with low crystallinity is more susceptible to bioabsorption as water and enzymes penetrate easier into the amorphous regions. Secondly, previous studies have established that macrophages are able to phagocytize PHA *in vitro* (Ali et al., 1994; Saad et al., 1999) and free radicals, acidic products or enzymes produced by these cells may also accelerate the degradation (Tracy et al., 1999). The size of the biomaterials determines the way macrophages respond to them (Xia & Triffitt, 2006). Macrophages may phagocytoze material particles with sizes smaller than a single-nucleated macrophage (usually around 10 µm in diameter). Large particles between 10 µm and several hundred micrometers in diameter may be engulfed by foreign body giant cells or multinucleated giant cells. For bulk materials in the form of medical devices such as scaffolds, foreign body giant cells will form and adhere to the surface of these materials. As seen in Fig. 9 D4, the number of inflammatory cells was the most concentrated at the interface of electrospun P(3HB-*co*-97mol%-4HB) suggesting their active part in the bioabsorption process. It is known that biodegradable materials will be degraded within phagosomes after phagocytosis, or eroded via extracellular resorption, with or without the involvement of foreign body giant cells. Also, after total resorption of the biodegradable materials, any associated inflammation will be resolved. Hence, accordingly, the number of macrophages surrounding the P(3HB-*co*-97mol%-4HB) decreased tremendously after 12 weeks. Thirdly, possibly the enzymatic degradation by lipase also contributed to the rapid bioabsorption of the P(3HB-*co*-97mol%-4HB). PHA can be enzymatically degraded by PHA depolymerases, but there is no evidence to date that these are present *in vivo* (Williams & Martin, 2002). P(4HB) was found to be also highly susceptible to lipase hydrolysis as opposed to P(3HB) (Mukai et al., 1993). Besides having good mechanical properties and biocompatibility, it is desirable for a medical implant to show good bioabsorption after its primary function has been fulfilled. The persistence of polymer at a wound healing site may lead to chronic inflammation as shown by the slowly degrading P(3HB) patches that elicited a long term (greater than two years) macrophage response (Malm et al., 1994). Hence the fast bioabsorption rates of the PHA containing 4HB have confirmed their potential in the application for medical implants.

4.2 (b) PLA-based nanofibrous scaffolds

The bioabsorption mechanisms of the PLA-based nanofibrous scaffolds are illustrated in Fig. 11. For the PLLA nanofibrous scaffold, it is believed that intracrystalline swelling induces preferential hydrolysis of the molecular chains in the amorphous regions between lamella crystals. This leads to the cleavage of the fiber strand and the decrease in the molecular weight. Following these events, the chain-end degradation at the edge of the cleaved fiber may occur and leads to the decrease in the crystallinity. The cleavage of the fiber strands may facilitate the delamination and subsequently the fragmentation of the scaffold. The inflammatory reaction at the early stage of implantation may be due to the acidic low-molecular-weight degradation products and the fragmented fibers. A different bioabsorption mechanism is proposed for the stereocomplexed PLA nanofibrous scaffold. It

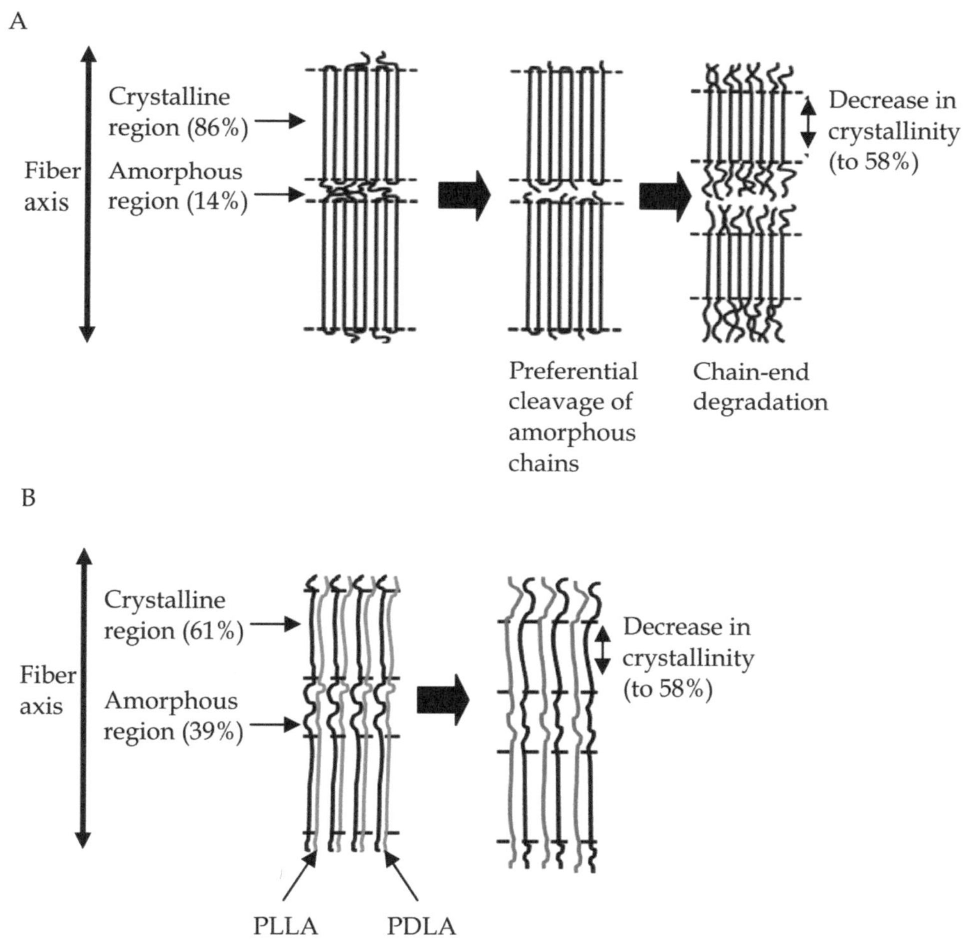

Fig. 11. Schematic representation of the structural changes of A. PLLA and B. stereocomplexed PLA nanofibers during implantation *in vivo*.

is assumed that a single strand of the stereocomplexed PLA nanofiber is composed of the PLLA and PDLA chains aligned in a side-by-side manner. This assumption is made according to the well known fact that the high melting temperature of the stereocomplexed PLA is caused by the strong molecular interaction between the PLLA and PDLA chains. This particular molecular arrangement may therefore suppress the hydrolysis of the molecular chains *in vivo*. As such, the morphology of the electrospun stereocomplexed PLA is retained. This also gives the reason why the inflammatory reaction of the stereocomplexed PLA nanofibrous scaffold is limited at the vicinity of the interfacial region between the scaffold and the surrounding tissues.

5. Conclusion

The overall aim of this research was to electrospin P(3HB-*co*-97mol%-4HB) for fabricating tissue-engineering scaffold with enhanced mechanical properties, bioabsorption and biocompatibility. Its performance as a nanofibrous scaffold for tissue engineering was compared with electrospun homopolymer P(3HB) and its copolymers containing 5mol%-3HHx and 7mol%-4HB as well as with electrospun PLA-based scaffolds. All of these nanofibrous scaffolds were implanted subcutaneously in rats to evaluate their tissue response. Thus far, this is the first study to evaluate the bioabsorption and tissue response of electrospun scaffolds containing 3HB, 3HHx and 4HB monomers implanted in rat models. SEM revealed that both the *in vitro* and *in vivo* surface erosion of the PHA nanofibrous scaffolds progressed dependently on the individual fiber dimensions and monomeric contents. However in the case of the PLA-based nanofibrous scaffolds, SEM showed that despite having similar fiber dimensions, the fragmentation of the PLLA proceeded faster than that of the stereocomplexed PLA. Hence, it is deduced that the molecular chain interactions within the fiber strands strongly influence the fragmentation of PLA-based nanofibers in the *in vitro* and *in vivo* experiments. Sterilization with ethylene oxide did not cause discoloration and damage to all nanofibrous scaffolds. The mechanical properties demonstrated by all nanofibrous scaffolds were comparable to those of human skin, thus suggesting that their structures are able to provide sufficient biomechanical support. For the PHA, nanofibrous scaffolds consisting of high 3HB content had higher degree of crystallinity and thus, they showed slower bioabsorption rate. Even though having rather comparable crystallinity content with the 3HB-rich scaffolds, the bioabsorption rate of PLA-based nanofibrous scaffolds was slightly faster possibly because their smaller fiber dimensions further enhanced hydrolysis. GPC revealed that the *in vitro* degradation of all nanofibrous scaffolds proceeded at a much slower rate in comparison to the bioabsorption *in vivo*. Thus, some enzymes present in the biological environment must have some influence in the bioabsorption of the nanofibrous scaffolds. Histological evaluation showed that subcutaneous implantations of the PHA nanofibrous scaffolds were well tolerated *in vivo* as the tissue response continued to be very mild throughout the course of the study. When compared with the PLA-based nanofibrous scaffolds, the PHA nanofibrous scaffolds showed better biocompatibility because their by-products of degradation did not further induce inflammatory reaction in the body of the rats. The current work also revealed that by changing the molar fraction of monomers in the PHA copolymers, it is possible to create tissue-engineering scaffolds that are tailor-made to meet the various needs in regenerating different cell type. Such versatility would definitely expand the potential of PHAs as biomaterials. Overall, the electrospun P(3HB-*co*-97mol%-4HB) outperformed all the P(3HB) copolymers and PLA-based scaffolds in terms of bioabsorbability and biocompatibility, hence proving it to be a promising new biomaterial for tissue engineering scaffold. In conclusion, this research lays the groundwork for a good understanding of the physical properties and tissue response of the electrospun P(3HB-*co*-97mol%-4HB) nanofibers in their future application as tissue engineering scaffold.

6. Acknowledgement

The authors thank Prof. Won-Ki Lee (Pukyong University, South Korea) and Dr. Toshihisa Tanaka (Shinshu University, Japan) for their kind supply of PDLA and P(3HB-*co*- 7mol%-4HB), respectively. They also thank Dr. Atsushi Mahara (National Cardiovascular Centre, Japan) and Dr. Sunao Murakami (AIST, Japan) for their help in the subcutaneous implantation of nanofiber samples in rats and their retrieval. The authors acknowledge the Asia Program Associate (APA) between Universiti Sains Malaysia and RIKEN Institute. This work was supported by a Grant-in Aid for Scientific Research (B) from the Ministry of Education, Culture Sports, Science and Technology (MEXT) of Japan (No. 19350075) (to T. Iwata) and by a grant for Ecomolecular Science Research II provided by RIKEN Institute.

7. References

Ali, SAM.; Doherty, PJ. & Williams, DF. (1994). Molecular biointeractions of biomedical polymers with extracellular exudates and inflammatory cells and their effects on the biocompatibility in vivo. Biomaterials, Vol. 15, (1994) pp. 779-785

Anderson, AJ. & Dawes, EA. (1990). Occurrence, metabolic role, and industrial uses of bacterial polyhydroxyalkanoates. *Microbial Review*, Vol. 54, (1990) pp. 450-472

Demir, MM.; Yilgor, I.; Yilgor, E. & Erman, B. (2002). Electrospinning of polyurethane fibers. *Polymer*, Vol. 43, (2002) pp. 3303-3309

Di Lorenzo, ML.; Raimo, M.; Cascone, E. & Martuscelli, E. (2001). Poly(3-hydroxybutyrate)-based copolymers and blends: influence of a second component on crystallization and thermal behavior. *Journal of Macromolecular Science: Physic*, Vol. 40, (2001) pp. 639-667

Doi, Y.; Kanesawa, Y.; Kawaguchi, Y. & Kunioka, M. (1992). Biodegradation of microbial polyesters in the marine environment. *Polymer Degradability and Stability*, Vol. 32, (1992) pp.173-177

Dong, H.; Nyame, V.; Macdiarmid, AG. & Jr. Jones, WE. (2004). Polyaniline/Poly(methyl methacrylate) coaxial fibers: The fabrication and effects of the solution properties on the morphology of electrospun core fibers. *Journal of Polymer Science Part B: Polymer Physics*, Vol. 42, (2004) pp. 3934-3942

FDA clears first of its kind suture made using DNA techonology. Available from: FDA News http://www.fda.gov/bbs/topics/NEWS/2007/NEW01560.html; 2007 February 12 [accessed 25.04.07]

Fong, H.; Chun, I. & Reneker, DH. (1999). Beaded nanofibers formed during electrospinning. *Polymer*, Vol. 40, (1999) pp. 4585-4592

Gogolewski, S. (1992). Resorbable polymers for internal fixation. *Clinical Materials*, Vol. 10, (1992) pp. 13-20

Griffith, LG. (2000). Emerging design principles on biomaterials and scaffolds for tissue engineering. *Annals of the New York Academy of Sciences*, Vol. 961, (2000) pp. 83-95

Hasirci, V.; Lewandrowski, K.; Gresser, J.; Wise, DL. & Trantolo DJ. (2001). Versatility of biodegradable biopolymers: degradability and an in vivo application. *Journal of Biotechnology*, Vol. 86, (2001) pp. 135-150

Ishii, D.; Lee, W.-K.; Kasuya, K. & Iwata, T. (2007). Fine structure and enzymatic degradation of Poly[(*R*) -3-hydroxybutyrate] and stereocomplexed poly(lactide) nanofibers. *Journal of Biotechnology*, Vol. 132, (2007) pp. 318-324

Ishii, D.; Tang, HY.; Mahara, A.; Murakami, S.; Yamaoka, T.; Lee, W.-K. & Iwata, T. (2009). *In vivo* tissue response and degradation behavior of PLLA and stereocomplexed PLA nanofibers. *Biomacromolecules*, Vol. 10, (2009) pp. 237-242

Jeong, EH.; Im, SS. & Youk, JH. (2005). Electrospinning and structural characterization of ultrafine poly(butylene succinate) fibers. *Polymer*, Vol. 4, (2005) pp. 9538-9543

Leibovich, SJ. & Ross, R. (1975). The role of macrophage in wound repair. A study with hydrocortisone and antimacrophage serum. *American Journal of Pathology*, Vol. 78, (1975) pp. 71-100

Li, WJ.; Cooper, JA. Jr.; Mauck, RL. & Tuan, RS. (2006). Fabrication and characterization of six electrospun poly(alpha-hydroxy ester)-based fibrous scaffolds for tissue engineering applications. *Acta Biomaterialia*, Vol. 2, (2006) pp. 377-385

Lo, CM.; Wang, HB.; Dembo, M. & Wang YL. (2000). Cell movement is guided by the rigidity of the substrate. *Biophysical Journal*, Vol. 79, (2000) pp. 144-152

Lyons, J.; Li, C. & Ko, F. (2004). Melt-electrospinning part I: Processing parameters and geometric properties. *Polymer*, Vol. 45, (2004) pp. 7597–7603

Malm, T.; Bowald, S.; Bylock, A.; Busch, C. & Saldeen, T. (1994). Enlargement of the right ventricular outflow tract and the pulmonary artery with a new biodegradable patch in transanular position. European Surgical Research, Vol. 26, (1994) pp. 298- 308

Marois, Y.; Zhang, Z.; Vert, M.; Deng, X.; Lenz, R. & Guidoin R. (2000). Mechanism and rate of degradation of polyhydroxyoctanoate films in aqueous media: a long term in vitro study. *Journal of Biomedical Material Research*, Vol. 49, (2000) pp. 216-224

Martin, DP. & Williams, SF. (2003). Medical applications of poly-4-hydroxybutyrate: a strong flexible absorbable biomaterial. *Biochemical Engineering Journal*, Vol. 16 (2003) pp. 97-105

Martin, DP.; Skraly, FA. & Williams, SF. (1999). Polyhydroxyalkanoate compositions having controlled degradation rates. PCT Patent Application No. WO 99/32536; 1999.

Martin, P. (1997). Wound healing-aiming for perfect skin regeneration. *Science*, Vol. 276, (1997) pp. 75-81

Mukai, K.; Doi, Y.; Sema, Y. & Tomita, K. (1993). Substrate specificities in hydrolysis of polyhydroxyalkanoates by microbial esterases. Biotechnology Letter, Vol. 15, (1993) pp. 601-604

Pelham, RJ. & Wang, YL. (1997). Cell locomotion and focal adhesions are regulated by substrate flexibility. *Proceedings of the National Academy of Science*, pp. 13661-13665, Vol. 94, 1997, USA.

Qu, XH.; Wu, Q.; Zhang, KY. & Chen, GQ. (2006). *In vivo* studies of poly(3-hydroxybutyrate-*co*-3-hydroxyhexanoate) based polymers: biodegradation and tissue reactions. *Biomaterials*, Vol. 27, (2006) pp. 3540-3548

Rappolee, DA.; Mark, D.; Banda, MJ. & Werb, Z. (1988). Wound macrophages express TGFalpha and other growth factors in vivo: analysis by mRNA phenotyping. *Science*, Vol. 241, (1988) pp. 708-712

Saad, B.; Neuenschwander. P.; Uhlschmid, GK. & Suter, UW. (1999) New versatile, elastomeric, degradable polymeric aterials for medicine. *International Journal of Biological Macromolecules*, Vol. 25, (1999) pp. 293-301

Saito, Y.; Nakamura, S.; Hiramitsu, M. & Doi, Y. (1996). Microbial synthesis and properties of poly(3-hydroxybutyrate-*co*-4-hydroxybutyrate). *Polymer International*, Vol. 39, (1996) pp. 169-174

Siew, EL.; Rajab, NF.; Osman, A.; Sudesh, K. & Inayat-Hussain, SH. (2007). *In vitro* biocompatibility evaluation of poly(3-hydroxybutyrate-*co*-4-hydroxybutyrate) copolymer in fibroblast cells. *Journal of Biomedical Research A*, Vol. 81A, (2007) pp. 317-325

Siew, EL.; Rajab, NF.; Osman, A.; Sudesh, K. & Inayat-Hussain, SH. (2008). Mutagenic and clastogenic characterization of poststerilized poly(3-hydroxybutyrate-*co*-4-hydroxybutyrate) copolymer biosynthesized by *Delftia acidovorans*. *Journal of Biomedical Materials Research Part A* (in press)

Su, F.; Iwata, T.; Tanaka, F. & Doi, Y. (2003). Crystal structure and enzymatic degradation of poly(4-hydroxybutyrate). *Macromolecules*, Vol. 36, (2003) pp. 6401-6409

Subbiah, T.; Bhat, GS.; Tock, RW.; Parameswaran, S. & Ramkumar, SS. (2005). Electrospinning of nanofibers. *Journal of Applied Polymer Science*, Vol. 96, (2005) pp. 557-569

Tang, HY.; Ishii, D.; Mahara, A.; Murakami, S.; Yamaoka, T.; Sudesh, K.; Samian, R.; Fujita, M.; Maeda, M. & Iwata, T. (2009). Scaffolds from electrospun polyhydroxyalkanoates copolymers: Fabrication, characterization, bioabsorption and tissue response. *Biomaterials*, Vol. 10, (2009) pp. 1307-1317

Taylor, GI. (1964). Disintegration of water drops in an electric field. *Proceeding of The Royal Society*, pp. 383-397, 1964, London

Tracy, MA.; Ward, KL.; Firouzabadian, L.; Wang, Y.; Dong, N.; Qian, R. & Zhang, Y. (1999). Factors affecting the degradation rate of poly(lactide-*co*-glycolide) microspheres *in vivo* and *in vitro*. *Biomaterials*, Vol. 20, (1999) pp. 1057-1062

Tsuji, H. (2002). Autocatalytic hydrolysis of amorphous-made polylactides: effects of L-lactide content, tacticity, and enantiomeric polymer blending. *Polymer*, Vol. 43, (2002) pp. 1789-1796

Tsuji, H. (2003). In vitro hydrolysis of blends from enantiomeric poly(lactide)s. Part 4: wellhomo- crystallized blends and nonblended films. *Biomaterials*, Vol. 24, (2003) pp. 537-547

Tsuji, H. (2005). Poly(lactide) Stereocomplexes: Formation, Structure, Properties, Degradation, and Applications. *Macromolecular Bioscience*, Vol. 5, (2005) pp. 567-597

Williams, SF. & Martin, DP. (2002) Applications of PHAs in medicine and pharmacy. In: Steinbüchel A, editor. Series of biopolymers in 10 volumes, vol. 4. Wiley/VCY/Verlag; 2002. p. 91-121

Xia, Z. & Triffitt, JT. (2006) A review on macrophages responses to biomaterials. *Biomedical Materials*, Vol. 1, (2006) pp. R1-R9

Yokouchi, M.; Chatani, Y.; Tadokoro, H.; Teranishi, K. & Tani, H. (1973). Structural studies of polyesters: 5. Molecular and crystal structures of optically active and racemic poly(β-hydroxybutyrate). *Polymer*, Vol. 14, (1973) pp. 267-272

Zong, X.; Bien, H.; Chung, CY.; Yin, L.; Fang, D.; Hsiao, BS.; Chu, B. & Entcheva, E. (2005). Electrospun fine-textured scaffolds for heart tissue constructs. *Biomaterials*, Vol. 26, (2005) pp. 5330-5338

Zong, X.; Ran, S.; Kim, KS.; Fang, D.; Hsiao, BS. & Chu, B. (2003). Structure and morphology changes during in vitro degradation of electrospun poly(glycolide-*co*-lactide) nanofiber membrane. *Biomacromolecules*, Vol. 4, (2003) pp. 416-423

Zuo, W.; Zhu, M.; Yang, W.; Yu, H.; Chen, Y. & Zhang, Y. (2005). Experimental study on relationship between jet stability and formation of beaded fibers during electrospinning. *Polymer Engineering and Science*, Vol. 45, (2005) pp. 704-709

11

Photocatalyst Nanofibers Obtained by Calcination of Organic-Inorganic Hybrids

Koji Nakane and Nobuo Ogata
University of Fukui
Japan

1. Introduction

Electrospinning (ES) is one of the most useful techniques to form nanofibers in a diameter of several hundred nanometers (Doshi & Reneker, 1995, Buchko et al., 1999, Huang et al., 2003). The diameter of the nanofibers produced by ES is at least one or two orders of magnitude smaller than those of conventional fiber production methods like melt or solution spinning. As a result, the electrospun nanofibers have high specific surface area (Yamashita, 2007). These nanofibers are well-suited to be used as chemical reaction fields (Nakane et al., 2005, 2007).

Much attention has been paid to the formation of both organic polymeric nanofibers and inorganic nanofibers using ES (Ramakrishna et al., 2005). Many kinds of inorganic nanofibers (SiO_2, Al_2O_3, ZrO_2, $NiCo_2O_4$, and so on) have been obtained by calcination of organic-inorganic hybrid precursor nanofibers formed by ES (Guan et al., 2004, Shao et al., 2004, Chronakis, 2005, Panda & Ramakrishna, 2007, Krissanasaeranee et al., 2008). The formation of TiO_2 nanofibers have been also reported by several research groups. Li and Xia formed anatase-type titanium oxide (TiO_2) nanofibers by the calcination of poly(vinyl pyrrolidone) (PVP)-Ti tetraisopropoxide (TTIP) hybrid nanofibers at 500°C in air (Li & Xia, 2003). The TiO_2 nanofibers obtained would be a useful material for a photocatalytic reaction, but their usage has not been investigated. Ethanol has been used as the solvent of the spinning solution to form the hybrid precursor nanofibers. Therefore, a spinneret could be stopped up by a solid material because ethanol will evaporate from the tip of the spinneret during the spinning. Furthermore, TTIP is very easily hydrolyzed, and thus a water-free condition is required for the use of TTIP on ES. Another groups also formed TiO_2 nanofibers by calcination of TiO_2-PVP and TiO_2-poly(vinyl acetate) precursors which were formed by ES using organic solvents such as ethanol and dimethylformamide (Kim et al., 2006, Nuansing et al., 2006, Kumar et al., 2007, Ding et al., 2008).

Li and Xia reported the formation of TiO_2 hollow-nanofibers (nanotubes) by ES of two immiscible liquids (TTIP-PVP ethanol solution and heavy mineral oil) through a coaxial, two-capillary spinneret, followed by selective removal of the cores and calcination in air (Li & Xia, 2004). The TiO_2 nanotubes with uniform and circular cross-sections were obtained by the method. Kobayashi et al. reported the preparation of TiO_2 nanotubes using the gelation (self-assembly with a rodlike fibrous structure) of an organogelator (It is not ES.) (Kobayashi et al., 2000, 2002). The organogelator is a cyclohexane derivative that was specially

synthesized by this research group. The outer and inner diameters of the TiO_2 nanotubes obtained were 150-600 nm and 50-300 nm, respectively. We considered that the TiO_2 nanotubes would be easily formed by using the ES process without organogelators.
In this study, we formed TiO_2 nanofibers (Nakane et al., 2007) and TiO_2 nanotubes (Nakane et al., 2007) by calcination of new precursor nanofibers of poly(vinyl alcohol) (PVA)-titanium compound hybrids. The precursor nanofibers were formed by using ES with water as a solvent. It is most safety method to use water as a solvent on ES. The photocatalytic reaction using the TiO_2 nanofibers obtained was also investigated.

2. Materals and methods

2.1 Formation of PVA-Ti compound hybrid precursor nanofibers and TiO_2 nanofibers

Two types of precursor nanofibers were formed using ES.
Precursor-1: PVA (degree of polymerization: 1500) 10 wt% aqueous solution was prepared. Titanium lactate (TL) $[(OH)_2Ti(C_3H_5O_2)_2]$ (5 g) was added to the PVA solution (10 g) to produce transparent PVA-TL mixed solution (spinning solution). TL was kindly gifted from Matsumoto Chemical Industry Co., Ltd., Japan (TC-310, content: TL 35-45%, 2-propanol 40-50%, water 10-20%). The mixed solution was loaded into a plastic syringe (2 ml) equipped with a needle. The solution extrusion rate was 0.015 ml/min. A voltage of 25 kV was applied to the needle, and the PVA-TL hybrid nanofibers were then deposited on a collector. The collector (copper plate) was grounded, and the distance between the tip of the needle and the collector was 10 cm. The PVA-TL hybrid nanofibers obtained were used as a precursor of TiO_2 nanofibers.
Precursor-2: Pure PVA nanofibers formed by ES were immersed in a titanium alkoxide [titanium tetraisopropoxide (TTIP)] (10wt%)-ethanol solution for 10 minutes. The treated nanofibers were washed in fresh ethanol and then PVA-TTIP hybrid precursor nanofibers were obtained.
These precursor nanofibers obtained were calcined up to a given temperature in an electric furnace (in air), and TiO_2 nanofibers were formed.

2.2 Apparatus and procedure

The structure of the nanofibers was observed by scanning electron microscope (SEM) (Hitachi S-2400, Japan). Thermogravimetric (TG) analysis was performed in air and a heating rate of 10 °C/min (Shimadzu DTG-60, Japan). X-ray diffraction (XRD) measurement was taken using a CuKα with a Ni filter (40 kV, 30 mA) (Shimadzu XRD-6100, Japan). The nitrogen adsorption isotherms (-196°C) of the TiO_2 nanofibers were measured by Micromeritics TriStar 3000, USA.
The photocatalysis of the TiO_2 nanofibers was evaluated using the photocatalytic decomposition of methylene blue (3,7-bis(dimethylamino) phenothiazin-5-ium chloride; $C_{16}H_{18}ClN_3S$) (Nakane et al., 2007). The TiO_2 nanofibers formed (5 mg) were dispersed in methylene blue (1×10^{-5} mol/l) aqueous solution (50 ml). Three ml of test liquid was taken from this solution and fed in a quartz cell. The test solution was irradiated with white light using an extra-high pressure mercury vapor lamp (Ushio Inc., Japan), and the absorbance at 665 nm, which is the maximum absorption wavelength of methylene blue, was measured by an absorptiometer (JASCO, CT-109, Japan). The decomposition rate of methylene blue was calculated from the absorbance.

3. Results and discussion

3.1 TiO_2 nanofibers obtained from PVA-TL hybrid nanofibers

Fig. 1 shows the SEM image of PVA-TL hybrid nanofibers (precursor-1). The fiber diameters of the PVA-TL nanofibers are 200-350 nm, and the fibers have a smooth surface without macropores. The specific surface area and the pore volume of the hybrid nanofiber are 1.80 m^2/g and 0.00764 cm^3/g, respectively. Thus, the hybrid nanofiber is considered a nonporous material.

Fig. 1. SEM image of PVA-TL hybrid nanofibers (precursor-1).

Fig. 2 shows TG curves of the pure PVA and the PVA-TL hybrid nanofibers. The weight residue of pure PVA becomes zero at 550°C, and that of the PVA-TL hybrid is 25% at 600°C. White residues (TiO_2) were obtained after measurement for PVA-TL hybrid.

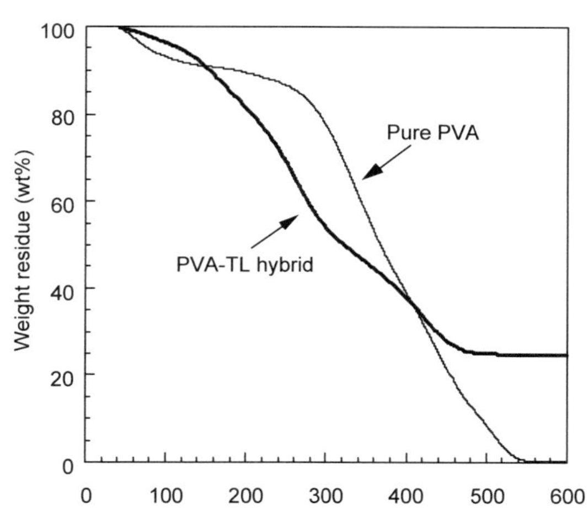

Fig. 2. TG curves of pure PVA and PVA-TL hybrid nanofibers.

Fig. 3 shows the SEM image of the TiO_2 nanofibers obtained by the calcination of PVA-TL hybrid nanofibers at 400°C for 5 hours. Compared to the image shown in Fig.1, the fiber diameter of the TiO_2 nanofibers is 70-80% of that of the PVA-TL hybrid nanofibers, with the space between the fibers made denser by calcination. The residues were brittle, but maintained the shape of the PVA-TL hybrid non-woven mat, although shrinkage occurred due to the calcination.

Fig. 3. SEM image of TiO_2 nanofibers obtained by calcination at 400°C for 5 hours in air.

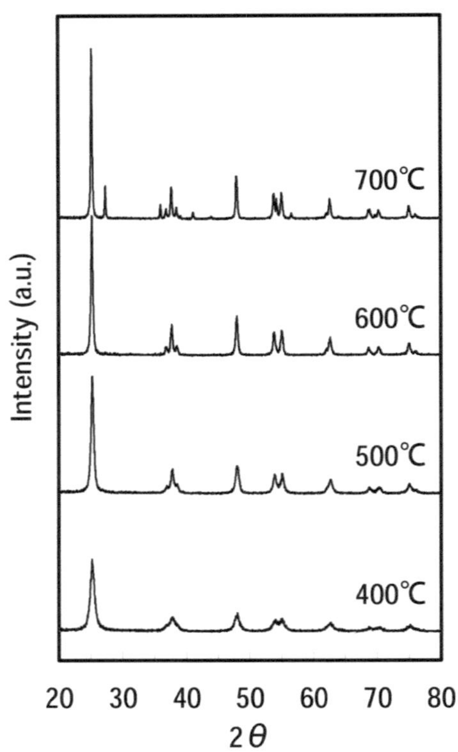

Fig. 4. XRD curves of TiO_2 nanofibers calcined at prescribed temperatures for 5 hours in air.

Fig. 4 shows XRD curves of the TiO_2 obtained by calcination of the PVA-TL nanofibers at 400-700°C for 5 hours. Anatase-type TiO_2 is formed at 400-600°C, and the peak intensities increase with calcination temperature. Rutile-type (rutile-anatase mixed) TiO_2 is formed at 700°C. It is well-known that anatase is superior to rutile for photocatalysis. Thus, a calcination temperature of 600-700°C would be an effective condition when using the TiO_2 nanofibers as a photocatalyst.

Fig.5 shows the relationship of the calcination temperature of the hybrid nanofibers and the pore characteristics (specific surface area and pore volume) obtained from the nitrogen adsorption isothermes (-196°C) of the TiO_2 nanofibers. The specific surface area and pore volume of the TiO_2 nanofibers decreases with increasing the calcination temperature. This is due to the sintering of the TiO_2 by the calcination. The average pore diameters (d_p), which were assumed to be cylindrical in shape, were based on the specific surface area (S) and pore volume (V) for each TiO_2 nanofiber: $d_p=4V/S$. The d_p of the TiO_2 nanotubes were 3.8 nm (calcination temperature: 400°C), 7.4 nm (500°C), 17.5 nm (600°C), and 44.0 nm (700°C). Consequently, the TiO_2 nanofibers obtained in this study are classified as mesoporous materials.

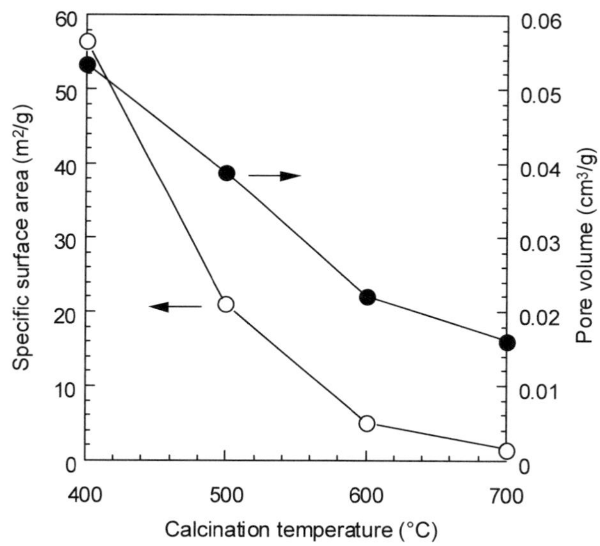

Fig. 5. Effect of calcination temperature upon the pore characteristics of the TiO_2 nanofibers.

3.2 TiO_2 nanotubes obtained from PVA-TTIP hybrid nanofibers

Fig.6 shows the SEM images of (a) pure PVA nanofibers formed by ES, and (b) PVA-TTIP hybrid precursor nanofibers (precursor-2). The fiber diameter of a PVA nanofiber is *ca.* 200 nm, and the fiber diameter of the precursor nanofiber becomes 1.3 times larger that that of the PVA nanofiber. Both fibers have a smooth surface. White residues (TiO_2) were obtained by calcination of the hybrid precursor nanofibers at 500 °C for 5 hours. The sample weight after calcination was 12.1% of the one before calcination.

Fig.7 shows the SEM image of the residue after calcination of the precursor nanofiber at 500°C for 5 hours. As can be seen from this image, hollow TiO_2 nanofibers (TiO_2 nanotubes: outer diameter, *ca.*440 nm, inner diameter, *ca.*270 nm) were formed. The nanotubes obtained are seemed to have a uniform diameter as compared to the diameter range reported by Kobayashi et al. (Kobayashi et al., 2000). We determined that a precursor nanofiber with a

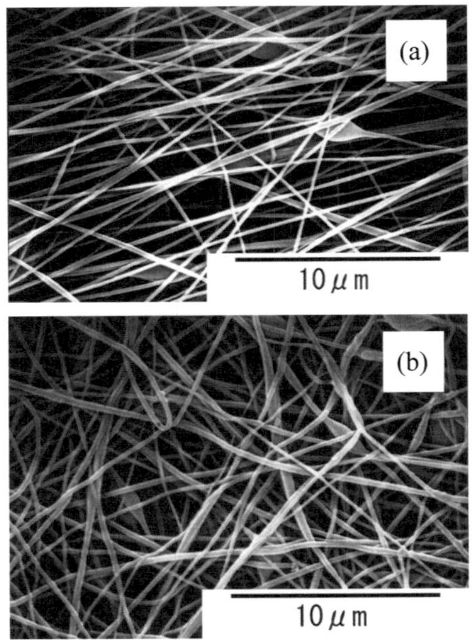

Fig. 6. SEM images of (a) pure PVA nanofibers and (b) PVA-TTIP hybrid nanofibers (precursor-2).

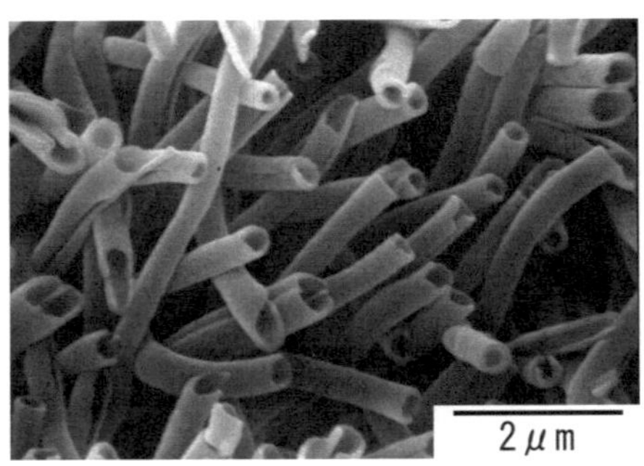

Fig. 7. SEM image of TiO_2 nanotubes formed by calcination of PVA-TTIP hybrid nanofibers.

skin-core structure would be obtained by our method; the skin layer is a PVA-TTIP hybrid, and the core is pure PVA. TTIP will penetrate into the PVA matrix when the PVA nanofibers are immersed in TTIP-ethanol solution, and a PVA-TTIP hybrid layer will be formed. The interaction between the PVA and TTIP could not be identified, but the hybrid would be formed by a coordination bond between the titanium and the oxygen of the hydroxyl group on the PVA molecules (Nakane et al., 2003). The structure of the TiO_2 nanotubes obtained would be reflected in the skin-core structure of the precursor nanofibers. Schematic illustration of the formation of TiO_2 nanotubes by our method is shown in Fig.8.

The diameter of the nanotube can be controlled by changing the diameter of the pure PVA nanofiber. However, nanotubes were not obtained when the outer diameter was *ca.*200 nm (Fig.9 (a), formed of non-hollow nanofibers) or *ca.*700 nm (Fig.9 (b), breaking of TiO_2 tube wall). Thus, the range of the outer diameter which can be formed in this study is 300–600 nm.

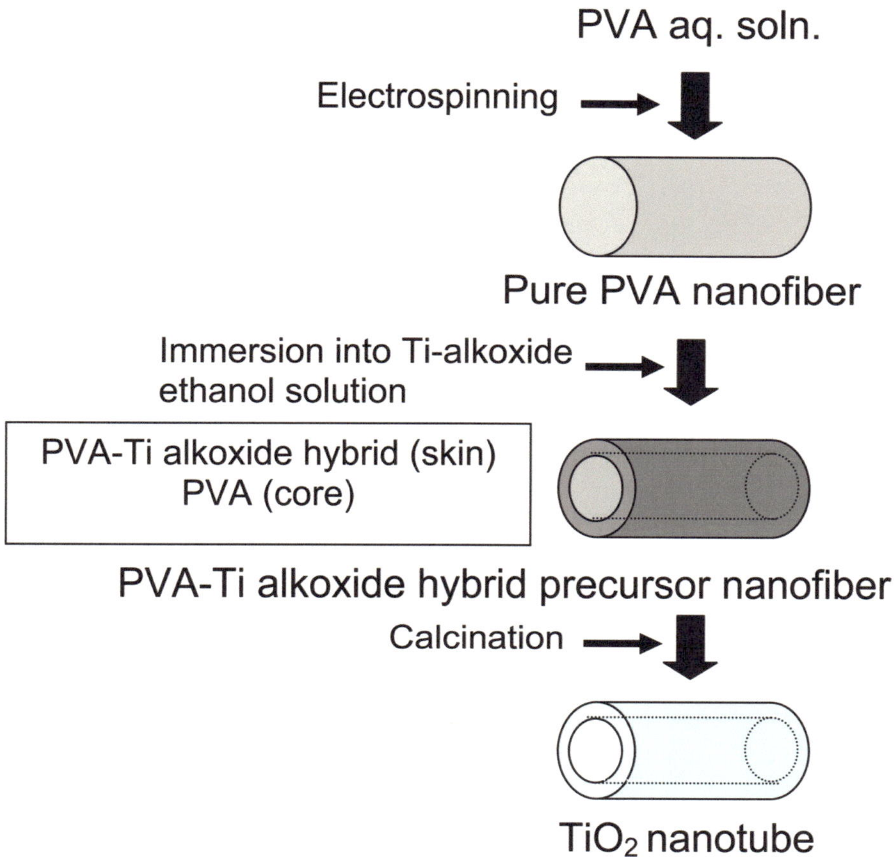

Fig. 8. Schematic illustration of the formation of TiO_2 nanotubes by our method.

Fig.10 shows the XRD curves of the residues (TiO_2) obtained by calcination of the precursor nanofibers at 400-800°C for 5 hours. Anatase type TiO_2 is mainly formed at 400-600°C, and the peak intensities increase with an increase in the calcination temperature. Rutile type (rutile-anatase mixed) TiO_2 is formed above 600°C.

Fig.11 shows the nitrogen adsorption isothermes (-196°C) of the TiO_2 nanotubes (calcination temperature range: 400-700°C). The adsorption amount of the TiO_2 nanotubes decreases with an increase of the calcination temperature. This is due to the sintering of TiO_2. The specific surface areas of the TiO_2 nanotubes were obtained from Fig.10 using the B.E.T. equation. The areas are 75.9 m^2/g (calcination at 400°C), 38.8 m^2/g (500°C), 17.4 m^2/g (600°C) and 6.4 m^2/g (700°C) (the area of pure PVA nanofiber was 3.8 m^2/g). The average pore diameters of the TiO_2 nanotubes were 8.3 nm (400°C), 14.8 nm (500°C) and 21.4 nm (600°C). These pore sizes are not reflected the hollow size of the TiO_2 nanotubes, because the hollow size is several hundred nanometers. Therefore, it is likely that the TiO_2 nanotubes have mesopores on their nanotube wall (Fig.12). The mesopores would be through-holes formed by thermal decomposition of PVA in the precursor hybrid nanofibers. By the presence of the mesopores, the specific surface area of the TiO_2 nanotubes becomes larger, and the photocatalytic reaction using the TiO_2 nanotubes would occur effectively. The TiO_2 nanotube calcined at 700°C had a non-porous wall due to the sintering.

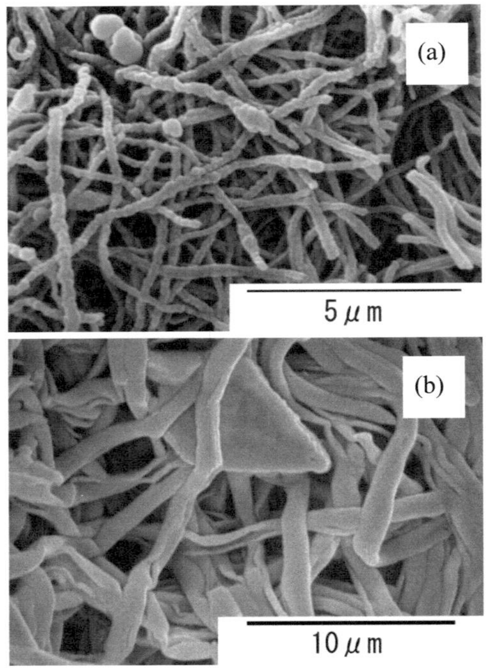

Fig. 9. SEM images of (a) non-hollow TiO_2 nanofibers and (b) wall-broken TiO_2 nanotubes.

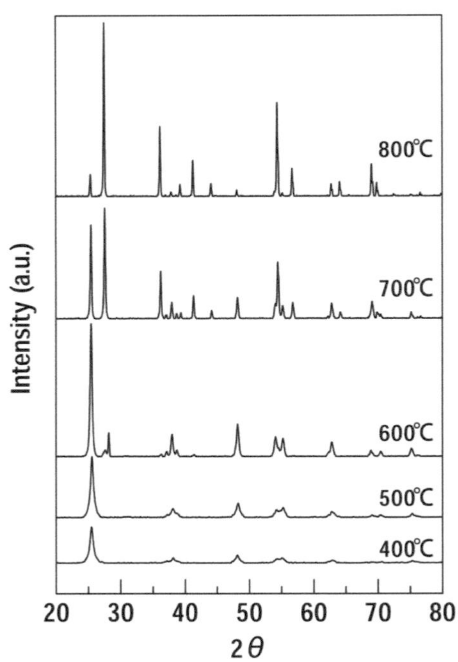

Fig. 10. XRD curves of TiO_2 nanotubes calcined at prescribed temperatures for 5 hours in air.

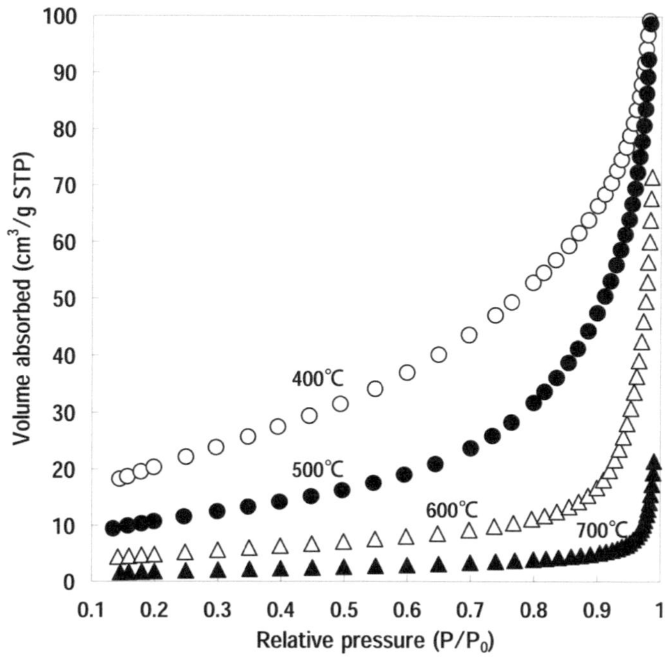

Fig. 11. Nitrogen adsorption isotherms (-196°C) of TiO_2 nanotubes calcined at prescribed temperatures for 5 hours in air.

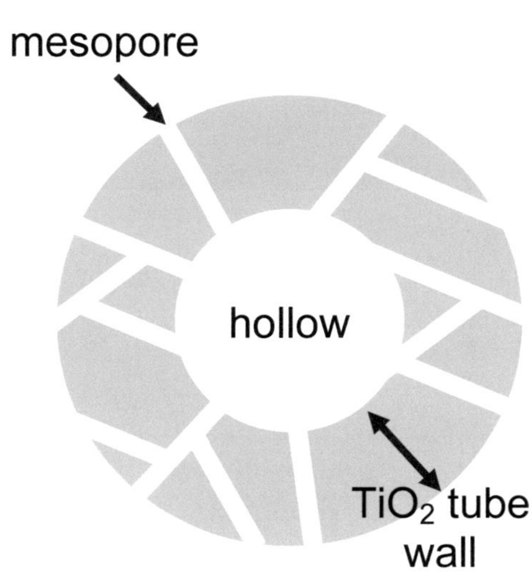

Fig. 12. Schematic illustration of TiO_2 nanotube obtained by our method (cross-section).

3.3 Photocatalysis of TiO_2 nanofibers and TiO_2 nanotubes

The photocatalysis of the TiO_2 nanofibers and nanotubes was investigated. Fig.13 (a) (b) show the relationship between the decomposition rate of methylene blue and the irradiation

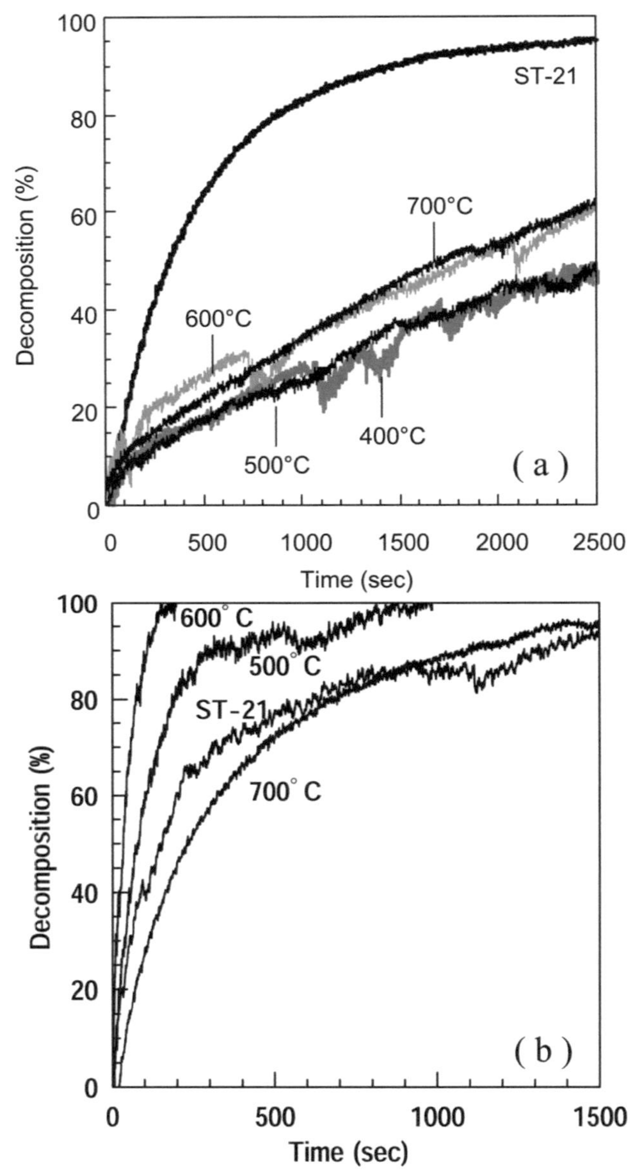

Fig. 13. Effect of irradiation time of white light upon the decomposition rate of methylene blue by the photocatalysis of TiO_2 nanofibers obtained by calcination of (a) PVA-TL and (b) PVA-TTIP hybrid nanofibers at prescribed temperatures.

time of white light for each TiO_2. Each figure includes the result of commercially available anatase type TiO_2 nanoparticles (ST-21, particle size: 20 nm, specific surface area: *ca.*50 m^2/g, Ishihara Sangyo Kaisha, LTD. Japan).

In Fig.13 (a), the photocatalysis of the TiO_2 nanofibers calcined at 600°C and 700°C is higher than that of the TiO_2 nanofibers calcined at 400°C and 500°C, but the major differences between each TiO_2 nanofiber are not observed. The crystallinity of anatase-type TiO_2 increases with calcination temperature, though the specific surface area becomes lower. The photocatalysis of the TiO_2 nanofibers would be affected by both the crystallinity and pore characteristics of the TiO_2 nanofibers. The TiO_2 nanofibers have good photocatalysis, but the properties of these TiO_2 nanofibers are inferior to that of ST-21. The specific surface area of the TiO_2 nanofiber calcined at 400°C (56.4 m^2/g) is higher than that of ST-21, but the ST-21 excels in photocatalysis. This would be due to the difference of the crystallinity of anatase.

In Fig.13 (b), the photocatalysis of the anatase type TiO_2 nanotubes (calcination at 500°C and 600°C) is clearly higher than that of ST-21, and the nanotube (calcination at 600°C) shows the highest photocatalysis in this experiment. Also, the photocatalysis of the rutile-anatase mixed TiO_2 nanotube (calcination at 700°C) is equivalent to that of ST-21. The specific surface area of ST-21 is higher than that of TiO_2 nanotubes, but the TiO_2 nanotubes excel in photocatalysis. At the present stage the reason why is uncertain, but the structure of the nanotube might contribute to the efficient photocatalysis of TiO_2. In other words, electron holes will be formed at the surface of the hollow when the TiO_2 nanotubes are irradiated. And the excited electron moves to inside the hollow because it might be the non-irradiation area (the electron density might be low). Thus the oxidation site and the reduction site would be separated, and the photocatalysis would proceed efficiently.

4. Conclusion

Titanium oxide (TiO_2) nanofibers were formed by calcination of poly(vinyl alcohol) (PVA)-Ti lactate hybrid precursor nanofibers in air. The fiber diameters of the PVA-Ti lactate hybrid nanofibers were 200-350 nm, and the fiber diameters of the TiO_2 nanofibers were 70-80% of those of the PVA-Ti lactate hybrid nanofibers. The specific surface area and average pore diameter of the TiO_2 nanofibers calcined at 500°C for 5 hours were 21.0 m^2/g and 7.4 nm, respectively.

TiO_2 hollow-nanofibers (nanotubes) were formed by calcination of PVA-Ti alkoxide hybrid nanofibers. The outer and inner diameters of the TiO_2 nanotubes calcined at 500°C for 5 hours were *ca.*440 nm and *ca.*270 nm, respectively. The specific surface area of the nanotubes was 38.8 m^2/g, and the existence of mesopores (average pore diameter, 14.8 nm) on the nanotube wall was indicated by the nitrogen adsorption isotherm (-196°C). The photocatalysis of the TiO_2 nanotubes was superior to that of commercially available anatase type TiO_2 nanoparticles.

The TiO_2 nanofibers and nanotubes have the advantage of being easily fixed on other materials, such as a refractory fabric, without a binder by the fiber length (TiO_2 powders such as ST-21 requires a binder in order to be fixed on other materials.). Fig.14 shows the SEM images TiO_2 nanofibers formed on carbon microfibers without a binder. These fibers are expected to be used as an air filter.

Fig. 14. SEM image of TiO_2 nanofibers formed on carbon microfibers (calcination temperature: 500°C).

5. Acknowledgements

The authors express their gratitude to Dr. Shinji Yamaguchi, Dr. Naoki Shimada and Ms. Kaori Yasuda for their helpful cooperation. This work was supported by the Ministry of Education, Culture, Sports, Science and Technology-Japan.

6. References

Buchko, C. J.; Chen, L. C.; Shen, Y. & Martin, D. C. (1999). Processing and microstructural characterization of porous biocompatible protein polymer thin films, *Polymer*, 40, 7397-7407

Chronakis, I. S. (2005). Novel nanocomposites and nanoceramics based on polymer nanofibers using electrospinning process - A review, *J. Mater. Proc. Tech.*, 167, 283-293

Ding, Y.; Zhang, P.; Long, Z.; Jiang, Y.; Xu, F. & Lei, J. (2008). Fabrication and photocatalytic property of TiO_2 nanofibers, *J. Sol-Gel Sci. Technol.*, 46, 176-179

Doshi, J. & Reneker, D. H. (1995). Electrospinning process and applications of electrospun fibers, *J. Electrostatics*, 35, 151-160

Guan, H.; Shao, C.; Liu, Y.; Yu, N. & Yang, X. (2004). Fabrication of $NiCo_2O_4$ nanofibers by electrospinning, *Solid State Commun.*, 131, 107-109

Huang, Z.-M.; Zhang, Y.-Z.; Kotaki, M. & Ramakrishna, S. (2003). A review on polymer nanofibers by electrospinning and their applications in nanocomposites, *Comp. Sci. Tech.*, 63, 2223-2253

Kim, ID.; Rothschild, A.; Lee, BH.; Kim, DY; Jo, SM.; & Tuller, HL. (2006). Ultrasensitive Chemiresistors Based on Electrospun TiO_2 Nanofibers, *Nano Lett.*, 6, 2009 -2013

Kobayashi, S.; Hanabusa, K.; Hamasaki, N.; Kimura, M. & Shirai, H. (2000). Preparation of TiO_2 hollow-fibers using supramolecular assemblies, *Chem. Mater.*, 12, 1523-1525

Kobayashi, S.; Hamasaki, N.; Suzuki, M.; Kimura, M.; Shirai, H. & Hanabusa, K. (2002). Preparation of helical transition-metal oxide tubes using organogelators as structure-directing agents, *J. Am. Chem. Soc.*, 124, 6550-6551

Krissanasaeranee, M.; Vongsetskul, T.; Rangkupan, R.; Supaphol, P. & Wongkasemjit, S. (2008). Preparation of ultra-fine silica fibers using electrospun poly(vinyl alcohol)/silatrane composite fibers as precursor, *J. Am. Ceram. Soc.*, 91, 2830-2835

Kumar, A.; Jose, R.; Fujihara, K.; Wang, J. & Ramakrishna, S. (2007). Structural and Optical Properties of Electrospun TiO_2 Nanofibers, *Chem. Mater.*, 19, 6536-6542

Li, D. & Xia, Y. (2003). Fabrication of titania nanofibers by electrospinning, *Nano Lett.*, 3, 555-560

Li, D. & Xia, Y. (2004). Direct fabrication of composite and ceramic hollow nanofibers by electrospinning, *Nano Lett.*, 4, 933-938

Nakane, K.; Ogihara, T.; Ogata, N. & Kurokawa, Y. (2003). Formation of composite gel fiber from cellulose acetate and zirconium tetra-*n*-butoxide and entrap-immobilization of β-galactosidase on the fiber, *J. Mater. Res.*, 18, 672-676

Nakane, K.; Ogihara, T.; Ogata, N. & Yamaguchi, S. (2005). Formation of lipase-immobilized poly(vinyl alcohol) nanofiber and its application to flavor ester synthesis, *Sen'i Gakkaishi*, 61, 313-316

Nakane, K.; Hotta, T.; Ogihara, T.; Ogata, N. & Yamaguchi, S. (2007). Synthesis of (*z*)-3-hexen-1-yl acetate by lipase immobilized in polyvinyl alcohol nanofibers, *J. Appl. Polym. Sci.*, 106, 863-867

Nakane, K.; Yasuda, K.; Ogihara, T.; Ogata, N. & Yamaguchi, S. (2007). Formation of poly(vinyl alcohol)-titanium lactate hybrid nanofibers and properties of TiO_2 nanofibers obtained by calcination of the hybrids, *J. Appl. Polym. Sci.*, 104, 1232-1235

Nakane, K.; Shimada, N.; Ogihara, T.; Ogata, N. & Yamaguchi, S. (2007). Formation of TiO_2 nanotubes by thermal decomposition of poly(vinyl alcohol)-titanium alkoxide hybrid nanofibers, *J. Mater. Sci.*, 42, 4031-4035

Nuansing, W.; Ninmuang, S.; Jarernboon, W.; Maensiri, S. & Seraphin, S. (2006). Structural characterization and morphology of electrospun TiO_2 nanofibers, *Mater. Sci. Eng. B*, 131, 147-155

Panda, PK. & Ramakrishna, S. (2007). Electrospinning of alumina nanofibers using different precursors, *J. Mater. Sci.*, 42, 2189-2193

Ramakrishna, S.; Fujihara, K.; Teo, WE.; Lim, TC. & Ma, Z. (2005). *An Introduction to Electrospinning and Nanofibers*, 22-62, World Scientific Publishing Co. Pte. Ltd., ISBN 981-256-454-3(pbk), Singapore

Shao, C.; Guan, H.; Liu, Y.; Gong, J.; Yu, N. & Yang, X. (2004). A novel method for making ZrO_2 nanofibres via an electrospinning technique, *J. Cryst. Growth*, 267, 380-384

Yamashita, Y. (2007). Electrospinning -The Latest in Nanotechnology-, 145-146, Sen-I Sya, ISBN 978-4-9902580-1-6, Osaka, Japan

12

Electrochemical and Adsorption Properties of Catalytically Formed Carbon Nanofibers

Liliana Olenic, Stela Pruneanu, Valer Almasan and Alexandru R. Biris
National Institute for Research and Development of Isotopic and Molecular Technologies
Romania

1. Introduction

After the development of high resolution electron microscopy, the carbon structures of nano dimensions could be explained and investigated in detail and have been used in many fields. These different nanoscale carbon structures have remarkable and unique chemical, physical and mechanical properties.

Carbon has long been known to exist as amorphous carbon and in two crystalline allotropic forms: graphite and diamond. Many other carbon based nanomaterials have been developed: fullerenes-discovered by Kroto et al., 1985; carbon nanotubes (CNTs): multi-wall carbon nanotubes (MWCNTs)-recognized discovery attributed to Iijima, 1991 and single-wall carbon nanotubes (SWCNTs) reported at the same time by Iijima & Ichihasi, 1993 and Bethune et al., 1993. The article by Iijima, 1991 which showed that carbon nanotubes were formed during arc-discharge synthesis of C60, has also brought a great interest for carbon nanofibers (CNFs).

The history of carbon nanofibers (nanofibers and nanotubes) also named nanofilaments, goes back in the 19th century. A method for growth of catalytically carbon filaments using iron catalyst and a carbon source gas was first patented by Hughes & Chambers, 1889.

Radushkevich & Lukyanovich, 1952 obtained hollow graphitic carbon fibers that were 50 nanometers in diameter. They were the first who mentioned *carbon nanofibers,* but for a long time these nanostructures were of no industrial importance (Peshnev et al., 2007).

The interest in the structure of these filaments and their properties emerged in the 1970s with the development of transmission electron microscopy, when the proposal of growth mechanism of Oberlin et al., 1976 was reported. They grew nanometer-scale carbon fibers by chemical vapour deposition (CVD). Tennent, 1987, presented a U.S. patent for graphitic, hollow core "fibrils".

The recent increasing scientific and industrial interest in carbon nanofilaments as one-dimensional nanomaterials, originates from their unusual application properties and similarities with carbon nanotubes.

Carbon nanofibers have been extensively studied: their synthesis and growth mechanism (Oberlin et al., 1976; Tibbetts et al., 1993, 1994; De Jong & Geus, 2000; Helveg et al., 2004; Cui et al., 2004a), their structure (Endo et al., 2002, 2003; Paredes et al., 2005; Eksioglu& Nadarajah, 2006; Lawrence et al., 2008;) and properties (Endo et al., 1993, 1995; Kavan & Dunsch, 2008; Charlier et al., 2008; Damnjanovic et al., 2005). CNFs have been recognized as a very promising material based on their nanostructure and properties.

In this chapter we focus on the electrochemical and adsorption properties of carbon nanofibers prepared by catalytic chemical vapour deposition (CCVD) method, underlying the results obtained by the authors.
We first present some of up to date literature concerning the CVD synthesis of CNFs. Thereafter, we resume the electrochemical and adsorption properties of the as prepared CNFs and we also show our own results. Our studies have evidenced that among all carbon nanostructures prepared by us, carbon nanofibers showed the best electrochemical characteristics. As a consequence, CNFs were successfully used as support for biologically active substances (amino acids, glucose oxidase, DNA). The application of nanofibers in the sensors area was also described. CNFs prepared by CCVD method have been successfully used for the construction of second-generation glucose biosensors. The enzyme and the redox mediator were easily co-immobilized on the surface of carbon nanofibers due to its high specific area. The linear response range of this glucose biosensor was between 1.7 and 7mM while the time required to reach 95% of the steady state, was around 30 seconds. We have used in an original manner an amperometric method to detect the changes in the specific activity of GOx, immobilized longer time on CNFs.
Finally, we summarize issues with respect to the research goals to be dealt with, in future work.

2. Application fields of carbon nanofibers

The application area of CNFs depends very much on the synthesis conditions, which strongly influence the properties of these nanostructures.
CNFs are the subject of extensive experimental and theoretical studies for specific applications, such as: adsorbent (including hydrogen storage material for fuel cells, lithium-ion secondary batteries and supercapacitors), catalysts or catalyst supports, polymer additives, template for fabrication of various nanostructures (Ju et al., 2008; Van der Lee et al., 2005; Kymakis & Amaratunga, 2002; Li et al, 2006). CNFs are of great interest for the development of nanoelectronics components (field effect transistors, diodes, electrochemical capacitors, electron sources) or analytical sensors (Baughman et al, 2002; Liu & Hu, 2002).
Carbon nanostructures such as vertically aligned carbon nanofibers (VACNFs) and nanocones produced by plasma enhanced CVD (PECVD) are nanomaterials of great interest due to their potential applications in areas such as: tips for scanning microscopy (Cui et al., 2004 b), field-emission devices (Fan et al., 1999), biological probes (Guillorn et al., 2002), interconnects for nanoelectronics and memory devices (Grobert et al., 1999). The incorporation of VACNFs as a nanostructured material into multiscale devices has often enhanced the performance of the combined system (Baker et al., 2005). Trace analysis and DNA hybridization detection with VACNF electrodes have been demonstrated by Koehne et al., 2004. Nanofiber arrays have been incorporated as vertically oriented diffusion barriers in microfluidic devices, to mimic cell functionality (Fletcher et al., 2004). The high aspect ratio and mechanical stability of VACNFs has also been proved useful for the parallel delivery of molecular species, including DNA, cells and tissues (McKnight et al., 2004).
The interest in nanosystems for biological applications is continuously growing, especially for fabrication of nanosensors, molecular probes, miniaturized biomedical devices and bioreactors (Huang et al., 2002; Hu et al., 2004; Brown et al., 2008). The adsorption of biological molecules on different carbon nanostructures may offer the possibility of

fabricating biosensors on a nanometer scale. Applications of nanotubes such as drug delivery into a single cell have been referred to (Kam et al., 2004).
Efficient adsorbents research for enzymes and microorganisms are very important for development of modern bioprocesses of hydrolysis, oxidation and isomerization. Highly stable heterogeneous biocatalysts were prepared by immobilization of enzymatic active substances on inorganic supports (enzymes-in particular glucoamylase; intracellular compartments and whole cells of microorganisms) (Kovalenko et al., 2009 a).
Multiple and specific applications require optimum preparation methods of CNFs. It is very important to identify and control the critical parameters for the optimization of the synthesis process and the application of nanocarbon products.

3. Synthesis methods of carbon nanofibers

Carbon nanofilaments have been synthesized by various methods, for example laser vapourization (Baker et al., 1997), arc discharge (Iijima, 1991), catalytic chemical vapour deposition (Zheng et al., 2004) and plasma-enhanced chemical vapour deposition. Several PECVD methods developed by Ren et al., 1999 have been used for growth of nanofibers, including microwave discharges (Woo et al., 2002), direct current (dc) or radio frequency (rf) glow discharges (Merkulov et al., 2002) and inductively coupled discharges (Delzeit et al., 2002).

Catalytic Chemical Vapour Deposition (CCVD) method

The most developed method for the synthesis of CNFs is CCVD method. The advantage of the method consists in the possibility to control the morphology and structure of nanocarbon products, to improve the alignment of nanofilaments and to obtain large amounts with high purity and low costs for all kind of applications. A variety of CVD processes have been used for carbon nanofilaments synthesis, which include catalytic thermal CVD, plasma-enhanced CVD, alcohol catalytic CVD, aerogel-supported CVD, laser-assisted CVD (Govindaraj & Rao, 2006) and thermal gradient CVD (Ling-Jun et al., 2009). Lower temperature for CNFs growth using heterogeneous metal catalysts, was suggested by other researchers (Poirier et al., 2001).
There are numerous experimental parameters that can be adjusted during the synthesis process of CNFs, by CCVD. By selecting the metal catalyst, carbon precursors (sources) and reaction conditions (thermal energy) one can control the structure, morphology and their relating properties. Therefore, right combination of these three components makes it possible to selectively synthesize various types of carbon nanofilaments, ranging from SWCNTs and MWCNTs to CNFs.
The synthesis of carbon nanofilaments by CCVD method is based on the catalytic decomposition of a gaseous or volatile compound of carbon source (methane, carbon monoxide or acetylene, C_2H_4, methanol/ethanol, benzene (Devaux et al., 2009), on a variety of transition metals (usually iron, cobalt, nickel and their alloys; palladium is rarely employed as a catalyst for solid carbon deposition-Atwater, et. al., 2009), either in a powdered or supported form as the catalytic entities (which also serve as nucleation sites for the initiation of nanocarbon growth), over the temperature range 400-1000^{0}C. The carrier gas is argon, hydrogen, nitrogen. Sometimes hydrogen is added to reduce the metallic oxides to metal.
Lupu et al., 2004a used the CCVD method in which the outer furnace was replaced by a high frequency induction heating. Various types of CNFs were obtained by using different

catalysts. This mode combined with the CCVD method allows a significant decrease of energy consumption and a shorter reaction time as compared with the heating mode with outer furnace. CNFs have been synthesized by decomposition of pure ethylene over Fe:Ni:Cu catalyst in a horizontal furnace. The catalyst was prepared from nitrate solutions by co-precipitation with ammonium bicarbonate and was calcined at 400^0 C for 4 h. The carbonaceous products were purified by extraction in HCl (37%) for 24 h, washed with distilled water, and dried at 150^0C for 3 h. A typical transmission electron microscope (TEM) image (Figure 1) of the sample shows nanofibers with "herringbone" structure and diameters ranging from 80 to 290 nm, similar to those reported in the literature. Their specific area was determined by the BET method and the value was between 170-242 m^2g^{-1}. The CNFs have been characterized by cyclic voltammetry and their adsorption properties for biologically active substances have been closely followed (Pruneanu et al., 2006; Olenic et al., 2009).

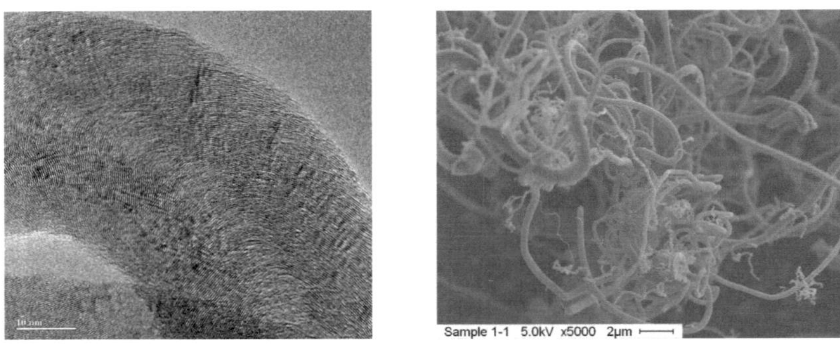

a b

Fig. 1. (a) HRTEM image of CNF (from ethylene at 600^0 C on Fe:Ni:Cu as catalyst); (b) SEM image of CNFs. Reprinted from ref. Olenic et al., 2009 with kind permission of Springer Science and Business Media.

In the synthesis of nanocarbon structures by CCVD method, the critical step is the catalyst preparation. Metal nanoparticles catalyst (optimum size between 0.4–5 nm) favours the catalytic decomposition of the carbon source gas in a temperature range of 600–1100^0 C. As was shown in the literature, the amorphous carbon is deposited from the thermal decomposition (pyrolysis) of the carbon source gas, whereas the carbon nanofibers are grown from the catalytic decomposition of the carbon source gas (Teo et al., 2003).
According to the growth procedure, CVD method includes the seeded catalyst method (Li et al., 1996) which uses the catalyst seeded on a substrate within a reactor (in this case the interactions between the catalyst and support (alumina, silica, silicon) dictates the growth mode (Randall et al., 2001); an advantageous one is the floating catalyst method which is a method wherein the carbon vapour and the catalytic metal particles both get deposited in the reaction chamber, without a substrate. (Martin-Gullon et al., 2006).
One of the CVD methods that has been developed is the synthesis of vertically aligned nanofibers bundles for specific applications. The synthesis of VACNFs arrays were all carried out in horizontal reactors (Cao et al., 2001). All the reported products by vertical floating catalyst method were randomly arranged CNFs (Perez-Cabero et al., 2003). There are few reports on aligned CNF bundles synthesized by floating catalyst procedure, in vertical reactors (Cheng et al, 2004).

VACNFs were also obtained by low-pressure inductively coupled PECVD (Caughman et al., 2003); isolated VACNFs were synthesized by Melechko et al., 2003.

When CNFs are prepared, crystallized structures are generally desired (amorphous carbon-free). The growth temperature affects the crystallinity: a too high temperature leads to the formation of pyrolytic amorphous carbon. This is the reason for preferring the highest deposition temperature without significant self-decomposition of the carbon source gas.

The growth mechanism leading to the formation of CNFs (reviewed by Teo et al., 2003) has been studied by many groups. Baker et al., 1972 proposed a growth mechanism for both nanofibers and nanotubes, which was later completed. Other models for growing CNFs were proposed by Oberlin et al., 1976, Koch et al., 1985, Zheng et al., 2004. Formation mechanism of large branched carbon nano-structures has been presented by Devaux et al., 2009.

Examination of synthesized CNFs by TEM and SEM reveals the basic microstructure of graphitic CNFs. There are two types of carbon nanotubes: single-wall and multi-wall and four types of carbon nanofibers that consist of stacked graphite layers, which can be arranged parallel (tubular-adopting the structure of a "multi-walled faceted nanotube"), perpendicular to the fiber axis (platelet-adopting the arrangement of a "deck of cards"), or herringbone structure (the graphite platelets are at a particular angle to the fiber axis), and amorphous type without crystalline structure. Most of carbon nanofibers and nanotubes synthesized by CCVD method are crystalline or partially crystalline and only a few of them are amorphous.

The herringbone structure seems to be favoured when the catalyst is an alloy. Herringbone-type CNFs with large diameter and a very small or completely hollow core have been synthesized through a CVD method (Terrones et al., 2001).

The only difference among the various forms of carbon nanofilaments is their chemical structure. Martin-Gullon, et al., 2006, present in detail a classification of nanofilaments depending on their structure.

The properties related to the morphology of CNFs depend on many factors, like: the chemical nature of the catalyst and the conditions of its pretreatment (Huang et al., 2009; Kovalenko et al., 2009 b), the composition and flow rate of a gas mixture and the temperature and duration of the synthesis (Endo et al., 2003; Chuang et al., 2008).

On the other hand, the electrical and optical properties of carbon nanostructures are largely dependent on their structures (Kataura et al., 1999; Yang et al., 2003).

The conducting properties of CNFs that can be varied from metal to semiconductor (depending on the structural parameters and doping with heteroatoms) are very important for practical applications (Ismagilov, 2009).

All CNFs products obtained by CCVD method contain impurities such as metal catalyst particles, amorphous carbon and carbon nanoparticles depending on the reaction conditions. Therefore, purification of carbon nanostructures is of great importance for technological applications.

A purification step is usually required before carbon nanofilaments can be used, especially for biomedical applications. Several purification methods are reported in the literature (Liu et al., 2007). Graphitization (or heat treatment) is one of the most effective methods to remove defects or impurities such as metallic compounds, which diminish the electrical and mechanical properties of conventional carbon nanofibers.

Huang et al., 2009 demonstrated that high purity CNFs can be formed by varying the synthesis temperature. Different types of CNFs were characterized by various techniques to understand their crystal structure, morphology, graphitization degree and thermal stability.

For more complex applications of carbon nanotubes, different functionalization methods have been introduced. Investigation of the interaction between carbon nanotubes and biological molecules are very important (Zhong et al., 2009).
McKnight et al., 2006 showed several approaches toward such site-specific functionalization along the nanofiber length, including physical and electrochemical coating techniques, chemical immobilization of DNA and enzyme species, and covalent attachment of biotin followed by affinity-based capture of streptavidin-conjugated molecules.

4. Electrochemical properties of carbon nanofibers

For many electrochemical applications, carbon is a well known material of choice. Among its practical advantages are: a wide potential window in aqueous solution, low background current, lack of corrosion processes at positive potentials and low costs.
The advantages of CNFs in the construction of biosensors, relate to their small size with large specific area, the promotion of electron transfer when used in electrochemical reactions and easy bio-molecules immobilization. DNA molecules can be covalently bound on the functionalized fiber surface (e.g. with carboxylic groups). In comparison with the classical carbon electrodes, CNFs show better electrodic behaviour including good conducting ability and high chemical stability. The electrochemical properties of CNFs paste electrodes have been largely studied. In most cases, CNFs were prepared as composite electrodes.
It is of interest to explore the properties of carbon nanocomposite electrodes to see if they might exhibit new properties, due to the high edge/surface area ratio of such materials.
Marken et al., 2001 have evaluated CNFs (obtained by ambient pressure CVD method) as novel electrode materials for electrochemical applications (porous, pressed onto a glassy-carbon substrate and non-porous, embedded in a solid paraffin matrix). They exhibit low BET surface areas and high electrochemical capacitances due to the fact that the spaces between the fibers allow the penetration of electrolyte solution. Capacitive currents tend to mask voltammetric currents during cyclic voltammetry. By comparison, when the spaces between CNFs are impregnated by an inert dielectric material (paraffin wax) the electrode has good conductivity and low capacitance. These materials were compared with other forms of nanostructured carbons: aerogel or activated charcoal.
Van Dijk et al 2001 prepared nanocomposite electrodes made of CNFs and black wax and used them for anodic stripping voltammetry of zinc and lead.
Maldonado et al., 2005 have prepared nondoped and nitrogen-doped (N-doped) CNFs films by the floating catalyst CVD method using precursors consisting of ferrocene and either xylene or pyridine to control the nitrogen content. CNF coated nickel-mesh was used as working electrode, to study the influence of nitrogen doping on the oxygen reduction reaction. The electrodes have significant catalytic activity for oxygen reduction in aqueous solutions (neutral to basic pH).
Yeo-Heung et al., 2006 tested the electrochemical actuation properties of carbon nanofiber–polymethylmethacrylate (CNF–PMMA) composite material. They characterized the CNF-PMMA actuator by impedance spectroscopy, at voltages up to 15V. The relationship between displacement and applied voltage was determined.
Roziecka et al., 2006 prepared ITO electrodes modified with hydrophobic CNFs–silica film, which was employed as support for liquid/liquid redox systems. The redox processes within the ionic liquid is coupled to ionic transfer processes at the ionic liquid/water

interface. Therefore, the CNFs electrode material was an excellent support for recording both the Faradaic and capacitive currents. The efficiency of the electrode process increases due to the use of the heterogeneous matrix.

Our group has studied the electrochemical properties of carbon nanofilaments (CNFs, MWCNTs and SWCNTs- unpublished data). Paste electrodes were prepared by mixing the carbon powder with silicon oil and then packing the resulting paste into the cavity of a PVC syringe (2.5 mm diameter). The electrical contact was ensured by a Pt wire, tightly inserted into the paste.

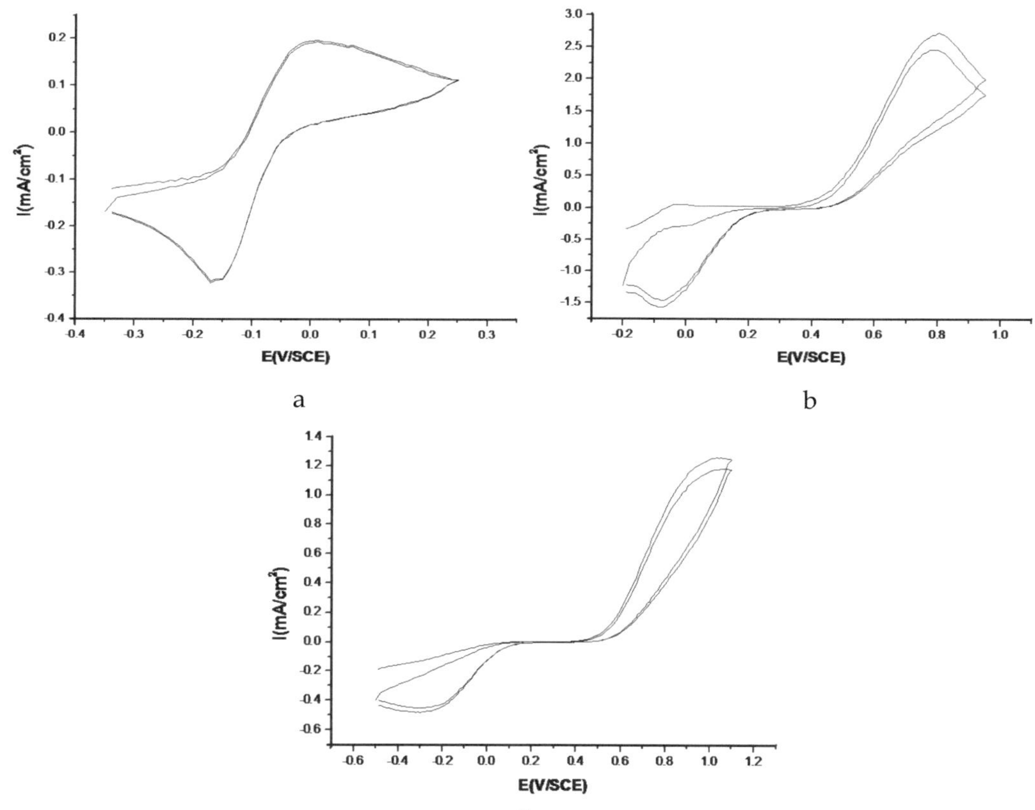

Fig. 2. Cyclic voltammograms recorded in solution of 10^{-2} M hydroquinone and 0.5M KCl for: a) CNFs; b) MWCNTs; c) SWCNTs paste electrode; all voltammograms were recorded with a sweep rate of 100 mVs^{-1}.

The electrochemical behaviour of these types of electrodes was investigated by cyclic voltammetry (100 mVs^{-1} sweep rate) using as redox mediator a solution of 10^{-2}M hydroquinone (Figure 2 a,b,c). From Figure 2a one can see that carbon nanofibers showed the best electrodic properties. The voltammograms exhibit two well-defined peaks, with the peak potential separation, ΔEp, around 150 mV. This value is higher than that generally obtained for a reversible redox system (60 mV/n, where n is the number of electrons transferred during the reaction).

For MWCNTs and SWCNTs paste electrode, the peak potential separation, ΔEp is considerable larger (850 mV and respectively 1100 mV), indicating a lower conductivity and a slow transfer of electrons.

Due to the excellent electrodic properties of CNFs paste electrode, Pruneanu et al., 2006 have studied the oxidation of calf thymus DNA. The interest in this kind of research is due to the fact that the electrochemical oxidation may mimic the biological oxidation mechanism, involving enzymes. All the four bases of DNA can be chemically oxidized; electrochemically, only guanine and adenine oxidation peaks can be recorded (thymine and cytosine have oxidation potentials larger than 1.2V vs. Ag/AgCl). In order to establish the exact position of purine oxidation potentials (adenine and guanine) the authors have registered differential pulse voltammetry (DPV) curves, in solution containing 10^{-3} M adenine hemisulphate and 10^{-3} M guanine hemisulphate (in 0.1M PBS pH 7+ 0.5M KCl, Figure 3). The two peaks that appeared around 0.9V vs. Ag/AgCl and 1.18V vs. Ag/AgCl were ascribed to guanine and adenine oxidation, respectively. The intensity of the peaks decreased after successive recordings, due to the irreversible character of the oxidation process.

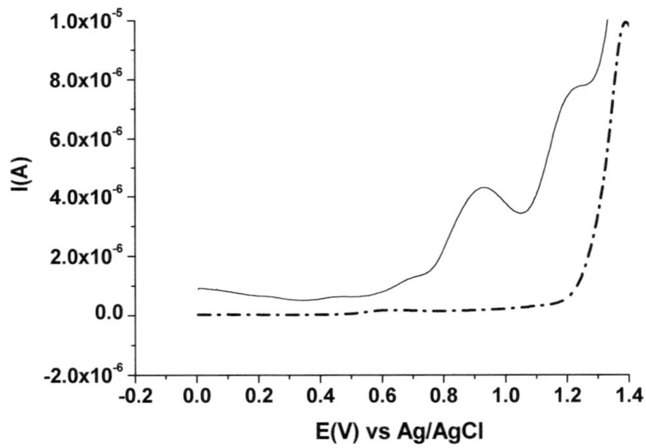

Fig. 3. DPVs recorded in a solution of 10^{-3} M adenine hemisulphate and 10^{-3} M guanine hemisulphate, in 0.1M PBS (pH 7) + 0.5M KCl.

The signals obtained from guanine or adenine oxidation can be used for the construction of a DNA biosensor. In Figure 4 one can see that the oxidation peak of adenine hemisulphate increases with the increase of solution concentration (10^{-7}....10^{-3} M).

Oxidation of calf thymus DNA (single stranded or double stranded DNA) at carbon nanofibers paste electrode was also studied by DPV (Figure 5). Prior experiments, calf thymus DNA was physically adsorbed on the electrode surface, by immersing it in DNA solution for about five minutes, under constant stirring. The two peaks corresponding to guanine and adenine oxidation were clearly recorded for single stranded DNA (Figure 5, straight line). In contrast, no signal was obtained when double stranded DNA was adsorbed at the electrode surface (Figure 5, dashed line). This may be explained by the fact that in double stranded DNA the purine bases are hidden between the double helix, so they have no free access to the electrode surface. In this case the transfer of electrons cannot take place.

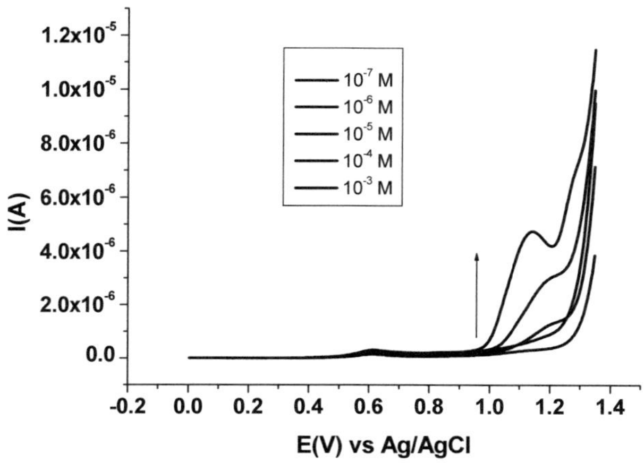

Fig. 4. DPVs recorded in solutions of adenine hemisulphate of different concentration: 10^{-7}.... 10^{-3} M in 0.1M PBS (pH 7) + 0.5M KCl.

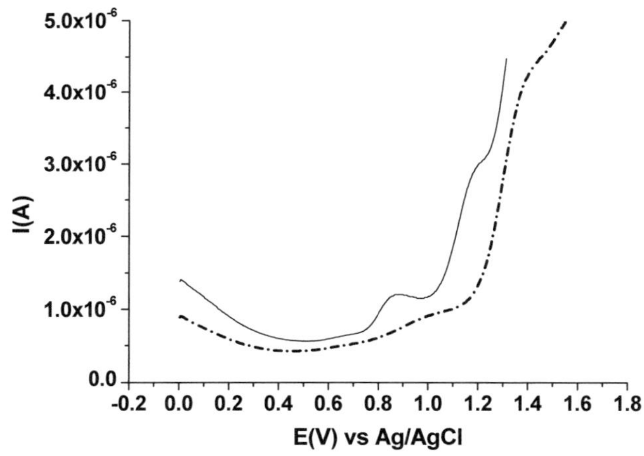

Fig. 5. DPVs of single-stranded DNA (straight line) and double-stranded DNA (dashed line) in solution of 0.1M PBS (pH 7) + 0.5M KCl (0.3 mgml^{-1} DNA)

Zhang et al., 2004 performed *I* –*V* measurements on individual VACNFs. They fabricated multiple Ti/Au ohmic contacts on individual fibers, having the contact resistance of only few kOhm. The measurements demonstrated that VACNFs exhibit linear *I* –*V* behaviour at room temperature. Between intergraphitic planes in VACNFs exists a dominant transport mechanism of electrons, along the length of the fiber.

VACNFs are increasingly used in bioelectrochemistry, due to the fact that they exhibit fast electron transfer to redox species from solution, or act as highly conducting substrates to connect redox enzymes to macro-sized electrodes. Their chemical stability combined with a high degree of biologically accessible surface area and nanoscale dimension make VACNFs ideal substrates for the development of scaffolds in biological detection. Additionally, their mechanical strength and narrow diameter allow easy cell penetration, making them suitable for intracellular electrochemical detection.

Baker et al., 2006a demonstrated the ability to use VACNFs as electrodes for biological detection. He also emphasized the importance of the surface functionalization, in order to control the overall electrochemical response. Functionalized VACNFs with the redox active protein *cytochrome c* were characterized by cyclic voltammetry (CV) measurements. Although the high surface area of the nanofibers allows the *cytochrome c* molecules to produce an increase of the electrochemical current, the high capacitive currents partially obscured this signal and partially offset the potential improvement in the signal-to-noise ratio.

VACNT arrays were successfully grown on planar graphitic carbon substrates, using a bilayer Al/Fe catalyst and water-assisted thermal CVD. Excellent voltammetric characteristics were demonstrated after insulating the arrays with a dielectric material (Liu et al., 2009).

A method for the development of an amperometric biosensor for interference-free determination of glucose was reported by Jeykumari & Narayan, 2009. The bienzyme-based biosensor was constructed with toluidine blue functionalized CNTs. The electrochemical behaviour of the sensor was studied by impedance spectroscopy, cyclic voltammetry and chronoamperometry. The excellent electrocatalytic activity of the biocomposite film allowed the detection of glucose under reduced over potential, with a wider range of determination and with a very good detection limit. The sensor showed a short response time, good stability and anti-interferent ability. The proposed biosensor exhibits good analytical performance in terms of repeatability, reproducibility and shelf-life stability.

Sadowska et al., 2009, functionalized MWCNTs with azobenzene and anthraquinone residues (chemical groups with redox activity) for potential application in catalysis and memory storage devices. Using the Langmuir–Blodgett method, the nanotubes containing electroactive substituents were transferred onto electrode substrates and characterized by cyclic voltammetry. The amount of electroactive groups per mg of nanotubes was calculated based on the cathodic current peak. A highly reproducible voltammetric response was obtained with a single nanotube layer or multiple nanotube/octadecanol layers. It is believed that such devices will be invaluable for future high-performance electrodes.

Minikanti et al., 2009 designed implantable electrodes as targets for wide frequency stimulation of deep brain structures. They have demonstrated by cyclic voltammetry and impedance spectroscopy, the enhanced performance of implantable electrodes coated with multi-wall carbon nanotube. The results were compared with those obtained for the more traditional stainless steel. They also investigated the surface morphology of aged electrodes due to the fact that implantable electrodes have to be mechanically stable and present high shelf life. The effect of superficial oxygen adsorption on the aged MWCNTs electrodes was observed through a modified cyclic voltammetric spectrum.

In the past few years, considerable interest was focused on the application of carbon based nanomaterials as electrodes for supercapacitors, due to their chemical inertness and easy processability. The capacitive behaviour of the CNFs was studied in term of charge-discharge curves and cyclic voltammetry.

Recently, carbon nanomaterials with various morphologies (carbon nanotubes, nanofibers, nanowires and nanocoils) have been intensively studied as negative electrode materials in

lithium-ion batteries (Zou et al., 2006). These nanofibers have low graphitic crystallinity. The experimental results showed that CNF electrodes had high reversibility with small hysteresis, in the insertion/extraction reactions of lithium-ion.
All these studies suggest that CNFs represent a new class of materials suitable for electrochemical applications.

5. Adsorption properties of carbon nanofibers

The biologically active substances can be attached to CNFs surfaces by physical adsorption (physisorption) or chemical immobilization.
For a long time, activated carbons (ACs) materials containing large surface area and well-developed porosity were successfully applied in various industrial processes including adsorption (gases and liquids), mixture separation, filtration, etc.
CNFs and activated CNFs have special properties, compared with activated carbon. Among these, we mention the high chemical reactivity due to the large fraction of active sites, available for chemical and physical interaction with different species.
Baker, 2007 noticed the use of nanofibers as adsorbents. He additionally emphasized that the functionality of carbon nanofiber surface has an important role. The raw graphitic materials are free of surface oxygen groups and therefore are hydrophobic in nature. CNFs surface can have a hydrophilic character after a normal activation procedure. The control of the acid-base properties of carbon nanofibers surface has an important impact on a variety of potential applications. The structural characteristics e.g. the infinite number of graphite layers and the weak Van der Waals forces are responsible for the high adsorption capacity observed for these nanostructures.
Bououdina et al., 2006 presented a review on hydrogen absorbing materials. The hydrogen is theoretically adsorbed on the surface of CNFs and then incorporated between the graphitic sheets. The structure of CNFs allows the physisorption of large amounts of hydrogen. The used catalyst was unsupported NiO powder. As regarding the catalyst, they noticed that at low temperatures (400^{0}C) Ni_3C is formed while metallic Ni is formed at high temperatures (500^{0}C). The usage of high temperature (700^{0}C) and Ni catalyst favour the formation of crystalline structure. The Ni_3C phase leads to the formation of herringbone structure while Ni favours the formation of platelet structure. They also noticed that at low temperature, the surface area of as-prepared CNFs increased about three times. The microstructural modifications of obtained carbon nanostructures bring great benefits, by correlating the catalytic phases (Ni_3C or Ni metal) with hydrogen uptake.
Lupu et al., 2004 b used palladium catalyzed CNFs for hydrogen adsorption.
CNFs based electrodes, grown into a porous ceramic substrate, show promising properties for applications in electrochemistry. Some aromatic compounds (hydroquinone, benzoquinone, and phenol - Murphy et al., 2003) are strongly adsorbed on the surface of carbon nanofiber composite electrode. The composite electrode has a high surface area due to the carbon nanofiber and shows promising properties for applications in electroanalysis.
Diaz et al., 2007 evaluated the performance of different nonmicroporous carbon structures (multi-wall carbon nanotubes, nanofibers, and high-surface-area graphites) as adsorbents for volatile organic compounds, hydrocarbons, cyclic, aromatic and chlorinated compounds. The evaluation was based on the adsorption isotherms, the values of heats of adsorption and values of free energy of adsorption. They observed that the adsorption of n-alkanes and

other polar probes on CNTs is less energetically favorable than the adsorption on flat graphite.
Cuervo et al., 2008 have evaluated the effect of the chemical oxidation, on the adsorption properties of CNFs. They discussed the adsorption of n-alkanes, cyclohexane and chlorinated compounds. They showed that the adsorption is a complex process, where morphological aspects are playing a key role. Both the capacity and adsorption strength decreased after the oxidative treatment of carbon nanofibers, especially in the case of chlorinated compounds. There is steric limitation in the adsorption process, after oxidation of nanofiber. In the case of aromatic compounds, the steric limitation is compensated by the interaction of aromatic rings with surface carboxylic groups. The absence of nucleophilic groups in the chlorinated compounds hinders their adsorption on the activated nanofibers.
Kovalenko et al., 2001 investigated the adsorption properties of catalytic filamentous carbon (CFC) with respect to biological adsorbates, like: L-tyrosine, bovine serum albumin, glucoamylase and non-growing bacterial cells of Escherichia coli, Bacillus subtilis and Rhodococcus sp. They have studied the influence of the surface chemical properties and textural parameters of CFC, on the adsorption. They used three independent methods for the calculation of the value of *accessible surface area*: comparative method, fractal method and external geometrical surface of granules. The conclusion was that the adsorption of biological adsorbates is mainly influenced by the *accessible surface area*. The roughness of the surface also affects the efficiency of the adsorption/desorption of bacterial cells.
Wei et al., 2007 presented in a review the biological properties of carbon nanotubes (the processing, chemical and physical properties, nucleic acid interactions, cell interactions and toxicological properties). The unique biological and medical properties of carbon nanostructured are of great interest in the last years. Finally, future directions in this area are discussed.
Li et al., 2005 prepared herringbone nanofibers that were subsequently oxidized, in order to create carboxylic acid groups on their surface. After that, they were functionalized with reactive linker molecules derived from diamines and triamines.
Surface functionalization is an important step to enhance wettability, dispersibility and surface reactivity of carbon nanostructures to help incorporation into composites and devices. There are two known strategies currently employed to modify carbon nanostructures surface: covalent functionalization and non-covalent wrapping of carbon nanostructures with surfactants, polymers or ceramic coatings.
The successful surface functionalization of vapour-grown carbon nanofiber materials has been extensively reported in literature. In particular, those having the platelet or herringbone structures are especially suitable for surface functionalization, due to the presence of edge-site carbon atoms.
A great advantage of carbon nanofibers is their compatibility with physiological cells and tissues; additionally, these fibers have excellent conductivity and high strength to weight ratios. The high conductivity is a promising property for electrical stimulation of neuronal cells and can be beneficial for studying the nerve functions and regeneration. The excellent electrical and mechanical properties of carbon nanofibers lead to promising potential applications as central and peripheral neural biomaterials (McKenzie et al., 2004).
Many supports as powders, beads or chips (polymers and resins, silica and silica-alumina composites and carbonaceous materials) have been studied for enzyme immobilization.

Immobilized enzymes are used as catalysts in fine chemicals and chemicals production. The immobilization of the enzymes on support brings important advantages over dissolved enzymes, e.g. the possibility of recovery and reuse, simple operation and improved stability. De Lathouder et al., 2004 functionalized ceramic monoliths with different carbon coatings and the biocatalyst (enzyme lipase) was adsorbed on the supports. They found that CNFs support have the highest adsorption capacity, preserve the activity of enzyme and have the highest stability during storage. The pore volume, surface area and the nature of surface groups of the supports influence the adsorption process of the different carbon types.
To investigate the interaction between carbon nanotubes and biomolecules, Bradley et al., 2004 used compact transistor devices with carbon nanotubes being the conducting channel and studied the interaction between nanotubes and streptavidin.
Olenic et al., 2009 have studied the adsorption properties of different bio-molecules onto the surface of CNFs, synthesized by CCVD method (Lupu et al 2004a). Few amino acids (alanine, aspartic acid and glutamic acid) and glucose oxidase (GOx) were adsorbed on CNFs and activated carbon (AC). Hydrophilic and hydrophobic properties of CNFs and AC surfaces were characterized by the pH value, the concentration of acidic/basic sites and by naphthalene adsorption. Carbon nanofibers with the "herringbone" structure (Figure 1) were purified in HCl. The specific area (170 m^2g^{-1}) was determined by BET method. The investigated carbon structures were weakly acidic mainly due to preparation and activation methods. The adsorption properties of CNFs and AC were different for various amino acids, depending on the molecular weight and acid-base functionalities of each amino acid. The interaction between GOx and CNF support was complex, depending on factors like steric hindrance or chemical groups attached to CNF surface. The filamentous morphology of CNF was responsible for the greater stability of adsorbed enzyme, compared with the enzyme used directly in solution.

Sample	BET surface (m^2g^{-1})	pH	Acidic values (meq g^{-1})	Basic values (meq g^{-1})	Naphthalene adsorption (nmol m^{-2})
CNFs	170	6.20	0.15	0.6	51.17
AC	1400	6.52	0.04	0.28	27.8

Table 1. pH, hydrophilic and hydrophobic properties of CNFs and AC. Reprinted from ref. Olenic et al., 2009 with kind permission of Springer Science and Business Media.

The data were fitted with the Langmuir adsorption isotherm. From the adsorption isotherms (Figures 6, 7) one can see that the adsorption of amino acids onto CNFs increases from alanine to aspartic acid; when the less hydrophobic AC was used as support, the adsorption of amino acids increased from aspartic acid to alanine and to glutamic acid. Glutamic acid adsorbed on CNFs doesn't obey the Langmuir equation, due to its hydrophobicity. GOx was also adsorbed on CNF and AC. In comparison with CNF, the adsorption process on AC does not obey the Langmuir equation. This means that the intermolecular interactions between adsorbate molecules are stronger than the interaction between the adsorbate molecules and support.

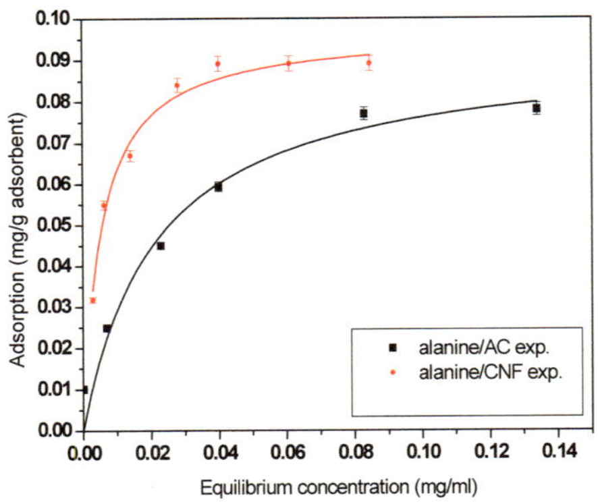

a

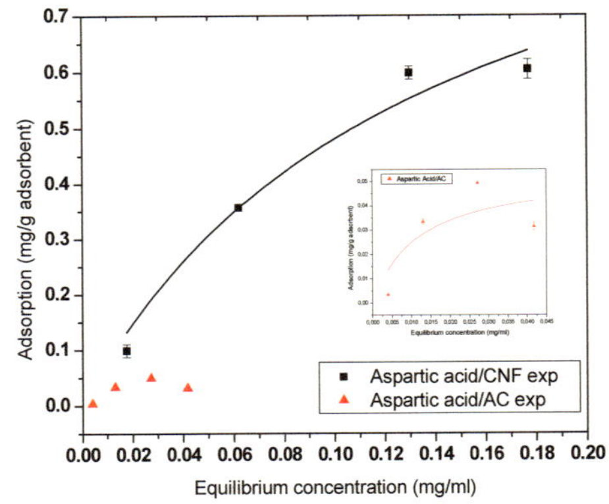

b

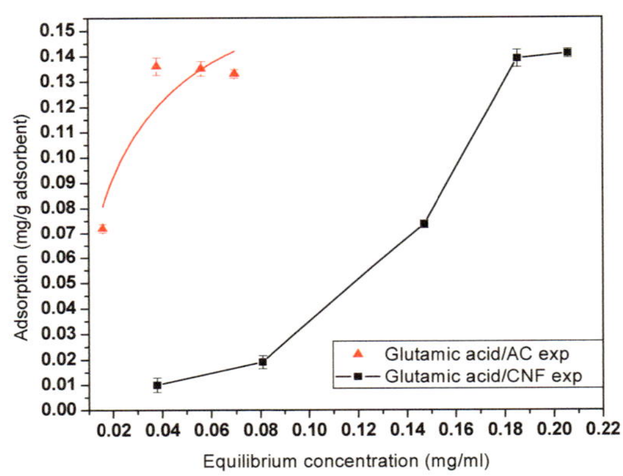

c

Fig. 6. The adsorption isotherms of alanine (a) aspartic acid (b) and glutamic acid (c) on CNFs and AC (error bars represent the standard deviation of the mean for 5 samples). Reprinted from ref. Olenic et al., 2009 with kind permission of Springer Science and Business Media.

Due to the fact that the *accessible surface area* (ASA) plays an important role in the adsorption of various bio-molecules, we have determined the ratio of ASA_{CNF}/ASA_{AC} by comparative method, for all adsorbate molecules. We have noticed that the adsorption of GOx on CNFs reaches saturation earlier than on AC (unpublished data).

Bio-molecules	Alanine	Glutamic acid	Aspartic acid
ASA_{CNF}/ASA_{AC}	1.02	0.027	5.66

Table 2. The ratios of ASA_{CNF}/ASA_{AC} for adsorbate molecules

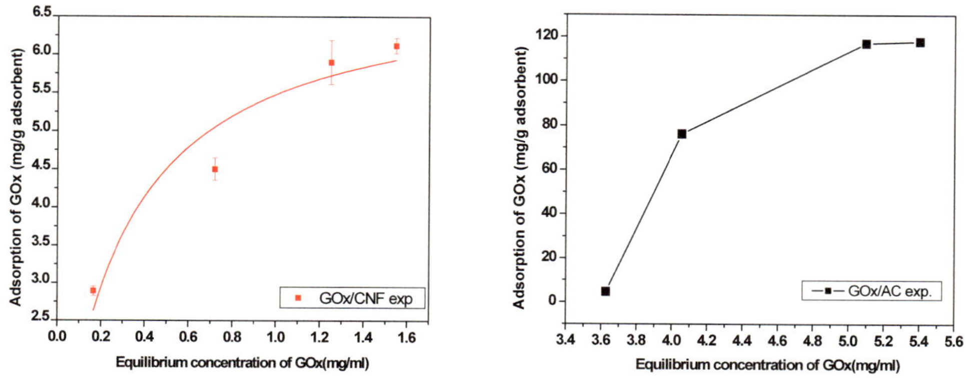

a b

Fig. 7. The adsorption isotherms of GOx on a-CNF and b-AC (error bars represent the standard deviation of the mean for 5 samples). Reprinted from ref. Olenic et al., 2009 with kind permission of Springer Science and Business Media.

Carbon nanofibers as sensors

CNFs represent a promising material to assemble electrochemical sensors and biosensors. The direct immobilization of enzymes onto the surface of CNFs was proved to be an efficient method for the development of a new class of sensitive, stable and reproducible electrochemical biosensors. Such sensors showed good precision, high sensitivity, acceptable stability and reproducibility.

CNFs can efficiently immobilize antigen/antibody on their surfaces and can be used in the preparation of amperometric immunosensors (Wohlstadter et al., 2003; O'Connor et al., 2004; Yu et al., 2005; Viswanathan et al., 2006). An amperometric immunosensor for separation-free immunoassay of carcinoma antigen-125, based on its covalent immobilization coupled with thionine on carbon nanofiber was prepared by Wu et al., 2007. The direct electrochemistry of NADH was studied at a glassy carbon electrode modified using CNFs (Arvinte et al., 2007).

VACNFs were also used for biosensing applications (Baker et al., 2006 b). The use of highly activated CNFs for the preparation of glucose biosensors, in comparison with SWCNT and graphite powder, is presented by Vamvakaki et al., 2006. They demonstrated that CNFs are far superior to carbon nanotubes or graphite powder as matrix for the immobilization of proteins and enzymes and for the development of biosensors. They characterized the buffer capacity and the electrochemical properties of supports. Carbon nanofiber-based glucose biosensors provide higher sensitivity, reproducibility and longer lifetime. This is due to the high surface area of nanofibers which together with the large number of active sites, offers the grounds for the adsorption of enzymes. In addition, they allow for both the direct electron transfer and increased stabilization of the enzymatic activity. These carbon nanofiber materials are thus very promising substrates for the development of a series of highly stable and novel biosensors.

Metz et al., 2006 demonstrated a method for producing nanostructured metal electrodes, by functionalization of CNFs with molecular layers bearing carboxylic acid groups, which then serve as a template for electroless deposition of gold.

CNFs have been incorporated into composite electrodes for use with *liquid|liquid* redox systems (Shul et al., 2005).

CNFs are very good materials for the interface between solid state electronics and biological systems. Integrated VACNFs, grown on electronic circuits, were used in a multiplex microchip for neural electrophysiology by Nguyen-Vu et al., 2005. The chip has multiple nanoelectrode arrays with dual function: either as electrical stimulation electrodes or as electrochemical-sensing electrodes. They tested the implantable electrodes in-vitro cell culture experiments.

Lee et al., 2004 provided the fabrication of high-density arrays of biosensor elements using functionalized VACNFs (with nitro groups). The surface of VACNFs was further modified by an electrochemical reduction reaction (nitro groups on specific nanostructures were reduced to amino groups). DNA was then covalently linked to only these nanostructures. DNA-modified nanostructures have excellent biological selectivity for DNA hybridization.

MWCNTs inlaid nanoelectrode array have ultrahigh sensitivity in direct electrochemical detection of guanine, in the nucleic acid target (Koehne et al., 2004).

Olenic et al., 2009 adsorbed the GOx on CNFs and prepared a glucose biosensor using potassium ferrocyanide as redox mediator (Figure 8 a). In order to detect the changes in the specific activity of GOx immobilized a long time on CNFs, an amperometric method was used in an original manner (Figure 8 b). The specific activity was determined by taking into consideration the decrease of the current in time. The proposed method is fast and very simple and demonstrates that not all the enzyme immobilized on nanofibers can catalyze the oxidation of glucose. The characteristics of biosensor are: linear range between 1.7 and 7 mM and sensitivity of 8.6 μA/mM. After 1 year, they have changed (linear range 1–3 mM and sensitivity 1.5 μA/mM).

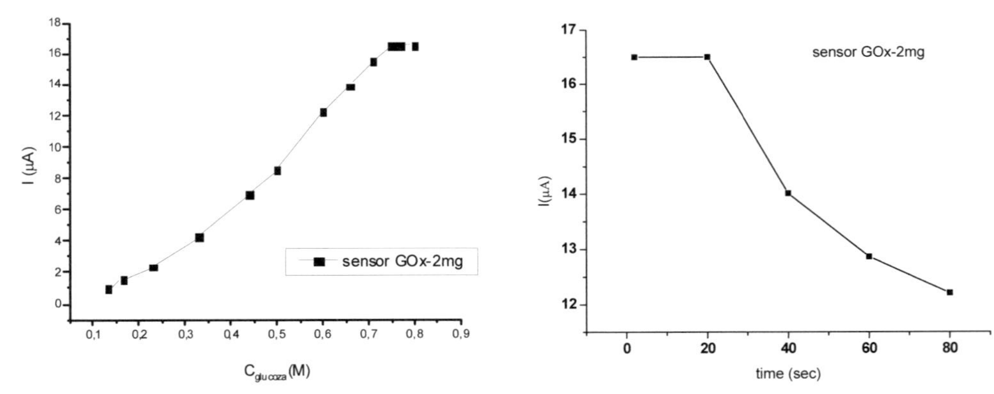

a b

Fig. 8. a-calibration curve of glucose biosensor; b- biosensor response during glucose consumption (the points represent the media of five determinations). Reprinted from ref. Olenic et al., 2009 with kind permission of Springer Science and Business Media.

The results presented in Table 3 shows that the enzymatic activity of GOx decreases in time.

Time	Current (μA)	Enzyme activity ($U\ mg^{-1}$)	Enzymatic activity decreased (%)
After preparation	96	157	0
After 12 months	38	64	59

Table 3. The decrease of GOx activity in time. Reprinted from ref. Olenic et al., 2009 with kind permission of Springer Science and Business Media.

We can conclude that the amount of enzyme required to prepare a high sensitive biosensor has to be larger than that adsorbed on CNFs, due to the fact that some of it does not participate to the reaction.

6. Conclusions and future research

A new synthesis technique of carbon nanofilaments in a cold wall reactor (CCVD method with inductive heating) has been achieved and improved in the laboratory where the authors are working. This method was a world premiere (Lupu et al., 2004).

Compared to the classical method, this technique is suitable for the synthesis of all types of high quality carbon nanofilaments. Its efficiency was proved by the reduction of the global synthesis time to one half and of the energetic consumption to a third. Nowadays, the method is used in many laboratories from Japan, China, USA, etc.

The obtained CNF's structures were electrochemically characterized by cyclic voltammetry. Additionally, single stranded and double stranded calf thymus DNA was physisorbed on the surface of a CNF's electrode. The oxidation peaks of adenine and guanine were recorded by differential pulse voltammetry. The authors also had in view the adsorbing properties of these nanostructures, in the presence of some biologically active substances (amino-acids and glucose oxidase). The nanomaterials have been used to obtain a glucose biosensor. A new simple and trustful method has been finalized which helps to determine the enzymatic activity of GOx. All the accomplished studies are genuine and they bring a great contibution to the literature in the field. The adsorption studies can contribute to the development of bio-technological processes, in the pharmaceutical industry and in clinical trials.

Further studies can be performed on CNFs with various morphological and structural characteristics, in order to see their influence on the adsorption and electrochemical properties. There is a possibility of enlarging the research area, by studying other biologically active substances and by simulation of their adsorption on nanostructured supports. Additionally, the study of direct oxidation (without redox mediator) of GOx and DNA on CNFs electrodes, would help in improving the construction of new types of biosensors.

Currently, the research in our laboratory is focused on the detection of new properties of the functionalized carbon nanostructures, for treatment of human and animal pancreatic cancer and other cancers in general.

7. Acknowledgements

Authors are thankful to the National Authority for Scientific Research, Romania for providing financial support for the work.

The authors are also thankful to Springer Science and Business Media for their kind permission to use the published data in this review and to Dr. G. A. Kovalenko, for helping gather relevant literature.

8. References

Arvinte, A.; Valentini, F.; Radoi, A.; Arduini, F.; Tamburri, E.; Rotariu, L.; Palleschi, G.& Bala, C. (2007). The NADH Electrochemical Detection Performed at Carbon Nanofibers Modified Glassy Carbon Electrode. *Electroanalysis,* Vol. 19, No. 14, (July 2007) pp. 1455 - 1459, Print ISSN: 1040-0397, Online ISSN: 1521-4109

Atwater, M. A.; Phillips, J.; Doorn, S. K.; Luhrs, C. C.; Fernandez, Y.; Menendez, J.A. & Leseman, Z. C. (2009). The production of carbon nanofibers and thin films on palladium catalysts from ethylene–oxygen mixtures. *Carbon,* Vol. 47, No.9, (August 2009) pp. 2269 -2280, ISSN 0008-6223

Baker, R.T.K.; Barber, M.A.; Harris, P.S.; Feates, F.S. & Waite, R.J. (1972). Nucleation and growth of carbon deposits from the nickel catalyzed decomposition of acetylene, *J. Catal.* Vol. 26, No.1, (July 1972) pp. 51–62. ISSN 0021-9517

Baker, R.T.K.; Kim, M.S.; Chambers, A.; Park, C. & Rodriguez, N. M. (1997). The relationship between metal particle morphology and the structural characteristics of carbon deposits, *Proceedings of Catalyst Deactivation, Book Series: Studies in Surface Science and Catalysis,* Vol.111, pp. 99-109, ISBN 0444826033, Oct., 1997 Cancun, Mexico Bartholomew C. H. & Fuentes G. A. (Eds.), Elsevier, Amsterdam, Netherlands

Baker, R. T. K. (2007). Nanofibers. In: *Concise Encyclopedia of Composite Materials,* Mortensen A., (Ed.) pp. 96-104, Elsevier, ISBN13: 9780080451268 ISBN10: 0080451268, Lausanne, Switzerland

Baker, S. E.; Tse, K.Y.; Hindin, E.; Nichols, B. M.; Clare, T. L. & Hamers, R. J. (2005). Covalent functionalization for biomolecular recognition on vertically aligned carbon nanofibers *Chem. Mater.,* Vol.17, No.20, (October 2005) pp. 4971-4978, Print ISSN 0897-4756, Web Edition ISSN 1520-5002

(a) Baker, S.E.; Tse, K.Y.; Lee, C.S. & Hamers, R.J. (2006). Fabrication and characterization of VACNFs electrodes for biosensing app.lications. *Diamond & Related Materials,* Vol. 15, No. 2-3, (February-March 2006) pp. 433 - 439, ISSN 0925-9635

(b) Baker, S. E.; Colavita, P. E.; Tse, K.Y. & Hamers, R. J. (2006). Functionalized Vertically Aligned Carbon Nanofibers as Scaffolds for Immobilization and Electrochemical Detection of Redox-Active Proteins. *Chem. Mater.,* Vol. 18, No. 18, (September 2006) pp. 4415-4422, Print ISSN: 0897-4756, Web Edition ISSN: 1520-5002

Baughman, R.H.; Zakhidov, A.A. & De Heer, W.A. (2002). Carbon nanotubes–the route toward applications. *Science,* Vol. 297, No. 5582, (August 2002) pp. 787–792, Print ISSN 0036-8075 Online ISSN 1095-9203

Bethune, D. S.; Kiang, C. H.; DeVries, M.S.; Gorman, G.; Savoy, R.; Vazquez, J. & Beyers, R. (1993). Cobalt-catalyzed growth of carbon nanotubes with single-atomic layerwalls, *Nature,* Vol. 363, No.6430, (June 1993) pp. 605-607, ISSN 0028-0836

Bououdina, M.; Grant, D. & Walker, G. (2006). Review on hydrogen absorbing materials—structure, microstructure, and thermodynamic properties. *Proceedings of International Journal of Hydrogen Energy,* Vol. 31, No. 2, pp. 177 - 182, ISSN 0360-3199 Donetsk, Ukraine, May 17-21, 2004 Pergamon-Elsevier Science LTD, Oxford, England

Bradley, K.; Briman, M.; Star, A & Gruner, G. (2004). Charge transfer from adsorbed proteins. *Nano Lett* , Vol. 4, No. 2, (February 2004) pp. 253–256, Printed ISSN 1530-6984, Online ISSN 1530-6992

Brown, S.; Jespersen, T.S. & Nygard J. (2008). A genetic analysis of carbon–nanotube-binding proteins. *Small,* Vol. 4, No.4, (April 2008) pp. 416–420, ISSN 1613-6810

Cao, A.; Zhu, H.W.; Zhang, X.F.; Li, X.S.; Ruan, D.B.; Xu, C. L.; Wei, B.Q.; Liang, J. & Wu, D.H., (2001). Hydrogen storage of dense-aligned carbon nanotubes *Chem.Phys. Lett.,* Vol. 342, No. 5-6, (July 2001) pp. 510-514 , ISSN 0009-2614

Caughman, J. B. O.; Baylor, L. R.; Guillorn, M. A.; Merkulov, V. I.; Lowndes, D. H., & Allard, L. F. (2003). Growth of vertically aligned carbon nanofibers by low-pressure inductively coupled plasma-enhanced chemical vapour deposition, *Appl.Phys. Lett.,* Vol. 83, No. 6, (August 2003) pp. 1207-1209, Print ISSN 0003-6951, Online ISSN 1077-3118

Charlier, J.-C. Eklund, P. C.; Zhu, J. & Ferrari, A.C. (2008). Electron and Phonon Properties of Graphene: Their Relationship with Carbon Nanotubes: In: *Carbon Nanotubes: Synthesis, Structure, Properties and App.lications,* Jorio, A., Dresselhaus, G & Dresselhaus, M.S. (Eds.), pp. 673-709, Springer, ISBN 978-3-540-72864-1, Berlin Heidelberg New York

Cheng, J.; Zhang, X.; Liu, F.; Tu, J.; Lu, H.; Sun, Y.& Chen, F. (2004). Long bundles of aligned carbon nanofibers obtained by vertical floating catalyst method. *Mater.Chem.Phys.,* Vol. 87, No. 2-3, (October 2004) pp. 241–245, ISSN 0254-0584

Chuang, C.C.; Liu, W.L.; Chen, W.J. & Huang, J.H. (2008). Temperature and substrate dependence of structure and growth mechanism of carbon nanofiber. *Applied Surface Science,* Vol. 254, No.15, (May 2008) pp. 4681–4687, ISSN 0169-4332

Cuervo, M. R. ; Asedegbega-Nieto, E.; Dýaz, E.; Vega, A.; Ordonez, S. ; Castillejos-Lopez, E. & Rodrýguez-Ramos, I. (2008). Effect of carbon nanofiber functionalization on the adsorption properties of volatile organic compounds. *Journal of Chromatography A,* Vol. 1188, No.2, (April 2008) pp. 264–273, ISSN 0021-9673

(a) Cui, H.; Yang, X; Simpson, M. L.; Lowndes, D. H. & Varela M (2004). Initial growth of vertically aligned carbon nanofibers. *Appl Phys Lett, Vol.* 84, No. 20, (May 2004) pp. 4077–4079, Print ISSN 0003-6951 Online ISSN 1077-3118

(b) Cui, H.; Kalinin, S.V.; Yang, X. & Lowndes, D. H., (2004). Growth of carbon nanofibers on tipless cantilevers for high resolution topography and magnetic force imaging. *Nano Lett.,* Vol. 4, No. 11, (November 2004) pp. 2157-2161, Printed ISSN 1530-6984, Online ISSN 1530-6992

Damnjanovic, M.; Milosevic, I.; Dobardzic, E.; Vukovic, T. & Nikolic, B. (2005). Symmetry Based Fundamentals of Carbon Nanotubes, In: *Applied Physics of Carbon Nanotubes: Fundamentals of Theory, Optics and Transport Devices,* Rotkin S.V. & Subramoney S. (Eds.), pp. 41-88, Springer, ISBN 10: 3-540-23110-2 ISBN 13: 978-3-540-23110-3 Berlin Heidelberg New York

De Jong, K. P. & Geus, J. W. (2000). Carbon Nanofibers: Catalytic Synthesis and Applications. *Catalysis Reviews-Sci.Eng.,* Vol. 42, No. 4, (November 2000) pp. 481–510, ISSN 0161-4940

De Lathouder, K. M.; Bakker, J. ; Kreutzer, M. T.; Kapteijn, F.; Moulijn, J. A. & Wallin, S. A. (2004). Structured reactors for enzyme immobilization: advantages of tuning the wall morphology. *Chem. Eng. Sci.*, Vol. 59, No.22-23, (September 2004) pp. 5027 - 5033, ISSN 0009-2509

Delzeit, L.; McAninch, I.; Cruden, B. A.; Hash, D.; Chen, B.; Han, J. & Meyyappan, M. (2002). Growth of multiwall carbon nanotubes in an inductively coupled plasma reactor. *J. Appl. Phys.* Vol. 91, No. 9, (May 2002) pp. 6027-6033, ISSN 0021-8979

Devaux, X.; Tsareva, S.Y.; Kovalenko, A.N.; Zharikov, E.V. & McRae E. (2009). Formation mechanism and morphology of large branched carbon nano-structures. *Carbon* Vol. 47, No. 5, (April 2009) pp. 1244-1250, ISSN 0008-6223

Díaz, E.; Ordóñez, S. & Vega A. (2007). Adsorption of volatile organic compounds onto carbon nanotubes, carbon nanofibers, and high-surface-area graphites. *Journal of Colloid and Interface Science*, Vol. 305, No. 1, (January 2007) pp. 7–16, ISSN 0021-9797

Eksioglu, B.& Nadarajah, A. (2006). Structural analysis of conical carbon nanofibers. *Carbon*, Vol. 44, No. 2, (February 2006) pp. 360–373, ISSN 0008-6223

Endo, M.; Takeuchi, K.; Igarashi S.; Kobori, K.; Shiraishi, M. & Kroto, H. W. (1993). The production and structure of pyrolytic carbon nanotubes (PCNTs), *J. Phys. Chem. Solids*, vol. 54, no.12 , (December 1993) pp. 1841-1848 , ISSN 0022-3697

Endo, M.; Takeuchi, K.; Kobori, K.; Takahashi, K.; Kroto, H. W.& Sarkar, A. (1995). Pyrolytic Carbon Nanotubes from Vapor-Grown Carbon Fibers, *Carbon*, Vol. 33, No.7, (February 1995) pp. 873-881 , ISSN 0008-6223

Endo, M.; Kim, Y. A. , Hayashi, T.; Fukai, Y.; Oshida, K.; Terrones, M.; Yanagisawa, T.; Higaki, S.; Dresselhaus, M. S. (2002). Structural characterization of cup-stacked-type nanofibers with an entirely hollow core. *Appl.Phys.Lett.*, Vol.80, No.7, (February 2002) pp. 1267–1269, ISSN 1077-3118

Endo, M.; Kim, Y. A.; Hayashi, T.; Yanagisawa, T.; Muramatsu, H., Ezaka, M., Terrones, H., Terrones, M. & Dresselhaus, M. S. (2003). Microstructural changes induced in "stacked cup" carbon nanofibers by heat treatment. *Carbon*, Vol. 41, No. 10 (April 2003) pp. 1941- 1947, ISSN 0008-6223

Fan, S.; Chapline, M. G.; Franklin, N. R.; Tombler, T. W.; Cassell, A. M. & Dai H. (1999). Self-oriented regular arrays of carbon nanotubes and their field emission properties. *Science*, vol. 283, No. 5401 (January 1999) pp. 512–514, Print ISSN 0036-8075, Online ISSN 1095-9203

Fletcher, B. L.; Hullander, E. D.; Melechko, A. V.; McKnight, T. E.; Klein, K. L.; Hensley, D. K.; Morrell, J. L.; Simpson, M. L. & Doktycz, M. J. (2004). Microarrays of biomimetic cells formed by the controlled synthesis of carbon nanofiber membranes. *Nano Lett.*, Vol. 4, No. 10, (October 2004) pp. 1809-1814, Print ISSN 1530-6984, Online ISSN 1530-6992

Govindaraj, A. & Rao, C. N. R. (2006). Synthesis, growth mechanism and processing of carbon nanotubes. In: *Carbon Nanotechnology: Recent Developments in Chemistry, Physics, Materials Science and Device App.lications*, Dai L. (Ed.), pp. 15-52, Elsevier ISBN 10: 0-444-51855-x, ISBN 13: 978-0-444-51855-2, Amsterdam, The Netherlands

Grobert, N.; Hsu, W. K.; Zhu, Y. Q.; Hare, J. P.; Kroto, H. W.; Walton, D. R. M.; Terrones, M.; Terrones, H.; Redlich, P; Ruhle, M.; Escudero, R.& Morales, F. (1999). Enhanced magnetic coercivities in Fe nanowires. *App.l Phys Lett.*, Vol. 75, No. 21, (Nov.1999) pp. 3363–3365, Print ISSN 0003-6951, Online ISSN 1077-3118

Guillorn, M. A.; McKnight, T. E.; Melechko, A.; Merkulov, V. I.; Britt, P. F.; Austin, D. W.; Lowndes, D. H. & Simpson, M. L. (2002), Individually addressable vertically aligned carbon nanofiber-based electrochemical probes. *J. App.l.Phys.* Vol. 91, No. 6 (March 2002) pp. 3824.-3828 ISSN 0021-8979

Helveg, S.; Lopez-Cartes, C.; Sehested, J.; Hansen, P.L.; Clausen, B.S.; Rostrup-Nielsen, J.R.; Abild-Pedersen, F. & Norskov, J. K. (2004). Atomic-scale imaging of carbon nanofibre growth. *Nature,* Vol. 427, No. 6973, (December 2003) pp. 426–429, ISSN 0028-0836

Hu, H.; Ni, Y.; Montana, V.; Haddon, C. & Parpura, V. (2004). Chemically functionalized carbon nanotubes as substrates for neuronal growth. *NanoLett.* Vol. 4, No. 3, (February 2004) pp. 507–511, ISSN 1530-6984

Huang, C.-W.; Wu, H.-C.; Lin, W.-H. & Li, Y.-Y. (2009). Temperature effect on the formation of catalysts for growth of carbon nanofibers. *Carbon,* Vol. 47, No. 3, (March 2009) pp. 795–803, ISSN 0008-6223

Huang, W.; Taylor, S.; Fu, K.; Lin, Y.; Zhang, D.; Hanks, T. W.; Rao, A. M. & Sun, Y-P. (2002). Attaching proteins to carbon nanotubes via diimide-activated amidation. *NanoLett.* Vol. 2, No. 4, (March 2002) pp. 311–314, ISSN 1530-6984

Hughes, T. V.& Chambers, C. R. (1889). Manufacture of carbon filaments. US Patent 405480

Iijima, S. (1991). Helical microtubules of graphitic carbon. *Nature,* Vol. 354, No. 6348, (November 1991) pp. 56-58, ISSN 0028-0836

Iijima, S. & Ichihashi, T. (1993). Single-shell carbon nanotubes of 1-nm diameter. *Nature,* Vol. 363, No. 6430, (June 1993) pp. 603-605, ISSN 0028-0836

Ismagilov, Z. R.; Shalagina, A E.; Podyacheva, O. Yu.; Ischenko, A. V.; Kibis, L. S.; Boronin, A. I.; Chesalov, Y. A.; Kochubey, D. I.; Romanenko, A. I.; Anikeeva O. B.; Buryakov, T. I. & Tkachev E. N. (2009) Structure and electrical conductivity of nitrogen-doped carbon nanofibers. *Carbon,* Vol. 47, No.8, (July 2009) pp. 1922–1929, ISSN 0008-6223

Jeykumari, D. R. S. & Narayanan, S. S. (2009). Functionalized carbon nanotube-bienzyme biocomposite for amperometric sensing. *Carbon,* Vol. 47, No. 4, (April 2009) pp. 957-966, ISSN 0008-6223

Ju, Y.W.; Choi, G.R.; Jung, H.R. & Lee, W.J. (2008). Electrochemical properties of electrospun PAN/MWCNT carbon nanofibers electrodes coated with polypyrrole. *Electrochimica Acta,* Vol. 53, No. 19, (August 2008) pp. 5796–5803, ISSN 0013-4686

Kam, N. W. S.; Jessop, T.C.; Wender, P. A. & Dai, H.J. (2004). Nanotube molecular transporters: internalization of carbon nanotube-protein conjugates into mammalian cells. *J.Am. Chem.Soc.,* Vol. 126, No.22, (June 2004) pp.6850-6851, ISSN 0002-7863

Kataura, H.; Kumazawa, Y.; Maniwa, Y.; Umezu, I.; Suzuki, S.; Ohtsuka, Y. & Achiba, Y. (1999). Optical properties of single-wall carbon nanotubes. *Proceedings of Synth. Metals,* Vol. 103, No. 1-3, pp. 2555–2558, ISSN 0379-6779, Montpellier, France, July 12-18, 1999, Elsevier Science SA, Lausanne, Switzerland

Kavan, L. & Dunsch, L. (2008). Electrochemistry of Carbon Nanotubes, In: *Carbon Nanotubes Advanced Topics in the Synthesis, Structure, properties and App.lications,* Jorio, A., Dresselhaus, G. & Dresselhaus, M. S. (Eds.), pp. 567-604, Springer, ISBN 978-3-540-72864-1, Berlin Heidelberg New York

Koch, A. J. H. M.; Debokx, P. K.; Boellaard, E.; Klop, W. & Geus, J. W. (1985). The formation of filamentous carbon on iron and nickel catalysts: II. Mechanism *J. Catal.,* Vol. 96, No.2, (December 1985) pp. 468-480, ISSN 0021-9517

Koehne, J.; Li, J.; Cassell, A. M.; Chen, H.; Ye, Q.; Ng, H. T.; Han, J. & Meyyapp.an, M. (2004) The fabrication and electrochemical characterization of carbon nanotube nanoelectrode arrays *J. Mater. Chem.*, Vol. 14, No. 4, (November 2003) pp. 676-684, ISSN 0959-9428

Kovalenko, G. A.; Kuznetsova, E. V.; Mogilnykh, Y. I.; Andreeva, I. S.; Kuvshinov, D. G. & Rudina, N. A. (2001). Catalytic filamentous carbons for immobilization of biologically active substances and nongrowing bacterial cells. *Carbon,* Vol. 39, No.7, (June 2001) pp. 1033-1043, ISSN 0008-6223

(a) Kovalenko, G.A.; Perminova, L.V.; Chuenko, T.V. & Rudina, N.A. (2009). Adsorptive immobilization of enzymatic active substances on alumina–silica foam coated by carbon nanofibers. *Carbon,* Vol. 47, No. 2, (February 2009) pp. 420–427, ISSN 0008-6223

(b) Kovalenko, G. A.; Rudina, N.A.; Chuenko, T.V.; Ermakov, D. Y. & Perminova, L.V. (2009). Synthesis of catalytic filamentous carbon by the pyrolysis of alkanes on alumina–silica foam supporting nickel nanoparticles, *Carbon,* Vol. 47, No. 2, (February 2009) pp. 428–435, ISSN 0008-6223

Kroto, H. W.; Hearth, J. R.; O`Brien, S. C.; Curl, R. F. & Smalley, R. E. (1985). C60 buckminsterfullerene. *Nature,* Vol. 318, No. 6042, (November 1985) pp. 162-163, ISSN 0028-0836

Kymakis, E.; & Amaratunga, G. A. J. (2002). Single-wall carbon nanotube/conjugated polymer photovoltaic devices. *Appl. Phys. Lett.*, Vol. 80, No. 1, (January 2002) pp. 112-114, Print ISSN 0003-6951 Online ISSN 1077-3118

Lawrence, J. G., Berhan, L. M., & Nadarajah, A. (2008). Structural transformation of vapour grown carbon nanofibers studied by HRTEM, *J.Nanopart.Res.*, Vol. 10, No. 7, (October 2008) pp. 1155–1167, Print ISSN 1388-0764, Online ISSN 1572-896X

Lee, C.S.; Baker, S. E.; Marcus, M. S.; Yang, W. S.; Eriksson, M. A. & Hamers, R. J. (2004). Electrically Adresable Biomolecular Functionalization of Carbon Nanotube and Carbon Nanofiber Electrodes, *Nano Letters*, Vol. 4, No. 9, (September 2004) pp. 1713-1716, Print ISSN 1530-6984, Online ISSN 1530-6992.

Li, C. P.; Chen, Y. & Fitzgerald, J.F. (2006). Substitution reactions of carbon nanotube template, *App.l.Phys.Lett.,* Vol. 88, No. 22, (May 2006) pp. 223105-1-223105-4 Print ISSN 0003-6951 Online ISSN 1077-3118

Li, J.; Vergne, M. J.; Mowles, E. D.; Zhong, W. H.; Hercules, D. M. & Lukehart, C. M. (2005). Surface functionalization and characterization of graphitic carbon nanofibers (GCNFs). *Carbon,* Vol. 43, No. 14 , (November 2005) pp. 2883–2893, ISSN 0008-6223

Li, W. Z.; Xie, S. S.; Qian, L. X.; Chang, B. H.; Zou, B. S.; Zhou, W. Y.; Zhao, R. A. & Wang, G. (1996), Large-Scale Synthesis of Aligned Carbon Nanotubes, *Science,* Vol. 274, No. 5293, (December 1996) pp. 1701-1703, Print ISSN 0036-8075, Online ISSN 1095-9203

Ling-jun, G.; Dong-sheng, Z.; Ke-zhi, L.; He-jun, L.; Chuang, W. & Qian-gang, F. (2009). Bamboo-shaped carbon nanofibers obtained in pyrolytic carbon by thermal gradient chemical vapour deposition. *Carbon,* vol. 47, No. 10 (August 2009) pp. 2528 –2555, ISSN 0008-6223

Liu, P.F. & Hu, J. H. (2002), Carbon nanotube powder microelectrodes for nitrite detection. *Sensors Actuators B Chem.*, Vol. 84, No. 2-3, (May 2002) pp. 194-199, ISSN 0925-4005

Liu, X.; Baronian, K. H. R. & Downard, A. J. (2009). Direct growth of vertically aligned carbon nanotubes on a planar carbon substrate by thermal chemical vapour deposition. *Carbon,* Vol. 47, No. 2, (February 2009) pp. 500–506, ISSN 0008-6223

Liu, Y.; Gao, L.; Sun, J., Zheng, S.; Jiang, L.; Wang, Y.; Kajiura, H., Li, Y. & Noda, K. (2007). A multi-step strategy for cutting and purification of single-walled carbon nanotubes. *Carbon*, Vol. 45, No. 10 (September 2007) pp. 1972–1978, ISSN 0008-6223

(a) Lupu, D.; Biris, A. R.; Jianu, A.; Bunescu, C.; Burkel, E.; Indrea, E.; Mihailescu, G.; Pruneanu, S.; Olenic, L. & Misan, I. (2004). Carbon nanostructures produced by CCVD with induction heating. *Carbon,* Vol. 42, No. 3, (January 2004) pp. 503-507, ISSN 0008-6223

(b) Lupu, D.; Biris, A.R.; Misan, I.; Jianu, A.; Holzhüter, G. & Burkel, E. (2004). Hydrogen uptake by carbon nanofibers catalyzed by palladium. *International Journal of Hydrogen Energy* Vol. 29, No. 1, (January 2004) pp. 97-102, ISSN 0360-3199

Maldonado, S. & Stevenson, K. J. (2005). Influence of Nitrogen Doping on Oxygen Reduction Electrocatalysis at Carbon Nanofiber Electrodes. *J.Phys.Chem. B,* Vol. 109, No. 10 (March 2005) pp. 4707-4716, Printed ISSN 1520-6106

Marken, F.; Gerrard, M. L.; Mellor, I. M.; Mortimer, R. J.; Madden, C. E.; Fletcher, S.; Holt, K.; Foord, J.S.; Dahm, R.H. & Page, F. (2001). Voltammetry at carbon nanofiber electrodes. *Electrochem. Commun.*, Vol. 3, No. 4 (April 2001) pp. 177-180, ISSN 1388-2481

Martin-Gullon, I.; Vera, J.; Conesa, J. A.; Gonzalez, J. L. & Merino, C. (2006). Differences between carbon nanofibers produced using Fe and Ni catalysts in a floating catalyst reactor. *Carbon,* Vol. 44, No. 8 (July 2006) pp. 1572–1580, ISSN 0008-6223

McKenzie, J. L.; Waid, M.C.; Shi, R. Y. & Webster, T. J. (2004). Decreased functions of astrocytes on carbon nanofiber materials. *Biomaterials,* Vol. 25, No. 7-8, (March-April 2004) pp. 1309–1317, ISSN 0142-9612

McKnight, T. E.; Melechko, A. V.; Hensley, D. K.; Mann, D. G. J.; Griffin, G. D. & Simpson, M. L. (2004). Tracking Gene Expression after DNA Delivery Using Spatially Indexed Nanofiber Arrays *Nano Lett.*, Vol. *4,* No. 7, (July 2004) pp. 1213-1219, ISSN 1530-6984

McKnight, T.E.; Peeraphatdit, C.; Jones, S. W.; Fowlkes, J. D.; Fletcher, B. L.; Klein, K. L.; Melechko, A. V.; Doktycz, M J. & Simpson, M. L. (2006). Site-Specific Biochemical Functionalization along the Height of Vertically Aligned Carbon Nanofiber Arrays *Chem. Mater.*, Vol. 18, No. 14, (July 2006) pp. 3203-3211, ISSN 0897-4756

Melechko, A.V.; McKnight, T.E.; Hensley, D. K.; Guillorn, M. A.; Borisevich A. Y.; Merkulov, V. I.; Lowndes, D.H. & Simpson, M. L. (2003). Large-scale synthesis of arrays of high-aspect-ratio rigid vertically aligned carbon nanofibres, *Nanotechnology,* Vol.14, No. 9, (September 2003) pp.1029–1035, Print ISSN 0957-4484, Online ISSN 1361-6528

Merkulov, V. I.; Hensley, D. K.; Melechko, A. V.; Guillorn, M. A.; Lownde,s D. H. & Simpson, M. L. (2002). Control Mechanisms for the Growth of Isolated Vertically Aligned Carbon Nanofibers, *J. Phys. Chem. B*, Vol. 106, No. 41, (September 2002) pp. 10570-10577, ISSN 1089-5647

Metz, K.M.; Tse, K. Y.; Baker, S. E.; Landis, E. C. & Hamers, R. J. (2006). Ultrahigh-Surface-Area Metallic Electrodes by Templated Electroless Deposition on Functionalized Carbon Nanofiber Scaffolds. *Chem..Mater.,* Vol. 18, No. 23, (November 2006) pp. 5398-5400, ISSN 0897-4756

Minnikanti, S.; Skeath, P. & Peixoto, N. (2009). Electrochemical characterization of multi-walled carbon nanotube coated electrodes for biological applications. *Carbon*, Vol. 47, No. 3, (March 2009) pp. 884–893, ISSN 0008-6223

Murphy, M. A.; Wilcox, G. D.; Dahm, R. H. & Marken, F. (2003). Adsorption and redox processes at carbon nanofiber electrodes grown onto a ceramic fiber backbone *Electrochemistry Communications,* Vol. 5, No. 1, (January 2003) pp. 51–55, ISSN 1388-2481

Nguyen-Vu, T.-D.B.; Chen, H.; Cassell, A.; Koehne, J.; Purewal, H.; Meyyapp.an, M.; Andrews, R. & Li, J. (2005). Carbon Nanofiber Nanoelectrode Array for Closed-Loop Electrical Stimulation, *10th Annual Conference of the International FES Society,*- Montreal, Canada, July 2005, http://www.ifess.org/ifess05/IFESS%202005.htm

Oberlin, A., Endo, M. & Koyama, T. (1976). Filamentous growth of carbon through benzene decomposition. *J.Cryst.Growth,* Vol. 32, No. 3, (March 1976) pp. 335–349, ISSN 0022-0248

O'Connor, M.; Kim, S.N.; Killard, A.J.; Forster, R.J.; Smyth, M.R.; Papadimitrakopoulos, F. & Rusling, J.F. (2004). Mediated amperometric immunosensing using single walled carbon nanotube forests. *Analyst,* Vol. 129, No. 12, (November 2004) pp. 1176-1180, ISSN 0003-2654

Olenic, L.; Mihailescu, G.; Puneanu, S.; Lupu, D.; Biris, A.R.; Margineanu, P.; Garabagiu, S. (2009). Investigation of carbon nanofibers as supp.ort for bioactive substances. *J. Mat.Sci.: Mat.in Med.* Vol.20, No.1, (August 2008) pp. 177-183, ISSN 0957-4530

Paredes, J. I.; Burghard, M.; Martinez-Alonso, A. & Tascon, J. M. D. (2005). Graphitization of carbon nanofibers: visualizing the structural evolution on the nanometer and atomic scales by scanning tunneling microscopy. *Appl Phys A,* Vol. 80, No. 4, (February 2005) pp. 675–682, ISSN 0947-8396

Perez-Cabero, M.; Rodriguez-Ramos, I. & Guerrero-Ruiz, A. (2003). Characterization of carbon nanotubes and carbon nanofibers prepared by catalytic decomposition of acetylene in a fluidized bed reactor *J. Catal.*, Vol. 215, No. 2, (April 2003) pp. 305-316, ISSN 0021-9517

Peshnev, B. V.; Nikolaev, A. I.; Pilipeiko, A. Yu. & Estrin R. I. (2007). Structural and Physicochemical Characterization of Carbon Nanofibers by the COMPAS Method. *Solid Fuel Chemistry* ,Vol. 41, No. 1, (February 2007) pp. 52–57 ISSN 0361-5219

Poirier, E; Chahine, R & Bose, TK. (2001) Hydrogen adsorption in carbon nanostructures. *Int J Hydrogen Energy,* Vol. 26, No. 8, (August 2001) pp. 831–835, ISSN 0360-3199

Pruneanu, S.; Ali, Z.; Watson, G.; Hu, S.-q.; Lupu, D.; Biris, A.R.; Olenic, L. & Mihailescu, G. (2006). Investigation of Electrochemical properties of carbon nanofibres prepared by CCVD method, *Part. Sci. Technol.*, Vol. 24, No. 3, (July-September 2006) pp. 311-320, ISSN 1548-0046

Radushkevich, L.V. & Lukyanovich, V.I. (1952). Carbon structure formed under thermal decomposition of carbon monoxide on iron, *Zh. Fiz. Khim.*, Vol. 26 , No. 1, (1952) pp. 88-95, ISSN 0044-4537

Randall, L.; Wal, V.; Ticich, T. M. & Curtis, V. E. (2001). Substrate–supp.ort interactions in metal-catalyzed carbon nanofiber growth. *Carbon,* Vol. 39, No.15, (December 2001) pp. 2277–2289, ISSN 0008-6223

Ren, Z. F.; Huang, Z. P.; Wang, D. Z.; Wen, J. G.; Xu, J. W.; Wang, J. H.; Calvet, L. E.; Chen, J.; Klemic, J. F. & Reed, M. A. (1999). Growth of a single freestanding multiwall carbon nanotube on each nanonickel dot. *Appl. Phys. Lett.*, Vol. 75, No. 8, (August 1999) pp. 1086-1088, Print ISSN 0003-6951, Online ISSN 1077-3118

Rozniecka, E.; Niedziolka, J.; Sirieix-Plenet, J. ; Gaillon, L.; Murphy, M. A. ; Marken, F. & Opallo, M. (2006). Ion transfer processes at the room temperature ionic liquid vertical bar aqueous solution interface supp.orted by a hydrophobic carbon nanofibers - silica composite film. *Journal of Electroanalytical Chemistry*, Vol. 587, No. 1, (February 2006) pp.133–139, ISSN 0022-0728

Sadowska, K.; Roberts, K. P.; Wiser, R.; Biernat, J. F.; Jablonowska, E. & Bilewicz, R. (2009). Synthesis, characterization, and electrochemical testing of carbon nanotubes derivatized with azobenzene and anthraquinone. *Carbon*, Vol. 47, No. 6, (May 2009) pp. 1501-1510, ISSN 0008-6223

Shul, G.; Murphy, M. A.; Wilcox, G. D.; Marken, F. & Opallo, M. (2005). Effects of carbon nanofiber composites on electrode processes involving liquid vertical bar liquid ion transfer. *J Solid State Electrochem*, Vol. 9, No. 12, (December 2005) pp. 874–881, Print ISSN 1432-8488, Online ISSN 1433-0768

Tennent, H. G. (1987). Carbon fibrils, method for producing same and compositions containing same. US Patent 4,663,230 Publication date: May 1987

Teo, K. B. K.; Singh, C.; Chhowalla, M. & Milne, W. I. (2003). *Catalytic synthesis of carbon nanotubes and nanofibers* In: Encyclopedia of Nanoscience and Nanotechnology. Nalwa, H.S., (ed.) American Scientific Publishers, pp. 665-686. ISBN 1-58883-001-2 CA, USA

Terrones, H.; Hayashi, T.; Munoz-Navia, M.; Terrones, M.; Kim, Y. A.; Grobert, N. Kamalakaran, R.; Dorantes-Davila, J.; Escudero, R.; Dresselhaus, M. S. & Endo, M. (2001). Graphite cones in palladium catalysed carbon nanofibers. *Chem Phys Lett.*, Vol. 343, No.3–4, (August 2001) pp. 241–250, ISSN 0009-2614

Tibbetts, G. G.; Gorkiewics, D. W. & Alig, R. L. (1993). A new reactor for growing carbon fibers from liquid-and vapor-phase hydrocarbons. *Carbon*, Vol. 31, No.5, pp. 809–814, ISSN 0008-6223

Tibbetts, G. G.; Bernado, C. A.; Gorkiewics, D. W. & Alig, R. L. (1994). Role of sulfur in the production of carbon fibers in the vapour phase. *Carbon*, Vol. 32, No. 4, (1994) pp. 569–576, ISSN 0008-6223

Wei, W.; Sethuraman, A.; Jin, C.; Monteiro-Riviere, N. A. & Narayan, R. J. (2007). Biological Properties of Carbon Nanotubes. *Journal of Nanoscience and Nanotechnology*, vol. 7, No. 4-5 (April-May 2007) pp. 1284–1297, ISSN 1533-4880

Wohlstadter, J. N.; Wilbur, J. L.; Sigal, G. B.; Biebuyck, H. A.; Billadeau, M. A.; Dong, L. W.; Fischer, A. B.; Gudibande, S. R.; Jameison, S. H.; Kenten, J. H.; Leginus, J.; Leland, J. K.; Massey, R. J. & Wohlstadter, S. J. (2003). Carbon nanotube-based biosensor. *Adv. Mater.*, Vol. 15, No. 14, (July 2003) pp. 1184-1187, ISSN 0935-9648

Woo, Y. S.; Jeon, D. Y.; Han, I. T.; Lee, N. S.; Jung J. E. & Kim J. M. (2002). In situ diagnosis of chemical species for the growth of carbon nanotubes in microwave plasma-enhanced chemical vapor deposition. *Diamond Relat. Mater.*, Vol. 11, No. 1, (January 2002) pp. 59-66, ISSN 0925-9635

Wu, L.; Yan, F. & Ju, H. (2007). An amperometric immunosensor for separation-free immunoassay of CA125 based on its covalent immobilization coupled with thionine on carbon nanofiber. *Journal of Immunological Methods*, Vol. 322, No. 1-2, (April 2007) pp. 12–19, ISSN 0022-1759

Van der Lee, M. K.; Van Dillen, A. J.; Bitter, J. H. & de Jong, K. P. (2005). Deposition precipitation for the preparation of carbon nanofiber supported nickel catalysts. *J. Am. Chem. Soc.*, Vol. 127, No. 39, (October 2005) pp. 13573-13582, ISSN 0002-7863

Van Dijk, N.; Fletcher, S.; Madden, C. E. & Marken, F. (2001). Nanocomposite electrodes made of carbon nanofibers and black wax. Anodic stripping voltammetry of zinc and lead *Analyst*, Vol. 126, No. 11, (August 2001) pp. 1878–1881, ISSN 0003-2654

Vamvakaki, V.; Tsagaraki, K. & Chaniotakis, N. (2006). Carbon nanofiber-based glucose biosensor. *Anal. Chem.*, Vol. 78, No. 15, (August 2006) pp. 5538- 5542, ISSN 0003-2700

Viswanathan, S.; Wu, L. C.; Huang, M.R. & Ho, J. A. A. (2006). Electrochemical immunosensor for cholera toxin using liposomes and poly(3,4-ethylenedioxythiophene)-coated carbon nanotubes. *Anal. Chem.*, Vol. 78, No. 4, (February 2006) pp. 1115- 1121, ISSN 0003-2700

Yang, X.; Guillorn, M. A.; Austin, D.; Melechko, A. V.; Cui, H.; Meyer H. M.; Merkulov V. I.; Caughman, J. B. O.; Lowndes, D. H. & Simpson, M.L. (2003). Fabrication and Characterization of Carbon Nanofiber-Based Vertically Integrated Schottky Barrier Junction Diodes. *Nanoletters*, Vol. 3, No. 12, pp. 1751-1755, (November 2003) ISSN 1530-6984

Yeo-Heung, Y.; Miskin, A.; Kang, P.; Jain, S.; Narasimhadevara, S.; Hurd, D.; Shinde, V.; Schultz, M. J.; Shanov, V.; He, P., Boerio, F. J.; Shi, D.& Srivinas, S. (2006). Carbon Nanofiber Hybrid Actuators: Part I - Liquid Electrolyte-based; *J. Intelligent Material Systems and Structures*, Vol. 17, No. 2, (2006) pp. 107-116, ISSN 1045-389X

Yu, X.; Kim, S. N.; Papadimitrakopoulos, F. & Rusling, J. F. (2005). Protein immunosensor using single-wall carbon nanotube forests with electrochemical detection of enzyme labels. *Mol. Biosyst.*, Vol. 1, No. 1, (May 2005) pp. 70-78, ISSN 1742-206X

Zhang, L.; Austin, D.; Merkulov, V. I.; Meleshko, A. V.; Klein, K. L.; Guillorn, M. A., Lowndes, D. H. & Simpson, M. L. (2004). Four-probe charge transport measurements on individual vertically aligned carbon nanofibers. *Appl.Phys. Lett.*, vol. 84, no. 20, (May 2004) pp. 3972-3974, Print ISSN 0003-6951, Online ISSN 1077-3118

Zheng, G. B. ; Kouda, K. ; Sano H.; Uchiyama, Y.; Shi, Y. F. & Quan, H. J.(2004). A model for the structure and growth of carbon nanofibers synthesized by the CVD method using nickel as a catalyst. *Carbon*, Vol. 42, No. 3, (January 2004) pp. 635–640, ISSN 0008-6223

Zou, G.; Zhang, D.; Dong, C.; Li, H.; Xiong, K.; Fei, L. & Qian, Y. (2006). Carbon nanofibers: Synthesis, characterization, and electrochemical properties. *Carbon*, Vol. 44, No. 5, (April 2006) pp. 828–832, ISSN 0008-6223

13

Synthesis of Carbon Nanofibers by a Glow-arc Discharge

Marquidia Pacheco, Joel Pacheco and Ricardo Valdivia
Instituto Nacional de Investigaciones Nucleares, Mexico

1. Introduction

The carbon nanofibers (CNF) consist of graphite platelets arranged in diverse orientations with respect to the fiber axis and present distinctive and special functional properties; these structures have a large number of edges and remarked chemical interaction that favor the absorption capacity [1], they also have a high-catalytic activity which can be used as solid carbon supports for other catalytic reactions [2], [3].

Because all these remarkable features, CNF are quite appropriate for health [4,5] and atmospheric pollutants treatment [6-8], on-chip interconnect integration [9,10] and they can also be used like chemical or biochemical sensing on molecular scale [11] .

To appreciate, a little bit more, the vast world of carbon nanostructures, M.Monthioux and V. Kuznetsov describe, from a carefully point of view [12], some amazing data about the history of carbon nanostructures; in particular they mention a patent of *Thomas Alba Edison* in 1892, dealing on the synthesis of carbon filaments for an incandescent lamp, employing a thermal decomposition of gaseous methane. However, such patent can not be considered as the first evidence for the growth of carbon nanotubes nor nanofibers, since the resolution of the available optical microscopes were scarcely able to image filaments smaller than few micrometers in diameter. Thanks to the subsequent invention of the transmission electron microscope (TEM), in 1953 first TEM images of CNF were published [13].

At the end of the fifties and during the sixties, many laboratories and companies begin to be interested on CNF, for example, R. Bacon had synthesized CNF of about 200nm by the electric arc technique [14]. Later, during the 70's, *A. Oberlin, M. Endo and co-workers* have obtained CNF of about 7nm with the chemical vapour deposition (CVD) technique [15-17].

Afterward, new techniques of CNF synthesis were constantly reported in literature. Nowadays, they exist a large quantity of methods to synthesize carbon nanofibers, the most common is the CVD method; this gas-phase process, generally, operate at lower temperatures, the experiment is carried out in a flow furnace at atmospheric pressure. In perhaps the simplest experimental setup, the catalyst is placed in a ceramic boat which is then put into a quartz tube. A reaction mixture consisting of, for example, acetylene and argon is passed over the catalyst bed for several hours at temperatures ranging from 500 to 1100°C [18-23].

Another technique to vapour-grown CNF production is based on a 'floating catalyst' carried in the gas stream inside a continuous flow reactor [24]. Supported catalysts have been used

for CNF synthesis to achieve control of fiber width. Silica-supported Fe, Ni or Co catalysts produce narrow tubular CNFs from CO/H_2 and ethylene/H_2mixtures [25]. A similar technique but on water-soluble supports (Na_4SiO_4, Na_2SiO_3, and Na_2CO_3) is also used [26].
A relative new technique to synthesize CNF, is based on an electrospinning process, were a polymer solution is delivered from syringe to a collector with two electrodes with several shapes [27, 28].
Concerning plasma processes for carbon nanostructures synthesis, the electric arc technique has been used to synthesize mostly carbon nanotubes [13, 29,30] but also CNF [31-33].
The CVD traditional technique is now enhanced with plasma technology; it has two advantages compared with the conventional CVD method: low process temperature and more flexible film properties [34-38].
A plasma technique that is also used for carbon nanostructures synthesis is the plasma torch; however a relatively low yield of these structures is produced [39]. Its principal limitation could be the very short residence time (lower than 1ms) of catalyst particles in the plasma because of high axial velocity of plasma jets (axial velocity can attain 300m/s).
The work herein presented is dedicated to the investigation of CNF synthesis using combination energy for two types of plasma (glow discharge and electric arc discharge) in order to diminish energy consumption. Our aim is then, to propose an alternative to the existing plasma methods for CNFs synthesis. Here, we propose an approach consisting in a high frequency glow-arc discharge working under low power consumption (<300W) and low flow rates of hydrocarbon injection (< 0.1 lpm). This method has additional advantages such as very short reaction times, simple to operate (it requires neither substrates nor special pretreatments) and low cost arrangement (expensive vacuum equipment is not involved).

2. Principle.

The present work deals with the growth of CNF by a glow-arc HF discharge; the alternating electric field across the electrodes and the temperature in the gap, provide the energy and dynamics necessary for the dissociation of carbon coming from an hydrocarbon gas (CH_4 + He) and from graphite electrodes to form CNF assisted by the transition metal catalysts (Ni, Y).
Although the conversion of the carbon and the catalyst to carbon nanostructures by arc discharge and laser method is very efficient, the energy consumption to vaporize the carbon becomes their principal drawback. To atomize a mol of graphite a quantity of 716.6kJ of energy is needed compared with 80kJ/mol if methane gas is used into the plasma discharge. Consequently, we propose an alternating process, by using not a high power DC discharge, but a high frequency and low power discharge assisted by methane as carbon-containing gas.
The method here exposed is characterized to have two discharge regimes: a glow discharge and an electric arc discharge. Glow discharge is characterized to have very energetic electrons contrary to other species in plasma with lower kinetic energies, a very characteristically mauve light, under our conditions, is emitted at, approximately, ambient temperature. This discharge is used to pre-heat catalysts to react with the carbon containing gas (CH4) when the electric arc is established. Under this electric arc regime, the sublimation of carbon occurs and the subsequent reaction with the catalyst takes place. The electric arc regime is characterized to be close to local thermodynamical equilibrium, it means that electrons and heavy particles have almost the same kinetic energy or temperature, which easily can reach 4000 K up to 12000 K.

The advantage to employ an alternating current (AC) and HF discharge is to obtain quasi instantaneous breakdown voltage peaks, getting enough specific energy to dissociate and to ionize the carbon supply, and consequently, the CNF synthesis is obtained at lower energy consumption.

During the processes, a particle agglomeration takes place in the inter-electrode gap; this effect has been observed by some authors [40-42]. They have found that the particles flowing through the gap depend on the alternating electric field intensity, gravitational force and friction forces described by the Newton motion equation. The amplitude of oscillation decreases with increasing frequency in the electric field vector. Higher frequencies are then, more appropriate because lower amplitude of particles oscillation results in lower probability of their precipitation to the electrodes, and therefore small power consumption is needed to attain the carbon sublimation.

To characterize some physical parameters during the plasma processes, a spectroscopic technique is applied: the optical emission spectroscopy (OES), which examines the wavelengths of photons emitted by the plasma species, during their transition from an excited state to a lower energy state. Each element emits a characteristic set of impulsions at specific wavelengths according to its electronic structure; by observing these wavelengths the elemental composition of the sample can be determined; furthermore, additional parameters like electronic and rotational temperatures (molecules temperature) and electron density, for example, can be obtained. This technique has the enormous advantage to be a non intrusive method.

3. Experimental set-up.

The plasma discharge was generated inside a reactor where the pressure could be controlled by a very simple pumping system, within a 10-100 kPa range. The electric energy is provided by a high frequency resonant converter specially designed for this purpose (see figure 1(a)) [43]. The AC current can be adjusted from 10mA up to 2A peak to peak. The frequency of the applied voltage will affect the voltage at which breakdown occurs, in stable condition the breakdown voltage is 200V. In general, if DC voltage is used, a higher breakdown voltage will be required [44]. The curves of the electric evolution from glow-toarc discharge as well as the traces of voltage versus current in both regimes were plotted using the real data obtained during the experiments carried out with the setup described in figure 1(a). These results are shown in figure 1(b) [45]. By multiplying the voltage by the current, the instant power applied to the plasma is determined and the *rms* power values are 51W for glow discharge and 96.7 W for the arc discharge. In the electric diagnostic, it is clear that the arc-regime discharge consumes more power than the glow-regime discharge. The difference, however, is small; in fact, the arc regime requires 45.7W more than the glow regime because the arc discharge exists almost 100% of the time, and the glow regime presents discontinuities. The reason is that some hydrogen atoms are free when the CH_4 molecule bonds are broken, and it causes an increment of the electric conductivity of the plasma requiring, then, less power to sustain the arc regime. The temperature dependence of the electric conductivity of various gases (H_2, He, N_2 and Ar) shows that H_2 presents more conductivity than He; this is due to the higher ionization potential of He [46].

The plasma reactor has a special optical port for measurement. The optical signal of the plasma discharge was guided by an optical fiber mounted in a XY electromechanical system to achieve a vertical and horizontal scanning of the plasma discharge with very precise incremental steps (0.1mm). The final end of the optical fiber is focused into the entrance slit of the monochromator, equipped with an 1800 grooves/mm holographic grating.

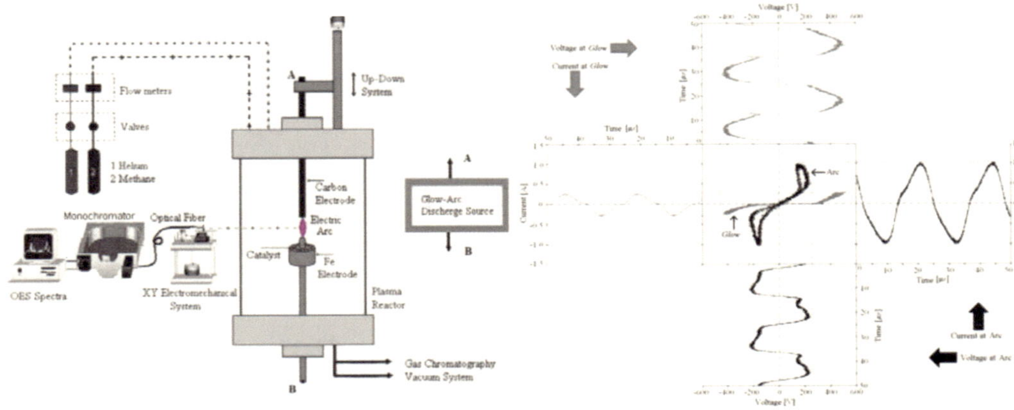

(a) (b)

Fig. 1. (a) Experimental set-up for CNF synthesis, (b) Electric waveforms during glow-arc process [45]

The inferior electrode plays at the same time the role of powder catalyst container. The catalysts mixture (34%at.Ni/10.32%at.Y/55.68%at.C) is disposed in the lower electrode (see figure 2). Two configurations of the lower electrode (containing the catalysts) were tested; in figure 2 a perforated structure with 1mm diameter holes is showed and a simplest configuration electrode consisting in a cylindrical carbon electrode with a co-axial hole drilled at its centre, filled with graphite powder and catalyst.

Fig. 2. Glow discharge and detailed view of the lower electrode.

The upper pure carbon electrode (10mm diameter, 100mm long) can be automatically adjusted to conserve a constant gap of 5mm. All the operating conditions were performed under stable conditions at 42 kHz of frequency and the power source was varied from 150 to 400 watts. Methane was used as carbon containing gas at a feed rate of 0.1 l/min. Helium flowing at 0.3l/min was used as a plasma gas.

A great advantage of using an alternating electric field is that a fine control of two separated regimes (glow and arc discharge) can be accomplished. Figure 2 shows a glow discharge required for the catalyst heating during a previous time (just a few seconds) to produce some vapours used to react with the carbon containing gas (CH_4) when the electric arc is established (figure 3). Under this regime, the sublimation of carbon occurs and the subsequent reaction with the catalyst takes place.

Fig. 3. Electric arc discharge

4. Results and discussion

4.1 Optical emission spectroscopy diagnostic

OES is one of the most used techniques for plasma diagnosis. The optical emission measurements of the Swan Band are commonly used in the diagnosis of high or even in low temperature processes by introducing carbon materials in the process.

For plasmas in departure from local thermal equilibrium, the rotational, vibrational and excitation temperatures from electrons can differ from those of the heavy species temperatures. Taking into account the relation between the rotational and translational states, the rotational temperature is derived generally from the temperature of heavy particles. Then, the temperature of heavy particles can be obtained from the measurement of the rotational temperature using the $C_2(0,0)$ Swan band situated between 513nm and 517nm. The evolution of the optical spectrum of the plasma in the Swan band for both atmospheres (He and He + CH_4), is presented in figure 12. It is clear from this figure, that the methane gas

accelerates the glow-to-arc discharge process; the transition evolution is reduced five times. In addition, the spectrum intensity under He-CH_4 atmosphere is higher than under He atmosphere. Applying the method described elsewhere [47, 48], temperature values of 6180K and 4830K are obtained under He and He-CH_4 atmospheres, respectively.
These are sufficient high temperatures to accomplish the catalyst, the carbon and the methane sublimation. Unfortunately, the diagnostic can not be prolonged until the final of the experiments, because dusty plasma rapidly fills out the reactor, and consequently, the spectral lines could not be clearly distinguished from the background, leading to a relatively large uncertainty under these conditions.

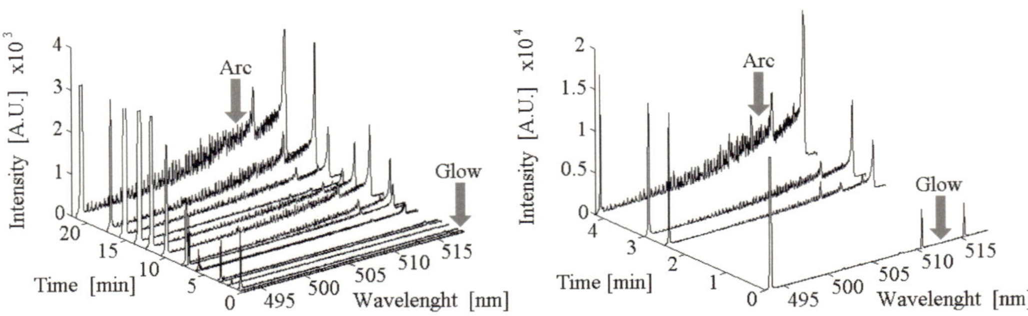

(a) (b)

Fig. 4. OES evolution from glow-to-arc under atmosphere of: (a) He, (b) He-CH_4

4.2 Morphological analysis

It is worth to note that the main product accumulation is formed in both electrodes because of the alternative current (figure 5), having, a major percentage of this product accumulated in the upper electrode due to the convection effect. In a dc electric arc discharge, usually employed to generate carbon nanotubes, only the cathode accumulates the carbon product in a more compact way (figure 6) [29].

Fig. 5. Carbonaceous depot in both electrodes when ac current is used

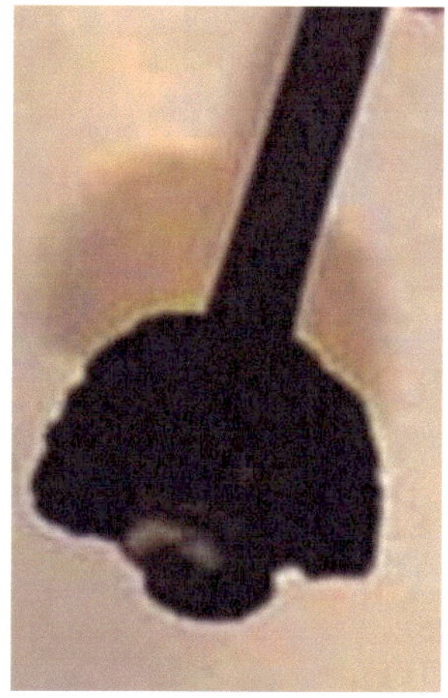

Fig. 6. Carbonaceous depot in cathode, when dc current is used

The main product was collected from the electrodes and the soot-like deposits from the reactor wall, then it was subjected to ultrasonic treatment in iso-propilic alcohol for 10 minutes and finally it was analyzed by a JEOL JSM-5900LV scan electronic microscopy.
Samples studied are mostly composed by carbon nanofibers (figure 7). An additional purification experiment with toluene solution during 5 min leads in a final product quite free of amorphous carbon, as can be shown in figure 8. Very similar morphological structure was obtained by Matsuura and co-workers [49], in a plasma reactor with twelve-phase alternating current discharge and 3 kW of power.
To further characterize the CNF microstructure, an analysis was done with JEOL 2010 Transmission Electron Microscope. Images of the samples are displayed in figure 9 and 10, where homogeneous nanofibers were observed; the CNF diameters vary from 30nm to 200nm. Figure 9 shows a CNF with diameter of about 120nm. Figure 10 shows a thicker CNF with relatively periodic joints, as those observed by [50], with 30nm of external diameter and an inner diameter of about 15nm.

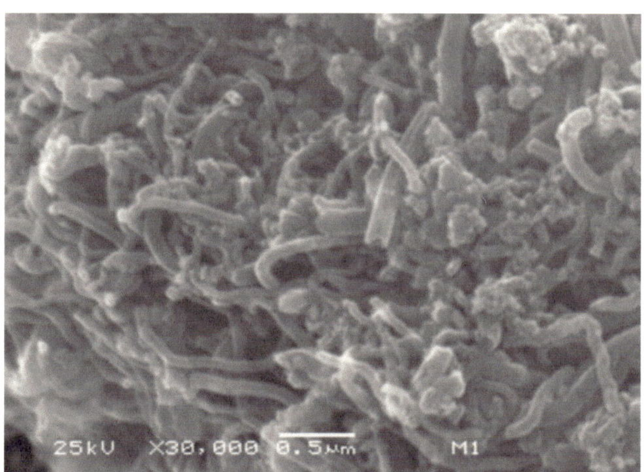

Fig. 7. SEM image detailing classical "spaghetti" morphology.

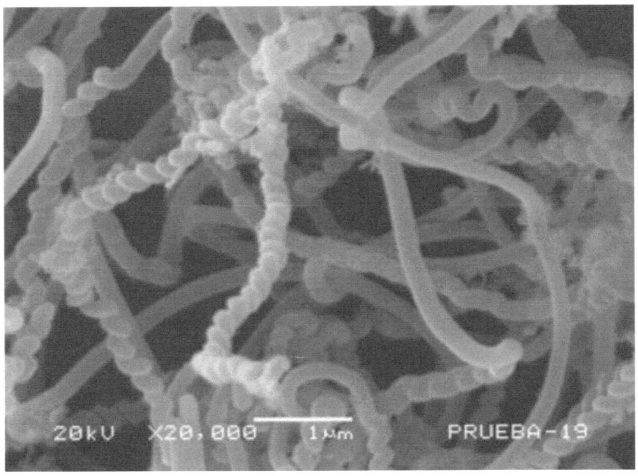

Fig. 8. SEM image detailing some braided nanofibers (purified sample).

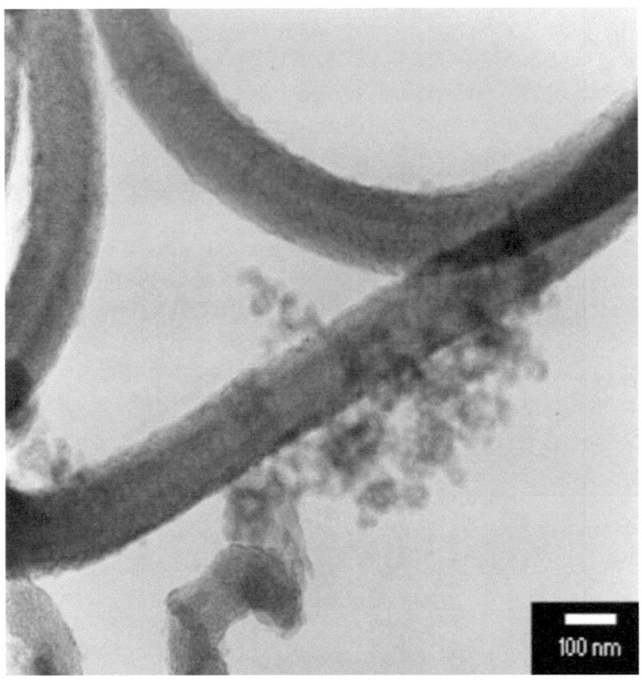

Fig. 9. CNF with relatively high diameter

4.3 XRD results.

In figure 11, four X-ray diffraction patterns are superposed. Relatively high CNFs quantity is corroborated with these patterns. Each pattern corresponds to samples obtained under very specific operational conditions. Sample a) is the catalyst-graphite mixture before their exposure to the plasma; this X-ray pattern shows a rich crystalline structure. Sample b) was obtained at low applied power (158W). After the electronic microscopy study (SEM and

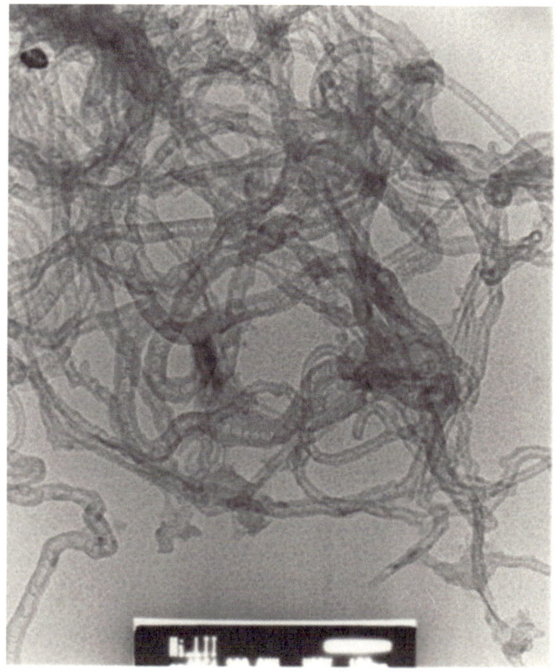

Fig. 10. CNF with periodic joints

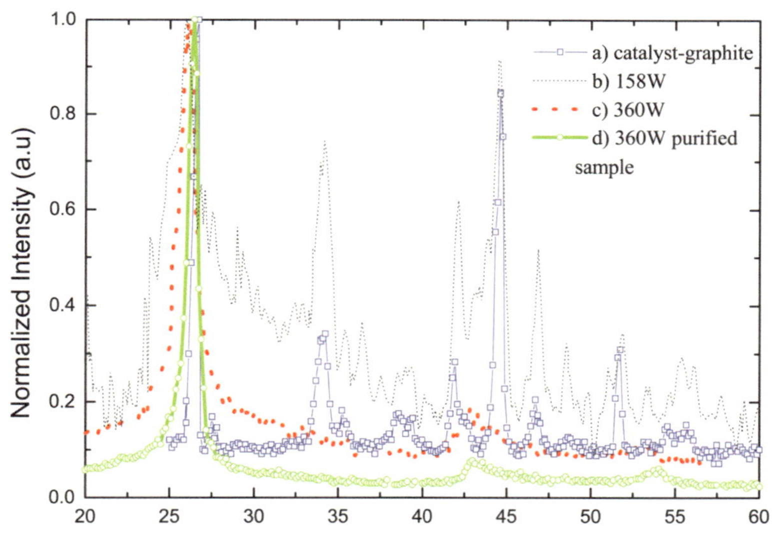

Fig. 11. X-ray diffraction patterns

TEM) it was found that the CNFs were not representative, however the X-ray pattern still shows a polycrystalline structure. The sample c) was obtained under 360W of applied power. The X-ray spectrum exhibits few defined peaks indicating a reduced crystalline structure of CNFs [51]. The most intense peaks are (0 0 2) and (1 0 0) peaks respectively situated at 26.25° and 42.20° in a 2θ system. The 26.25° angle corresponds to the interplanar spacing d_{002} of carbon nanofibers and nanotubes [52]. Finally, spectrum d) corresponds to a purified sample. Peak (0 0 2) is more intense than peaks found in other samples.

4.4 Raman scattering results

To support our analysis obtained by SEM, TEM and XRD techniques a fourth one was applied. The samples were also analyzed by the Raman scattering technique which is mostly used to characterize the crystalline structure.

The main criterion used in literature [50, 53] to reveal the carbon nanostructures quality by Raman scattering technique, is the ratio between the peaks G to D. The G peak is located around 1590 cm^{-1} and attributes C-C elongated vibration of graphite layers, indicating a well graphitized carbon nanostructure. Imperfect graphite structure is characterized by the D peak, near 1349 cm^{-1}, and it is also associated with the existence of amorphous carbon fragments rather than structure imperfections. The peak B situated at 159 cm^{-1} usually represents the radial breathing mode (RBM) in monowall carbon nanotubes. The formation of nanofibers instead of nanotubes could be explained by the presence of hydrogen in the plasma discharge that will terminate the dandling bonds at the edges of stacked graphite platelets [54].

From figure 12 it is deduced that the G/D ratio is around 1.41 corresponding to a high quality of samples [55 - 57].

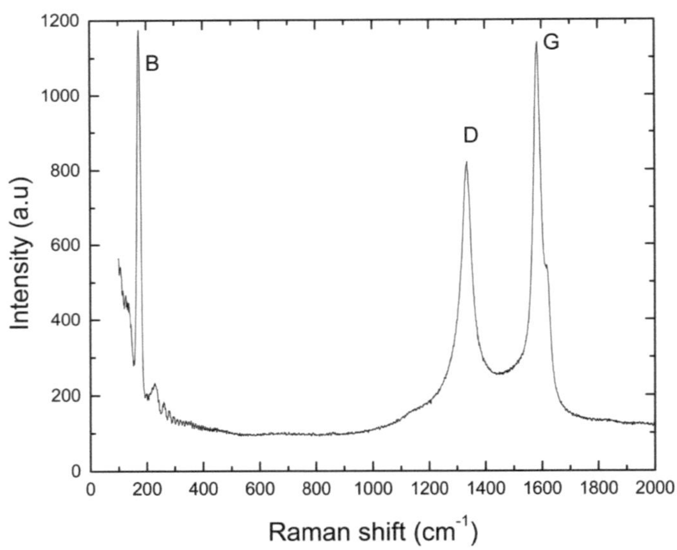

Fig. 12. Raman spectrum for sample obtained at 360W, showing the peaks B, D and G.

4.5 Power influence

To study the influence of the power input in the CNF synthesis several values of power input were tested and the obtained products were analyzed by SEM technique. Results of these tests are schematized in figure 13 which shows the CNFs evolution in function of power, that higher CNF yields are obtained at 360W; under this experimental condition the plasma remains very stable. To increase the power capacity several module reactors could be assembled into an array. The simplicity of its electric circuitry and adaptability to an AC glow-arc discharge are some of the most attractive features of this method. Modular plasma discharge working in an array has been already reported by Kuo and Koretzky [58, 59].

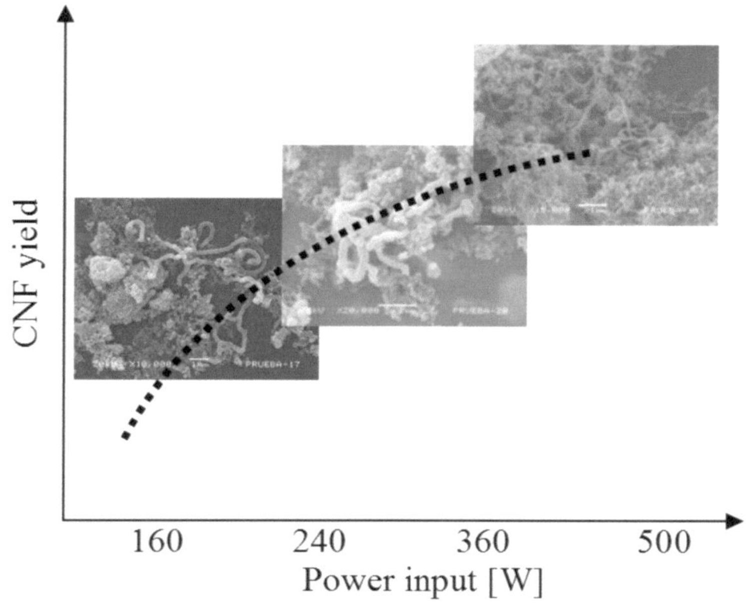

Fig. 13. Qualitatively CNFs yield in function of power input

4.6 Preliminary results of NO_x adsorption by CNF

To determine the energy of activation in CNF and, then, the process of sorption, CNF samples were contaminated with NO_X. Contaminated and uncontaminated CNF, were analyzed by thermogravimetry (figure 14), that usually consists in weight lost in function of temperature determination.

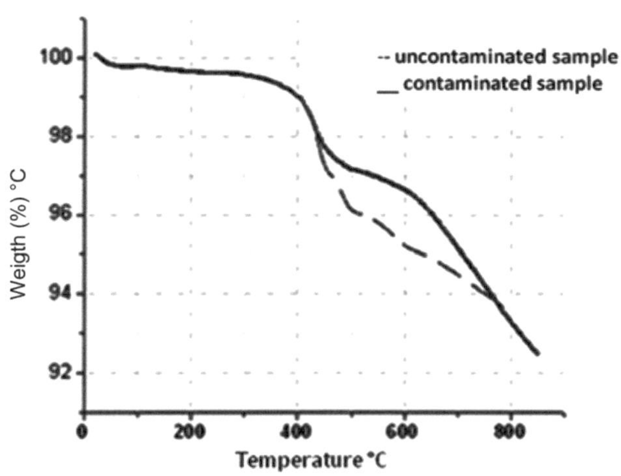

Fig. 14. CNF uncontaminated and contaminated with NO_x

By following the next procedure is possible to obtain the energy activation. Equation (1) represents a first order kinetics adsorption

$$\ln\left(-\ln(1-\alpha)\right)=\ln\left(\frac{A}{\beta}\frac{RT^2}{E_a^{des}}\right)-\frac{E_a^{des}}{R}\left(\frac{1}{T}\right) \tag{1}$$

Where,
T is temperature
R = 8.134 J/mol-K,
A: pre-exponencial factor (s^{-1})
β: heating velocity (K/min)
E_a^{des} : energy activation
α is the weight loss in function of T, more specifically,

$$\alpha=\frac{m_o-m_T}{m_o-m_f} \tag{2}$$

m_o is initial weight at T, m_T is the weight in function of T and m_f is final weight.
From data of figure 15, by plotting $ln(-ln(1\text{-}\alpha))$ versus $1/T$, is possible to obtain the activation energy of uncontaminated CNF (figure 15a) and from these contaminated with NOx (figure 15b).

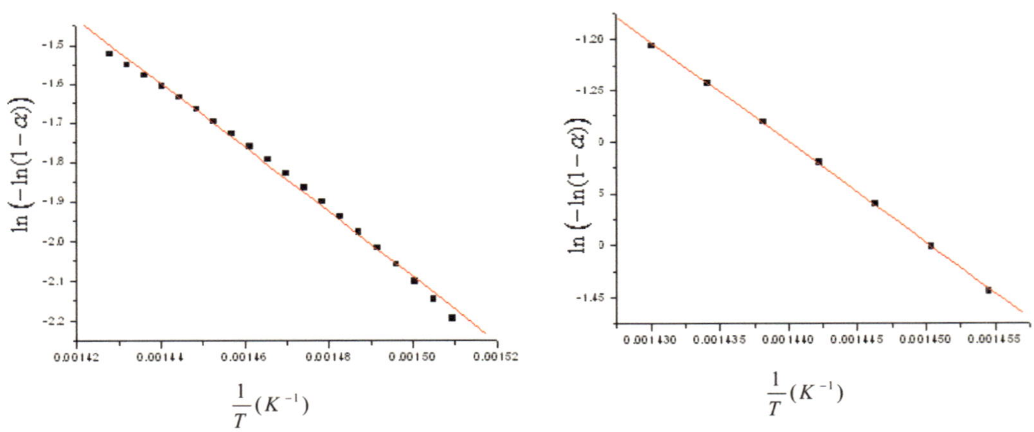

(a) (b)

Fig. 15. (a) Uncontaminated CNF, (b) Contaminated CNF

For the uncontaminated and contaminated samples the values of energy activation respectively are: E_a^{des} = 68.12 $\frac{kJ}{mol}$ and E_a^{des} = 80.98 $\frac{kJ}{mol}$. These, relatively low values corresponds to a physical absorption.
These results are similar to values obtained by some others authors (for carbon nanostructures the energy activation is between 10 KJ/mol-100KJ/mol [60,61]). The advantage of the physical adsorption, confirmed by thermogravimetric analysis, is that NOx, could be removed from CNF fluid bed by employing physical means such as a pressure camera.

An additional experiment was effectuated to test the capacity of adsorption of CNF, consisting in passing a constant flux of 400ppm of NOx during few minutes, through a CNF bed. By employing a NOx sensor (PG250) it was possible to determinate the removal rate being of about 87%. It is worth to note that additional experiments must be done, in order to confirm the life time of the CNF as support in a fluid bed.

5. Conclusion

A simple technique for CNFs synthesis is reported, the duration of processes is lower than 5 minutes and it requires neither preheating nor high flux of carrier gas. The synthesis has been achieved by the decomposition of methane in an AC low energy plasma discharge. The formed CNFs, exhibited a diameter of about 80nm with relatively no impurities. This purity allows the CNFs to be used as a catalyst support for subsequent applications in polymer composite formation or polluted gas absorbers.
The power input of the plasma discharge is an important parameter in the process, an optimization of the CNF synthesis was obtained at about 360W. A great advantage of using a high frequency electric field consists in controlling the power transferred during the glow discharge, and electric arc modes.
By comparing the energy consumptions for this AC plasma discharge with others different configurations, it is clearly shown that a CNFs synthesis can be produced with minimal energy consumption when this kind of AC glow-arc discharge is used. 800 kJ are needed to produce 1g of CNFs.
Preliminary experimental results shows that CNF obtained have a potential to be used as toxic gas adsorbers.
To increase the power and CNFs production, these modular plasma reactors can be connected in series or parallel configuration. The advantage of using a carbon-containing gas, instead of carbon consumable electrodes, resides in the small amount of energy that is needed to atomize it. All these attributions, would favor the implementation of a novel device for producing research quantities of CNF with a low cost and simplicity.

6. Acknowledgements

The supports obtained from the ININ (Mexican Institute of Nuclear Research), CONACyT (the Mexican Council for Technological Education contracts SEP-2004-C01-46959 and PCP), and the ICyTDF (Council for science and technology of México City) are gratefully acknowledged. The authors would also thank M. Durán, M. Hidalgo, F.Ramos, N. Estrada, S. Velazquez, C. Torres, A. Juanico, M.L Jiménez for their assistance in experimental tests and analysis. To M.I. Martínez and J. Pérez del Prado for their valuable help in the microscopy analysis and to L. Escobar-Alarcon for the Raman analysis.

7. References

[1] J. M. Blackman, J. W. Patrick, A. Arenillas, W. Shi, and C. E. Snape, *"Activation of carbon nanofibres for hydrogen storage"* Carbon, vol. 44, no. 8, pp. 1376–1385, Jul. 2006.
[2] P. Serp, M. Corrias, and P. Kalck, *"Carbon nanotubes and nanofibers in catalysis"*, Appl. Catal. A, Gen, vol. 253, no. 2, pp. 337–358, (2003).
[3] C. Park and R. T. K. Baker, *"Catalytic behavior of graphite nanofibers supported nickel particles. The effect of chemical blocking on the performance of the system"* J. Phys. Chem. B, vol. 103, no. 13, pp. 2453–2459,(1999).

[4] S.V. Mikhalovsky, , L.I. Mikhalovska, V.G. Nikolaev, V.I Chernyshev, V.V. Sarnatskaya, K.Bardakhivskaya, A.V. Nikolaev, L.A. Sakhno, *"Use Of High Surface Nanofibrous Materials In Medicine"*, Chapter in Nanoengineered Nanofibrous Materials, NATO Science Series, Mathematics, Physics and Chemistry Vol 169, 243-254, (2004)

[5] G. Tejral, N. R. Panyala, J. Havel, *"Carbon nanotubes: toxicological impact on human health and environment"*, J Appl Biomed, 7: 1–13, (2009).

[6] T. Masciangioli and W. X. Zhang, *"Environmental technologies at the nanoscale"* Environ. Sci. Technol., vol. 37, no. 5, pp. 102–108, (2003).

[7] V. L. Colvin, *"The potential environmental impact of engineered nanomaterials"* Nat. Biotechnol 21 (10), 1166–1170, (2003).

[8] A.B.M Danna, S.E Iyuke, A. Fakhrul-Razi, T. G. Chuah, M.A.Atieh, M.F Al-Khatib, *"Synthesis and Characterization of Carbon Nano-Structures for Methane Storage"*, Environmental Informatics Archives, Volume 1, 597-605, (2003).

[9] Q.Ngo, S. Krishnan,A.M Cassell, Y. Ominami, J. Li, M. Meyyappan, C.Y.Yang *"Electrical Characterization of Carbon Nanofibers for On-chip Interconnect Applications"*, Proceedings of 2005 5th IEEE Conference on Nanotechnology, Nagoya Japan, July 2005

[10] Q. Ngo, A. M. Cassell, A. J. Austin, J. Li, S. Krishnan, M. Meyyappan, C. Y. Yang, *"Characteristics of Aligned Carbon Nanofibers for Interconnect Via Applications"*, IEEE Electron Device Letters, 27 (4), 221- 224, (2006)

[11] T. Gao and T. H. Wang, *"Synthesis and properties of multipod-shaped ZnO nanorods for gas-sensor applications"* Appl. Phys. A, Solids Surf., vol. 80, no. 7, pp. 1451–1454, (2005).

[12] M. Monthioux, and V. Kuznetsov, *"Who should be given the credit for the discovery of carbon nanotubes?"*, Carbon 44, 1621-1622, (2006)

[13] W.R. Davis, R.J. Slawson, G.R. Rigby, *"An unusual form of carbon"*, Nature 171, 756, (1953).

[14] R. Bacon, *"Growth, Structure, and Properties of Graphite Whiskers."* J. Appl. Phys, 31(2), 283-290 (1960).

[15] A. Oberlin, M. Endo, and T. Koyama, *"Filamentous growth of carbon through benzene decomposition"*, J. Cryst. Growth, 32, 335 (1976).

[16] M. Endo, T. Koyama, Y. Hishiyama, *"Structural Improvement of Carbon Fibers Prepared from Benzene"*, Jap. J. Appl. Phys., 15, 2073-2076 (1976).

[17] V. Ivanov, J.B. Nagy, P. Lambin, A. Lucas, X.B. Zhang, X.F. Zhang, D. Bernaerts, G. van Tendeloo, S. Amelinckx, J. van Landuyt; *"The study of carbon nanotubules produced by catalytic method"*, Chem. Phys. Lett. 223(4) (1994), 329–335.

[18] N. M. Rodriguez, *"A review of catalytically grown carbon nanofibres"*, J. Mater. Res. 8, 3233-3250 (1993).

[19] R. T. K. Baker, *"Catalytic Growth of Carbon Filaments"* Carbon 27(33), 315-323 (1989).

[20] N. M. Rodriguez, A. Chambers, and R. T. K. Baker, *"Catalytic Engineering of Carbon Nanostructures"*, Langmuir 11(1010), 3862-3866 (1995).

[21] C. Park, N. M. Rodriguez, and R. T. K. Baker, *"Carbon deposition on Iron-Nickel during interaction with carbon monoxide-hydrogen mixtures"*, J. Catal. 169 (1), 212-227 (1997).

[22] S.A.Manafi and S. H. Badiee, *"Production of carbon nanofibers using a CVD method with lithium fluoride as a supported cobalt catalyst"*, Research Letters in Materials Science Article ID 850975, 5 pages, (2008)

[23] A.M. Benito, Y. Maniette, E. Munoz, M.T. Martinez, *"Carbon nanotubes production by catalytic pyrolysis of benzene"*, Carbon 36(5), 681–683, (1998).

[24] G.G. Tibbetts, D.W. Gorkiewicz, R.L. Alig; *"A new reactor for growing carbon-fibers from liquid-phase and vapor-phase hydrocarbons"*, Carbon 31(5), 809–814, (1993).

[25] P. E. Anderson and N. M. Rodriguez, "*Influence of the support on the structural characteristics of carbon nanofibers produced from the metal-catalyzed decomposition of ethylene*", Chem. Mater. 12, 823-830 (2000)
[26] E.S. Steigerwalt and C. M. Lukehart, "*Preparation of graphitic carbon nanofibers with the use of water-soluble supports*", J. Nanosci. Nanotech. 2, 25–28, (2002)
[27] D.Sun, L. Wang, D. Wu, and L. Lin, "*Electrospun Nanofibers Bundles*" Proceedings of the 1st IEEE International Conference on Nano/Micro Engineered and Molecular Systems January 18 - 21, 2006, Zhuhai, China
[28] L. Y. Yeo, J. R. Friend, "*Electrospinning carbon nanotube polymer composite nanofibers*", Journal of Experimental Nanoscience, Vol. 1, No. 2, June 2006, 177–209.
[29] M. Pacheco « *Synthèse des Nanotubes de Carbone par Arc Electrique*, PhD Thesis, Université Paul Sabatier, (2003)
[30] E. Dervishi; Z. Li; Y. Xu; V. Saini ; A. R. Biris; D. Lupu; A. S. Biris, "*Carbon Nanotubes: Synthesis, Properties, and Applications*", Particulate Science and Technology, 27 (2), 107 - 125, (2009).
[31] H. Huang, H. Kajiura, Y. Murakami, M.Ata, "*Metal sulfide catalyzed growth of carbon nanofibers and nanotubes*". Carbon 41, 615–618, (2003).
[32] S. Seraphin, S. Wang, D. Zhou, J.Jiao, "*Strings of spherical carbon clusters grown in a catalytic arc discharge*", Chem Phys Lett. 228, 506–512, (1994).
[33] H. Kajiura, H. Huang, S. Tsutsui, Y. Murakami, M.Miyakoshi, "*High purity fibrous carbon deposit on the anode surface in hydrogen DC arc-discharge*", Carbon 40, 2423– 2428, (2002)
[34] J. B. O. Caughman, L. R. Baylor, M. A. Guillorn, V. I. Merkulov, D. H. Lowndes, and L. F. Allard, "*Growth of vertically aligned carbon nanofibers by low-pressure inductively coupled plasma-enhanced chemical vapor deposition*", Appl. Phys. Lett. 83, 1207 (2003);
[35] V. I. Merkulov, A. V. Melechko, M. A. Guillorn, D.H. Lowndes, M. L. Simpson, "*Alignment mechanism of carbon nanofibers produced by plasma-enhanced chemicalvapor deposition*", Appl. Phys. Lett., 79 (18), 2970-2972 (2001).
[36] H. Le Poche, J. Dijon and T. Goislard de Monsabert, "*Radio-frequency plasma system to selectively grow vertical field-aligned carbon nanofibers from a solid carbon source*", Carbon , 45 (15), 2904-2916, (2007).
[37] A. Caillard, C. Charles, R.W. Boswell, P. Brault, "*Synthesis of Carbon Nanofibers and Pt-Nanocluster-Based Electrochemical Microsystems by Combining Low-Pressure Helicon Plasma Techniques*" IEEE Transactions on Plasma Science, 36 (4), 882 - 883, (2008)
[38] R. C. Smith, J. D. Carey, C. H. P. Poa, D. C. Cox, S. R. P. Silva "*Electron field emission from room temperature grown carbon nanofibers*", J. Appl. Phys. 95, 3153 (2004).
[39] D. Harbec, J. L. Meunier, L. Guo, R. Gauvin, N. El Mallah, "*Carbon nanotubes from the dissociation of C_2Cl_4 using a DC thermal plasma torch,*" J. Phys. D, Appl. Phys., 37, 15, pp. 2121-2126, 2004
[40] A. A. Howling, C. Hollenstein, P.-J. Paris, "*Direct visual observation of powder dynamics in RF plasma-assisted deposition*", Appl. Phys. Lett., 59, no. 12, pp. 1409–1411, 1991.
[41] T. Nitter, T. Aslaksen, F. Melandso, O. Havnes, "*Levitation and dynamics of a collection of dust particles in a fully ionized plasma sheath*", IEEE Trans. Plasma Sci., vol. 22, no. 2, pp. 159–172, Apr. 1994.
[42] M. Lackowski, A. Jaworek, A. Krupa, "*Current–voltage characteristics of alternating electric field charger*" , J. Electrost., vol. 58, no. 1/2, pp. 77–89, May 2003.
[43] J. O. Pacheco-Sotelo, R. Valdivia-Barrientos, M. Pacheco-Pacheco, J. F. Ramos-Flores, M. A. Durán-García, J. S. Benitez-Read, "*A universal resonant converter for equilibrium and nonequilibrium plasma discharges*", IEEE Trans. Plasma Sci., vol. 32, no. 5,pp. 2105–2112, Oct. 2004.

[44] E. E. Kunhardt, *"Generation of large-volume, atmospheric-pressure, nonequilibrium plasmas"*, IEEE Trans. Plasma Sci., vol. 28, no. 1, pp. 189–200, Feb. 2000.

[45] R. Valdivia-Barrientos, J. Pacheco-Sotelo, M. Pacheco-Pacheco, F. Ramos Flores, A. Cruz-Azocar, M. L. Jiménez López, J. S. Benitez-Read, R. López-Callejas, *"Optical and electrical diagnosis of a High-Frequency Glow-Arc Discharge and its Application to the Synthesis of Carbon Nanofibers"*, IEEE Transaction on Plasma Science, Vol. 35, No. 5, pp 1550-1558, 2007.

[46] M. I. Boulous, P. Fauchais, and E. Pfender, *Thermal Plasmas: Fundamentals and Applications*, vol. I. New York: Plenum, p. 9. 273, 1994.

[47] M. Pacheco, J. Pacheco, M. Valdivia, L. Bernal, R. Valdivia, A. Huczko, et al.,*"Synthesis of carbon nanostructures by using thermal plasma torch"*, Braz.J.Phys., vol. 34, no. 4B, pp. 1684–1688, Dec. 2004.

[48] H. Lange, A. Huczko, M. Sioda, M. Pacheco, M. Razafinimanana, A. Gleizes, *"On self absorption method for determination of C_2 in carbon arc plasma"*, Progress Plasma Process. Mater., P. Fauchais, Ed. NewYork: Begell House Inc., 2003.

[49] T. Matsuura, K. Taniguichi, and T. Watanabe, *"A new type of arc plasma reactor with twelve-phase alternating current discharge for synthesis of carbon nanotubes"* 17th Int. Symp. Plasma Chem., Toronto, ON, Canada, Aug. 7–12, 2005, Paper ID 634.

[50] J. Yu, Q. Zhang, J. Ahn, S. F. Yoon, R. Y. J. Li, B. Gan, K. Chew, K. H. Tan, *"Synthesis of Carbon nanostructures by microwave plasma chemical vapor deposition and their characterization"*, Mat. Sci. Eng., vol. B90, no. 1, pp. 16–19, Mar. 2002.

[51] C. T. Hsieh, J.M. Chen, R. R. Kuo, and Y. H. Huang, *"Formation and field emission properties of carbon nanofibers by a simplified thermal growth"*, Rev. Adv. Mater. Sci., vol. 5, no. 5, pp. 459–463, 2003.

[52] E. G. Rakov, S. N. Blinov, I. G. Ivanov, E. V. Rakova, and N. G. Digurov, *"Continuos process for obtaining carbon nanofibers"*, Russ. J. Appl. Chem., vol. 77, no. 2, pp. 187–191, Feb. 2004.

[53] X. Zhao, S. Inoue, M. Jinno, T. Suzuki, and Y. Ando, *"Macroscopic oriented web of single-wall carbon nanotubes"*, Chem. Phys. Lett., vol. 373, no. 3/4, pp. 266–271, May 2003.

[54] L. Delzeit, I. McAninch, B. A. Cruden, D. Hash, B. Chen, J. Han *"Growth of multiwall carbon nanotubes in an inductively coupled plasma reactor"* , J. Appl. Phys., vol. 91, no. 9, pp. 6027–6033, May 2002.

[55] T. Nozaki, Y. Kimura, K. Okazaki, *"Carbon nanotubes deposition in glow barrier discharge enhanced catalytic CVD"*, J. Phys. D, Appl. Phys., vol. 35, no. 21, pp. 1–6, Nov. 2002.

[56] G. S. Choi, Y. S. Cho, S. Y. Hong, J. B. Park, K. H. Son, D. J. Kim, *"Carbon nanotubes synthesized by Ni-assisted atmospheric pressure thermal chemical vapor deposition"*, J. Appl. Phys., vol. 91, no. 6, pp. 3847–3854, Mar. 2002.

[57] C. Pham-Huu, R. Vieira, B. Louis, A. Carvalho, J. Amadou, T. Dintzer, and M. J. Ledoux, *"About the octopus-like growth mechanism of carbon nanofibers over graphite supported nickel catalyst"*, J. Catal., vol. 240, no. 2, pp. 194–202, Jun. 2006.

[58] S. P. Kuo, E. Koretzky, and L. Orlick, *"Design and electrical characteristics of a modular plasma torch"*, IEEE Trans. Plasma Sci., vol. 27, no. 3,pp. 752–758, Jun. 1999.

[59] E. Koretzky and S. P. Kuo, *"Characterization of an atmospheric pressure plasma generated by a plasma torch array"* Phys. Plasmas, vol. 5, no. 10, pp. 3774–3780, Oct. 1998.

[60] M.A. Keane, "*Interfacial Applications in Environmental Engineering*", Marcel Dekker, New York., 2003,

[61] C.-H Wu, *"Adsorption of reactive dye onto carbon nanotubes: Equilibrium, kinetics and thermodynamics"*, Journal of Hazardous Materials, 144, 1, 93-100, 2007.

14

Morphology and Dispersion of Pristine and Modified Carbon Nanofibers in Water

Jian Zhao
Qingdao University of Science and Technology, Qingdao, China, and Department of Chemical and Materials Engineering, University of Cincinnati, Cincinnati, OH, USA

1. Introduction

Carbon nanofibers have been of great interest due to their extraordinary mechanical and electronic properties. Carbon nanofibers (CNF) are different from carbon nanotubes in that they have many more walls of crystalline carbon and usually have more structural defects than carbon nanotubes. The cost of preparing carbon nanofibers is significantly less than carbon nanotubes due to the synthesis techniques used, defects and the remaining amorphous carbon.

Carbon nanofibers are suitable for a range of applications such as reinforcing fillers, field emitters and nanoelectronic devices etc. (Dresselhaus, Dresselhaus et al. 2001; Safadi, Andrews et al. 2002; Gong, Li et al. 2005; Li, Zhao et al. 2005) Unfortunately, the advantages of carbon nanofibers have not been realized because of the difficulty of obtaining fully dispersed nanofibers. Although hundreds of papers have been published describing enhanced dispersion of carbon nanofibers by surface modification, plasma treatment and functionalization of the sidewalls and fiber tips, quantitative measurement of the degree of dispersion remains challenging and the nature of the dispersed entities remains unknown.

Scattering methods is an ideal tool to provide structural information about nanofiber morphology. In this chapter, we review several approaches that are used to assist dispersion, including surface modification, PEG-functionalization and plasma treatment Small angle light scattering is utilized as a primary tool to assess the morphology of the carbon nanofibers and quantify dispersion of the carbon nanofibers treated through these approaches. A simplified tube or fiber model is introduced to assist in further understanding the morphology. The chapter is divided into three sections.

The first section focuses on dispersion of untreated and acid-treated carbon nanofibers suspended in water. Analysis of Light scattering data provides the first insights into the mechanism by which surface treatment promotes dispersion. Both acid-treated and untreated nanofibers exhibit hierarchical morphology consisting of small-scale aggregates (bundles) that agglomerate to form fractal clusters that eventually precipitate. Although the morphology of the aggregates and agglomerates is nearly independent of surface treatment, their time evolution is quite different. Acid oxidation has little effect on bundle morphology. Rather acid treatment slows agglomeration of the bundles. The second section discusses the morphology and dispersion of solubilized carbon nanofibers. Light scattering data indicate that PEG-functionalized nanofibers are dispersed at small rod-like bundle (side-by-side aggregate) level. PEG-functionalization of the carbon fibers leads to solubilization not by disrupting small-scale size-by-side bundles, but by inhibiting formation of the large-scale agglomerates. The third section focuses on dispersion of plasma-treated carbon nanofibers. Comparison of untreated and plasma-treated nanofibers shows that plasma treatment facilitates dispersion of nanofibers. The chapter will conclude with a summary.

1.1 Thermodynamics of nanophase carbon

In this section, the origin of the dispersion problem, mainly with respect to thermodynamics is explored. Several factors make the dispersion of nanophase carbon particularly troublesome. These factors are dominated by strong attraction between carbon species of both enthalpic and entropic origin. In addition, the low dimensionality of carbon nanotubes leads to an enhancement of these attractive forces.

The origin of the attractive forces between graphitic structures is well known. Due to the extended pi electron system, these systems are highly polarizable, and thus subject to large attractive van der Waals forces. These forces are responsible for the secondary bonding that holds graphitic layers together. In the case of carbon nanofibers, these forces lead to so called "bundles," extended structures formed by side-by-side aggregation of the nanofibers.

When suspended in a polymer, an attractive force between filler particles also arises due to pure entropic factors. (Bechinger, Rudhardt et al. 1999) Polymer chains in the corona region of the colloidal filler suffer an entropic penalty since roughly half of their configurations are precluded. Therefore there is a depletion of polymer in the corona. This depletion results in an osmotic pressure forcing the filler particles together. This effective attraction is intrinsic to colloids dispersed in polymers.

Finally, the linear structure of carbon nanotubes leads to a cooperative effect that enhances the forces described above. Whereas spherical particles touch at a point, rods interact along a line. As a result the above forces are augmented by filler geometry.

1.2 Structure and small-angle scattering:

Small-angle scattering is a powerful technique for characterization of fractal objects. Small angle scattering (SAS) is the collective name given to the techniques of small angle neutron (SANS), x-ray (SAXS) and light (SALS, or just LS) scattering. Fractal objects are geometrically self-similar under a transformation of scale. (Schaefer 1988) This self-similarity is implicit in the power-law functions. In a scattering experiment, however, self-similarity is manifest in a power-law relationship between intensity I and wave vector q.

$$I(q) \propto q^{-P} \tag{1}$$

In scattering experiments, the scattered intensity I(q), which is proportional to the scattering cross section per unit volume $d\Sigma/Vd\Omega$, is measured as a function of scattering angle θ. This angle is related to the wave vector, q.

$$q = 4n\pi/\lambda \sin(\theta/2) = 2\pi/d \tag{2}$$

where λ is the wavelength of the incident beam in the media, θ is the scattering angle, and d is the length scale probed in the experiment.

The scattered intensity, I(q), then is expressed as:

$$I(q) \propto NP^2(q)S(q) \tag{3}$$

where N is the number density of individual scatterers, P(q) is a form factor related to the shape and scattering crosssection of the scatterers and S(q) is the structure factor related to correlations between the scatterers.

In the scattering experiment, a beam of electromagnetic radiation strikes a sample. The radiation is elastically scattered by the sample. A detector records the scattered beam. The resulting scattering pattern can be analyzed to provide information about the size and shape etc. of nanoparticles. Scattering techniques effectively probe an object on different length scales as determined by q^{-1}. (Schaefer, Bunker et al. 1989)

In our study, we use small angle light scattering as a primary tool to investigate dispersion of nanofibers. The dispersion efficiency was determined using a low-angle light scattering photometer–a Micromeritics Saturn Digitizer (www.micromeritics.com). Light scattering data are reported in reciprocal space (intensity vs. wave vector, q). Data in this form are directly available. Light scattering covers the regime 10^{-6} Å^{-1} < q < 10^{-3} Å^{-1}. The q-range corresponds to length-scales (~q^{-1}) from 100 μm at low q to 1000 Å at high q. A scattering curve can be fitted over two-level regimes by a unified function related to the aggregated bundles and agglomerate structure respectively.

2. Acid-treated and As-received nanofibers

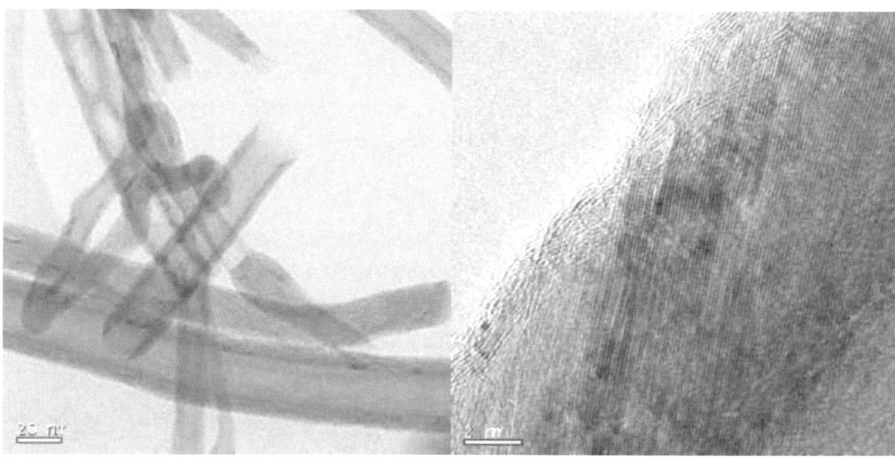

Fig. 1. TEMs of unmodified carbon nanofibers PR19HT. Graphitic layers are visible at both magnifications. The low-resolution image shows a variety of tube shapes and morphologies including concentric cylinders and stacked cones. No metallic catalyst was observed. The bars are 20 nm and 2 nm.

Applied Sciences, Inc. (ASI) made all the nanofiber samples used in this research using full scale chemical vapor deposition (CVD). A 3:1 concentrated H_2SO_4:HNO_3 mixture is commonly used for surface modification. (Chen, Hamon et al. 1998; Chen, Rao et al. 2001) After such acid treatment, nanofibers form relatively stable colloidal solutions in water. Dispersions have been characterized by atomic force microscopy (AFM), UV/visible-NIR spectra etc. (Shaffer, Fan et al. 1998; Ausman, Piner et al. 2000) The evolution of the dispersed state under quiescent conditions following sonication, however, remains unknown. Here, we use light scattering to quantify the state dispersion of as-received and acid-treated carbon nanofibers as a function of time. To understand the state of aggregation of the nanofibers, the size distribution from the light scattering data is determined using the maximum entropy (ME) method. (Potton, Daniell et al. 1988; Morrison, Corcoran et al. 1992; Boukari, Long et al. 2000) We used the Irena code (http://www.uni.aps.anl.gov/~ilavsky/irena.html) developed by Ilavsky and Jemian to obtain the maximum-entropy solution. (Jemian, Weertman et al. 1991; Ilavsky 2004)

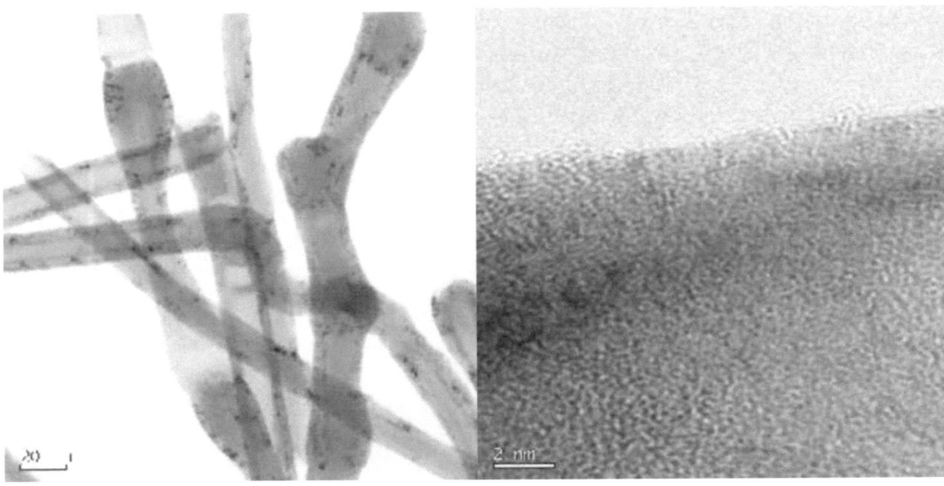

Fig. 2. TEMs of acid-treated nanofibers PR19LHT: More defects on the walls are evident and breakage of graphite layers. The bars are 20 nm and 2 nm.

2.1 Transmission Electron Microscopy (TEM)

The raw Pyrograf®-III PR19HT powder consists of loosely aggregated nanofibers. Some nanofibers are curved with open ends. A representative HRTEM image of the original Pyrograf®-III PR19HT (Figure 1) shows the graphite structure with the interlayer spacing d = 0.34 nm. No iron catalyst particles are found by TEM. Defects on the walls of nanofibers are occasionally observed in pristine nanofibers.

	Time	5min	1 hr	2 hr	5 hr	8 hr	24 hr	32hr	44hr
Low q	Rg (μm)	4.8	4.6	4.5	4.3	3.5	8.4	11.9	14.4
	P	1.04	1.07	1.09	1.01	1.00	1.29	1.45	1.78
	G	9.5	8.7	8.5	7.2	4.6	14.0	30.5	69.1
	105 B	6.33	9.89	2.75	5.28	1.94	0.94	0.45	0.006
High q	Rg (μm)	0.80	0.83	0.78	0.75	0.86	0.79	0.77	0.87
	P	2.10	2.13	2.15	2.03	2.09	2.00	2.08	2.07
	G	0.87	1.15	0.72	0.42	1.30	0.97	1.04	1.61
	108 B	3.25	2.72	2.82	4.01	3.35	6.09	3.14	3.29

Table 1. Guinier radii and exponents as a function of time for acid-treated carbon nanofibers PR19HT.

Bright-field, high-resolution TEM images of the acid-treated nanofibers PR19HT are shown in Figure 2. More defects and even serious damage are found after the acid treatment. Disruption of outer graphitic layers is also observed. The stripping of the altered outer graphite layers after strong oxidation can give rise to thinning of nanofibers. These observations are consistent with the literature. (Shaffer, Fan et al. 1998; Monthioux, Smith et al. 2001)

2.2 Light scattering investigation

There is considerable experimental evidence for the presence of carboxylic acid bound to carbon nanotubes through acid treatment. (Liu, Rinzler et al. 1998; Hu, Bhowmik et al. 2001) These carboxylic groups result in improved dispersion of carbon nanotubes in polar solvents. The carboxylated carbon nanofibers are stable in water for days. In the absence of sonication, however, tubes eventually aggregate and precipitate. We use light scattering to monitor this process.

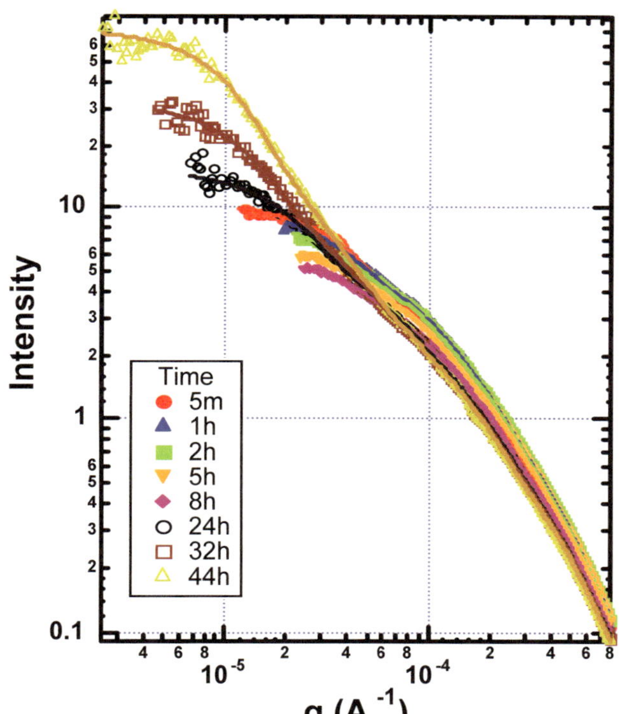

Fig. 3. Dispersion of acid-treated nanofibers PR19HT in water during two-day suspension. The suspensions were sonicated at 10W for five minutes before data were taken using light scattering in batch mode. Minimal change is observed at large q indicating minimal change in morphology below 1 μm. The micron-size entities originally present simply aggregate into larger structures in a hierarchical fashion. The lines are two-level unified fits. The unified parameters are collected in Table 1.

Figure 3 shows the light scattering profiles as a function of time for acid-treated nanofibers PR19HT in water at a concentration of 5.0×10^{-6} g/ml. The data were obtained in the batch mode with no circulation or sonication. Initially the scattered intensity at low q decreases up to 8 hours. Below we argue that this decrease is due to settling with minimal agglomeration. After 8 h, however, the intensity increases substantially at low-q, consistent with agglomeration. In dilute solution the intensity at q -> 0 is proportional to molecular weight, so the data in Fig. 3 imply an increase of molecular weight by a factor of ten between 8 and 44 hours.

Except for the data near 8 h, two "knees" are observed indicating two length scales. The knee region is referred to as Guinier scattering. The curvature in the Guinier regime defines a length scale (Guinier radius or radius-of-gyration, Rg, in the case of independent scatterers). Each Guinier knee is followed by a quasi power-law regime. The curves were fit using Beaucage's Unified Model to extract Rg, the power-law exponents, P, and the Guinier prefactors, G, and power-law prefactor, B, associated with each length scale. (Beaucage, Schaefer et al. 1994) These parameters are displayed in Table 1 for the two structural levels. The slope near –2 (P = 2) on a log-log plot around q = 0.002 $Å^{-1}$ could arise from a hollow tube since the wall of such an object is a two-dimensional on scales larger than its wall thickness and shorter than the radius. Such a slope, however, can also arise from more complex aggregated structures. (Schaefer, Brown et al. 2003; Schaefer, Zhao et al. 2003) The crossover length scale ($q^{-1} \cong 1$ μm) between the two power-law regimes corresponds to the largest radius of the tube aggregates. Minimal change in Rg and P is observed for $q > 10^{-4}$ $Å^{-1}$, indicating minimal change in morphology with time on length scales below ~1 μm.

The prefactor, G, derived from high-q structural level (level 1) decreases up to 8 hours, indicating that the number and/or molecular weight of the small-scale entities is decreasing up to 8 h. After 8 h, however, the data indicate that these small-scale structure cluster form large-scale objects, which we call agglomerates. All the carbon precipitates after several days. Chen et al observed similar behavior in the region $2 \times 10^{-4} < q < 2 \times 10^{-3}$ for single-walled nanotube suspensions. (Chen, Saltiel et al. 2004)

We also monitored dispersion behavior of untreated carbon nanofibers PR19HT although such a suspension is quite unstable in water even with aid of ultrasound. The data for the untreated sample (Figure 4) show similarities and differences when compared to the treated sample. Although less visible, two structural levels are present and the length scales are similar to the treated case in Figure 3. For the as-received sample, however, the large-scale agglomerates are observed immediately (5 min) whereas they form after 8 h in the treated case. Acid-treatment retards this agglomeration. The overall intensity also shows a nearly monotonic trend with time to lower values consistent with precipitation being the dominant process. The extracted Rgs (Table 2) are virtually unchanged during the precipitation process. The similarity of the length scales for the treated and untreated samples shows that acid treatment does not change the gross morphology of the nanotube aggregates, but only inhibits agglomeration of these smaller scale aggregates. (Zhao, Schaefer et al. 2005)

Figure 5 compares the scattering profile for acid-treated and as-received carbon nanofibers at 5 min after sonication. The intensity at low q (prefactor, G) for the acid-treated sample is one order of magnitude smaller than that for the untreated one, indicating smaller entities in former suspension. Detailed analysis shows that the low-q Guinier radius is 4.8 μm for acid-treated nanofibers, compared to 21.3 μm for untreated nanofibers. These observations are consistent with improved dispersion due to the presence of oxygen-containing functional groups on the surface.

In principle, the morphology of both the small and large-scale objects can be inferred from the power-law exponents, P, since P is the fractal dimension, D, of the objects giving rise to

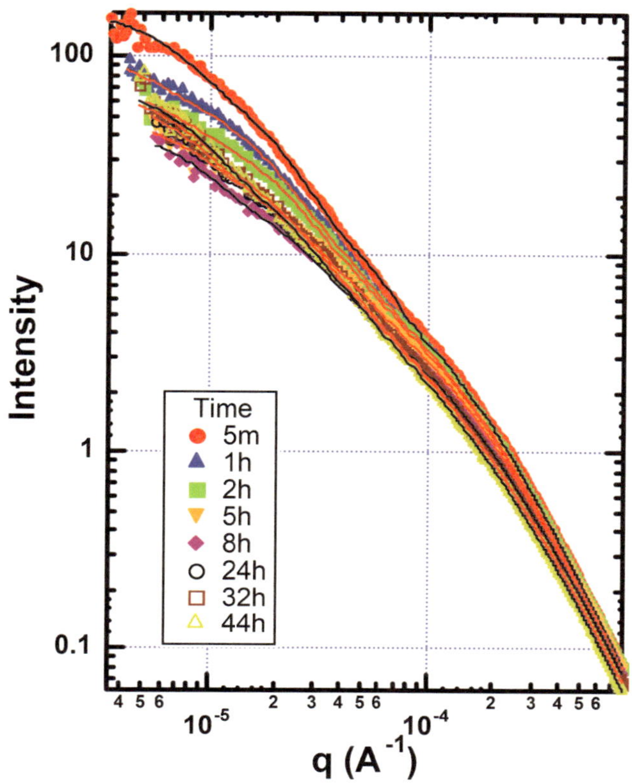

Fig. 4. Evolution of the light scattering profile of un-modified nanofibers PR19HT for two days following dispersion by sonication. The suspensions were sonicated at 10 W for five minutes before the observations began. The measurements were taken in the batch mode, so the sample was undisturbed during the course of the experiment. Note that the intensity at small q is a factor of 10 larger than Fig. 3, implying larger aggregated structures. The lines are two-level unified fits.

	Time	5min	1 hr	2 hr	5 hr	8 hr	24 hr	32hr	44hr
Low q	Rg (μm)	21.3	20.9	20.8	19.4	18.6	19.5	20.1	20.2
	P	1.48	1.44	1.43	1.22	1.23	1.32	1.35	1.40
	G	160.2	110.1	74.2	51.5	44.9	65.2	70.1	74.3
	106 B	0.59	4.23	4.14	29.72	23.24	10.67	8.30	4.43
High q	Rg (μm)	0.86	0.84	0.88	0.83	0.87	0.83	0.83	0.82
	P	2.01	2.07	2.15	2.14	2.08	2.00	1.98	2.10
	G	1.53	1.24	1.46	0.92	1.00	0.67	0.66	0.69
	108 B	4.53	2.93	1.67	1.77	2.56	3.90	4.75	1.89

Table 2. Guinier radii and exponents as a function of time for as-received carbon nanofibers PR19HT.

the scattering. D = 1 implies a linear objects and D > 1 indicates more branched or flexible structures. (Schaefer, Brown et al. 2003; Schaefer, Zhao et al. 2003) For our data, however, the scattering entities are polydisperse and the power-law regions extend over a very limited q range, so this approach is unworkable. An alternative is to use the relationship

$$M_w = R_z^D ,$$

where Mw is the weight-average molecular weight, Rz is the weight-squared-average radius and D is the fractal dimension of the object. Since G ~ Mw and Rz ~ Rg, D can be extracted from the slope of a log-log plot of G vs. Rg. Such a plot is shown in Figure 6 for the two structural levels for both the treated and untreated samples. Except for the low-q data for the treated sample, the data imply D ≥ 6, which is unphysical. When the slope is greater than 3 we interpret the evolution of the scattering profile as due to precipitation of carbon, because, to first approximation, G simply decreases at fixed Rg. For the untreated samples, therefore, both structural levels evolve by precipitation.

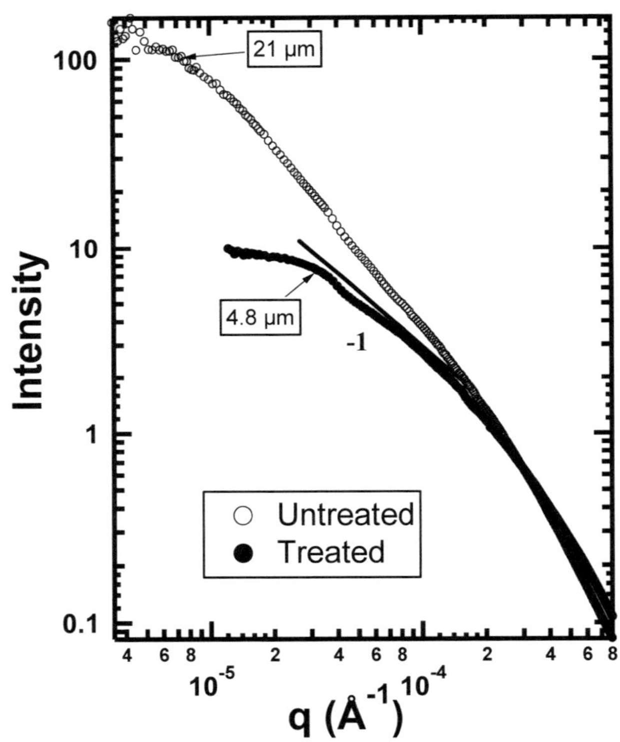

Fig. 5. Comparison of the scattering profiles for untreated and acid-treated carbon nanofibers PR19HT 5 min after sonication. A substantial population large-scale clusters is present only for the untreated sample.

For the treated sample, the small-scale objects precipitate for the first 8 h. and then begin to agglomerate. The latter conclusion is reached because D = 1.7 ± 0.15 for the large-scale structure (Figure 6), consistent with a fractal morphology characteristic of agglomerates formed by kinetic growth. This number is also consistent with the value of P in Table 1 for the 44-h treated sample. For the other times, the power-law region is insufficient to compare the measured Ps and D from Figure 6. For the treated sample, precipitation dominates agglomeration up to 8 h and agglomeration dominates precipitation after 8 h.

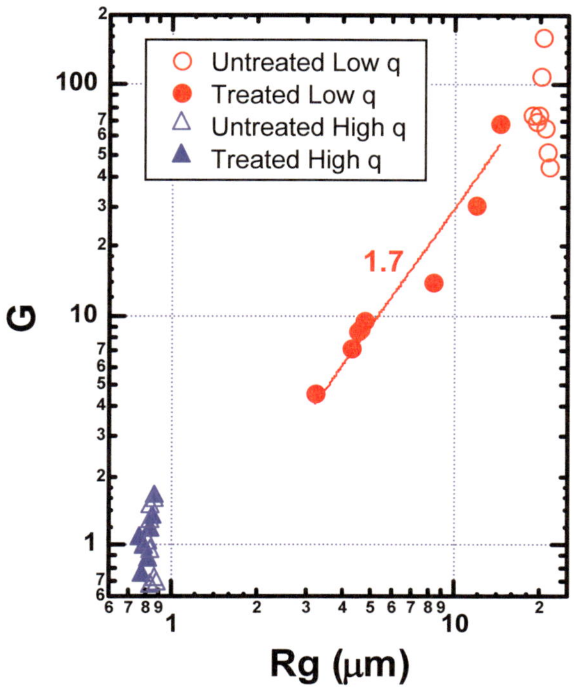

Fig. 6. Relationship between the Guinier prefactors and the Guinier radius for two structural levels observed in Figures 4 and 5. The low-q result for the treated sample is consistent with agglomeration to form a fractal cluster with fractal dimension D = 1.7. In the other cases, the large change in G with minimal change in Rg is consistent with sedimentation.

2.3 Size distribution

In order to further understand the morphology evolution, the data were analyzed to estimate the bundle size distributions using the maximum-entropy method. To extract the size distribution a particle shape must be assumed. Electron microscopy shows that carbon nanofibers are tube-like with some fibers showing more rod-like character. The high-q feature of the data should arise from this one-dimensional morphology.

We investigated both rod and tube models as shown in Figures 7 and 8, which compare the two models for the 8-h acid-treated sample. Figure 5.1.7 shows the fits to the light scattering data. Figure 8 shows the corresponding volume distributions assuming both rod and tube form factors. In the tube case, distributions calculated for different tube-wall thicknesses are shown. For the tube model the fits to the scattering profiles are virtually independent of wall thickness, although the volume distributions show a change in amplitude consistent with the increase of molecular weight with wall thickness, the solid rod being the limiting case. Fortunately, the position and shape of the distribution does not depend strongly on the assumed form factor.

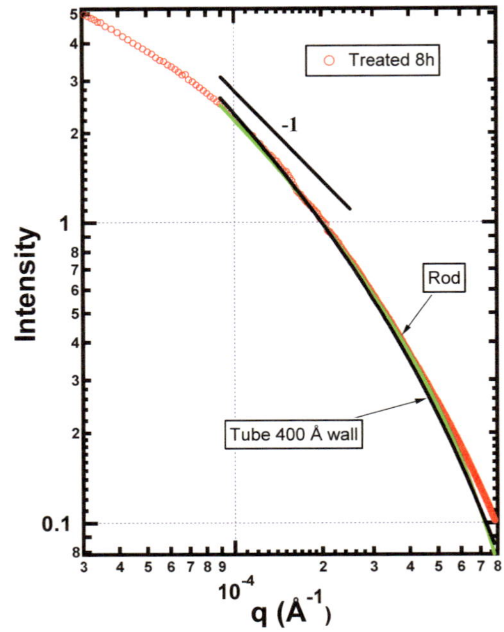

Fig. 7. Maximum-entropy fits to the 8-h treated data assuming rods and tubes. The tube fit is virtually independent of the assumed tube wall thickness. Both rods and tube distributions fit equally well.

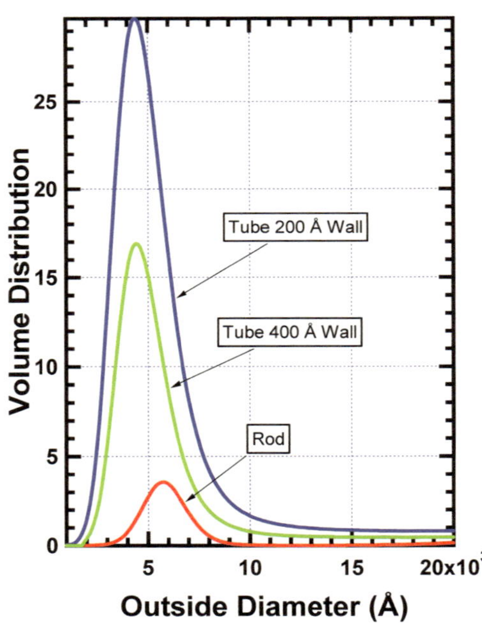

Fig. 8. Volume distributions assuming rods and tubes for the 8-h acid treated data. Both models show diameters somewhat larger than that observed by TEM, indicating some (side-by-side) aggregation.

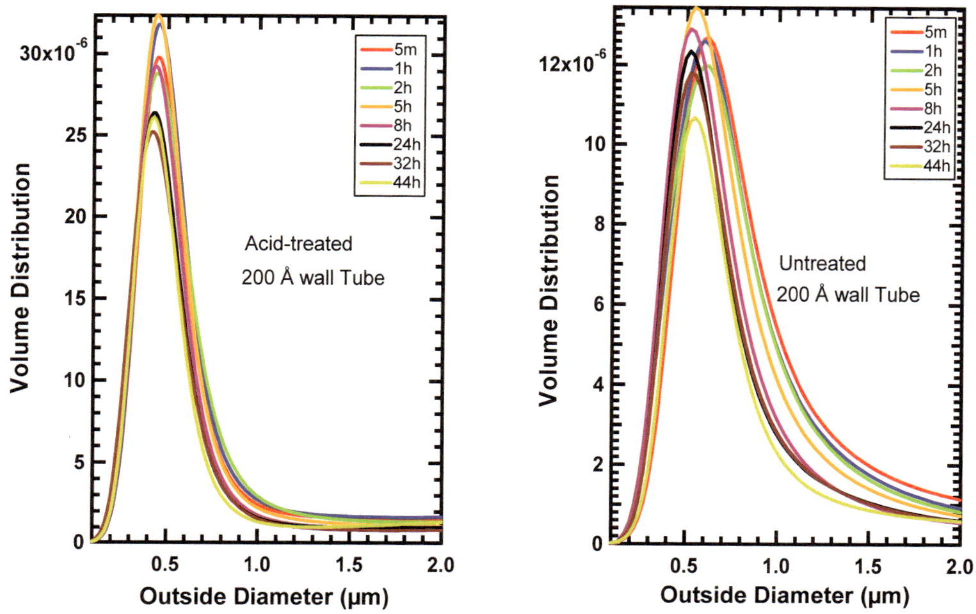

Fig. 9. (a) High-q volume distributions as a function of time for acid-treated nanofibers. The curves are obtained by ME method from light scattering data in the high q region. The particle shape used is tube. The ordinate is arbitrary in that the contrast is not known.
b) High-q volume distributions for untreated nanofibers. The curves are obtained by ME method from light scattering data in the high q region. The particle shape used is tube. Although the ordinate is in arbitrary units, the distributions can be quantitatively compared to Figure 9.

Based on the fits, it is impossible to distinguish between rods and tubes or for that matter more complex structures. Both models give similar results with a peak in the diameter distribution around 0.5 μm. In all cases, the diameter at the peak is considerably larger than the tube diameters seen in TEM. The scattering entities are not individual tubes, but bundles thereof.

Interestingly, the size distributions (Figures 9) extracted from these high-q data show minimal change with time implying that the short-scale morphology is maintained during agglomeration and precipitation. Both the peak position and the tails to larger sizes indicate that dispersion is not complete, based on the tube diameters seen in TEM. The bias to larger sizes is likely due to side-by-side fiber aggregates that are never disrupted. Because light scattering is weighted by volume, it doesn't take much aggregation to produce such tails on the size distributions.

Comparison of Figures 9 (a) and (b) shows that the large-diameter wing of the distribution is suppressed in the acid-treated sample. That is, acid treatment breaks up the larger aggregates. Since the contrast is not known, the volume distributions (ordinate in Figure 9) are on an arbitrary scale. Nevertheless, comparison of the distributions in Figures 9 is meaningful. Based on this comparison, the volume missing from the large-diameter wings of the treated distributions shows up at around 0.4 μm, which is still substantially larger than the largest individual fibers seen in TEM.

The simplified rod and tube models used here were developed by Justice and Schaefer. (Schaefer, Justice et al. 2005) These models approximate the exact rod and tube models in various power-law regimes and give the proper crossover length scales. Exact models, however, display oscillations in the power-law regimes, which are suppressed in the simplified models. This simplification is of minimal consequence when dealing with polydisperse distributions but it does accelerate the ME fitting code.

Determination of the size distribution for the low-q data is more challenging. In fact, we are not able to extract reasonable size distributions from the low-q portion data using a fractal aggregate model. The process of dispersion and precipitation, however, can be inferred from time evolution of Rg and G extracted from the low-q unified fits. These parameters are found in Table 1.

Figures 10 and 11 show Rg and G derived from low-q region as function of time for acid-treated and untreated nanofibers. For the untreated fibers, G decreases abruptly at constant Rg, consistent with precipitation. After 10 h both Rg and G stabilize. For the treated case, G and Rg increase with time consistent with agglomeration. It is interesting that after 44 h, these parameters approach that of the untreated fibers. These observations are consistent with the fact that acid treatment slows agglomeration and precipitation, but ultimately the fate of the treated fibers is the same as that of the untreated.

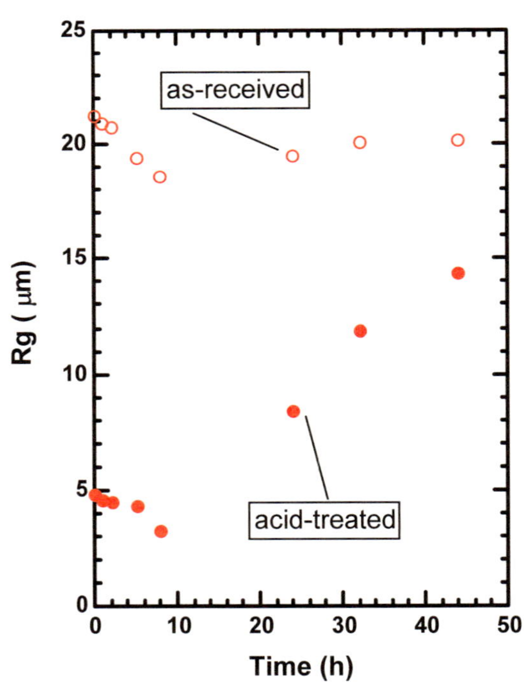

Fig. 10. Rg derived from low q region as a function of time for untreated and acid-treated nanofibers.

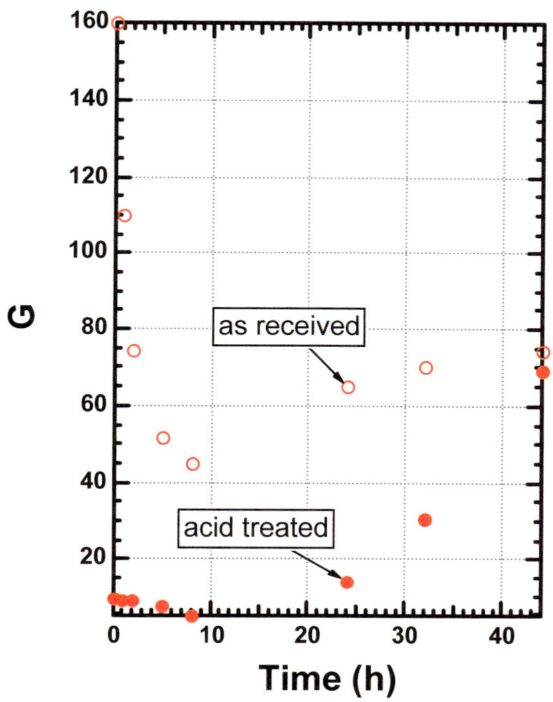

Fig. 11. G derived from low q region as a function of time for untreated and acid-treated nanofibers. At a given concentration G is proportional to the molecular weight.

2.4 Effect of post production processing

In order to optimize their structural features for different applications, these carbon nanofibers are usually processed upon production. Such post production processing can render nanocarbons to possess more desirable strength and electrical conductivity. (http://www.apsci.com/ppi-pyro3.html) For instance, heat treatment (HT) (up to 3000 °C) is performed to graphitize chemical vapor deposited (CVD) carbon present on the surface of as-grown carbon nanofibers. Thus, the heat-treated nanofibers possess highly electrical conductivity. Meanwhile, iron catalysts are removed in this process.

Post production thermal processing also alters surface properties of nanofibers, and thus probably influence their dispersion behavior. The acid-treated nanofibers PR19HT were studied in previous sections. Here we investigate dispersion of another type of acid-treated nanofibers PR19PS. PS means pyrolytically stripped carbon nanofibers. Typically, polyaromatic hydrocarbons are removed during this processing.

Presented in Figure 12, HRTEM images of as-received nanofibers PR19PS show that the naofibers are wrapped with a smooth CVD layer. Defects are occasionally found on the surface. Figure 13 shows typical TEM images of acid-treated PR19PS. Nanofibers with serious damage are observed. The CVD layer becomes rugged and defects are much more obvious along sidewalls after acid treatment.

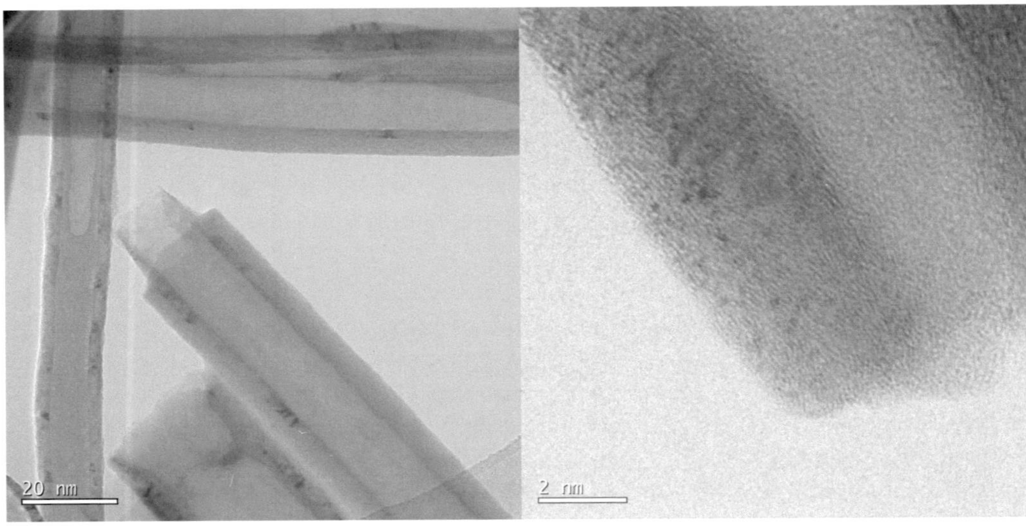

Fig. 12. TEMs of unmodified carbon nanotubes PR19PS. CVD layers are visible at both magnifications. The low-resolution image shows a variety of tube shapes and morphologies including concentric cylinders and stacked cones. The bars are 20 nm and 2 nm.

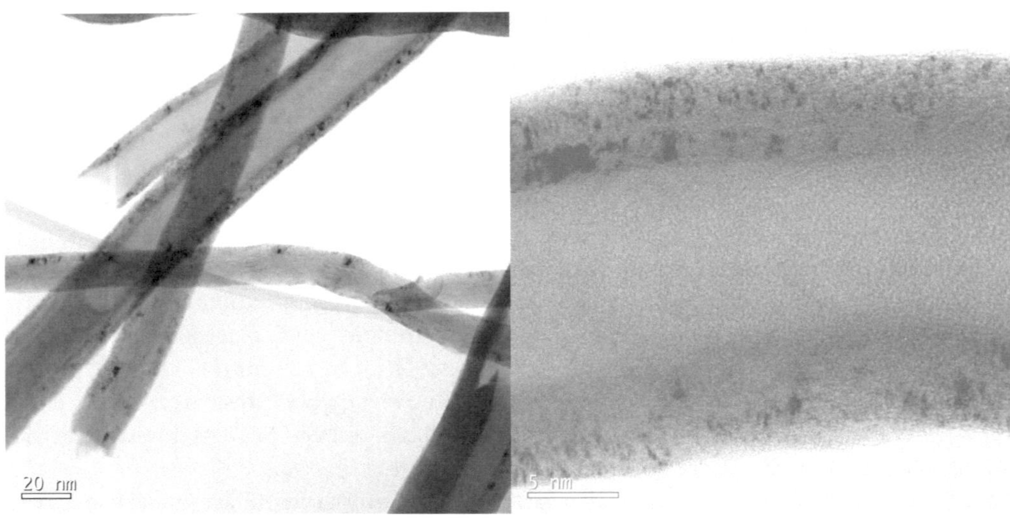

Fig. 13. TEMs of acid-treated PR19PS: More defects on the walls are evident. The bars are 20 nm and 2 nm.

We also measured dispersion behavior of acid-treated nanofibers PR19PS suspended in water using light scattering. Figure 14 shows the light scattering profiles as a function of time for acid-treated nanofibers PR19PS in water at a concentration of 5.0×10^{-6} g/ml. The data were obtained in the batch mode with no circulation or sonication. The overall intensity for acid-treated PR19PS shows similar tendency as acid-treated PR19HT, with an initial decrease followed by substantial increase. The maximal decrease in the intensity at low q is observed at 21 hr for acid-treated PR19PS whereas it occurs at 8hr for acid-treated PR19HT. That is, acid-treated PR19PS shows slower precipitation process with minimal agglomeration during the initial period. The considerable increase in the intensity at low q indicates agglomeration. Figure 15 compares scattering profiles of acid-treated PR19HT at 44 hr and PR19PS at 120 hr. Minimal difference in intensity at low q for both cases implies slower agglomeration for acid-treated PR19PS.

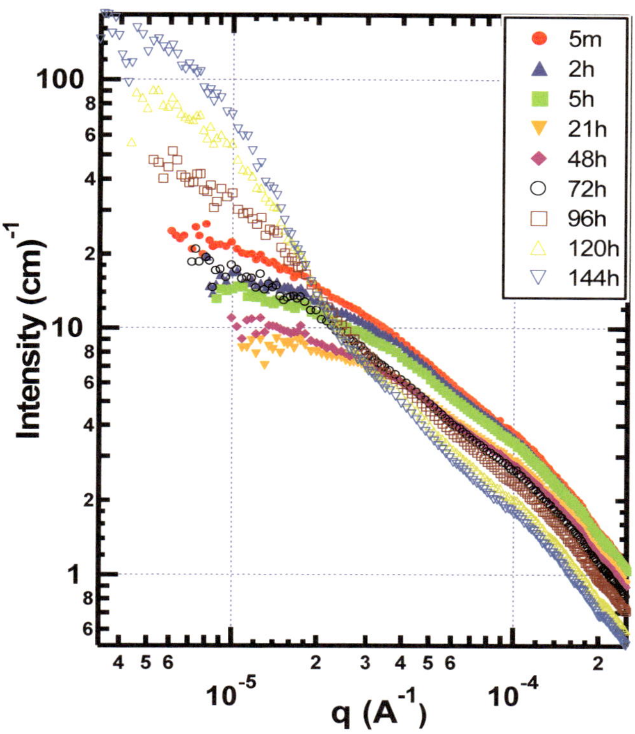

Fig. 14. Dispersion of acid-treated nanofibers (PR19PS) in water during four-day suspension. The suspensions were sonicated at 10W for five minutes before data were taken using light scattering in batch mode.

The scattering curves consist of one quasi power-law regimes and one Guinier regime. The curves were fit using Beaucage's Unified Model to extract Rg, the power-law exponents, P, and the Guinier prefactors, G, and power-law prefactor, B.(Beaucage, Schaefer et al. 1994) These parameters are listed in Table 3.

Figures 16 and 17 show Rg and G as a function of time for acid-treated nanofibers PR19PS and PR19HT. Following initial decrease due to precipitation, Rgs for acid-treated PR19PS nanofibers show a gradual increase over time whereas Rg for acid-treated PR19HT increase dramatically. It takes 120 hr for treated PR19PS to approach the Rg at 44 hr for treated PR19HT. At this Rg ($\cong$ 15 μm), G derived from treated PR19PS is much larger than that from treated PR19HT, indicating that more carbon entities are present in treated PR19PS suspension. Acid-treated PR19PS is suspended longer in water and agglomerates slowly.

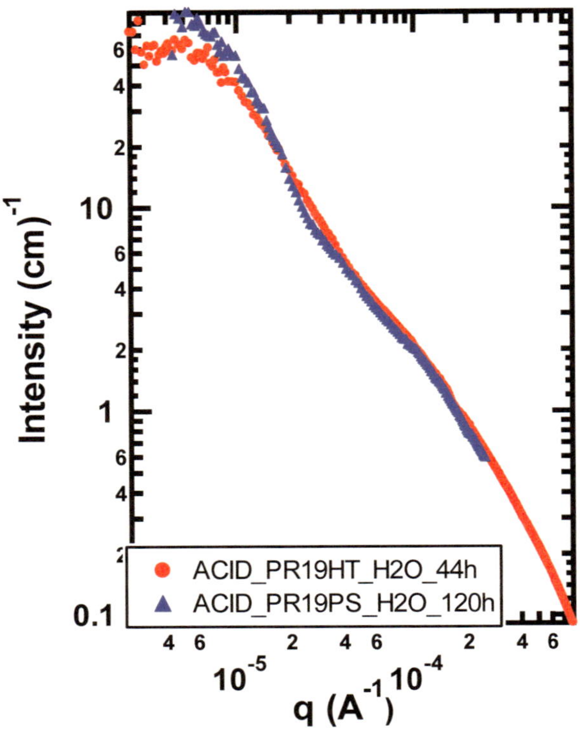

Fig. 15. Comparison of the scattering profiles for acid-treated carbon nanofibers PR19HT and PR19PS.

Time	5min	2 hr	5 hr	21 hr	48 hr	72 hr	96 hr	120 hr	144 hr
Rg (μm)	7.4	4.9	4.5	3.5	4.4	7.6	12.4	15.9	16.6
P	1.18	1.25	1.26	1.35	1.24	1.13	1.11	1.15	1.18
G	23.34	16.47	14.06	9.03	10.61	18.52	48.42	97.81	158.92
105 B	6.67	3.54	3.04	1.25	3.28	7.66	7.64	4.70	2.95

Table 3. Guinier radii and exponents as a function of time for acid-treated carbon nanofibers PR19PS.

Due to their graphete nature, the sidewalls of nanofibers PR19HT possess a chemical stability. However, the presence of CVD layers on the surface of PR19PS makes the chemistry available for surface modification less restrictive. Thus, PR19PS is more vulnerable to oxidative attack. Acid treatment creates denser surface acid sites that slow precipitation and agglomeration.

3. PEG-functionalized nanofibers

The presence of oxygen-containing groups on nanofibers after acid treatment provides rich chemistry for the attachment of functional groups to the surface of nanofibers. The simplest

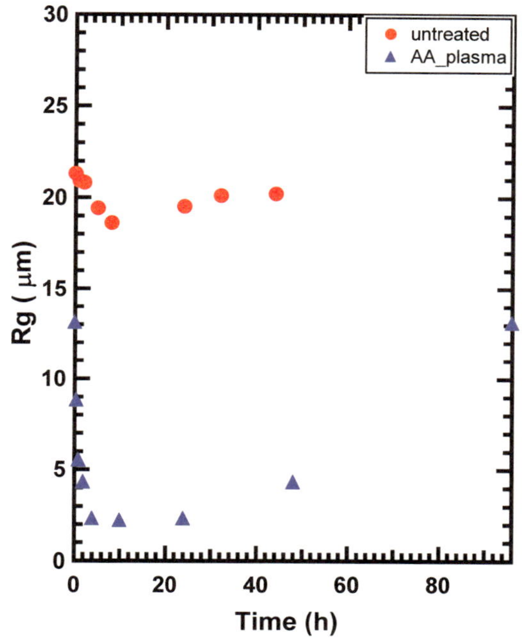

Fig. 16. Rg derived from low q region as a function of time for acid-treated nanofibers PR19PS and PR19HT.

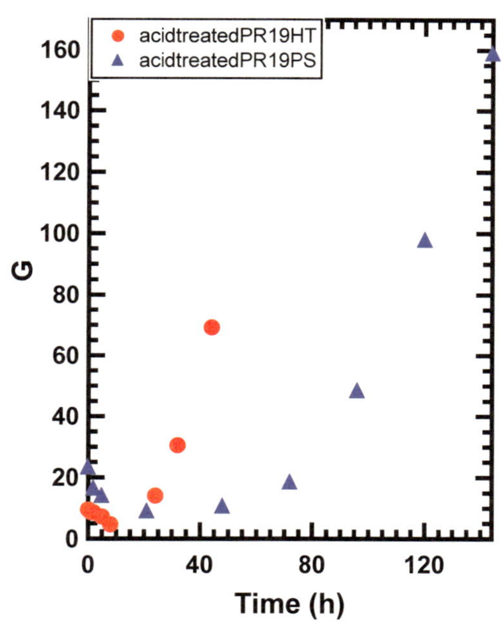

Fig. 17. G derived from low q region as a function of time for acid-treated nanofibers PR19PS and PR19HT. At a given concentration G is proportional to the molecular weight.

route to the dissolution of the nanofibers is to directly react an amine with the carboxylated nanofibers. A zwiterion is produced through a simple acid-base reaction. (Huang, Fernando et al. 2003)

Through the acid-base zwitterion interaction, diamine-terminated oligomeric poly(ethylene glycol) (PEG) was reacted with the carboxylic acids, which are bound to nanofibers via acid treatment. This process yields a dark-colored homogeneous solution that is stable in water for months. The dispersion was characterized using light scattering.

Figure 18 shows the light scattering profiles as a function of time for PEG-functionalized nanofibers (PR19HT) in water at a concentration of 5.0 × 10-6 g/ml. The data were obtained in the batch mode with no circulation or sonication. The scattering curve does not change with time. No Rg (radius of gyration) can be extracted in limiting regime due to the absence of a crossover at low q. The slope near –1 in this regime indicates a rod-like character on large scales. The slope near –2 (P = 2) on a log-log plot around q = 0.003 Å -1 may arise from a hollow tube.

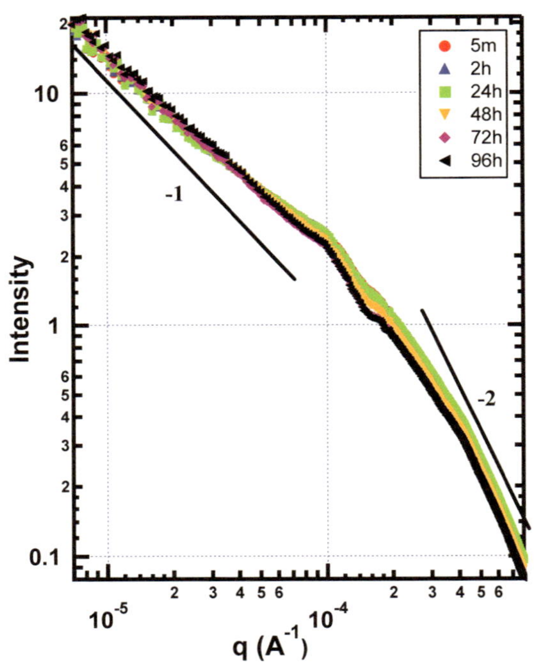

Fig. 18. Evolution of the light scattering profile of PEG-functionalized nanofibers (PR19HT) for four days following dispersion by sonication. The suspensions were sonicated at 10 W for five minutes before the observations began. The measurements were taken in the batch mode, so the sample was undisturbed during the course of the experiment.

Figure 19 compares the scattering profiles for PEG-functionlized, acid-treated and as-purchased carbon nanofibers at 44h after sonication. Compared to PEG-functionalized sample, Guinier region at low q is observed for untreated and acid-treated nanofibers, indicating the presence of large-length-scale structures formed by the small-scale structures at high q. In untreated state, as described above, the suspension is quite unstable and fibers precipitate. The large-scale objects we call agglomerates are observed immediately after sonication. The Guinier prefactors G derived from low-q region using Beaucage's Unified Model decrease abruptly right after sonication at fixed Rg, consistent with precipitation being dominant process. Such agglomeration is observed after 24 h for acid-treated case. After 44 h, the scattering curve of acid-treated sample approaches that of untreated fibers. The presence of oxygen-containing groups on the surface of nanofibers through acid treatment can not stop agglomeration and precipitation, but retard these processes.

Solubility is a measure of an equilibrium between the dissolved phase and aggregated (or agglomerated) phase. A long solubilizing chain on the functionalities should be favorable to obtain soluble nanofibers. Compared to the acid-treated and untreated sample, which shows strong agglomeration and no rod-like behavior, PEG-functionalized nanofibers do not show any agglomeration over time. Attachment of water-soluble oligemeric PEG on nanofibers does shift the equilibrium towards the dissolved phase, preventing agglomeration. (Zhao & Schaefer 2008)

In order to further understand the effect of the functionalization on the morphology, the data were analyzed by fitting with a simplified tube or rod form factor. The crossover length scale at high q between the two power-law regimes corresponds to the largest radius of the individual fibers or fiber aggregates. We argue below that the scattering objects are side-by-side fiber aggregates or bundles.

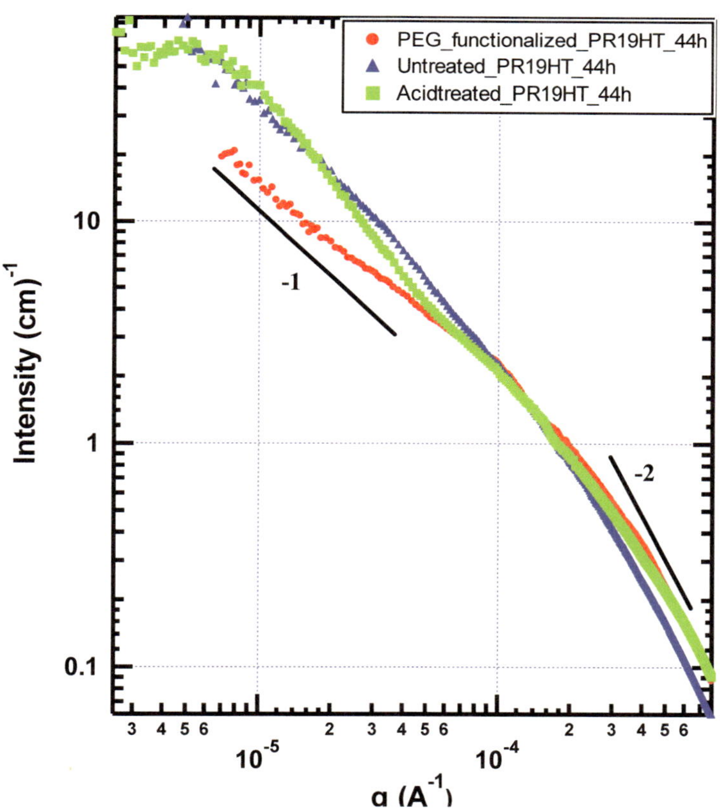

Fig. 19. Comparison of the scattering profiles for PEG-functionalized, untreated and acid-treated carbon nanofibers 44 hr after sonication.

The data do not fit with a rod form factor although they do follow a tube-like form factor reasonably well. We fit the scattering curve of PEG-functionalized nanofibers at 72 h using a simplified tube form factor (Figure 20). The simplified tube form factor approximates an exact tube model but suppress the oscillations that are found for exact models.(Justice, Wang et al. 2007) The simplified tube model fits the data reasonably well except at intermediate q, where it seems to underestimate the size of the relevant size scale (tube outer radius).

Based on the fit to the simplified tube form factor, one might naively conclude that dispersion is down to the level of individual tubes. However, the outer diameter of 0.47 μm (calculated from the radius of 2350 Å) derived from the simplified tube model is substantially larger than the largest individual fibers seen in TEM. That is, scattering entities are not individual tubes, but side-by-side fiber bundles. A bundled structure also accounts for the deviation from the simplified tube form factor. Nevertheless, fitting with the simplified tube model implies that these fiber aggregates are rigid and isolated. Time-evolution of dispersion behavior of PEG-nanofibers shows that fiber bundles remain well dispersed and do not form large-scale agglomerates over weeks.

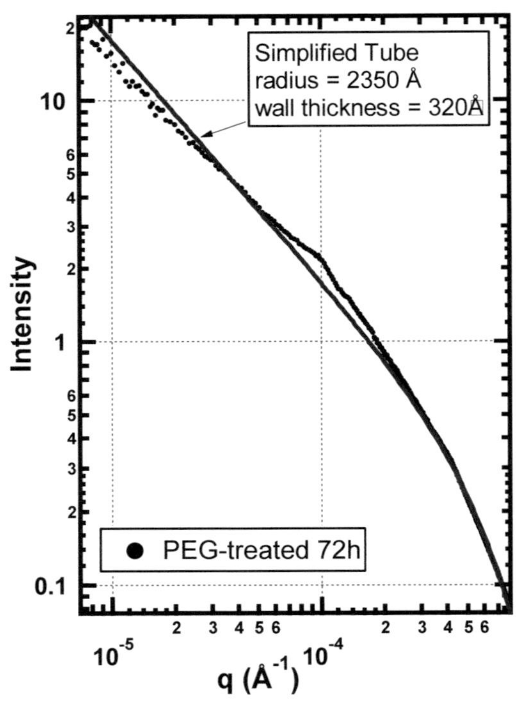

Fig. 20. Light scattering data for PEG-functionalized nanofibers compared with the simplified tube model.

4. Plasma-treated nanofibers

Plasma polymerization has been an active area of research over the past decades. (van Ooij, Zhang et al. 1999) This technique has been used in surface and interface engineering for improving adhesion, hydrophobicity and hydrophilicity, corrosion resistance or for surface

etching. Low temperature plasma polymerization, a room temperature and environmentally benign process, can be used for surface modification and thin film deposition on almost all substances. Deposition of ultrathin films on nanoparticles by plasma treatment has been achieved. (Shi, Lian et al. 2002; Shi, Lian et al. 2003)
To enhance dispersion ability of carbon nanofibers in water, acrylic acid is selected as a monomer for plasma polymerization. (Shi, He et al. 2002)Bright-field and high-resolution TEM images of the plasma-coated nanofibers are shown in Figure 21. An ultrathin film amorphous layer can be clearly seen covering the surface of the carbon nanofibers. The thin film is uniform with a thickness of approximately 2-7 nm surrounding the entire nanofiber surface. In our experiment, carbon nanofibers are treated by plasma polymerization. Brief plasma treatment deposits a coating of highly crosslinked polymer on the fibers, which substantially improves their compatibility with water, and thus assists dispersion of carbon nanofibers.
We measured dispersion efficiency of AA-plasma-treated nanofibers suspended in water and made a comparison with untreated nanofibers. Figure 22 shows the light scattering profiles as a function of time for acrylic acid (AA) plasma-treated nanofibers PR19HT in water at a concentration of $5.0 \times 10\text{-}6$ g/ml. The data were obtained in the batch mode with no circulation or sonication. For untreated nanofibers, as described above, the overall intensity shows a nearly monotonic, gradual decrease consistent with precipitation being the dominant process. After plasma treatment, the intensity at low q tends to drop abruptly with time. At 10 hr, the maximal decrease in the intensity is observed. After 10 hr, however, the intensity increases substantially, consistent with agglomeration.

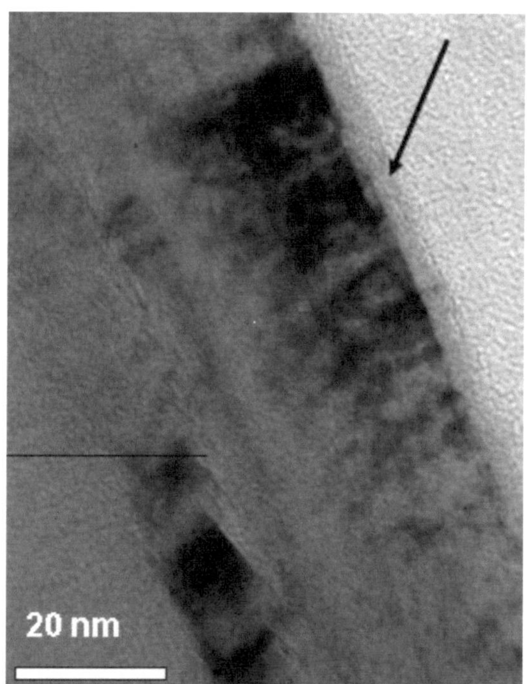

Fig. 21. TEM image of plasma-treated carbon nanofibers PR19LHT: The bars are 10 nm. An ultrathin film of acrylic acid was observed on the surface of the nanofibers.

	Time	5min	30min	1hr	2 hr	4hr	8hr	24hr	48 hr	96 hr
Low q	Rg (μm)	13.1	8.8	5.5	4.3	2.3	2.2	2.3	4.3	13.1
	P	1.58	1.43	1.39	1.22	1.21	1.20	1.20	1.47	1.70
	G	70.78	31.13	14.52	7.68	3.48	1.92	2.90	10.35	36.40
	106 B	70.31	3.26	4.41	21.87	13.81	10.84	15.50	1.47	17.06
High q	Rg (μm)	0.67	0.72	0.71	0.65	0.71	0.73	0.68	0.69	0.67
	P	1.93	2.07	1.90	1.97	1.97	1.90	1.95	1.94	2.10
	G	0.73	0.55	0.58	0.37	0.48	0.58	0.42	0.43	0.32
	108 B	6.58	1.14	11.34	5.69	5.67	29.6	5.55	4.92	1.64

Table 4. Guinier radii and exponents as a function of time for plasma-treated carbon nanofibers.

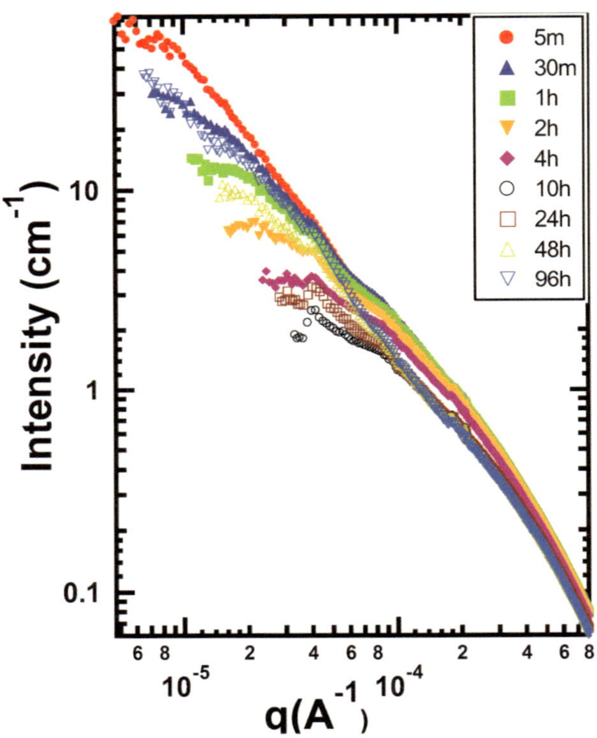

Fig. 22. Evolution of the light scattering profile of plasma-treated nanofibers in water for four days following dispersion by sonication. The suspensions were sonicated at 10W for five minutes before the observations began. The measurements were taken in the batch mode.

The scattering curves consist of two power-law regimes and two Guinier regimes that define two "length scales". Each Guinier regime is followed by a quasi power-law regime. The curves were fit using Beaucage's Unified Model to extract Rg, the power-law exponents, P, and the Guinier prefactors, G, and power-law prefactor, B, associated with each length scale.

These parameters are displayed in Table 4 for the two structural levels. The high q data share similarity with that for the untreated sample. These data imply minimal change in morphology with time on length scales below ~1 μm.

The decrease in the scattered intensity at low q up to 10 hours is pronounced and ascribed to precipitation. After 10 hr, however, the large-scale agglomerates gradually form.

Figure 23 compares the scattering profile for plasma-treated and as-purchased carbon nanofibers PR19HT at 10 hr after sonication. Compared to the untreated sample, the intensity at low q (G) for the treated sample is much smaller, indicating small entities in the suspension. The extracted length scale Rg at 10 hr, 2.2 μm is consistent with much smaller agglomerates compared to the untreated case. After 10 hours there is evidence for agglomeration. Plasma treatment retards this agglomeration.

Figures 24 and 25 show Rg and G derived from low-q region as a function of time for plasma-treated and untreated nanofibers. In both cases, G decreases during the first ten hours, consistent with precipitation. After 10 h, G increases with time consistent with agglomeration (increased Rg) for the plasma-treated sample, whereas both Rg and G show minor change for the untreated sample. Rgs extracted from the plasma-treated sample is considerably smaller than those in the untreated case, indicating improved dispersion.

After plasma treatment, the carbon stays suspended much longer although all the fibers precipitate finally. The clusters are much easier to break up and more difficult to agglomerate. Plasma treatment improves compatibility with water, thus slowing agglomeration and precipitation.

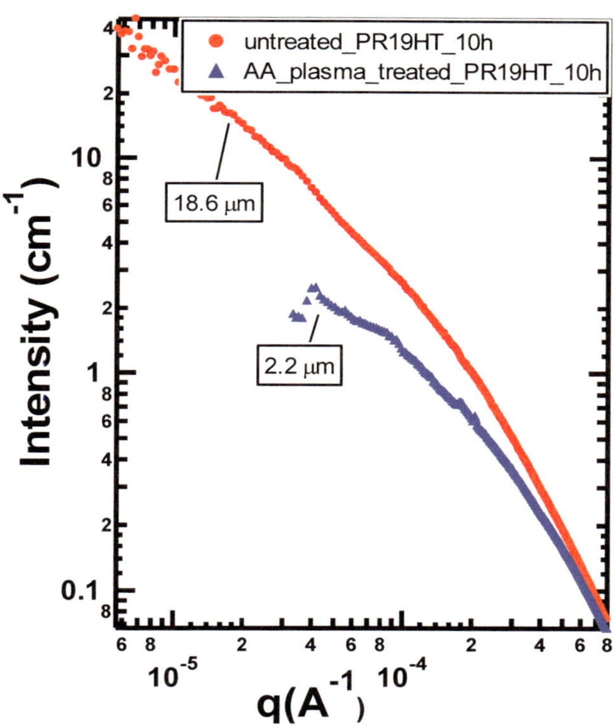

Fig. 23. Comparison of the scattering profiles for untreated and plasma-treated carbon nanofibers 8 h after sonication. A substantial population of large-scale clusters is present only for the untreated sample.

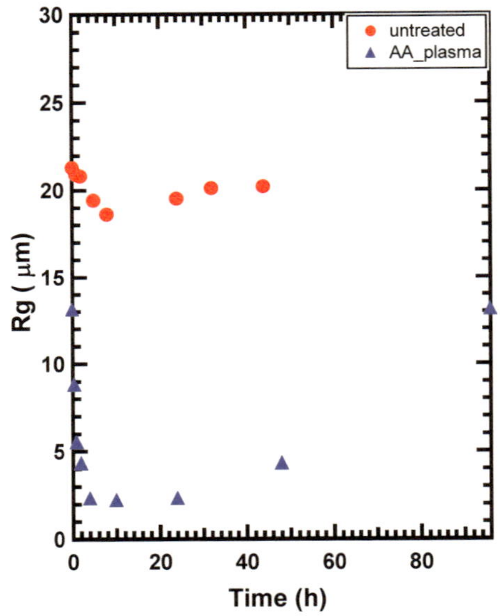

Fig. 24. Rg derived from low q region as a function of time for untreated and plasma-treated nanofibers.

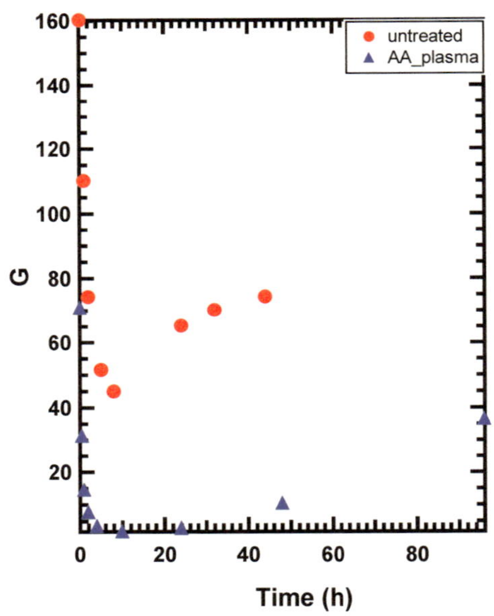

Fig. 25. G derived from low q as a function of time for untreated and plasma-treated nanofibers. At a given concentration region G is proportional to the molecular weight.

5. Summary

Dispersion of nanotubes in suspensions has been investigated using light scattering. Functionalization, plasma treatment and surfactants were used to assist dispersion. Improved dispersion in solutions was achieved. The main conclusions are summarized as follows.

We compare dispersion behavior of acid-treated and as-received carbon nanofibers suspended in water under quiescent conditions. Both samples show a hierarchical morphology consisting small-scale aggregates and large-scale agglomerates. The aggregates could be side-by-side bundles of individual nanofibers or more complex structures. In any case these objects agglomerate to form large-scale fractal clusters. Acid treatment shifts the small-scale size distributions to smaller bundle sizes. In the absence of surface treatment these bundles agglomerate immediately after sonication. In the acid-treated case, by contrast, it takes many hours for the agglomerates to form. Thus acid treatment assists dispersion primarily by retarding large-scale agglomeration not by suppressing small-scale aggregation. Post production processing affects dispersion. Acid-treated PR19PS shows slower agglomeration and precipitation than acid-treated PR19HT.

Dispersion behavior of PEG-functionalied nanofibers suspended in water in a quiescent condition was investigated. Comparison with untreated and acid-treated carbon nanofibers show that PEG-functionalization completely prevents formation of large-scale agglomerates that consist of small scale side-by-side aggregates. The presence of PEG oligomer has little effect on the small-scale bundle size distributions. Prevention of agglomeration is the primary mechanism by which functionalization leads to solubilization of nanofibers.

Nanofibers are plasma-treated using acrylic acid as a monomer. The plasma-treated nanofibers show greater tendency to suspend. The presence of COOH on the nanofibers could alter the surfaces of carbon nanofibers towards hydrophilicity, thus improving dispersion of nanofibers in water.

6. References

Ausman, K. D., R. Piner, et al. (2000). "Organic solvent dispersions of single-walled carbon nanotubes: Toward solutions of pristine nanotubes." Journal of Physical Chemistry B 104(38): 8911-8915.

Beaucage, G., D. W. Schaefer, et al. (1994). "Multiple Size Scale Structures in Silica Siloxane Composites Studied by Small-Angle Scattering." Abstracts of Papers of the American Chemical Society 207: 144-149.

Bechinger, C., D. Rudhardt, et al. (1999). "Understanding depletion forces beyond entropy." Physical Review Letters 83(19): 3960-3963.

Boukari, H., G. G. Long, et al. (2000). "Polydispersity during the formation and growth of the Stober silica particles from small-angle X-ray scattering measurements." Journal of Colloid and Interface Science 229(1): 129-139.

Chen, J., M. A. Hamon, et al. (1998). "Solution properties of single-walled carbon nanotubes." Science 282(5386): 95-98.

Chen, J., A. M. Rao, et al. (2001). "Dissolution of full-length single-walled carbon nanotubes." Journal of Physical Chemistry B 105(13): 2525-2528.

Chen, Q., C. Saltiel, et al. (2004). "Aggregation behavior of single-walled carbon nanotubes in dilute aqueous suspension." Journal of Colloid and Interface Science 280: 91-97.

Dresselhaus, M. S., G. Dresselhaus, et al. (2001). Carbon Nanotubes: Synthesis, Structure, Properties and Applications. Berlin, Springer.

Gong, Q., Z. Li, et al. (2005). "Synthesis and characterization of in situ grown carbon nanofiber/nanotube reinforced carbon/carbon composites." Carbon 43: 2426-2429.

Hu, H., P. Bhowmik, et al. (2001). "Determination of the acidic sites of purified single-walled carbon nanotubes by acid-base titration." Chemical Physics Letters 345(1-2): 25-28.

Huang, W. J., S. Fernando, et al. (2003). "Solubilization of single-walled carbon nanotubes with diamine-terminated oligomeric poly(ethylene glycol) in different functionalization reactions" Nano letters 3(4): 565-568.

Ilavsky, J. (2004). Particle Size ditribution from USAX, Irena SAS Modeling Macros Manual, UNICAT,Argonne Illinois, USA.

Jemian, P. R., J. R. Weertman, et al. (1991). "Characterization of 9cr-1movnb Steel by Anomalous Small-Angle X-Ray-Scattering." Acta Metallurgica Et Materialia 39(11): 2477-2487.

Justice, R. S., D. H. Wang, et al. (2007). "Simplified tube form factor for analysis of small-angle scattering data from carbon nanotube filled systems." Journal of Applied Crystallography 40: S88-S92.

Li, P., T. Zhao, et al. (2005). "Deuterated water as super solvent for short carbon nanotubes wrapped by DNA." Carbon 43: 2701-2703.

Liu, J., A. G. Rinzler, et al. (1998). "Fullerene pipes." Science 280(5367): 1253-1256.

Monthioux, M., B. W. Smith, et al. (2001). "Sensitivity of single-wall carbon nanotubes to chemical processing: an electron microscopy investigation." Carbon 39(8): 1251-1272.

Morrison, J. D., J. D. Corcoran, et al. (1992). "The Determination of Particle-Size Distributions in Small-Angle Scattering Using the Maximum-Entropy Method." Journal of Applied Crystallography 25: 504-513.

Potton, J. A., G. J. Daniell, et al. (1988). "Particle-Size Distributions from Sans Data Using the Maximum- Entropy Method." Journal of Applied Crystallography 21: 663-668.

Safadi, B., R. Andrews, et al. (2002). "Multiwalled carbon nanotube polymer composites: synthesis and characterization of thin films " J. Appl. Poly. Sci. 84: 2660-2669.

Schaefer, D. W. (1988). "Fractal models and the structure of materials." Materials Research Society Bulletin 13(2): 22.

Schaefer, D. W., J. M. Brown, et al. (2003). "Structure and dispersion of carbon nanotubes." Journal of Applied Crystallography 36: 553-557.

Schaefer, D. W., B. C. Bunker, et al. (1989). "Fractals and Phase-Separation." Proceedings of the Royal Society of London Series a- Mathematical Physical and Engineering Sciences 423(1864): 35-53.

Schaefer, D. W., R. S. Justice, et al. (2005). "Large-Scale Morphology of Dispersed Layered Silicates " Materials Research Society symposia proceedings 840: Q3.3.1-3.3.6.

Schaefer, D. W., J. Zhao, et al. (2003). "Morphology of Dispersed Carbon Single-Walled Nanotubes." Chemical Physics Letters 375(3-4): 369-375.

Shaffer, M. S. P., X. Fan, et al. (1998). "Dispersion and packing of carbon nanotubes." Carbon 36(11): 1603-1612.

Shi, D. L., P. He, et al. (2002). "Plasma deposition and characterization of acrylic acid thin film on ZnO nanoparticles." Journal of Materials Research 17(10): 2555-2560.

Shi, D. L., J. Lian, et al. (2002). "Plasma deposition of Ultrathin polymer films on carbon nanotubes." Applied Physics Letters 81(27): 5216-5218.

Shi, D. L., J. Lian, et al. (2003). "Plasma coating of carbon nanofibers for enhanced dispersion and interfacial bonding in polymer composites." Applied Physics Letters 83(25): 5301-5303.

van Ooij, W. J., N. Zhang, et al. (1999). Fundamental and Applied Aspects of Chemically Modified Surfaces, Royal Society of Chemistry.

Zhao, J. and D. W. Schaefer (2008). "Morphology of PEG-Functionalized Carbon Nanofibers in Water." Journal of Physical Chemistry C 112: 15306 - 15310.

Zhao, J., D. W. Schaefer, et al. (2005). "How Does Surface Modification Aid in the Dispersion of Carbon Nanofibers?" Journal of Physical Chemistry B 109: 23351-23357.

15

Non-Catalytic, Low-Temperature Synthesis of Carbon Nanofibers by Plasma-Enhanced Chemical Vapor Deposition

Shinsuke Mori and Masaaki Suzuki
Tokyo Institute of Technology, Japan

1. Introduction

Plasma-enhanced chemical vapour deposition (PECVD) has some unique advantages of allowing low-temperature growth of vertically aligned carbon nanotubes (CNTs) and less crystalline carbon nanofibers (CNFs) (Meyyappan et al., 2003; Melechko et al., 2005). In the conventional PECVD methods for CNTs/CNFs synthesis, metal catalyst particles are used because the CNFs/CNTs are grown by the following steps: (i) adsorption and decomposition of the reactant molecules and their fragments formed in the plasma on a surface of catalyst, (ii) dissolution and diffusion of carbon species through the metal particle, and (iii) precipitation of carbon on the opposite surface of the catalyst particle to form the nanofibers structure (Baker & Harris, 1978; Melechko et al., 2005). Hofmann et al. (2003) have suggested that the rate-determining step for the growth of CNF at a low temperature is not the diffusion of carbon through the catalyst particle bulk, as was proposed by Baker et al. (Baker & Harris, 1978) and is generally accepted for high-temperature conditions, but the diffusion of carbon on the catalyst surface. In this surface diffusion model, carbon atoms adsorbed at the top surface of the metal particles diffuse along the surface, where their motion is much faster than bulk diffusion, and then segregate at the bottom of the particles, forming graphitic planes. These graphitic basal planes are parallel to the metal surface, and the orientation angle between the graphite basal planes and the tube axis is not zero. As a result, although CNFs grown at a higher temperature (> 500 ^{o}C) consist of several graphitic basal planes oriented parallel to the fibre axis with a central hollow region (shell structures; they can be called carbon nanotubes), CNFs grown at a lower temperature consist of stacked cone-segment shaped graphite basal plane sheets (fish-bone, herring-bone, stucked-cone, or stacked-cup structures) or the basal planes oriented perpendicular to the fibre axis (platelets structures) and CNFs with large orientation angles are often not hollow (Fig. 1). For the practical application of CNTs/CNFs, their low-temperature synthesis by PECVD is attractive to achieve the direct deposition of CNTs/CNFs on various substrates involving materials with low melting points. So far, several studies on the low-temperature (< 400°C) synthesis of CNFs/CNTs by PECVD with various discharge systems using hydrocarbons have been reported, such as the RF discharge of CH_4 (Boskovic et al., 2002), the DC discharge of C_2H_2/NH_3 (Hofmann et al., 2003), the AC discharge of $C_2H_2/NH_3/N_2/He$ (Kyung et al., 2006), the microwave discharge of CH_4/H_2 (Liao and Ting, 2006), and a combination of ECR C_2H_2 plasma with ICP N_2 plasma (Minea et al., 2004) while few attempts at low-temperature PECVD of CNFs/CNTs using CO as the carbon source have been made (Han et al, 2002; Plonjes et al., 2002).

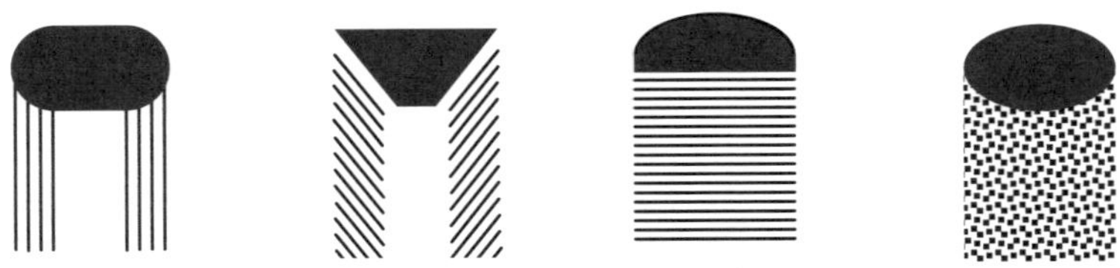

Fig. 1. Schematic cross-sectional illustrations of carbon nanofibers grown by catalytic CVD

The preparation of catalyst particle often limits to lower the process temperature because high-temperature treatment is usually necessary for the activation of catalyst. The elevated temperature is also needed to create the metal particles because metal particles are usually created by breaking up a thin metal film on a substrate into small islands on annealing at elevated temperatures (Merkulov et al., 2000). At the early stage of our CNF synthesis study, the vertically aligned CNFs could be synthesized on a Fe catalyst layer using a $CO/Ar/O_2$ discharge system at extremely low temperatures (Room temperature - 180 °C) (Mori et al., 2007, 2008, 2009a). In our subsequent study on the low-temperature activation of metal catalyst particles, it was found that the CNF growth process is not controlled by the catalyst particle, and that, surprisingly, CNFs can be grown even if no catalyst is used in the $CO/Ar/O_2$ plasma system at the optimal growth conditions (Mori & Suzuki, 2009b, 2009c). From the viewpoint of process simplification and product purification, this catalyst-free synthesis is attractive. In this chapter, therefore, we describe only non-catalytic PECVD of CNFs grown at a low-temperature (< 180 °C) in a $CO/Ar/O_2$ discharge system.

2. Synthesis

The CNFs were grown using a DC plasma-enhanced CVD system (DC-PECVD) and a microwave plasma-enhanced CVD system (MW-PECVD). In both systems, a low-temperature $CO/Ar/O_2$ plasma was used. In general, the advantages of low-temperature plasma CVD using CO instead of hydrocarbons as the carbon source gas are as follows: (1) the deposition of amorphous carbon is suppressed even at low temperatures (Muranaka et al., 1991; Stiegler et al., 1996); (2) the CO disproportionation reaction, $CO+CO \rightarrow CO_2+C$, is thermodynamically favorable at low temperatures; (3) vibrationally excited molecules are formed which enhance reactions at low temperature, such as $CO(v)+CO(w) \rightarrow CO_2+C$ (Plonjes et al., 2002; Capitelli 1986; Mori et al., 2001); (4) C_2 molecules are known to be formed effectively through the reactions $C + CO + M \rightarrow C_2O + M$ and $C + C_2O \rightarrow C_2 + CO$ and can be precursors for the deposition of functional carbon materials (Caubet & Dorthe, 1994; Ionikh et al., 1994; McCauley et al., 1998).

2.1 DC-PECVD system

Figure 2(a) shows a schematic diagram of the experimental apparatus for the DC-PECVD system. The quartz discharge tube has a 10-mm inner diameter, and there are two electrodes spaced 5 cm apart and connected to the DC power supply in the discharge tube; one of them is a stainless-steel rod cathode with a diameter of 6 mm and the other is a stainless-steel rod anode with a diameter of 1.5 mm. In this study, borosilicate glass pieces (4 x 4 x 0.2 mm^3) were used as substrates which were placed on the cathode. Before CNF synthesis, the surfaces of substrates were cleaned with ethanol and no catalysts were used in the synthesis. The parameters for the CNFs deposition process were as follows: CO flow rate: 20 sccm, Ar flow rate: 20 sccm; O_2 flow rate: 0-1.0 sccm; total pressure: 800 Pa; discharge current: 2 mA. The substrate temperature, *Ts*, was monitored by a thermocouple placed below the substrate while it would be lower than the upper surface and CNF temperature. Although the substrate was heated up by the discharge, the temperature, *Ts*, of all the samples in this system remained as low as 90 °C.

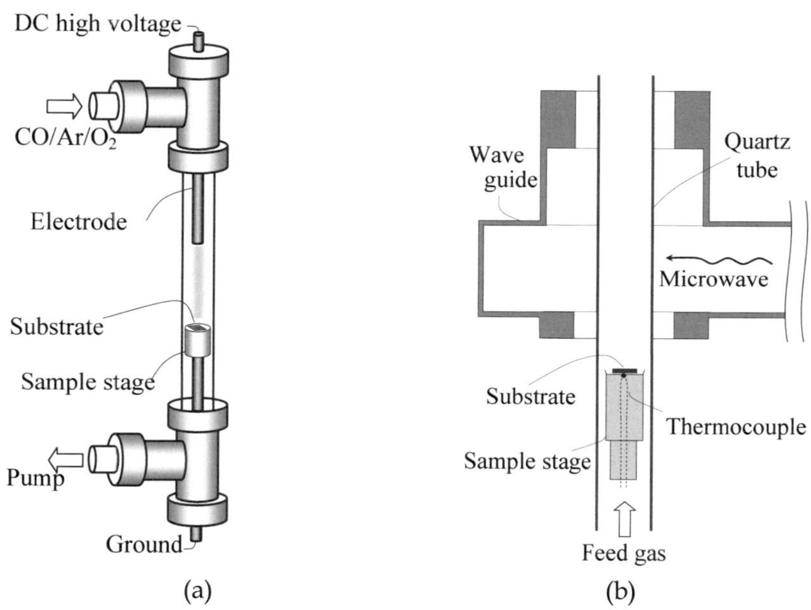

Fig. 2. Schematic diagram of plasma reactor: (a) DC-PECVD and (b) MW-PECVD system

2.2 MW-PECVD system

Figure 2(b) shows a schematic diagram of the MW-PECVD system in which the CNFs were grown. This system comprises a modified ASTeX DPA25 plasma applicator with a quartz discharge tube of 10-mm inner diameter. Borosilicate glass, silicon single-crystal wafers, CaF_2, and polycarbonate plates (4 mm × 4 mm) were used as substrates, and the substrate was placed 52 mm below the center of the waveguide. Before CNF synthesis, the surfaces of substrates were cleaned with ethanol and no catalysts were used in this system. The conditions for CNF deposition process were as follows: CO flow rate, 10 sccm; Ar flow rate, 30 sccm; O_2 flow rate, 0-1.0 sccm; total pressure, 400 Pa; and microwave power, 80 W. The substrate temperature, *Ts*, was monitored by a thermocouple placed below the substrate. In the present configuration, the substrate temperature was automatically increased to about 150 °C when plasma irradiation was applied. However, this temperature was unstable. Therefore, in order to achieve steady temperature condition, *Ts* above 150 °C was controlled using a nichrome wire heater equipped with a temperature controllerand maintained stably at 180 °C throughout the MW-PECVD process.

3. Properties

The carbon deposits growing on the substrate were observed by scanning electron microscopy (Hitachi S-4500, KEYENCE VE-8800) and transmission electron microscopy (JEOL JEM-2010F) and analyzed by Raman spectroscopy (JASCO NRS-2100).

3.1 DC-PECVD system

Figure 3 shows scanning electron microscope (SEM) images of the carbon deposits with different additional O_2 gas compositions. The morphology of carbon deposits is strongly affected by the O_2/CO ratio. Without the addition of oxygen, pillar-like carbon films were formed. When a small amount of O_2 was added to the CO plasma, the morphology of the carbon films changed to a cauliflower-like structure (O_2/CO ~ 1/1000) and a fibrous structure (O_2/CO = 2/1000 ~ 5/1000). At higher O_2 flow rates, however, the deposition rate decreased and the fibrous structure was no longer observed.

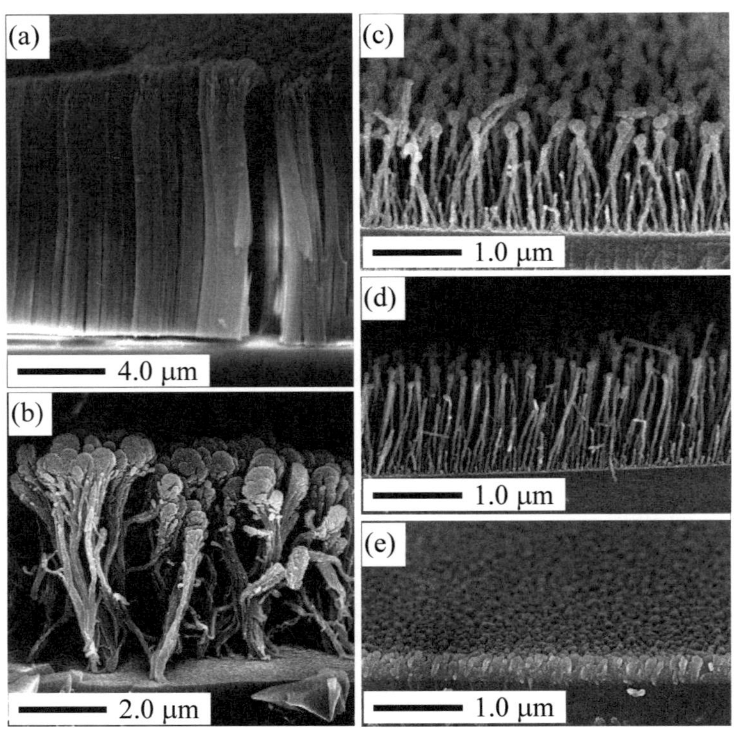

Fig. 3. SEM images of carbon materials synthesized with different O_2/CO ratio without catalyst at 90 °C. O_2/CO ratio; (a) O_2/CO = 0; (b) O_2/CO = 1/1000; (c) O_2/CO = 2/1000; (d) O_2/CO = 4/1000; (e) O_2/CO = 7/1000: Growth time: (a), (c) 1 h; (b), (d), (e) 2h

Figure 4 shows transmission electron microscope (TEM) images of CNFs synthesized at O_2/CO = 3/1000. Under this condition, the diameter of the CNFs was about 10-50 nm. The surface of the CNFs was not so smooth. In the high-magnification images, the lattice structure of the crystallized carbon layers is clearly visible. In most of the thinner fibers, the layers were perpendicular to the fiber axis, and it was revealed that the CNFs had a platelet structure [Figs. 4(a) and 4(b)]. That structure has already been reported by some researchers in their catalytic-grown CNFs using carbon monoxide as the carbon source gas (Murayama & Maeda, 1990; Rodriguez et al., 1995; Yoon et al., 2005). In the rest of the fibers, those layers were not clearly seen because their directions were random relative to the fiber axis and they overlapped each other [Figs. 4(d)]. However, it can be said that the crystallinity of the carbon fibers was quite high in spite of the low growth temperature.

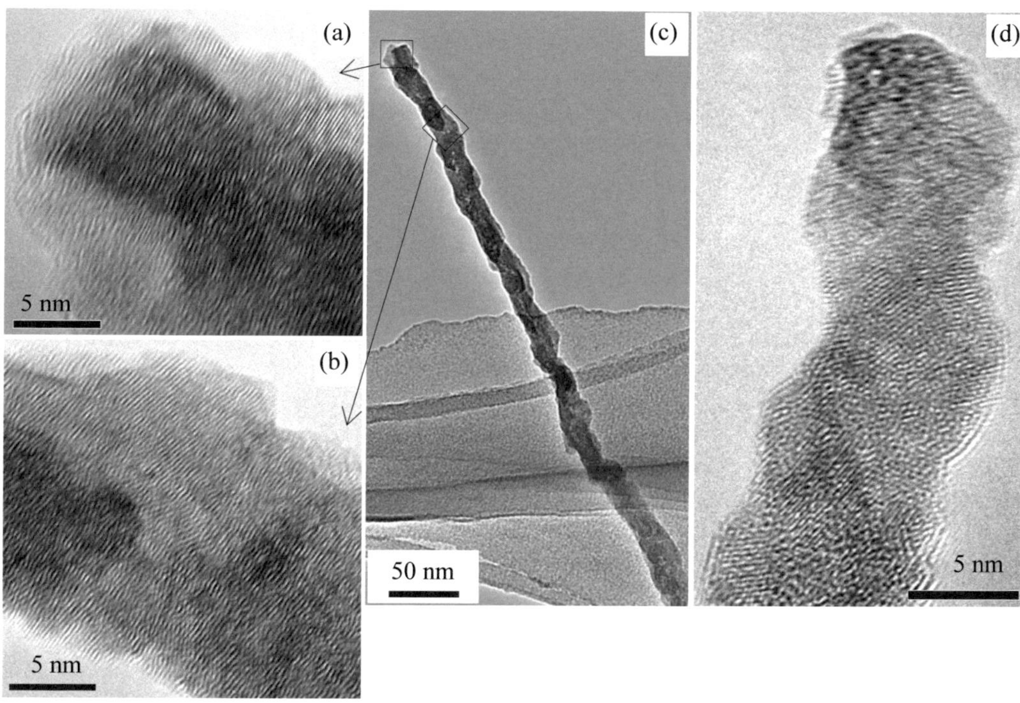

Fig. 4. TEM images of carbon nanofibers synthesized without catalyst at 90 °C. O_2/CO = 3/1000. Growth time: 2 h. (a)-(c) platelets CNFs, (d) randomly oriented CNFs

The Raman spectra of carbon deposits were examined as shown in Fig. 5 and it was found that there was no appreciable difference between the Raman spectra in the present non-catalytic study and previous one in which Fe catalyst was used (Mori & Suzuki, 2008): (1) the spectra for all the samples present two peaks of carbon material: the rather sharp G-band peak at approximately 1590 cm^{-1}, which indicates the presence of crystalline graphene layers, and the broad D-band peak at 1350 cm^{-1}, which indicates the existence of defective graphene layers; (2) the D-band decreased with increasing O_2/CO ratio while the G-band was almost unchanged. Therefore, from the Raman spectroscopic analysis, it is concluded that the deposition of amorphous carbon is selectively suppressed by the addition of O_2.

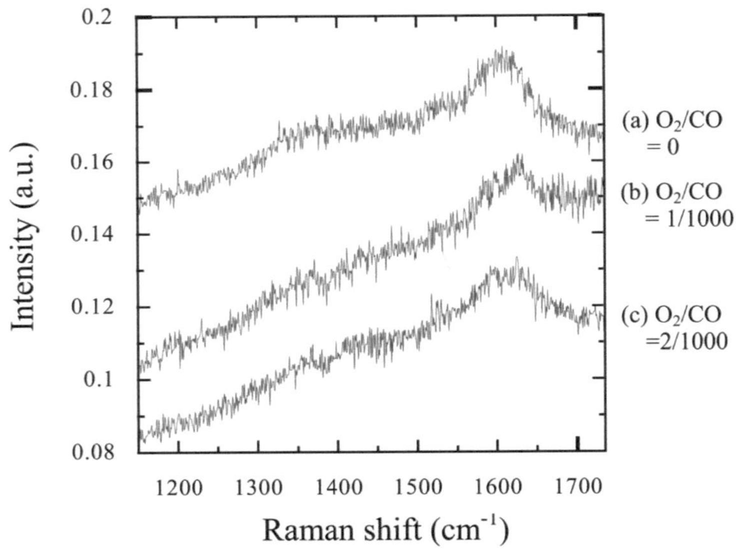

Fig. 5. Raman spectra of the carbon deposits prepared at O_2/CO ratios of; (a) 0/1000; (b) 1/1000; and (c) 2/1000.

3.2 MW-PECVD system

Figure 6 shows SEM images of the carbon deposits grown by MW-PECVD on the glass substrates after 10 minutes of deposition with different levels of O_2 gas supplementation. In the absence of added oxygen, pillar-like carbon films was formed. When a small amount of O_2 was added to the CO plasma, the morphology of the carbon films changed to fibrous structure. At higher O_2 flow rates, however, the deposition rate decreased and no carbon deposits could be observed. While O_2/CO window for CNFs formation is shifted towards a higher O_2 concentration side, the influence of oxygen addition on the morphology of carbon deposits was almost the same as that seen for DC-PECVD system. This is probably due to the fact that in microwave plasma the generation rates of the precursors for carbon deposition, i.e., C and C_2 are much higher than those in DC plasma. Although CNFs synthesized by DC-PECVD are straight, MW-PECVD grown CNFs are slightly waved.

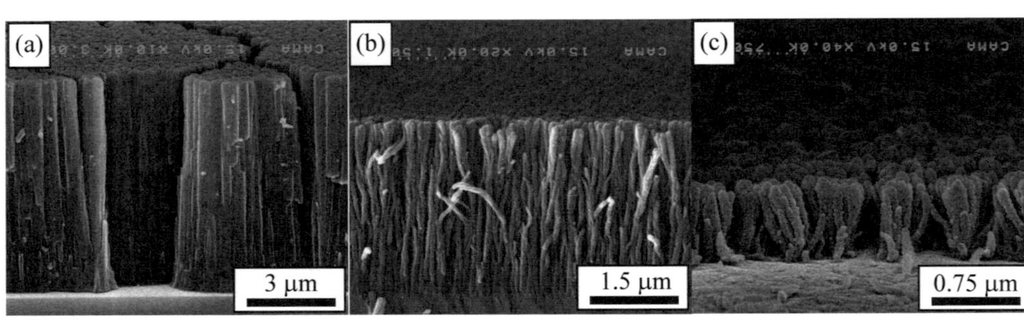

Fig. 6. SEM images of the carbon deposits on the glass substrates at O_2/CO rations of: (a) 0/1000; (b) 7/1000; and (c) 9/1000

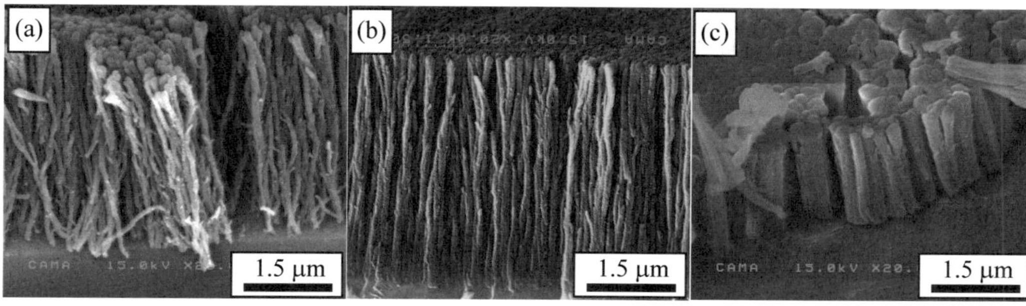

Fig. 7. SEM images of carbon deposits on different substrate materials.
Substrates: (a) CaF_2; (b) Si; (c) polycarbonate

Figure 7 shows SEM images of CNFs grown on different material substrates. The morphologies of the CNFs grown on Si and CaF_2 substrates were almost the same as those grown on the glass substrates. However, CNFs grown on the polycarbonate showed a different morphology. The diameters of the CNFs were increased, fiber-bundling was evident, and the fiber length was diminished. The high affinity that exists between the precursor species and organic materials may result in the formation of large nuclei on the substrates and result in the growth of CNFs with large diameter.

Figure 8 shows TEM images of the CNFs. The diameters of CNFs were 50-100 nm and no tubular structure was evident (Fig.8(a)). The surfaces of the CNFs were covered with the branching fibers and their nuclei, whose diameters are 5-10 nm. The high-magnification image of the CNF edge is shown in Fig. 8(b). Although it is not clearly seen because they overlapped and their directions were random in relation to the fiber axis, the lattice images of crystallized carbon were partially observed especially in the branching fibers.

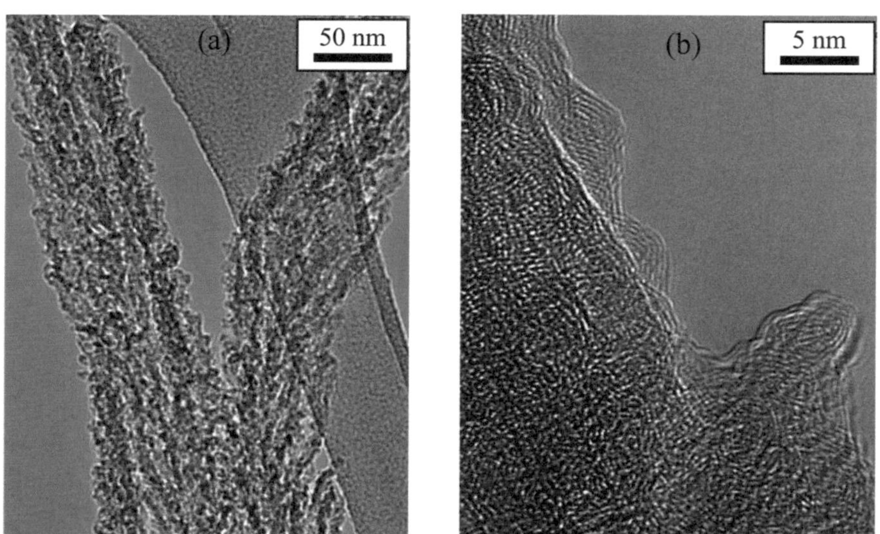

Fig. 8. TEM images of CNFs grown on the glass substrates at an O_2/CO ratio of 7/1000. (a) Low-magnification TEM image of two bundling CNFs; (b) high-magnification TEM image of the CNF surface.

The Raman spectra for the carbon materials formed on the glass substrate by MW-PECVD system were also examined. As shown in Figure 9, the rather sharp G-band peak and the broad D-band peak were observed and the D-band peak at 1350 cm^{-1}decreased with increasing O_2 flow rate, which is similar to the DC-PECVD results.

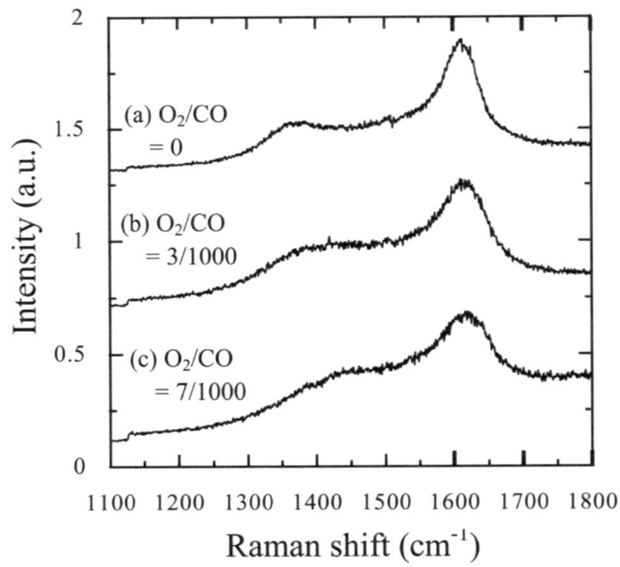

Fig. 9. Raman spectra of the carbon deposits prepared at O_2/CO ratios of; (a) 0/1000; (b) 3/1000; and (c) 7/1000.

4. Reaction mechanism and growth model

In order to infer the reaction mechanism, the plasma emission was monitored by a spectrometer (Ocean Optics, HR4000). The typical emission spectra from CO/Ar/O_2 plasma were shown in Fig. 10. A strong C_2 high-pressure band and CO Angstrom bands ($B^1\Sigma+ \rightarrow A^1\Pi$) and also a weak C atom spectrum at 247.9 nm can be seen. Interestingly, instead of C_2 swan bands ($d^3\Pi_g \rightarrow a^3\Pi_u$), which are well known as the most prominent bands of C_2 in hydrocarbon discharge and combustion flames, C_2 high-pressure bands ($d^3\Pi_g$, v=6 $\rightarrow a^3\Pi_u$) were observed in this system, which are known to be predominant compared to other C_2 band systems under certain CO discharge conditions (Caubet et al., 1994).

Figure 11 shows the influence of oxygen fraction on the emission intensities of CO Angstrom band, C_2 HP band, and C atom spectra. From this figure, the contribution of C_2 molecules to the CNF synthesis is suggested, because it is only C_2 molecules that the emission intensity shows a substantial change when the amount of oxygen increases. As for atomic carbon, the emission intensity is not influenced by the addition of oxygen and it is thought that there is no substantial change in the amount of C atom concentration.

It is more clearly suggested from Fig. 12 in which the normalized growth rate of CNFs and emission intensity of CO, C and C_2 by those without oxygen addition are plotted as a function of oxygen fraction. Although the normalized emission intensity of C atom spectra are not influenced by the addition of oxygen, that of C_2 HP band and normalized growth

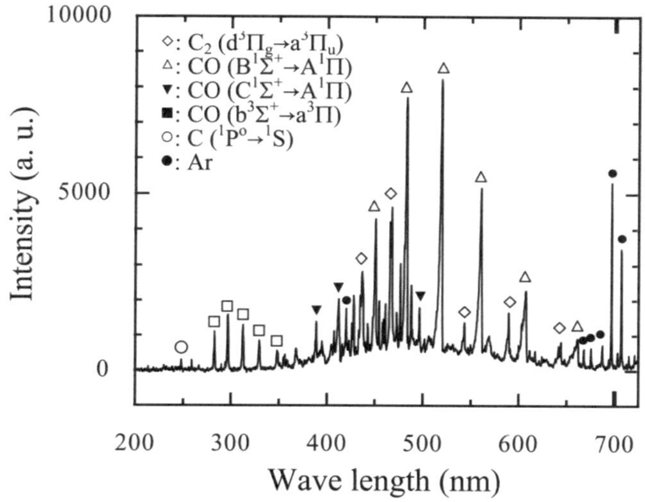

Fig. 10. Typical Emission Spectra of CO/Ar Plasma from the cathode region ($O_2/CO = 0$)

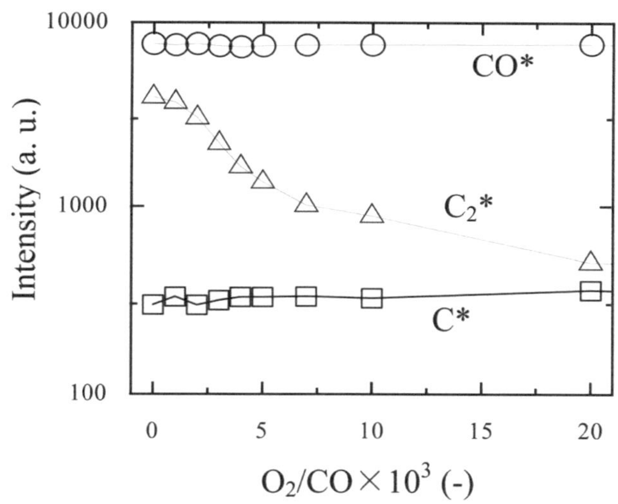

Fig. 11. Emission intensity of CO*, C_2*, and C* as a function of O_2/CO ratio

rates decrease drastically with increasing additional oxygen fraction and show a good correlation between them. In general, the change in the precursor density and the increase in the etching ability are thought to be the reasons why CNFs disappears as the amount of oxygen increases. However, as shown in the previous study, it cannot be thought that an increase in the etching ability is the reason for the disappearance of the CNFs in this case. When the small amount of hydrogen was added to the CO/Ar plasma, the CNFs disappear but the carbon deposits are not removed. As for the change in the spectrum, it is only C_2 molecules that the emission intensity shows a substantial change when the amount of hydrogen increases (Mori & Suzuki, 2008). In other words, even if the etching ability is low, the suppression of C_2 molecule formation results in the disappearance of fibrous structure. The carbon etching ability of hydrogen is much lower than that of oxygen (Mucha et al.,

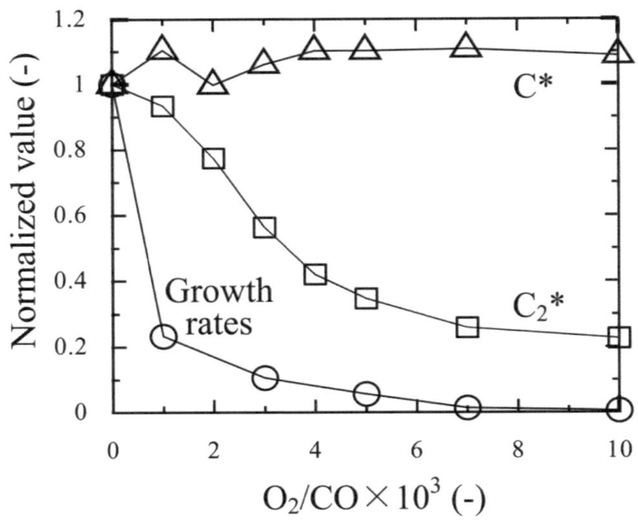

Fig. 12. Normalized growth rates of CNFs and emission intensities of C_2* and C* as a function of O_2/CO ratio. In this figure, growth rates and emission intensities are normalized using their values at $O_2/CO = 0$

1989). Therefore, it is reasonable to assume that, instead of C atoms, C_2 molecules play an important role as the main precursors of CNF synthesis in this system.

Although the growth mechanism for the formation of fibrous structure without catalyst metal particles has not been clearly understood yet, the nucleation and island growth process can be a key to control the morphology of carbon materials and the dispersed nuclei play the role of growth site for the CNFs. This is because CNFs grown on the polycarbonate has different morphology as shown in Fig. 7. The diameter of CNF grown on the polycarbonate is increased and fiber bundling arises and the length of fiber is diminished. The large affinity between precursor species and organic materials would result in the formation of large nuclei on the substrates and result in the growth of CNFs with large diameter (Mori & Suzuki, 2009c). In addition, the termination of the CNF surface by the oxygen would also be the key for the linear growth because the FTIR spectra of CNF includes a lot of oxygen containing groups, which is similar to C_3O_2 polymers and C_3O_2 polymers was known to have a linear structure (Mori & Suzuki, 2009a; Blake et al., 1964). It would be also ascribed to the anisotropic crystallization at the tip of the fiber due to the anisotropic action normal to the substrates in the cathode sheath region such as ion bombardment. Furthermore, C_2 molecules, which are known to be formed effectively in CO plasma, may crystallize anisotropically more readily than C atoms, thereby enhancing the linear growth of the CNFs.

Based on the results described above, the following model was postulated for the non-catalytic CNF growth mechanism. At first, precursor species such as C and C_2 appear mainly in the negative glow area, in which strong emission from C_2 molecules are observed, and diffuse to the substrate. Then, the Volmer–Weber island growth occurs on the substrate. During this nucleation process, if the affinity between the substrate and precursor species is high, then the diameters of the nuclei become large and the fibrous structure disappears.

Next, the carbon film grows up from the nucleus. During this nucleation and growth process, the precursor species deposited on the substrate are etched by the oxygen selectively, and only the crystallized carbons on the nuclei remain (Muranaka et al., 1991; Stiegler et al., 1996). Because of the etching action of oxygen on the nuclei, amorphous carbon, which tends to grow isotropically, is selectively removed, and only the crystallized carbon (graphitic carbon) remains anisotropically (Muranaka et al., 1991; Stiegler et al, 1996; Mucha et al., 1989). Moreover, because of the surface diffusion of the etchant species from the substrate and/or that of the precursor species to the substrate, the growth rate on the nucleus near the substrate is slower than at the tip area, which results in anisotropic growth from the nuclei. After some growth of the nuclei, the local deposition of precursor species occurs around the tip area, which leads the anisotropic growth. This is because the sticking probability of the precursor species onto the CNF surface is so high (Traskelin et al., 2008) that almost all of the precursor species are deposited around the tip area and rarely reach the lower fiber surface. In this anisotropic growth process, if the O_2/CO ratio is optimal, the diameter of CNFs does not increase as the fiber grows, and a fibrous structure can be achieved. However, if the O_2/CO ratio is slightly lower than its optimum value, the diameter of the CNFs increases as the fiber grows and the CNFs attach to adjacent fibers, which finally extinguishes the fibrous structure. On the other hand, if the O_2/CO ratio is slightly higher than its optimum value, the diameter of CNFs decreases as the fiber grows or nuclei can not grow further. The additional oxygen leads not only to amorphous carbon etching but also to the suppression of C_2 molecules formation by scavenging C_2O radicals (Mori & Suzuki, 2009d), which results in the disappearance of the fibrous structure and suppression of the carbon deposition. This optimum O_2/CO window is so narrow that this phenomenon would not have been observed until now.

5. Conclusion

The morphology of carbon deposits in the $CO/Ar/O_2$ plasma system is strongly affected by the O_2/CO ratio and, at the optimal O_2/CO ratio, vertically aligned CNFs were synthesized at a extremely low temperature in the absence of catalyst. The optimum O_2/CO ratio is crucial for the synthesis of CNFs without catalyst, as it suppresses the isotropic deposition of carbon materials by etching the amorphous carbon selectively and facilitates the anisotropic linear growth of carbon deposits by allowing the formation of crystallized carbon. The spectroscopic study reveals that there is a strong correlation between C_2 formation and CNF growth and C_2 molecules play an important role as the main precursors of CNF synthesis in the $CO/Ar/O_2$ PECVD system. The nucleation and island growth process is a key to control the morphology of carbon materials and the dispersed nuclei play the role of growth site for the CNFs. In addition, the termination of the CNF surface by the oxygen would also be an important factor for the linear growth of carbon materials without catalyst. From the viewpoint of process simplification and product purification, this catalyst-free synthesis is attractive. As is well known, high-temperature treatment is usually necessary for the activation of catalyst in the conventional CNFs growth method; the non-catalytic synthesis could lead to the development of a more viable process for the direct growth of CNFs onto low-temperature substrates like plastics, because no catalyst preparation step would be necessary.

6. Acknowledgments

We thank Mr. Katsuaki Hori, Mr. Akira Genseki, and Mr. Jun Koki of the Center for Advanced Material Analysis in Tokyo Institute of Technology for assistance with the SEM and TEM observations.

7. References

Baker, R.T.K. & P.S. Harris (1978). The formation of filamentous carbon, *Chemistry and Physics of Carbon*, Vol. 14, p. 83, Marcel Dekker, New York

Blake, A.R., W.T. Eeles & P.P. Jennings (1964). Carbon suboxide polymers, *Trans. Faraday Soc.*, Vol. 60, pp. 691-699

Boskovic, B.O., V. Stolojan, R.U.A. Khan, S. Haq & S.R.P. Silva (2002). Large-area synthesis of carbon nanofibres at room temperature, *Nat. Mater.*, Vol. 1, pp. 165-168

Capitelli, M. (1986). *Nonequilibrium Vibrational Kinetics*, Springer, ISBN 3-540-16250-X, Berlin

Caubet, P. & G. Dorthe (1994). Origin of C_2 high-pressure bands observed in the products of a microwave discharge through CO, *Chem. Phys. Lett.*, Vol. 218, pp. 529-536

Han, J.H., S.H. Choi, T.Y. Lee, J.B. Yoo, C.Y. Park, T.W. Jeong, H.J. Kim, Y.J. Park, I.T. Han, J.N. Heo, C.S. Lee, J.H. Lee & W.K. Yi (2002). Field emission properties of modified carbon nanotubes grown on Fe-coated glass using PECVD with carbon monoxide, *Physica B Condens. Matter*, Vol. 323, pp. 182-183

Hofmann, S., C. Ducati, J. Robertson & B. Kleinsorge (2003). Low-temperature growth of carbon nanotubes by plasma-enhanced chemical vapor deposition, *Appl. Phys. Lett.*, Vol. 83, pp. 135-137

Ionikh, Yu Z., I.N. Kostyukevich & N.V. Chernysheva (1994). Excitation of swan bands of a C_2 molecule in steady-state and decaying plasma in a He-CO mixture, *Opt. Spectrosc.*, Vol. 76, pp. 361-367

Kyung, S.-J., Y.-H. Lee, C.-W. Kim, J.-H. Lee & G.-Y. Yeom (2006). Field emission properties of carbon nanotubes synthesized by capillary type atmospheric pressure plasma enhanced chemical vapor deposition at low temperature, *Carbon*, Vol. 44, pp. 1530-1534

Liao, K.-H. & J.-M. Ting (2006). Characteristics of aligned carbon nanotubes synthesized using a high-rate low-temperature process, *Diamond Relat. Mater.*, Vol. 15, pp. 1210-1216

McCauley, T.G., D.M. Gruen & A.R. Krauss (1998). Temperature dependence of the growth rate for nanocrystalline diamond films deposited from an Ar/CH_4 microwave plasma, *Appl. Phys. Lett.*, Vol. 73, pp. 1646-1648

Melochko A.V., V.I. Merkulov, T.E. McKnight, M.A. Guillorn, K.L. Klein, D.H. Lowndes & M.L. Simpson (2005). Vertically aligned carbon nanofibers and related structures; Controlled synthesis and directed assembly, *J. Appl. Phys.*, Vol. 97, pp. 041301-1-39

Merkulov, V.I., D.H. Lowndes, Y.Y. Wei, G. Eres & E. Voelkl (2000). Patterned growth of individual and multiple vertically aligned carbon nanofibers, *Appl. Pys. Lett.*, Vol. 76, pp. 3555-3557

Meyyappan, M, L. Delzeit, A. Cassell & D. Hash (2003). Carbon nanotubes growth by PECVD; a review, *Plasma Source Science Technol.*, Vol. 12, pp. 205-216

Minea, T.M, S. Point, A. Granier & M. Touzeau (2004). Room temperature synthesis of carbon nanofibers containing nitrogen by plasma-enhanced chemical vapor deposition, *Appl. Phys. Lett.*, Vol. 85, pp. 1244-1246

Mori, S., H. Akatsuka & M. Suzuki (2001). Carbon and oxygen isotope separation by plasma chemical reactions in carbon monoxide glow discharge, *J. Nucl. Sci. Technol.*, Vol. 38, pp. 850-858

Mori, S., M. Fukuya & M. Suzuki (2007). Synthesis of carbon nanofibers using carbon monoxide DC plasma at room temperature, *Trans. Mater. Res. Soc. Jpn.*, Vol. 32, pp. 513-516

Mori, S. & M. Suzuki (2008). Effect of oxygen and hydrogen addition on the low-temperature synthesis of carbon nanofibers using a low-temperature CO/Ar DC plasma, *Diamond Relat. Mater.*, Vol. 17, pp. 999-1002

Mori, S. & M. Suzuki (2009a). Characterization of carbon nanofibers synthesized by microwave plasma-enhanced CVD at low-temperature in a $CO/Ar/O_2$ system, *Diamond Relat. Mater.*, Vol. 18, pp. 678-681

Mori, S. & M. Suzuki (2009b). Non-catalytic low-temperature synthesis of carbon nanofibers by plasma-enhanced chemical vapour deposition in a $CO/Ar/O_2$ DC discharge system, *Appl. Phys. Exp.*, Vol. 2, pp. 015003-1-3

Mori, S. & M. Suzuki (2009c). Catalyst-free low-temperature growth of carbon nanofibers by microwave plasma-enhanced CVD, *Thin Solid Films*, Vol. 517, pp. 4264-4267

Mori, S. & M. Suzuki (2009d). The role of C_2 in low temperature growth of carbon nanofibers, *J. Chem. Eng. Jpn.*, in press

Mucha, J.A., D.L. Flamm & D.E. Ibbotson (1989). On the role of oxygen and hydrogen in diamond-forming discharges, *J. Appl. Phys.*, Vol. 65, pp. 3448-3452

Muranaka, Y., H. Yamashita, H. Miyadera (1991). Characterization of diamond films synthesized in the microwave plasmas of CO/H_2 and $CO/O_2/H_2$ systems at low temperatures (403-1023 K), *J. Appl. Phys.*, Vol. 69, pp. 8145-8152

Murayama, H & T. Maeda (1990). A novel form of filamentous graphite, *Nature*, Vol. 345, pp. 791-793

Plonjes, E., P. Palm, G.B. Viswanathan, V.V. Subramaniam, I.V. Adamovich, W.R.Lempert, H.L. Fraser & J.W. Rich (2002). Synthesis of single-walled carbon nanotubes in vibrationally non-equilibrium carbon monoxide, *Chem. Phys. Lett.*, Vol. 352, pp. 342-347

Rodriguez, N.M. A. Chambers & R.T.K. Baker (1995). Catalytic engineering of carbon nanostructures, *Langmuir*, Vol. 11, pp. 3862-3866

Stiegler, J., T. Lang, M. Nygard-Ferguson, Y.von Kaenel & E. Blank (1996). Low temperature limits of diamond film growth by microwave plasma-assisted CVD, *Diamond Relat. Mater.*, Vol. 5, pp. 226-230

Traskelin, P., O. Saresoja & K. Nordlund (2008). Molecular dynamics simulations of C_2, C_2H, C_2H_2, C_2H_3, C_2H_4, C_2H_5, and C_2H_6 bombardment of diamond (1 1 1) surfaces, *J. Nucl. Mater.*, Vol. 375, pp. 270-274

Yoon, S.-H., S. Lim, S.-h. Hong, W. Qiao, D.D. Whitehurst, I. Mochida, B. An & K. Yokogawa (2005). A conceptual model for the structure of catalytically grown carbon nanofibers, *Carbon*, Vol. 43, pp. 1828-1838

16

Carbon Nanotubes Reinforced Electrospun Polymer Nanofibres

Minoo Naebe, Tong Lin and Xungai Wang
Centre for Material and Fibre Innovation, Deakin University, VIC 3217, Australia

1. Introduction

With the rapid development in nanoscience and nanotechnology, there is an ever increasing demand for polymer fibres of diameters down to a nanometre scale having multiple functionalities. Electrospinning, as a simple and efficient nanofibre-making technology, has been used to produce polymer nanofibres for diverse applications. Electrospun nanofibres based on polymer/carbon nanotube (CNT) composites are very attractive multifunctional nanomaterials because they combine the remarkable mechanical and electronic properties of CNTs and the confinement-enhanced CNTs alignment within the nanofibre structure, which could greatly improve the fibre mechanical, electrical and thermal properties. In this chapter, we summarise recent research progress on electrospun CNTs/polymer nanofibres, with an emphasis on fibre mechanical properties and structure-property attributes. Outlook towards the challenge and future directions in this field is also presented.

2. Carbon nanotubes

Carbon nanotubes (CNTs) which were first reported by Oberlin *et al.* [1] and later by Iijima [2] have remained the focus of intense academic investigation since early 90's. They are still at the forefront of research in many areas of nanoscience and technology due to their spectacular combination of mechanical, electrical and thermal characteristics.

a. Structure of carbon nanotubes

Carbon nanotube is a new form of carbon which has a seamless hollow cylindrical structure, consisting of carbon hexagons with both ends capped with fullerene molecule. There are two general types of carbon nanotubes, namely single walled nanotubes (SWNTs) and multi walled nanotubes (MWNTs). SWNTs consist of one single layer of hexagonal carbon atom rolled into tubular form the diameter of which ranges from 0.4 to over 3 nm, while MWNTs have several concentric cylinders with diameter from 1.4 to over 100 nm [3]. SWNTs are varied in chiral angle i.e. the angle at which the atoms of CNTs are twisted about the main axis of the CNTs. According to their chiral angle and diameters, SWNTs are classified to three forms of structures including armchair, zigzag and intermediate [4] as shown in Figure 1. The chiral angle governs the electrical conductivity. While the armchair structure exhibits very high conductivity, the zigzag and the intermediate forms show semi-conductivity.

Fig. 1. Schematic illustration of various carbon nanotube structural forms, from left to right: armchair structure, zigzag structure and intermediate or chiral structure [3] [Copyright AAAS].

Several techniques have been developed to synthesise SWNTs and MWNTs. The most used methods are carbon arc discharge [2], laser ablation of carbon [5] and chemical vapour deposition (on catalytic particles) [6]. Either vacuum or process gas is used in the production. The defining issue in selecting an appropriate method is the ability to produce nanotubes on a large scale. Advances in catalysis and continuous growth processes are making CNTs more commercially feasible.

b. Mechanical properties

CNTs combine outstanding mechanical, electronic, thermal properties, and low density. However, their extraordinary mechanical properties due to carbon-carbon sp2 bonds and cylinder structure set them apart from many other different materials and other forms of carbons. Soon after Ijima's discovery and before the large scale production of CNTs, computer simulation was used to calculate the rigidity of SWNTs [7]. The calculated Young's modulus was 1500 GPa. However, later studies predicted that Young's modulus of CNTs is approximately 1 TPa [8]. It was not until 1997 that the first direct mechanical measurement was carried out on arc-MWNTs using atomic force microscopy and an average Young's modulus value of 1.28 TPa was obtained [9]. The highest Young's modulus and tensile strength measured for MWNTs produced by chemical vapour deposition method is 0.45 TPa and 4 GPa, respectively. The highest measured tensile strength for arc-MWNTs is 63 GPa [10]. Mechanical measurements on SWNTs did not commence until the late 1990's due to difficulties in handling them [11]. The highest measured values for Young's modulus and tensile strength of SWNTs are 1.47 TPa and 52 GPa, respectively [12]. The tensile strength of SWNTs could be more than 5 times higher than that of a steel fibre with the same diameter, yet only one-sixth of its density [13][14].

c. Electronic properties

The remarkable electronic properties of CNTs make them particularly attractive for the creation of miniaturized electronic component. While the resistivity values for high quality graphite, the measured resistivity copper is approximately 0.40 and 0.017 μΩm, respectively, and the measured resistivity of CNTs falls into the range of 0.05 μΩm ~ 10 mΩm [15]. Due to structural defects catalytically produced CNTs are expected to have a higher resistivity, similar to those of disordered carbon (i.e of the order of 10~100 μΩm). Earlier studies suggested that electronic properties of CNTs could vary widely from tube to tube and they could be metallic or semiconducting depending on their structure (*e.g.* tube chirality) and diameter [16][17][18][19]. However, recent studies showed that the electronic properties of a given nanotube were not only specified by the diameter and chirality of nanotubes but also depended on their chemical environment particularly gas exposure history [20][21]. Exposure to air or oxygen significantly influences CNTs electrical resistance. Semiconducting CNTs can be converted into metals through room temperature exposure to oxygen [20].

d. Carbon nanotubes/ polymer composites

Because of the exceptional properties and large aspect ratio, incorporation of nanotubes into polymer matrix has been proven to be a promising approach leading to structural materials and composites with excellent mechanical and physical properties. Various researches have been conducted and comprehensive reviews on mechanical properties of carbon nanotubes/polymer composites have been given by Coleman *et al.* [22][23]

The significant challenges in processing of CNTs/polymer composites lie in the uniform dispersion and orientation of nanotubes within the polymer matrix. CNTs tend to aggregate to form tight bundles due to strong van der Waals interactions and small size [24][25]. Dense and entangled CNTs network in the bundles prevent the CNTs from dispersing uniformly within the polymeric resins. As a result of uneven dispersion, the physical and mechanical properties of the composite material are considerably lower than the expected.

The difficulties in obtaining uniform dispersion were highlighted in the literature [26][27][28][29][30][31][32][33] and several techniques have been developed to improve the CNT dispersion. Ultrasonication has been used to de-aggregate CNT bundles and force the nanotubes to disperse uniformly throughout the material [27]. Other investigators have utilised solution-evaporation methods with high-energy sonication [26], surfactant-assisted processing through the formation of a colloidal intermediate [34][35] and melt spinning [36], as well as mechanical stretching of nanotube/polymer composites [37][38]. Others considered deposition of carbon nanotubes suspension under a magnetic field [39] and onto chemically modified substrate [40]. The most commonly used techniques to fabricate CNT/polymer composites are solution casting [41][42], melt processing [43][44], electrospinning [45] and *in-situ* polymerisation [46][47]. Electrospinning is an efficient processing method to produce CNT/polymer nanofibres with the CNTs orienting to the axes of the as-spun nanofibres [48][49]. There is a fast growing interest in applying this technique to produce nanofibres using various polymers [48]. This chapter summarises the research and development of electrospun carbon nanotube/polymer nanofibres with an emphasis on processing conditions, fibre morphology, mechanical properties and applications.

One of the interesting features for CNTs is the nucleation crystallisation of surrounding polymer molecules when the CNTs are dispersed into some polymer matrices. It has been observed that the introduction of CNTs to some polymer matrices alters the crystallisation dynamics of polymers [42][50]. In other words, the presence of CNTs induces crystallisation of host polymer. As a result of the crystalline polymer layer formed around the embedded nanotubes, the interaction between the CNT and polymer matrix is enhanced leading to improved mechanical properties.

Since 1998, several publications have reported the nucleation crystallinity of polymer in CNT/polymer composites [51][52][53][54][55][56][57]. The transmission electron microscopy (TEM) studies by McCarthy *et al.* [51] identified that carbon nanotubes were coated with a thin layer polymer when they were dispersed in a semi-conjugated polymer, poly(m-phenylenevinylene-co-2,5-dioctyloxy-p-phenylenevinylene). The nucleation crystallisation of proteins such as streptavidin around carbon nanotubes was also reported [52][54][56]. Also, polycarbonate has been found to form a thin layer around MWNTs. A clearly observable polymer sheath was identified at the fracture surface of a composite by scanning electron microscope (SEM) [54]. The nucleation crystallinity of other polymers, such as polypropylene (PP), was also studied [50][58][59][60][61][62][63]. However, no direct evidence for the formation of PP coating on the nanotubes was presented, except that the differential scanning calorimetry (DSC) measurements revealed an increase in the crystallinity due to the presence of CNTs.

It was also reported that the presence of CNTs in polyvinyl alcohol (PVA) improved the composite mechanical properties more than in other polymer systems [42][64][65][66][67][68][69]. The nucleation crystallisation of PVA was confirmed by microscopic investigations of fractured PVA composite film, which showed that a thick PVA coating covered the nanotubes [69]. It has been established that the presence of ordered polymer coating is the main reason leading to the enhanced mechanical strength [66].

3. Electrospinning and electrospun nanofibres

Electrospinning is a relatively low cost, fast and versatile method to produce continuous nanofibres mainly from polymer solutions. This technique has not been well studied until last decade even though it was invented in 1934 [70].

a. Basic electrospinning principle

A basic electrospinning setup, as shown in Figure 2a, consists of a container for polymer solution, a high-voltage power supply, spinneret (needle) and an electrode collector. During electrospinning, a high electric voltage is applied to the polymer solution and the electrode collector leading to the formation of a cone-shaped solution droplet at the tip of the spinneret, so called "Taylor cone" [71]. A solution jet is created when the voltage reaches a critical value, typically 5-20 kV, at which the electrical forces overcome the surface tension of the polymer solution. Under the action of the high electric field, the polymer jet starts bending or whipping around stretching it thinner. Solvent evaporation from the jet results in dry/semidry fibres which randomly deposit onto the collector forming a nonwoven nanofibre web in the most cases (Figure 2b).

Extensive research has been carried out on various aspects of electrospinning including operating parameters (e.g. applied voltage, feeding rate, distance between the nozzle and collector), material properties (e.g. viscosity, surface tension, conductivity), spinningability of many different polymers [72][73][74][75][76][77][78][79][80][81][82][83][84][85][86][87][88], process modelling [88] [89] [90] [91] [92] [93] [94] [95], nanofibre characterisations and morphology [87].

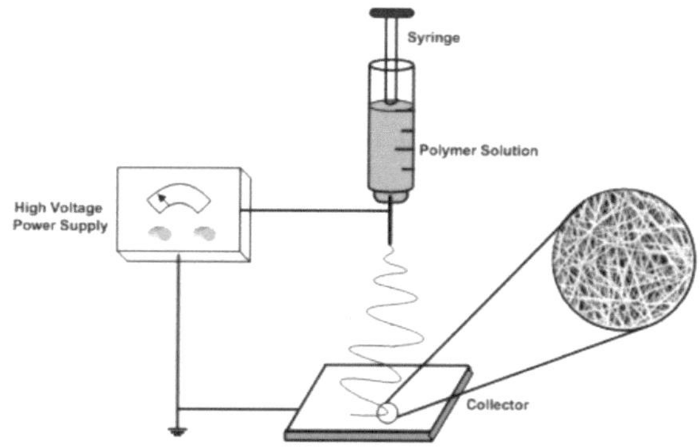

(a)

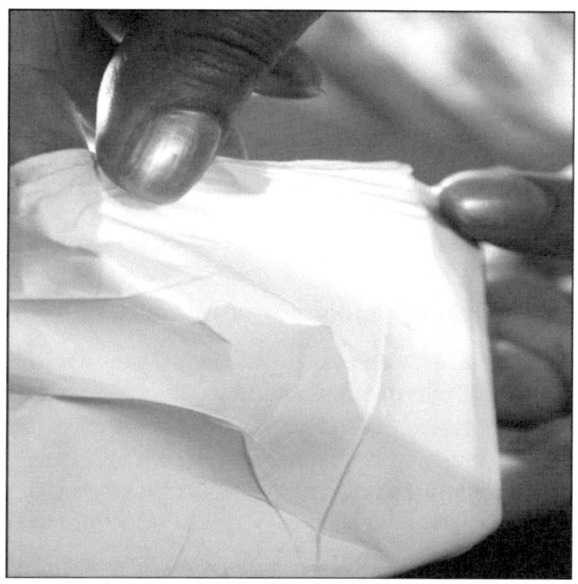

(b)

Fig. 2. (a) Basic apparatus for electrospinning, (b) A photo of typical electrospun nanofibre mat.

b. Electrospun nanofibres

Fibres obtained from electrospinning vary from uniform fibres to fibres with an irregular beads-on-string structure. The morphology of the electrospun fibres are dependent upon a number of factors including the polymer solution parameters such as molecular weight, molecular weight distribution, electrical conductivity, surface tension, viscosity and solvent, and the operating parameters such as electrical field, the distance from the nozzle tip and the collector and the flow rate of the polymer, as well as ambient conditions [84][96][97]. The diameter of electrospun fibres can be in the range between several microns to tens of nanometres. The small fibre diameter and large aspect ratio lead to extremely high surface-to-volume (weight) ratio, which makes the electrospun nanofibre desirable for many applications [98]. Recently, several review articles have been published on electrospinning [99][100][101], which demonstrate the great potential of electrospun nanofibres in diverse application fields.

c. Mechanical properties of electrospun polymer nanofibres

Although several experimental investigations have been carried out on the mechanical properties of nanofibre mats, only a few studies have been reported on the stress-strain behaviour of single electrospun nanofibres [102][103][104]. These studies have demonstrated that single electrospun nanofibres have promising mechanical properties. Gu *et al.* [102] calculated the Young's modulus of a single electrospun polyacrylonitrile (PAN) fibre from the force displacement curves obtained by bending a single fibre attached to an atomic force microscopy (AFM) cantilever. High Young's modulus of up to 50 GPa was reported suggesting the orientation of PAN molecular chains as evidenced by X-ray diffraction study. In a study by Bellan *et al.* [104], the average Young's modulus of electrospun poly (ethylene oxide) (PEO) nanofibres measured by AFM technique was reported to be 7.0 GPa, which was significantly larger than that of PEO bulk materials. While electrospinning induced molecular orientation was attributed to the high stiffness of the nanofibre, no data on the molecular orientation of single nanofibre was provided. Lee *et al.*[103] reported that the Young's modulus of PVP nanofibres containing TiO_2 nanoparticles was only 0.9 GPa, which was measured with an AFM using the three point bending method.

4. Electrospun carbon nanotube/polymer composite nanofibres

CNT/polymer composites have been fabricated by a number of processing methods including melt processing, solution processing and *in situ* polymerisation [105]. However, it is difficult to control the orientation of CNTs within the polymer matrix. This is particularly important for the reinforcement of polymers and for meeting the expectation for nanocomposites with significantly enhanced mechanical properties.

It has been established that electrospinning a polymer solution containing well-dispersed carbon nanotubes leads to nanocomposite fibres with the embedded carbon nanotubes orienting parallel to the nanofibre axis due to large shear forces in a fast fibre-drawing process [106]. It has been demonstrated that electrospinning is a potential method for aligning and debundling CNTs [107]. The improved CNT alignment within the fibres plus low cost and fairly simple spinning process has made this technique promising for producing CNT/polymer composite nanofibres.

The reported electrospun CNT/polymer composites including corresponding solvent and CNT concentration are listed in table 1. While CNTs have potential to be embedded into various polymer matrices, some polymers cannot be easily electrospun into nanofibres.

a. Carbon nanotube dispersion

A common method for preparing CNT/polymer solution for electrospinning involves making nanotube dispersion and polymer solution separately and then mixing them together. In general, interest has been focused on achieving homogenous nanotube dispersion in a polymer solution, which will affect the orientation and distribution of CNTs in the resultant nanofibres.

Polymers	Solvents	CNTs	CNTs (wt%)	Diameter (nm)	Tensile strength* (MPa) (Modulus, MPa)	Ref
PAN	--------	SWNTs/ MWNTs	1	180	--------	[108]
PAN	DMF	MWNTs	1-20	50~300	285~312 (6.4~14.5GPa)	[109]
PAN	DMF	MWNTs	1.5,7	20~140	------	[110]
PAN	DMF	SWNTs	1-10	50~400	20~30	[111]
PAN	DMF	MWNTs	1	200~2000	--------	[112]
PAN	DMF	SWNTs	1~4	50~200	(140 GPa)	[45]
PAN	DMF	MWNTs	2-20	100-300	37-80 (2-4.4GPa)	[113]
PVA	Water	MWNTs	4.5	295~429	4.2~12.9	[114]
PVA	Water	SWNTs	10	315~447	5.9~6.0	[115]
PEO	Ethanol/Water	SWNTs	3	----	(0.7~1.7GPa)	[116]
Polymethyl methacrylate	Chloroform	MWNTs	0.5-2	200~6000	------	[117]
Polymethyl methacrylate	DMF	MWNTs	1-5	100~800	------	[118]
Polyurethane	DMF	SWNTs	1	50~100	10~15	[119]
Polycaprolactone	Chloroform/ Methanol	MWNTs	7-15	100~550	--------	[120]
Polylactic acid	DMF	SWNTs	1-5	1000	-------	[121]
Regenerated silk fibroin	Formic acid	SWNTs	0.5-5	147	2.8-7.4 (180~705)	[122]
Regenerated silk	Formic acid	SWNTs	1	147~153	13.9~58.0 (633.8~6549.3)	[123]
Polybutylene terephthalate	1,1,1,3,3,3 Hexafluoro-2-propanol	MWNTs	5	250~3500	(1.79 GPa)	[124]
Polycarbonate	Chloroform	MWNTs	4	350	------	[125]
Nylon 6,6	Formic acid	MWNTs	2-20	150~200	-------	[126]
Polystyrene	DMF/ tetrahydrofuran	MWNTs	0.8,1.6	300,4500	----	[127]

Polyvinyl alcohol (PVA); Dimethylformamide (DMF); * Unit for modulus values is MPa unless otherwise stated.

Table 1. Electrospun CNT/polymer composite nanofibres and their mechanical strength

The stable dispersion of CNTs can be achieved by using surfactants (e.g. sodium dodecyl sulphate), large amphiphilic polymers (e.g. polyvinyl pyrrolidone) and natural macromolecules (e.g. polysaccharide, Gum Arabic) which can be adsorbed onto the hydrophobic nanotubes [116][128]. Nevertheless, the most common method for dispersing individual nanotubes is ultrasonication treatment. While ultrasonication is a valuable technique to overcome the entanglement of nanotubes and break up the agglomerates, it could also introduce defects and irregularity into the CNTs [128].

Purity is also an important factor affecting the composite quality. CNTs are typically purified by chemical methods to remove amorphous carbon and metal catalyst. The purification treatment also facilitates the dispersion of the CNTs in solvents as well as improves the nanotube-matrix interaction [15]. It was demonstrated that the caps of SWNTs could be removed by treatment with hot nitric acid due to the formation of carboxylic acid and hydroxyl groups at the nanotube ends [129][130]. Apart from this, the purified CNTs can be functionalised to get better dispersability and additional functions [131]. There are several approaches to chemical functionalisation of nanotubes [105][132]. Although functionalised CNTs showed improved dispersion, the electronic and photonic properties of nanotubes would be altered as well [133].

b. Structure and morphology

Alignment of individual nanotubes within the host polymer is a crucial step, in particular with the applications for reinforcement. Electrospinning is expected to make CNTs align with the fibre axis during the fibre formation process because of the high stretching ratio. Strategies to make the best use of this capability have been employed to improve the dispersion of CNTs in polymer solution using surface functionalised CNTs and ultrasonication treatment.

Salalha *et al.* [134] proposed a theoretical model for the behaviour of rodlike particles in the electrospinning jet and predicted the possibility for CNTs to be aligned along the streamlines of an electrospun jet. They fabricated electrospun MWNT/PEO nanocomposite fibres [25].They found that PEO crystals were highly aligned along the fibre axis during the electrospinning process. The MWNTs were individually embedded in the nanofibres, mostly aligned along the fibre axis. Nevertheless, twisted, bent and non-uniform nanotubes were also observed (Figure 3a). By comparing the MWNT dispersions with the raw nanotubes under TEM, they suggested that over-sonication of nanotubes for dispersion might damage nanotubes, preventing them from being well orientated. This finding is consistent with what Kearns and Shambaugh reported [135].

In other similar studies [116][134], well dispersed SWNTs were incorporated into PEO nanofibres by electrospinning. It was shown that nanotube alignment within the nanofibres depended strongly on the quality of the initial dispersion. Ko *et al.* [45] also found that a better alignment of CNTs was formed in polyacrylonitrile (PAN) than in polylactic acid (PLA) nanofibres, indicating the matrix dependency of nanotube alignment (Figure 3b). The improved orientation also resulted in a better distribution of carbon nanotubes within nanofibres. Vibration electrospinning employing in-built ultrasonic generator has also been used to align CNTs in nanofibres [136].

There have been several studies on controlling the spatial orientation of electrospun CNT/polymer composite nanofibres [109] [137] [138] [139] [140]. Modification of electrospinning setup has been a common approach to produce aligned electrospun composite nanofibres. This includes employing rotating collectors such as a disk or mandrel to collect nanofibres aligned in the rotating direction.

Ge *et al.* [109] developed aligned electrospun MWNT/PAN composite nanofibre sheets by collecting the nanofibres onto a winder with a surface velocity larger than the velocity of electrospun nanofibres. They observed highly oriented CNTs within the nanofibres and attributed this to the structural formation during the electrospinning process and the slow relaxation of CNTs [109]. Using a similar setup, Jose *et al.* [137] also fabricated aligned electrospun Nylon-6 nanofibres containing surface modified MWNTs. The MWNTs showed high degree of alignment in the nanofibres. They also demonstrated that MWNT/nylon-6

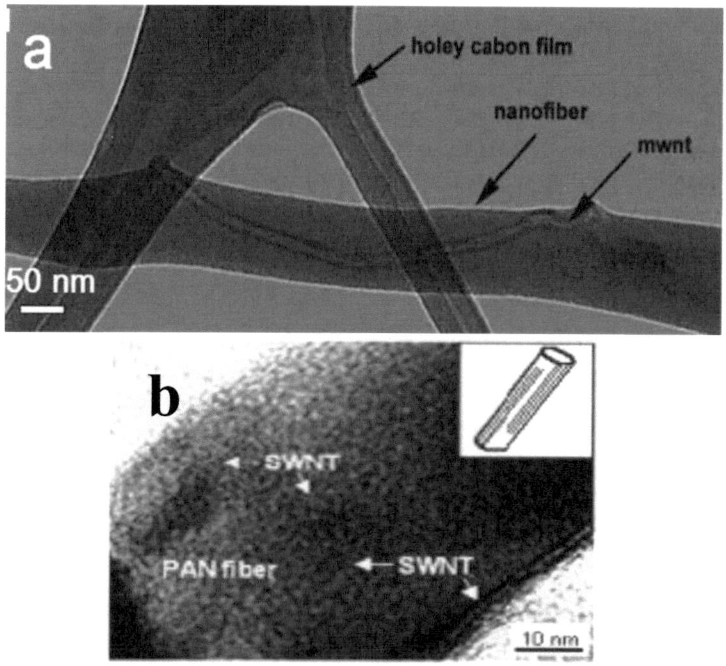

Fig. 3. a) TEM image of protrusion of MWNTs from the fibre [128] [Copyright ACS]. b) High resolution TEM image showing a uniform distribution and alignment of SWNTs in PAN fibre with the diameter of about 50nm [45] [Copyright Wiley-VCH].

nanofibres had significantly smaller diameter, ranging from 250 nm to 750 nm, than the neat nylon-6 fibres and the fibre diameter reduced with the increase in the concentration of CNTs. This is consistent with what Ra *et al.* reported for electrospun MWNT/PAN nanofibres [141].

Huang *et al.* [142] used a modified rotating disk collector including two separate, parallel aluminium plates with sharp edges to fabricate CNTs/polyvinylidene difluoride (PVDF). They found that the interfacial interaction between SWNTs and PVDF and the application of extensional forces had a strong synergistic effect on crystalline structures of PVDF inducing highly oriented crystallites at only 0.01 wt% of nanotubes. However, the effect of the MWNTs on crystal orientation was low and was attributed to the detrimental effect of the MWNTs on the preferred orientation of PVDF chains.

Zhang *et al.* [138] fabricated uniaxially aligned electrospun nanofibres of PAN and Triton X-100 grafted MWNTs using two collectors including i) parallel copper sheets [143] and ii) a slowly rotating drum with multi electrodes [144]. While large scale aligned MWNT/PAN can be produced using the multi-electrode drum at a slow rate of 1 rpm no aligned nanofibres were observed at high rate of drum rotation, i.e. 600 rpm. This was attributed to lack of enough time for nanofibres to be attracted across the axes of the drum at higher rotation rate.

Yee *et al.* [139] modified the design of a disk by using two separate parallel aluminium electrodes attached to a rotating disk to collect well-aligned MWNT/PVDF nanofibres where both electric field and mechanical force contributed to nanofibre alignment. It was found that the alignment of nanofibres collected by the modified disk was about the same as that by the conventional disk. However, nanofibres collected by modified rotating disk showed a uniform lateral distribution across a relatively larger distance. This was mainly because the specific electric field distribution created by aluminium electrodes coupled with the repelling force from the residual charges on the electrospun nanofibres. Aligned MWNT/nylon-6 nanofibres were also electrospun using perpendicular rotating disks [140]. In addition to rotating collectors, a secondary collector plate (as simple as a cardboard) was found to be an effective way of fabricating aligned electrospun SWNT/regenerated silk nanofibres of less than100 nm in diameter [123].

c. Mechanical properties

Most studies on CNT/polymer composites have been driven by improving the mechanical strength. This is of particular importance for electrospun nanofibres, because the relatively low bulk mechanical properties hinder their applications in some areas. Due to the small size, measuring the tensile properties for individual electrospun nanofibres is difficult. A few experimental investigations on mechanical properties of electrospun CNT/polymer nanofibres have been reported [45][124][145][146]. In these studies, atomic force microscopy (AFM) has been used as a tool to study the mechanical behaviour of single electrospun composite nanofibres.

Ko *et al.* [45] were among the first to study the mechanical properties of CNT reinforced electrospun nanofibres. They measured the elastic modulus of electrospun SWNT/PAN nanofibres. The fibre modulus obtained was 140 GPa which was well beyond the value of 60 GPa measured for PAN fibres [45]. Nevertheless, other mechanical properties such as strength and strain at break have not been measured. Liu *et al.* [145] studied the stress-strain behaviour of electrospun MWNT/ poly (methyl methacrylates) (PMMT) single nanofibres. Using AFM and environmental scanning electron microscope, tensile strength as high as 80 MPa and Young's modulus of 267 MPa were reported for composite nanofibres, which are significantly higher than that of pure PMMA nanofibres. The significant improvement in modulus and strength was due to the good dispersion and orientation of nanotubes within the polymer and to the strong interfacial adhesion due to the nanotube surface modification. Recently, Almecija *et al.* [146] developed a method to deposit electrospun nanofibres across pre-prepared trenches on silicon substrates. Measuring both force and displacement on SWNT/PVA strained nanofibres. They then developed a model in which the tensile properties of single nanofibres were generated using force-displacement data. The maximum strength and modulus obtained by this method was 2.6 and 85 GPa, respectively.

A novel method to study the mechanical deformation of electrospun composite nanofibres has been described by Kim *et al.* [125]. They performed *in-situ* tensile tests on single electrospun composite nanofibres by electron beam induced thermal stress under TEM. In this method, the bombardment of electron beam onto the fibres resulted in local thermal expansion and hence initiated tensile deformation. The strain rate can be controlled by adjusting the electron beam flux on the fibres [26] [147]. As the strain increased, the fibre elongated and with further increase of strain the necking occurred mainly at the end of a MWNT embedded into fibre. This was due to the slippage of the MWNTs in the direction of the applied tensile stress taking place beyond a certain critical stress value. A combination between CNT slippage mechanism and stress concentration (provided by the nanopores on the fibre surface), as well as the transfer of mechanical load from polymer matrix to MWNTs (due to highly aligned nanotubes along the fibre axis) were considered to be the main reason for the enhancement of the critical fracture strain and the toughening the electrospun composite nanofibre [125]. More recently Singh and co-workers [148] developed a novel characterisation device for testing individual electrospun polymer nanofibres. The tool consists of a nano manipulator, a transducer and associated probes and is operated inside a scanning electron microscope (Figure 4a). The three-plate capacitive transducer independently measures forces and displacement with a resolution of micronewton and nanometre, respectively. The tensile test of an electrospun polyaniline fibre (diameter ~1 μm) demonstrated the capabilities of the system. Engineering stress versus strain curves exhibited two distinct regions (Figure 4b); the Young's modulus of the latter region was approximately 5.9 GPa. Failure at the probe-specimen weld occurred at ~67 MPa, suggesting a higher yield stress for polyaniline microfibres when compared with the bulk counterpart.

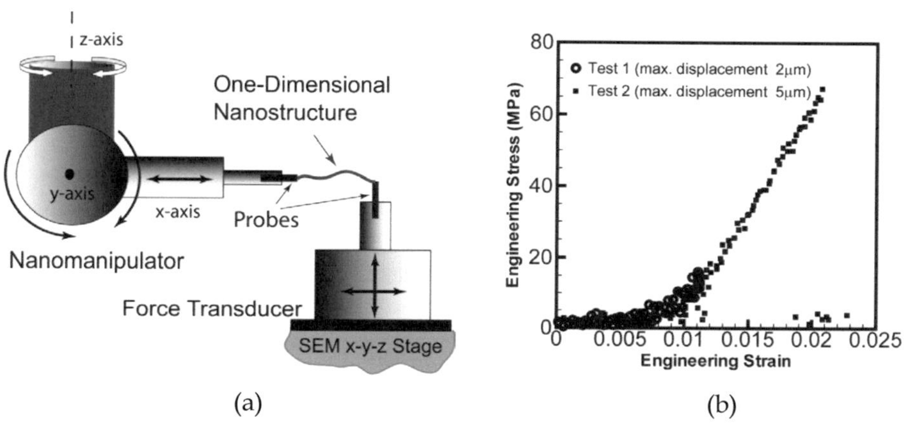

Fig. 4. (a) Schematic of the nano mechanical characterisation device showing the magnified view of the fibre specimen between two probes and the available degrees of freedom. (b) Engineering stress and strain for *test 1* (circles) and *test 2* (squares); the latter experiment involved loading the specimen to a maximum displacement of 5 μm; failure at the probe-specimen weld occurred prior to reaching maximum displacement [148] [Copyright AIP].

Ye *et al.* [147] used TEM to study the reinforcement and rupture behaviour of SWNT/PAN and MWNT/PAN nanofibres. Under tension, two-stage rupture behaviour of the composite nanofibres including crazing of polymer matrix and pulling-out of carbon nanotubes was observed. The involved mechanism was explained by the fact that CNTs reinforce the polymer fibres by hindering crazing extension, hence reducing the stress concentration and dissipating the energy. Two major factors, the distribution of CNTs in the polymer matrix and interfacial adhesion between nanotubes and polymers, determine the reinforcement effect of CNTs in polymer fibres.

Elastic deformation of MWNTs in electrospun MWNTs/PEO and MWNTs/PVA nanofibres were studied by Zhou and co-workers [149]. The degree of elastic deformation was found to increase as the modulus of the polymer matrix increased. A simplified model was also proposed to estimate the elastic modulus ratio of MWNTs and polymers. To confirm the validity of the model, the results were compared with that from AFM measurement.

As for CNT reinforced electrospun nanofibre mats, several research groups studied the mechanical behaviour of electrospun nanofibre membranes. These studies have demonstrated that CNTs improve the mechanical properties of electrospun polymer nanofibres as long as they are well dispersed into the polymer matrix. For example stress-strain analysis showed that the tensile strength of SWNT reinforced polyurethane (PU) nanofibre membrane was enhanced by 46% compared to pure PU nanofibre mat [119]. However, this value was further increased by 104% for PU membranes containing ester-functionalized SWNTs. This improvement in the mechanical strength was attributed to improved dispersion of the SWNTs as well as enhanced interfacial interaction of nanotubes with the PU matrix because of modified nanotube surface [119]. Recently, Yoon *et al.* [150] reported enhancement in mechanical strength of CNT reinforced nanofibres caused by better nanotube-polymer adhesion and good dispersion of SWNT because of the plasma treatment of nanotubes. Uniform dispersion of amino functionalised MWNTs and nanotube alignment in nylon 6 led to increased mechanical properties of electrospun MWNT/nylon-6 nanofibre mat [137] [151].

The upper limit of CNT concentration in electrospun nanofibres is also confined by the extent of CNT dispersion. Hou *et al.* [113] reported thick sheets of electrospun PAN nanofibres containing well-aligned MWNTs with concentrations from 0 to 35 wt%. It was shown that the presence of MWNTs improved the modulus and tensile strength of the composite nanofibre sheet. The tensile modulus increased with increasing the concentration of MWNTs in nanofibres. However, the tensile strength of nanofibres increased with an increase in the concentration of MWNTs up to 5wt% and then started to reduce for higher MWNTs content. This was attributed to poor dispersion of the MWNTs and poor interfacial cohesion between the MWNTs and the polymer matrix at higher concentrations. Meanwhile, strain to break reduced with increasing the MWNT concentration. Similar findings have also been reported by other research groups [152] [153].

The importance of fibre alignment on the mechanical properties has been well established. In a study by Jeong et al. [154], aligned electrospun MWNT/PVA membranes have been reported. The tensile strength of these membranes increased from 5.8 MPa to 12.9 MPa by adding 1wt% of MWNTs. In a recent study, however, Blond et al. [155] achieved a higher level of reinforcement. They produced aligned SWNT/PVA nanofibre membrane with the strength of up to 40 MPa using a rotating drum collector followed by mechanical stretching.

It has been demonstrated that CNTs nucleate crystallisation in CNT/polymer composite films [50][57][66][67]. The presences of crystalline polymer coating around the nanotubes significantly enhance the stress transfer and therefore the mechanical properties of composites [42]. It is normally believed that crystallisation of polymers is a slow process involving orientation of polymer molecules and solidification. Therefore, nucleate crystallisation of polymer should occur mainly in composite films that normally take a long time for evaporation of solvent during the film casting process, and a fast drying and solidification process, such as in electrospinning, could hinder the nucleation crystallisation because the polymer molecules have not sufficient time to orient around nanotubes. In a recent study, Naebe *et al.* [114] revealed that the nucleation crystallisation indeed happened in CNT reinforced electrospun PVA nanofibres. They demonstrated that the increased PVA crystallinity due to the presence of CNTs resulted in considerable improvement in the strength of composite nanofibres. Later, other researchers [123] also demonstrated the occurrence of nucleation crystallisation in other CNT-polymer systems with improved in tensile properties.

Post-electrospinning treatment, using methanol for instance, was found to be an effective way to increase the mechanical properties of electrospun PVA nanofibres [156]. Naebe *et al.* [114] performed a series of post-spinning treatments on MWNT/PVA composite nanofibres including soaking in methanol and crosslinking with glutaric dialdehyde. These treatments induced the crystallinity of nanofibres as well as established a crosslinked PVA network. They showed that the tensile strength of MWNT/PVA composite nanofibres was significantly improved by applying post–electrospinning treatments. This was attributed to the increased polymer crystallinity due to the combined effect of post-spinning and nucleation crystallization of polymer matrix induced by the nanotubes. Similar results were found for SWNT reinforced PVA electrospun nanofibres [115]. In a similar study, Gandhi *et al.* [123] showed that post-spinning treatment with methanol and stretching significantly increased the strength and toughness of electrospun silk nanofibres containing only 1% CNTs. Methanol increased the polymer crystalline structure whereas stretching assisted in aligning them in the nanofibres.

d. Influence of polymer types

Different types of polymers, including semi-crystalline, amorphous and elastomeric polymers, have been used to fabricate CNT-containing composite nanofibres [119] [125] [127] [128] [145]. It was revealed that flow-induced crystallisation might have occurred during electrospinning of semi-crystalline polymers, and the polymer crystals were oriented along the fibre axis [128] [134]. On the other hand, it was shown that nanotubes aligned well during electrospinning of CNT/polymer nanofibres. Since the presence of oriented polymer crystals has a significant influence on mechanical properties, it is complicated to evaluate the real contribution of CNTs regarding the improvement in the mechanical performance of electrospun composite nanofibres.

With the amorphous polymers, only a few studies on CNT/polymer nanofibres have been reported [125] [127] [145]. Although enhanced mechanical properties were reported for the nanofibres, the role played by polymer morphologies (i.e. crystalline, amorphous, and rigid) was not fully understood.

e. Influence of carbon nanotube types

SWNTs and MWNTs differ from one another in their size and dispersability in solution and polymer matrix as well as in mechanical and electrical properties [3]. However, few papers have reported on the influence of CNT types on the structure-property relationship of electrospun nanofibres.

Dror et al. [128] and Salalha et al. [134] studied the effect of SWNTs and MWNTs on the formation of electrospun PEO nanofibres. On the basis of X-ray diffraction, it was demonstrated that while the PEO crystal orientation in electrospun nanofibres was not affected by the inclusion of SWNTs, the incorporation of MWNTs into PEO matrix had a detrimental effect on the degree of the crystal orientation. Nevertheless, no data on mechanical properties of CNT/PEO nanofibres was reported. Electrospun MWNT/PVA and SWNT/PVA nanofibres have been reported [114] [115]. It was observed that the SWNTs and MWNTs induced different crystal phases in the PVA. With the same CNT concentration, the tensile strength of MWNT/PVA nanofibres showed no significant difference to that of SWNT/PVA ones.

f. Electric and thermal properties

The formation of electrospun CNT/polymer nanofibres has been explored for possible improvement in the electrical and thermal properties of polymer. As for electrical conductivity, most polymers possess a very low conductivity and the presence of CNTs provides a platform for inherently conducting polymer nanofibres suitable for many applications. Incorporation of CNTs into polymer nanofibres was found to increase the electrical conductivity of composite nanofibres [109]. The electrical properties of electrospun MWNT/PAN composite fibres were investigated by two independent groups [109] [141]. Ge *et al.* [109] developed highly orientated PAN nanofibre mats containing MWNTs. At a concentration of 10 wt% MWNTs, the composite nanofibres started to form the percolating network. Due to highly anisotropic orientation of the composite nanofibre structure, the electrical conductivity enhanced to ~1.0 S/cm at a concentration of 20 wt% MWNTs. Ra *et al.* [141] achieved a rather high conductivity with carbonised MWNT/PAN nanofibres. While carbonised PAN nanofibres without CNTs did not reveal anisotropy in electrical conductivity, a high anisotropy in electrical conductivity was observed for the carbonised MWNT/PAN nanofibres. The conductivity parallel to the spinning direction was about three times higher than that perpendicular to the spinning direction at only 2.5 wt% of MWNT. The authors claimed that the direction dependency of conductivity is an indication of CNT alignment along the nanofibre axis, which was further supported by the TEM observation.

Electrospun MWNT/nylon composite nanofibres were also prepared and the electrical properties were examined as a function of the filler concentration [126]. The MWNT/nylon nanofibres were electrospun on the ITO coated glass and a metal coated glass electrode was placed on the composite fibre sheet. The filler concentration was varied from 0 to 20 wt% and the I~V characteristics were examined. As shown in Figure 5, the I~V curve indicates a non-ohmic behaviour, which changed with the filler concentration. Similar electrical behaviour was also reported for SWNT/PVDF [157] and MWNT/PEO [158] composite nanofibres.

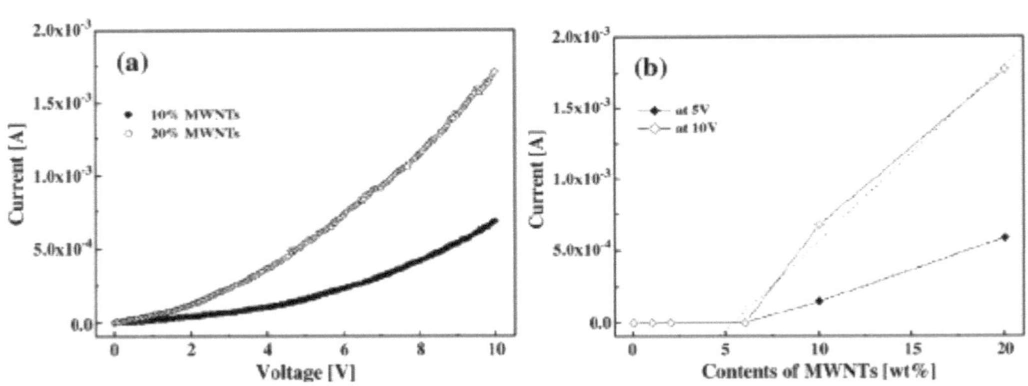

Fig. 5. (a) I~V characteristics for the nylon electrospun nanofibres loaded with 10 and 20 wt% CNTs. (b) Plot of the current as a function of the CNTs wt.% at 5 and 10 V [126] [Copyright Elsevier Science].

In an attempt to define the parameters that determine the conductivity of the nanofibre mats, McCullen *et al.* [152] performed a study on electrospun MWNT/PEO nanofibre. Electrical conductivity measurements of the randomly deposited nanofibre mats showed that by increasing the concentration of MWNTs the electrical conductivity increased remarkably. Above a percolation threshold of about 0.35 % of MWNTs in PEO, the conductivity increased by a factor of 10^{12} and then became approximately constant as the concentration of MWNTs was further increased. Maximum conductivity was obtained at about 1 wt % loading of MWNTs. The addition of only 1 % CNTs to silk nanofibres was found to increase the conductivity of nanofibres mat significantly [123].

In a rather different approach to studying the electrical conductivity of polymer nanofibres, Kang *et al.* [159] prepared MWNT/silk protein nanofibre mat. The electrical conductivity of the electrospun mat was found to be significantly higher than the plain silk protein nanofibres (from ~10^{-15} to ~10^{-4} S/cm) regardless of the dip-coating time. It was hypothesised that CNTs not only deposited on the surface of electrospun mat but also adsorbed by nanofibres due to strong interaction between the oxidised MWNTs and the peptide groups of silk protein.

Sundaray and co-workers [117] described the electrical conductivity of single electrospun MWNT/PMMA composite nanofibres. Alignment of MWNTs in the direction of the fibre axis was confirmed by bright field TEM images. The room temperature DC electrical conductivity of an electrospun MWNT/PMMA fibre showed a ten-orders increase compared to pure PMMA fibre. Percolation threshold of the composite nanofibre was well below the 0.05% w/w of CNTs loading and the conductivity increased with increase in MWNT concentration.

Not many papers reported on the thermal properties of electrospun CNT/polymer composite nanofibres. Thermal analysis has been carried out on the electrospun composite nanofibres to understand the relationship between the presence of carbon nanotubes and thermal properties. It was indicated that the presence of CNTs enhanced the thermal stability of polymer nanofibres.

The effect of heat treatment on SWNT/PAN composite fibres was investigated using TEM by Ko *et al.* [45]. SWNT/PAN was found to keep its shape but its microstructure changed significantly after the heat treatment. PAN lost hydrogen and oxygen during heat treatment and the shrinkage led to SWNTs sticking out of the fibres. Thermal properties of MWNT/PAN was investigated by Ge *et al.* [109] using thermal gravimetric analysis (TGA) and thermal mechanical analysis (TMA). They found that the thermal stability of MWNT/PAN nanofibres increased when compared to pure PAN nanofibres. It was attributed to the structural changes occurred in the nanofibres due to the presence of the carbon nanotubes, although the driving force behind the structural change has yet to be determined. An increased Tg was also found for MWNT/PAN composite nanofibres due to the formation of charge-transfer complexes which restricted molecular segment motions at the interface between the nanotube and PAN. The thermal expansion coefficient (CTE) of the MWNT/PAN composite nanofibres also increased [109]. A similar trend in thermal stability was also reported for MWNT reinforced polybutylene terephthalate (PBT) [124], PVA [114] and nylon-6 [151] composite nanofibres.

g. Applications

Electrospun nanofibres have a broad range of applications due to the combination of simplicity of fabrication process and their unique features. While several reviews on polymer nanofibre applications have been published [99][100][101][160], the works on CNT/polymer nanofibres have been mainly focused on developing a fundamental understanding of the fibre structure property relationships. Conducting electrospun CNT/polymer nanofibres have been demonstrated to be attractive for a large variety of potential applications, such as in optoelectronic and sensor devices [161]. For example, electrochemical biosensors were fabricated using electrospun MWNT/polymer composite nanofibres [162] [163]. In a recent study, the electrospun MWNT/poly(acrylonitrile-co-acrylic acid) nanofibres were found to enhance the maximum current of glucose oxide electrode and the enzyme electrode could be used several times without significant decrease in current [162]. Electrospun PVA nanofibres containing chitosan grafted MWNTs also exhibited sensory ability to hydrogen peroxide and potassium ferricyanide [163]. This nanofibre-based sensor demonstrated more sensitive response and intense current as well as faster electric charge transport than those of film-based sensors. Other potential applications of electrospun CNTs/polymer nanofibres include tissue engineering scaffolds, composite reinforcement, drug carriers for controlled release and energy storage. Given the advantages of CNT/polymer nanofibres in mentioned fields above, the number of investigations on these topics is very small.

5. Concluding remarks

The use of the electrospinning technique to incorporate carbon nanotubes (CNTs) into polymer nanofibres has been shown to induce alignment of the nanotubes within the polymer matrix, leading to significant improvements in fibre strength, modulus and electrical conductivity. To realise their commercial applications, considerable work is still required. This includes a thorough understanding of the structure–property relationship for various electrospun polymer nanofibres, the effective incorporation of carbon nanotubes into polymer fibres with a high loading content, and large scale production of composite nanofibres of consistent and high quality but at a low cost [164] [165] [166]. Core-shell CNT/polymer nanofibres are also a subject that warrant further research [167].

6. References

[1] A. Oberlin, M. Endo, T. Koyama, *Journal of Crystal Growth* 32, 335 (1976).
[2] S. Iijima, *Nature* 354, 56 (1991).
[3] R. H. Baughman, A. A. Zakhidov, W. A. Heer, *Science* 297, 787 (2002).
[4] S. M. Bachilo *et al., Science* 298, 2361 (2002).
[5] P. N. T. Guo, A.G. Rinzler, D. Toma´ nek, D.T. Colbert, R.E. Smalley, *J.Phys.Chem.* 99, 10694 (1995).
[6] Z. F. Ren *et al., Science* 282, 1105 (1998).
[7] G. Overney, W. Zhong, D. Tomanek, *Z. Phys.D: At., Mol. Clusters* 27, 93 (1993).
[8] J. P. Lu, *J. Phys. Chem. Solids* 58, 1649 (1997).
[9] E. W. Wong, P. E. Sheehan, C. M. Lieber, *Science* 277, 1971 (1997).
[10] M. Yu, O. Lourie, M. Dyer, T. Kelly, R. Ruoff, *Science* 287, 637 (2000).
[11] J. P. Salvetat *et al., Phys Rev Lett* 82, 944 (1999).
[12] M. F. Yu, B. S. Files, S. Arepalli, R. S. Ruoff, *Phys Rev Lett* 84, 5552 (2000).
[13] S. Berber, Y. K. Know, D.Tomanek, *Physics Review Letters* 84, 4613 (2000).
[14] R. Andrews *et al., Applied Physics Letters* 75, 1329 (1999).
[15] P. J. F. Harris, *International Materials Reviews* 49, 31 (2004).
[16] T. W. Ebbesen *et al., Nature* 382, 54 (1996).
[17] W. Mintmire, B. I. Dunlap, C.T.White, *Phys Rev Lett* 68, 631 (1992).
[18] N. Hamada, S.-I.Sawada, A.Oshiyama, *Phys Rev Lett* 68, 1579 (1992).
[19] R. Saito, M. Fujita, G. Dresselhaus, M.S.Dresselhaus, *Phys Rev,B* 46, 1804 (1992).

[20] P. G. Collins, K. Bradley, M. Ishigami, A. Zettl, *Science* 287, 1801 (2000).
[21] J. Kong *et al.*, *Science* 287, 622 (2000).
[22] J. N. Coleman, U. Khan, W. J. Blau, Y. K. Gunko, *Carbon* 44, 1624 (2006).
[23] J. N. Coleman, U. Khan, Y. K. Gunko, *Advanced Materials* 18, 689 (2006).
[24] R.Tucknott, S. N. Yaliraki, *Chemical Physics* 281, 455 (2002).
[25] Y.Dror *et al.*, *Langmuir* 19, 7012 (2003).
[26] D. Qian, E. C. Dickeya, R. Andrews, T. Rantell, *Applied physics Letters* 76, 2868 (2000).
[27] M. J. Biercuk *et al.*, *Applied Physics letters* 80, 2767 (2002).
[28] L. Liu, A. H. Barber, S. Nuriel, H. D. Wagner, *Advanced Functional Materials* 15, 975 (2005).
[29] S. Kumar *et al.*, *Macromolecules* 35, 9039 (2002).
[30] S. L. Ruan, P. Gao, X. G. Yang, T. X. Yu, *Polymer* 44, 5643 (2003).
[31] M. C. Paiva *et al.*, *Carbon* 42, 2849 (2004).
[32] X.-L. Xie, Y.-W. Mai, X.-P. Zhou, *Materials Science and Engineering* 49, 89 (2005).
[33] N. Grossiord, J. Loos, O. Regev, C. E. Koning, *Chem. Mater.* 18, 1089 (2006).
[34] B. Vigolo *et al.*, *Science* 290, 1331 (2000).
[35] M. S. P. Shaffer, A. H. Windle, *Advanced Materials* 11, 937 (1999).
[36] S. Kumar, H. Doshi, M. Srinivasarao, J. O. Park, D. A. Schiraldi, *Polymer* 43, (2002).
[37] L. Jin, C. Bower, O. Zhou, *Applied Physics Letters* 73, 1197 (1998).
[38] C. Bower, R. Rosen, L. Jin, J. Han, O. Zho, *Applied Physics Letters* 74, 3317 (1999).
[39] B. W. Smith *et al.*, *Applied Physics Letters* 77, 663 (2000).
[40] R. R. Schlittler *et al.*, *Science* 292, 1136 (2001).
[41] B. E. Kilbride *et al.*, *Journal of Applied Physics* 92, 4024 (2002).
[42]M. Cadek, J. N. Coleman, V. Barron, K. Hedicke, W. J. Blau, *Applied Physics Letters* 81, 5123 (2002).
[43] O. Meincke *et al.*, *Polymer* 45, 739 (2004).
[44] J. K. W. Sandler *et al.*, *Polymer* 45, 2001 (2004).
[45] F. Ko *et al.*, *Advanced Materials* 15, 1161 (2003).
[46] J. Gao *et al.*, *Journal of American Chemical Society* 126, 16698 (2004).
[47] C. Zhao *et al.*, *Polymer* 46, 5125 (2005).
[48] D. Li, Y. Xia, *Advanced Materials* 16, 1151 (2004).
[49] Y. Dzenis, *Science* 304, 1917 (2004).
[50] B. P. Grady, F. Pompeo, R. L. Shambaugh, D. E. Resasco, *Journal of Physical Chemistry* 106, 5852 (2002).
[51] B. McCarthy *et al.*, *Journal of Physical Chemistry B* 106, 2210 (2002).
[52] H. J. Barraza, F. Pompeo, E. A. O'Rear, D. E. Resasco, *Nano letters* 2, 797 (2002).
[53] F. Balavoine *et al.*, *Angewandte Chemie-International Edition* 38, 1912 (1999).
[54] W. Ding *et al.*, *Nano letters* 3, 1593 (2003).
[55] K. Keren, R. S. Berman, E. Buchstab, U. Sivan, E. Braun, *Science* 302, 1380 (2003).
[56] C. Richard, F. Balavoine, P. Schultz, T. W. Ebbesen, C. Mioskowski, *Science* 300, 775 (2003).
[57] K. P. Ryan *et al.*, *Composites Science and Technology* 67, 1640 (2007).
[58] A. R. Bhattacharyya *et al.*, *Polymer* 44, 2373 (2003).
[59] L. Valentini, J. Biagiotti, J. M. Kenny, S. Santucci, *Composites Science and Technology* 63, 1149 (2003).

[60] L. Valentini, J. Biagiotti, J. M. Kenny, S. Santucci, *Journal of Applied Polymer Science* 87, 708 (2003).
[61] L. Valentini, J. Biagiotti, J. M. Kenny, M. A. L. Manchado, *Journal of Applied Polymer Science* 89, 2657 (2003).
[62] E. Assouline *et al.*, *Journal of Polymer Science: Part B: Polymer Physics* 41, 520 (2003).
[63] J. Sandler *et al.*, *Journal of Macromolecular Science, Physics* B42, 479 (2003).
[64] J. N. Coleman *et al.*, *Applied Physics Letters* 82, 1682 (2003).
[65] O. Probst, E. M. Moore, D. E. Resasco, B. P. Grady, *Polymer* 45, 4437 (2004).
[66] J. N. Coleman *et al.*, *Polymer* 47, 8556 (2006).
[67] M. Cadek *et al.*, *Nano Letters* 4, 353 (2004).
[68] A. B. Dalton *et al.*, *Nature* 423, 703 (2003).
[69] J. N. Coleman *et al.*, *Advanced Functional Materials* 14, (2004).
[70] A.Formalas. (1934).
[71] GI.Taylor, *Proc R Soc London,Ser A* 313, 453 (1969).
[72] M. Bognitzki *et al.*, *Advanced Materials* 12, 637 (2000).
[73] M. Bognitzki *et al.*, *Advanced Materials* 13, 70 (2001).
[74] C. J. Buchko, L. C. Chen, Y. Shen, D. C. Martin, *Polymer* 40, 7397 (1990).
[75] Z. H. Chen *et al.*, *Macromolecules* 34, 6156 (2001).
[76] A. Theron, E.Zussman, AL.Yarin, *Nanotechnology* 12, 384 (2001).
[77] S. Megelski, JS.Stephens, JF.Rabolt, CD.Bruce, *Macromolecules* 35, 8456 (2002).
[78] J. M. Deitzel *et al.*, *Polymer* 43, 1025 (2002).
[79] B. Ding *et al.*, *Journal of Polymer Science: Part B: Polymer Physics* 40, 1261 (2002).
[80] J.-S. Kim, D.H.Reneker, *Polymer Composites* 20, 124 (1999).
[81] S. Koombhongse, W. L. WX, DH.Reneker, *Journal of Polymer Science: Part B: Polymer Physics* 39, 2598 (2001).
[82] A. MacDiarmid *et al.*, *Synthetic Metals* 119, 27 (2001).
[83] J. A. Matthews, G. E. Wnek, D. G. Simpson, G. L.Bowlin, *Biomacromolecules* 3, 232 (2002).
[84] D. Reneker, I.Chun, *Nanotechnology* 7, 216 (1996).
[85] X. Zong *et al.*, *Polymer* 43, 4403 (2002).
[86] H. Fong, W.-D. Liu, C.-S. Wang, R. Vaia, *Polymer* 43, 775 (2002).
[87] H. Fong, I. Chun, D. H. Reneker, *Polymer* 40, 4585 (1999).
[88] D. H. Reneker, A. L. Yarin, H. Fong, S. Koombhongse, *Journal of Applied physics* 87, 4531 (2000).
[89] M. M. Hohman, M. Shin, G. Rutledge, M. P. Brenne, *Physics of Fluids* 13, 2201 (2001).
[90] D. A. Saville, *Annual Review of Fluid Mechanics* 29, 27 (1997).
[91] J. J. Feng, *Physics of Fluids* 14, 3912 (2002).
[92] M. M. Hohman, M. Shin, G. Rutledge, M. P.Brenner, *Physics of Fluids* 13, 2221 (2001).
[93] A. M. Gan˜a´n-Calvo, *Journal of Fluid Mechanics* 335, 165 (1997).
[94] R. P. A. Hartman, D. J. Brunner, D. M. A. Camelot, J. C. M. Marijnissen, B. Scarlett, *Journal of Aerosol science* 30, 823 (1999).
[95] A. L. Yarin, S. Koombhongse, D. H. Reneker, *Journal of Applied Physics* 89, 3018 (2001).
[96] J.Doshi, D.H.Reneker, *Journal of Electrostatics* 35, 151 (1995).
[97] I.S.Chronakis, *Journal of Materials Processing Technology* 167, 283 (2005).
[98] Z. Huanga, Y.Z. Zhangb, M. Kotakic, S. Ramakrishnab, *Composites Science and Technology* 63, 2223 (2003).
[99] A. Greiner, J. H. Wendorff, *Angewandte Chemie International* 46, 5670 (2007).

[100] J. Fang, H. Niu, T. Lin, X. Wang, *Chinese Science Bulletin* 53, 2265 (2008).
[101] V. Thavasi, G. Singh, S. Ramakrishna, *Energy and Environmental Science* 1, 205 (2008).
[102] S.-Y. Gu, Q.-L. Wu, J. Ren, G. J. Vancso, *Macromol. Rapid Commun.* 26, 716 (2005).
[103] S.-H. Lee, C. Tekmenb, W. M. Sigmunda, *Materials Science and Engineering A* 398, 77 (2005).
[104] L. M. Bellan, J. Kameoka, H. G. Craighead, *Nanotechnology* 16, 1095 (2005).
[105] O. Breuer, U.Sundararaj, *Polymer composites* 25, (2004).
[106] P. Kannan, R. J. Young, S. J. Eichhorn, *Nanotechnology* 18, 235707(7pp) (2007).
[107] P. Kannan, R. J. Young, S. J. Eichhorn, *Small* 4, 930 (2008).
[108] H.Ye, H.Lam, N.Titchenal, Y.Gogotsi, F.Ko, *Applied Physics Letters* 85, 1775 (2004).
[109] J. J. Ge *et al.*, *Journal of American Chemical Society* 126, 15754 (2004).
[110] S.Kedem, J.Schmidt, Y.Paz, Y.Cohen, *Langmuir* 21, 5600 (2005).
[111] H.Lam, H.Ye, Y.Gogotsi, F.Ko, *Polymer Preprints* 45, 124 (2004).
[112] N.Titchenal *et al.*, *Polymer Preprints* 44, 115 (2003).
[113] H. Hou *et al.*, *Chemistry of Materials* 17, 967 (2005).
[114] M. Naebe, T. Lin, W. Tian, L. Dai, X. Wang, *Nanotechnology* 18, 225605 (2007).
[115] M. Naebe, T. Lin, M. P. Staiger, L. Dai, X. Wang, *Nanotechnology* 19, 305702 (2008).
[116] Y. Dror *et al.*, *Progress in Colloid and Polymer Science* 130, 64 (2005).
[117] B. Sundaray, V. Subramanian, T. S. Natarajan, *Appled Physics Letters* 88, 143114 (2006).
[118] H.S.Kim, J.H.sung, H.J.Choi, I.Chin, H.Jin, *Polmer Preprints* 46, 736 (2005).
[119] R. Sen *et al.*, *Nano Letters* 4, 459 (2004).
[120] K.Saeed, S.Y.Park, H.J.Lee, J.B.Baek, W.S.Huh, *Polymer* 47, 8019 (2006).
[121] F.Ko *et al.*, *Advanced Materials* 15, 1161 (2003).
[122] J.Ayutsede *et al.*, *Biomacromolecules* 7, 208 (2006).
[123] M. Gandhi, H. Yang, L. Shor, F. Ko, *Polymer* 50, 1918 (2009).
[124] G. Mathew, J. P. Hong, J. M. Rhee, H. S. Lee, C. Nah, *Polymer Testing* 24, 712 (2005).
[125] G. M. Kim, G. H. Michlera, P. Po¨tschke, *Polymer* 46, 7346 (2005).
[126] J. S. Jeong *et al.*, *Diamond & Related Materials* 15, 1839 (2006).
[127] C. Pan, L. Q. Ge, Z. Z. Gu, *Composites Science and Technology* 67, 3721 (2007).
[128] Y. Dror *et al.*, *Langmuir* 19, 7012 (2003).
[129] S.C.Tsang, Y.K.Chen, P.J.F.Harris, M.L.H.Green, *Nature* 372, 159 (1994).
[130] R.M.Lago, S.C.Tsang, K.L.Lu, Y.K.Chen, M.L.H.Green, *Chemical Communications*, 1355 (1995).
[131] V. N. Khabashesku, M. X. Pulikkathara, *Mendeleev Communications* 16, 61 (2006).
[132] T. Lin, V. Bajpai, T. Ji, L. Dai, *Australian Journal of Chemistry* 56, 635 (2003).
[133] K. Mylvaganam, L. C. Zhang, *Recent Patents on Nanotechnology* 1, 59 (2007).
[134] W. Salalha *et al.*, *Langmuir* 20, 9852 (2004).
[135] J.C.Kearns, R.L.Shambaugh, *Journal of Applied Polymer Science* 86, 2079 (2002).
[136] Y.-Q. Wan, J.-H. He, J.-Y. Yu, *Polymer International* 56, 1367 (2007).
[137] M. V. Jose *et al.*, *Polymer* 48, 1096 (2007).
[138] Q. Zhang, Z. Chang, M. Zhu, X. Mo, D. Chen, *Nanotechnology* 18, 115611 (2007).
[139] W. A. Yee *et al.*, *Polymer* 49, 4196 (2008).
[140] M. B. Bazbouz, G. K. Stylios, *European Polymer Journal* 44, 1 (2008).
[141] E. J. Ra, K. H. An, K. K. Kim, S. Y. Jeong, Y. H. Lee, *Chemical Physics Letters* 413, 188 (2005).
[142] S. Huang *et al.*, *Langmuir* 24, 13621 (2008).

[143] D. Li, Y. Wang, Y. Xia, *Nano Letters* 3, 1167 (2003).
[144] P. Katta, M. Alessandro, R. D. Ramsier, G. G. Chase, *Nano Letters* 4, 2215 (2004).
[145] L.-Q. Liu, D. Tasis, M. Prato, H. D. Wagner, *Advanced Materials* 19, 1228 (2007).
[146] D. Almecija, D. Blond, J. E. Sader, J. N. Coleman, J. J. Boland, *Carbon* 47, 2253 (2009).
[147] H. Ye, H. Lam, N. Titchenal, Y. Gogotsi, F. Ko, *Applied Physics Letters* 85, 1775 (2004).
[148] U. Singh *et al., Applied Physics Letters* 89, 73103 (2006).
[149] W. Zhou, Y. Wu, F. Wei, G. Luo, W. Qian, *Polymer* 46, 12689 (2005).
[150] O. J. Yoon *et al., Plasma Processes and Polymers* 6, 101 (2009).
[151] K. Saeed, S.-Y. Park, S. Haider, J.-B. Baek, *Nanoscale Research Letters* 4, 39 (2009).
[152] S. D. McCullen *et al., Macromolecules* 40, 997 (2007).
[153] S. D. McCullen *et al., Journal of Applied Polymer Science* 105, 1668 (2007).
[154] J. S. Jeong *et al., Thin Solid Films* 515, 5136 (2007).
[155] D. Blond *et al., Advanced Functional Materials* 18, 2618 (2008).
[156] L. Yao *et al., Chemistry of Materials* 15, 1860 (2003, 2003).
[157] C. Seoul, Y. T. Kim, C. K. Baek, *Journal of Polymer Science:Part B:Polymer Physics* 41, 1572 (2003).
[158] J. Y. Lim, C.K.Lee, S.J.Kim, I.Y.Kim, S.I.Kim, *Journal of Macromolecular Science,Part A: Pure and Applied Chemistry* 43, 785 (2006).
[159] M. Kang, H.-J. Jin, *Colloid and Polymer Science* 285, 1163 (2007).
[160] S. Tan, X. Huang, B. Wu, *Polymer International* 56, 1330 (2007).
[161] L. Dai, Ed., *Carbon Nanotechnology: Recent Developments in Chemistry, Physics, Materials Science and Device Applications* (Elsevier, Amsterdam, 2006).
[162] Z.-G. Wang, Y. Wang, H. Xu, G. Li, Z.-K. Xu, *Journal of Physical Chemistry* 113, 2955 (2009).
[163] W. Feng, Z. Wu, Y. Li, Y. Feng, X. Yuan, *Nanotechnology* 19, 05707 (2008).
[164] H. Niu, T. Lin, X. Wang, *Journal of Applied Polymer Science* 114, 3524 (2009).
[165] X. Wang, H. Niu, T. Lin, X. Wang, *Polymer Engineering and Science* 49, 1582 (2009).
[166] O. Jirsak *et al., International Pat.* WO 2005/024101, (2005).
[167] J. Liu, T. Wang, T. Uchida, S. Kumar, *Journal of Applied Polymer Science* 96, 1992 (2005).

17

On the Electron Transport in Conducting Polymer Nanofibers

Natalya A. Zimbovskaya
Department of Physics and Electronics, University of Puerto Rico-Humacao, CUH Station, Humacao, PR 00791;
Institute for Functional Nanomaterials, University of Puerto Rico, San Juan, PR 00931
Puerto Rico

1. Introduction

During the past decade, transport properties of conducting polymers such as doped polyacetylene and polyaniline-polyethylene oxides, were and still are intensively studied [1, 2]. These materials are significant mostly due to various possible applications in fabrication of nanodevices. Polymer-based devices should have advantages of low cost and flexible, controlled chemistry. Also, there are some unresolved problems concerning the physical nature of charge transfer mechanisms in conducting polymers, which make them interesting subjects for fundamental research. Chemically doped polymers are known to be very inhomogeneous. In some regions polymer chains are disorderly arranged, forming an amorphous, poorly conducting substance. In other places the chains are ordered and densely packed [3, 4]. These regions could behave as metallic-like grains embedded in the disordered environment. The fraction of metallic-like islands in bulk polymers varies depending on the details of the synthesis process. In practical samples such islands always remain separated by disordered regions and do not have direct contacts. In some cases, electronic states are delocalized over the grains, and electrons behave as conduction electrons in conventional metals. In these cases electrons motion inside the grains is diffusive with the diffusion coefficient $D = \frac{1}{3}v_F^2\tau$ (v_F is the Fermi velocity, and τ is the scattering time). In whole, electron transport in conducting polymers shows both metallic and nonmetallic features, and various transport mechanisms contribute to the resulting pattern.
An important contribution to the conduction in these substances is provided by the phonon-assisted electron hopping between the conducting islands and/or variable range hopping between localized electronic states. The effect of these transport mechanisms strongly depends on the intensity of stochastic nuclear motions. The latter increases as temperature rises, and this brings a significant enhancement of the corresponding contributions to the conductivity. The temperature dependence of the "hopping" conductivity $\sigma(T)$ is given by the Mott's expression [5]:

$$\sigma(T) = \sigma(0) \exp\left[-(T_0/T)^p\right] \tag{1}$$

where T_0 is the characteristic temperature of a particular material, and the parameter p takes on values 0.25, 0.33 or 0.5 depending on the dimensions of the hopping processes. Also, it was suggested that phonon-assisted transport in low-dimensional structures such as nanofibers and nanotubes, may be substantially influenced due to electron interactions [6, 7]. This results in the power-low temperature dependencies of the conductance $G(T)$ at low values of the bias voltage V ($eV < kT$, k being the Boltzmann constant), namely: $G \sim T^{\alpha}$. Experimental data for the conductance of some nanofibers and nanotubes match this power-low reasonably well, bearing in mind that the value of the exponent α varies within a broad range. For instance, α was reported to accept values about 0.35 for carbon nanotubes [8], and $\alpha \sim 2.2 \div 7.2$ for various polyacetylene nanofibers [9–11]. In general, hopping transport is very important in disordered materials with localized states. For this kind of transport phonons play part of a source of electrical conductivity. Accordingly, the hopping contribution to the conductivity always increases as temperature rises, and more available phonons appear. When polymers are in the insulating state, the hopping transport predominates and determines the temperature dependencies of transport characteristics.

In conducting state of conducting polymers free charge carriers appear, and their motion strongly contributes to the conductance. While moving, the charge carriers undergo scattering by phonons and impurities. This results in the conductivity stepping down. Metallic-like features in the temperature dependencies of dc conductivity of some polymeric materials and carbon nanotubes were repeatedly reported. For instance, the decrease in the conductivity upon heating was observed in polyaniline nanofibers in Refs. [12] and [13]. However, this electron diffusion is not a unique transport mechanism responsible for the occurrence of metallic-like behavior in the dc conductivity of conducting polymers. Prigodin and Epstein suggested that the electron tunneling between the grains through intermediate resonance states on the polymer chains connecting them, strongly contributes to the electron transport [14]. This approach was employed to build up a theory of electron transport in polyaniline based nanofibers [15] providing good agreement with the previous transport experiments [16]. Considering the electron tunneling through the intermediate state as a mechanism for the intergrain transport, we see a similarity between the latter and electron transport mechanisms typical for tunnel molecular junctions. In the case of polymers, metallic-like domains take on part of the leads, and the molecular bridge in between is simulated by intermediate sites. The effect of phonons on this kind of electron transport may be very significant. These phonons bring an inelastic component to the intergrain current and underlie the interplay between the elastic transport by the electron tunneling and the thermally assisted dissipative transport. Also, they may cause some other effects, as was shown while developing the theory of conduction through molecules [17–24].

2. Electron-resonance tunneling as a transport mechanism in conducting polymers

Here, we concentrate on the analysis of the electric current-voltage characteristics and conductance associated with the resonance tunneling transport mechanism. Considering the electron intergrain resonance tunneling, the transmission coefficient is determined with the probability of finding the resonance state in between the grains. The latter is estimated as $P \sim \exp(-L/\xi)$ (L is the average distance between the adjacent grains, and ξ is the localization length for electrons), and it takes values much greater than the transmission probability for sequental hoppings along the chains, $P_h \sim \exp(-2L/\xi)$ [14]. Nevertheless, the probability for existence of a resonance state at a certain chain is rather low, so only a few out of the whole set of chains connecting two grains are participating in the intergrain

electron transport. Therefore, one could assume that any two metallic domains are connected by a single chain providing an intermediate state for the resonance tunneling. All remaining chains can be neglected for they poorly contribute to the transport compared to the resonance chain. Within this approximation the "bridge" linking two islands is reduced to a single electron state. Realistic polymer nanofibers have diameters within the range 20÷100 nm, and lengths of the order of a few microns. This is much greater than a typical size of both metallic-like grains and intergrain separations, which take on values ~ 5÷10nm (see e.g. Refs. [16] and [25]). Therefore, we may treat a nanofiber as a set of working channels connected in parallel, any single channel being a sequence of grains connected with the resonance polymer chains. The net current in the fiber is the sum of currents flowing in these channels, and the voltage V applied across the whole fiber is distributed among sequential pairs of grains along a single channel. So, the voltage ΔV applied across two adjacent grains could be roughly estimated as $\Delta V \sim V L/L_0$ where L is the average separation between the grains, and L_0 is the fiber length. In practical fibers the ratio $\Delta V/V$ may take on values of the order of $10^{-2} \div 10^{-3}$.
In further current calculations we treat the grains as free electron reservoirs in thermal equilibrium. This assumption is justified when the intermediate state (the bridge) is weakly coupled to the leads, and conduction is much smaller than the quantum conductance $G_0 = 2e^2 = h$ (e, h are the electron charge, and the Planck constant, respectively). Due to the low probabilities for the resonance tunneling between the metallic islands in conducting polymers, the above assumption may be considered as a reasonable one. So, we can employ the well-known expression for the electron current through the molecular junction [26], and we write:

$$I = \frac{2en}{h}\int_{-\infty}^{\infty} dET(E)[f_1(E) - f_2(E)]. \tag{2}$$

Here, n is the number of the working channels in the fiber, $f_{1,2}(E)$ are Fermi functions taken with the different contact chemical potentials $\mu_{1,2}$ for the grains. The chemical potentials differ due to the bias voltage ΔV applied across the grains:

$$\mu_1 = E_F + (1-\eta)e\Delta V; \qquad \mu_2 = E_F - \eta e\Delta V. \tag{3}$$

In these expressions (2), (3) the parameter η characterizes how the voltage ΔV is divided between the grains, E_F is the equilibrium Fermi energy of the system including the pair of grains and the resonance chain in between, and $T(E)$ is the electron transmission function.
The general approach to the electron transport studies in the presence of dissipation is the reduced dynamics density-matrix formalism (see, e.g., Refs. [27] and [28]). This microscopic computational approach has the advantages of being capable of providing the detailed dynamics information. However, this information is usually more redundant than necessary, as far as standard transport experiments in conducting polymer nanofibers are concerned. There exists an alternative approach using the scattering matrix formalism and the phenomenological Buttiker dephasing model [29]. Adopting this phenomenological model we are able to analytically study the problem, and the results agree with those obtained by means of more sophisticated computational methods, as was demonstrated in the earlier works [30].

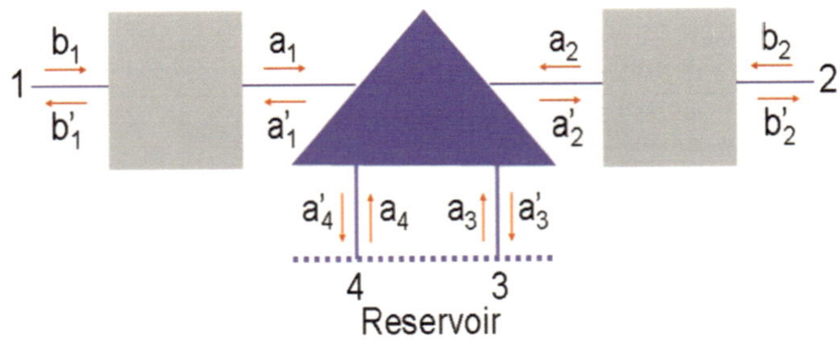

Fig. 1. Schematic drawing illustrating the intergrain electron transport in the presence of dissipation [29]. Rectangles correspond to the barriers separating the adjacent metallic-like islands from the intermediate state (the bridge), and the triange stands for a scatterer attached to the bridge.

Within the Buttiker model we treat the intergrain electron transport as a multichannel scattering problem. In the considered case the "bridge" between two adjacent grains inserts a single electron state. Therefore, an electron could be injected into the system (including two metallic-like domains and the intermediate "bridge" in between) and/or leave from there via four channels presented in the Fig. 1. The electron transport is a combination of tunneling through two barriers (the first one separates the left metallic island from the intermediate state and the second separates this state from the right island, supposing the transport from the left to the right). Inelastic effects are accounted for by means of a dissipative electron reservoir attached to the bridge site. The dissipation strength is characterized by a phenomenological parameter ϵ, which can take values within the range [0, 1]. When $\epsilon = 0$ the reservoir is detached from the bridge, which corresponds to the elastic and coherent electron transport. The greater is ϵ value the stronger is the dissipation. In the Fig. 1 the barriers are represented by the squares, and the triangle in between imitates a scatterer coupling the bridge to a dissipative electron reservoir.
Incoming particle fluxes (J_i) are related to those outgoing from the system (J_j') by means of the transmission matrix T [29, 30]:

$$J_j' = \sum_i T_{ji} J_i, \qquad 1 \leq i, j \leq 4. \tag{4}$$

Off-diagonal matrix elements $T_{ji}(E)$ are probabilities for an electron to be transmitted from the channel i to the channel j, whereas diagonal matrix elements $T_{ii}(E)$ are probabilities for its reflection back to the channel i. To provide charge conservation, the net particle flux in the channels connecting the system with the reservoir must be zero. So we have:

$$J_3 + J_4 - J_3' - J_4' = 0. \tag{5}$$

The transmission function $T(E)$ relates the particle flux outgoing from the channel 2 to the flux incoming to the channel 1, namely:

$$J_2' = T(E) J_1. \tag{6}$$

Using Eqs. (4) and (5) we can express the transmission function in terms of the matrix elements T_{ji}. The latter are related to matrix elements of the scattering matrix S, which expresses the outgoing wave amplitudes b'_1, b'_2, a'_3, a'_4 as linear combinations of the incident ones b_1, b_2, a_3, a_4 : $T_{ij} = |S_{ij}|^2$. In the considered case of a single site bridge the S matrix takes the form [15, 30]:

$$S = Z^{-1}\begin{pmatrix} r_1+\alpha^2 r_2 & \alpha t_1 t_2 & \beta t_1 & \alpha\beta t_1 r_2 \\ \alpha t_1 t_2 & r_2+\alpha^2 r_1 & \alpha\beta r_1 t_2 & \beta t_2 \\ \beta t_1 & \alpha\beta r_1 t_2 & \beta^2 r_1 & \alpha r_1 r_2 - \alpha \\ \alpha\beta t_1 r_2 & \beta t_2 & \alpha r_1 r_2 - \alpha & \beta^2 r_2 \end{pmatrix}. \tag{7}$$

where $Z = 1 - \alpha^2 r_1 r_2$, $\alpha = \sqrt{1-\epsilon}$, $\beta = \sqrt{\epsilon}$, $r_{1,2}$ and $t_{1,2}$ are the transmission an reflection coefficients for the barriers ($|t_{1,2}|^2 + |r_{1,2}|^2 = 1$).

When the bridge is detached from the dissipative reservoir $T(E) = |S_{12}|^2$. On the other hand, in this case we can employ a simple analytical expression for the electron transmission function [31]:

$$T(E) = 4\Delta_1(E)\Delta_2(E)|G(E)|^2, \tag{8}$$

where $\Delta_{1,2}(E) = -\text{Im}\Sigma_{1,2}(E)$. In this expression, self-energy terms $\Sigma_{1,2}$ appear due to the coupling of the metallic-like grains to the intermediate state (the bridge). The retarded Green's function for a single-site bridge could be approximated as follows:

$$G(E) = \frac{1}{E - E_1 + i\Gamma} \tag{9}$$

where E_1 is the site energy. The width of the resonance level between the grains is determined by the parameter $\Gamma = \Delta_1 + \Delta_2 + \Gamma_{en}$ (Γ_{en} describes the effect of energy dissipation). Further we consider dissipative effects originating from electron-phonon interactions, so, Γ_{en} is identified with Γ_{ph}.

Equating the expression (8) and $|S_{12}|^2$ we arrive at the following expressions for the tunneling parameters $\delta_{1,2}(E)$:

$$\delta_{1,2}(E) \equiv t_{1,2}^2 = \frac{2\Delta_{1,2}}{\sqrt{(E-E_1)^2+\Gamma^2}}. \tag{10}$$

Using this result we easily derive the general expression for the electron transmission function:

$$T(E) = \frac{g(E)(1+\alpha^2)[g(E)(1+\alpha^2)+1-\alpha^2]}{[g(E)(1-\alpha^2)+1+\alpha^2]^2} \tag{11}$$

where:

$$g(E) = 2\sqrt{\frac{\Delta_1\Delta_2}{(E-E_1)^2+\Gamma^2}}. \tag{12}$$

To simplify further analysis we approximate the self-energy terms $\Delta_{1,2}$ as constants: $\Delta_{1,2} \approx W_{1,2}^2/\sigma_{1,2}$. Here, $W_{1,2}$ are coupling strengths characterizing the coupling of the grains to the bridge and $\gamma_{1,2}$ characterize interatomic couplings inside the grains (leads). Simulating the leads by semiinfinite chains of identical sites, as was first suggested by D'Amato and Pastawski [32], one may treat the parameters $\gamma_{1,2}$ as coupling strengths between the nearest sites in these chains. Then the elastic electron transmission (ϵ = 0) shows a sharp peak at the energy E_1 (see Fig. 2a), which gives rise to a steplike form of the volt-ampere curve presented in Fig. 2b. When the reservoir is attached to the electronic bridge ($\epsilon \neq 0$), the peak in the transmission is eroded. The greater is the value of the parameter ϵ, the stronger is the erosion. When ϵ takes on the value of 0.7 the peak in the electron transmission function is completely washed out as well as the steplike shape of the $I-V$ curve. The latter becomes linear, corroborating the well-known Ohmic law for the sequential hopping mechanism. So, we see that the electron transmission is affected by stochastic nuclear motions in the environment of the resonance state. When the dissipation is strong (e.g. within the strong thermal coupling limit), the inelastic (hopping) contribution to the intergrain current predominates, replacing the coherent elastic tunneling. Typically, at room temperatures the intergrain electron transport in conducting polymers occurs within an intermediate regime, when both elastic and inelastic contributions to the electron transmission are manifested.

The described approach was succesfully employed to analyze experimental results on the electrical characterization of polyaniline-polyethylene oxides (PANi) nanofibers reported by Zhou et al [16], and Pinto et al [33]. In these experiments single-fiber electrical characterization was carried out at T ~$300K$. The experiments revealed that current-voltage curves for conducting nanofibers were non-Ohmic. Conducting samples included a 70-nm-diameter nanofiber (sample 1), a pair of 18-nm- and 25-nm-diameter fibers connected in parallel (sample 2) and a fiber whose diameter was steeply reduced in the middle from 70 nm to 20 nm (sample 3). The latter did exhibit a very asymmetric rectifying current-volltage characteristic. The expression (2) was employed to compute the current originating from electron resonance tunneling in the nanofibers [15]. To evaluate the number of working channels in a nanofiber it was assumed that for certain doping and crystallinity rates the number of channels is proportional to the fiber cross-sectional area. Also, it was taken into account that the contact of the fiber surfaces with atmospheric gases reduced the number of working channels. Accepting the value 5 nm for both average grain size and intergrain distance we estimated that the 70 nm fiber could include about 30 - 40 conducting channels, and we could not expect more than two working channels in the pair of fibers (sample 2). This gives the ratio of conductions ~ 15 - 20, which is greater than the ratio of cross-sectional areas of the samples $A_1/A_2 \approx 5$. The difference originates from the stronger dedoping of thinner fibers. Calculating the current, appropriate values for $W_{1,2}$ providing the best match between calculated and experimental $I-V$ curves were found applying the least-squares procedure. We estimated values of the coupling strengths $W_{1,2}$ in PANi fibers as 2.0 - $3.5meV$. As shown in the Fig. 2, computed volt-ampere characteristics match the curves obtained in the experiments reasonably well. The agreement between the presented theoretical results and experimental evidence proves that electron tunneling between metallic-like grains through intermediate states at the resonance chains can really play a significant part in the transport in conducting polymers provided that they are in the metallic state.

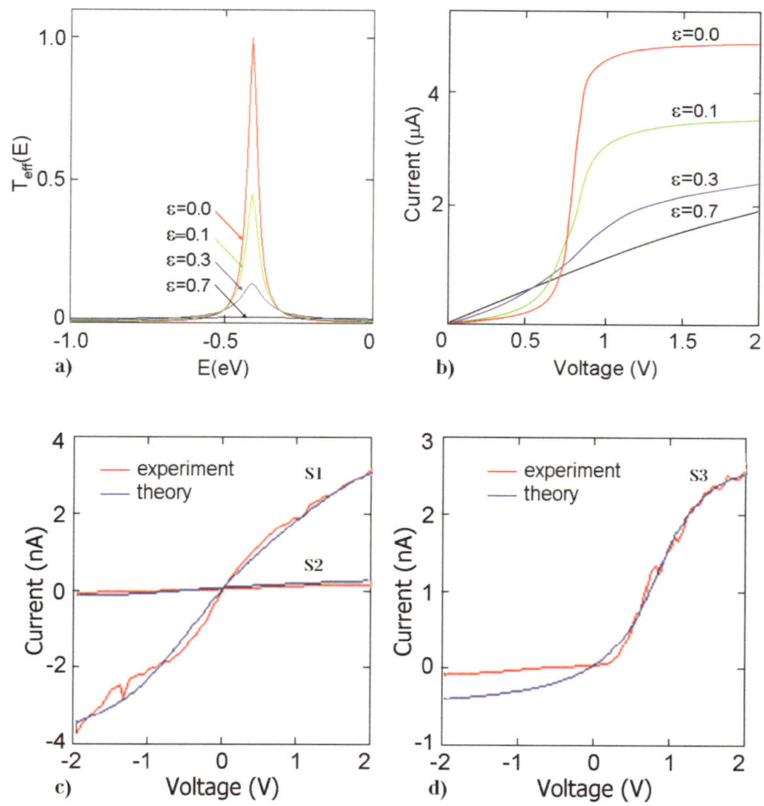

Fig. 2. Dissipation effect on the effective transmission function (a) and the current voltage characteristics (b). The curves are plotted at $T = 70K$, $\Delta_1 = \Delta_2 = 10meV$, $\epsilon = 0.0, 0.1, 0.3, 0.7$ from the top to the bottom. Calculated (dash-dot lines) and experimental (solid lines) current (nA) - voltage (V) characteristics for PANi nanofibers: (c) samples 1 (S1), and 2 (S2), and (d) sample 3 (S3).

However, we remark that the good fitting between the theory and the experiment demonstrated in the work [15] was achieved assuming that dissipation was very weak ($\alpha \approx 0.05$). This assumption could hardly be justified while one is observing electron transport in nanofibers at room temperature. So, the phenomenological approach employed in this section has some significant shortcomings. Its main disadvantage is that the dissipative effects are described in terms of a phenomenological parameter ϵ, whose dependence of characteristic factors affecting the transport (such as temperature, electron-phonon coupling strength and some others) remains unclear. It is necessary to modify the Buttiker's model to elucidate the relation of the parameter ϵ to the relevant energies characterizing electron transport in the considered systems. To obtain the desired expression for ϵ, we compare our results with those presented in Ref. [17]. In that work the inelastic correction δI to the coherent tunnel current via a single-site bridge is calculated using the nonequilibrium Green's functions formalism. The relevant result is derived in the limit of weak electron-phonon interaction when $\Gamma_{ph} \ll \Delta_{1,2}$. It is natural to assume that the dissipation strength ϵ is small within this limit, so we expand our expression for $T(E)$ in powers of ϵ. Keeping two first terms in this expansion, and assuming that $\Delta_1 = \Delta_2 \equiv \Delta$ we obtain:

$$T(E) \approx g^2(E)\Big(1 + \frac{\epsilon}{2}[1 - 2g^2(E)]\Big). \tag{13}$$

We employ this approximation to calculate the current through the bridge, and we arrive at the following expression for δI:

$$\delta I = \frac{e}{\hbar}\int_{-\infty}^{\infty} dE \frac{\epsilon\Delta}{\Gamma + \Gamma_{ph}} \rho_{el}(E) \frac{(E - E_1)^2 - 4\Delta^2}{(E - E_1)^2 + 4\Delta^2}\big[f_1(E) - f_2(E)\big]. \tag{14}$$

Here, $\rho_{el}(E)$ is the electron density of states at the bridge:

$$\rho_{el}(E) = -\frac{1}{\pi}\mathrm{Im}G(E). \tag{15}$$

Comparing the expression (14) with the corresponding result of Ref. [17], we find that these two are consistent, and we get [34]:

$$\epsilon = \frac{\Gamma_{ph}}{\Gamma + \Gamma_{ph}}. \tag{16}$$

When $\Delta \gg \Gamma_{ph}$ ($\epsilon \ll 1$) the bridge coupling to the dissipative reservoir is weak, and the elastic electron tunneling predominates.The opposite limit $\Delta \ll \Gamma_{ph}$ ($\epsilon \sim 1$) corresponds to the completely incoherent phonon-assisted electron transport.

3. Temperature dependencies of electron transport characteristics in conducting polymer nanofibers

Now, we turn to studies of temperature dependencies of the electron current and conductance associated with the intergrain electron tunneling. It is known that various conduction mechanisms may simultaneously contribute to the charge transport in conducting polymers, and their relative effects could significantly differ depending on the specifics of synthesis and processing of polymeric materials. The temperature dependencies of the resulting transport characteristics may help to identify the predominating transport mechanism for a particular sample under particular conditions. The issue is of a significant importance because the relevant transport experiments are often implemented at room temperature, so that the influence of phonons cannot be disregarded. Therefore, we study the effect of temperature (stochastic nuclear motions) on the resonance electron tunneling between metallic-like grains (islands) in polymer nanofibers. In these studies we consider the dissipative reservoir attached to the "bridge" site as a phonon bath, and we assume that the phonon bath is characterized by the continuous spectral density $J(\omega)$ of the form [35]:

$$J(\omega) = J_0 \frac{\omega}{\omega_c} \exp\left(-\frac{\omega}{\omega_c}\right) \tag{17}$$

where J_0 describes the electron-phonon coupling strength, and ω_c is the cut-off frequency of the bath, which determines the thermal relaxation rate of the latter. The expression for Γ_{ph} (E) was derived in earlier works [17, 36, 37] using nonequilibrium Green's functions approach. Using these results and the expression (17), we may present c_{ph} (E) as follows:

$$\begin{aligned}\Gamma_{ph} =& 2\pi J_0 \int d\omega \frac{\omega}{\omega_c} \exp\left(-\frac{\omega}{\omega_c}\right) \{N(\omega)[\rho_{el}(E-\hbar\omega)+\rho_{el}(E+\hbar\omega)] \\ &+ [1-n(E-\hbar\omega)]\rho_{el}(E-\hbar\omega) + n(E+\hbar\omega)\rho_{el}(E+\hbar\omega)\}.\end{aligned} \tag{18}$$

Here,

$$n(E) = \frac{\Delta_1 f_1(E) + \Delta_2 f_2(E)}{\Delta_1 + \Delta_2}, \tag{19}$$

and $N(\omega)$ is the Bose-Einstein distribution function for the phonons at the temperature T. The asymptotic expression for the self-energy term Γ_{ph} depends on the relation between two characteristic energies, namely: $\hbar\omega_c$ and kT (k is the Boltzmann constant). At moderately low or room temperatures $kT \sim 10 \div 30 meV$. This is significantly greater than typical values of $\hbar\omega_c$ ($\hbar\omega_c \sim 1 meV$ [17]). Therefore, in further calculations we assume $\hbar\omega_c \ll kT$. Under this assumption, the main contribution to the integral over ω in the Eq. (18) originates from the region where $\omega \ll \omega_c \ll kT/\hbar$, and we can use the following approximation [38]:

$$\Gamma_{ph}(E) = \frac{2\Gamma\Lambda(J_0, \omega_c, T)}{(E-E_1)^2 + \Gamma^2}. \tag{20}$$

Here, $\Gamma = \Delta_1 + \Delta_2 + \Gamma_{ph}$;

$$\Lambda = \frac{4J_0}{\hbar\omega_c}(kT)^2 \zeta\left(2; \frac{kT}{\hbar\omega_c} + 1\right) \tag{21}$$

where $\zeta(2; kT/\hbar\omega_c + 1)$ is the Riemann ζ function:

$$\zeta = (2; kT/\hbar\omega_c + 1) = \sum_{n=1}^{\infty} \frac{1}{(n + kT/\hbar\omega_c)^2}. \tag{22}$$

Under $\hbar\omega_c \ll kT$, we may apply the estimation $\Lambda \approx 4kTJ_0$.

Solving the equation (20) we obtain a reasonable asymptotic expression for Γ_{ph}:

$$\Gamma_{ph} = \frac{\Delta_1 + \Delta_2}{2} \frac{\rho^2(1+\sqrt{1+\rho^2})}{4\left(\frac{E-E_1}{\Delta_1+\Delta_2}\right)^2 + (1+\sqrt{1+\rho^2})^2}. \tag{23}$$

where $\rho^2 = 8\Lambda/(\Delta_1 + \Delta_2)^2$. Substituting this expression into Eq. (16) we arrive at the result for the dissipation strength ϵ:

$$\epsilon = \frac{1}{2} \frac{\rho^2(1+\sqrt{1+\rho^2})}{4\left(\frac{E-E_1}{\Delta_1+\Delta_2}\right)^2 + \frac{1}{2}(1+\sqrt{1+\rho^2})^3} \tag{24}$$

This expression shows how the ϵ depends on the temperature T, the electron-phonon coupling strength J_0, and the energy E. In particular, it follows from the Eq. (24) that ϵ reachs its maximum at $E = E_1$, and the peak value of this parameter is given by:

$$\epsilon_{max} = \frac{\sqrt{1+\rho^2}-1}{\sqrt{1+\rho^2}+1}. \tag{25}$$

The maximum value of the dissipative strength is determined with two parameters, namely, T and J_0. As illustrated in the Fig. 3, $\epsilon_{\max}$ increases when the temperature rises, and it takes on greater values when the electron-phonon interaction is getting stronger. This result has a clear physical sense. Also, as follows from the Eq. (24), the dissipation parameter exhibit a peak at $E = E_1$ whose shape is determined by the product kTJ_0. When either J_0 or T or both enhance, the peak becomes higher and its width increases. The manifested energy dependence of the dissipation strength allows us to resolve the above mentioned difficulty occurring when the inelastic contribution to the electron transmission function is estimated using the simplified approximation of the parameter ϵ as a constant. When the energy dependence of ϵ is accounted for, the peak in the electron transmission at $E = E_1$ may be still distinguishable when $\epsilon_{\max}$ takes on values as big as 0.5.

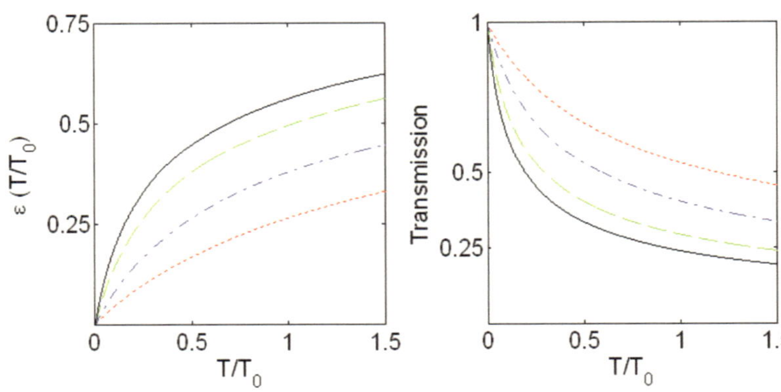

Fig. 3. Temperature dependencies of the maximum dissipation parameter ϵ (left panel) and the electron transmission (right panel). The curves are plotted at $T_0 = 50K$, $\Delta_1 = \Delta_2 = 4meV$, $E = E_1 = 0$, $J_0 = 9.0meV$ (solid line), $6.0meV$ (dashed line), $3meV$ (dash-dotted line), $1.5meV$ (dotted line).

The obtained result enables us to analyze the temperature dependencies of the electric current and conductance of the doped polymer fibers assuming that the resonance tunneling predominates in the intergrain electron transport in the absence of phonons. Current-voltage characteristics and voltage dependencies of the conductance $G = dI/dV$ computed using the expressions (2), (11), (24) are presented in the Fig. 4. We see that at low values of the applied voltage the electron-phonon coupling brings an enhancement in both current and conductance, as shown in the top panels of the Fig. 4. The effect becomes reversed as the voltage grows above a certain value (see Fig. 4, the bottom panels). This happens because the phonon induced broadening of the intermediate energy level (the bridge) assists the electron transport at small bias voltage. As the voltage rises, this effect is surpassed by the scattering effect of phonons which resists the electron transport. When the electron-phonon coupling strengthens, the I-V curves lose their specific shape typical for the elastic tunneling through the intermediate state. They become closer to straight lines corresponding to the Ohmic law. At the same time the maximum in the conductance originating from the intergrain tunneling gets eroded. These are the obvious results discussed in some earlier works (see e.g. Ref. [30]). The relative strength of the electron-phonon interaction is determined by the ratio of the electron-phonon coupling constant J_0 and the self-energy terms describing the coupling of the intermediate state (bridge) to the leads $\Delta_{1,2}$. The effect of phonons on the electron transport becomes significant when $J_0 > \Delta_{1,2}$. Otherwise, the coherent tunneling between the metallic-like islands prevails in the intergrain electron transport, and the influence of thermal phonon bath is weak. Again, we may remark that J_0 and T are combined as kTJ_0 in the expression (24). Therefore, an increase in temperature at a fixed electron-phonon coupling strength enhances the inelastic contribution to the current in the same way as the previously discussed increase in the electron-phonon coupling.

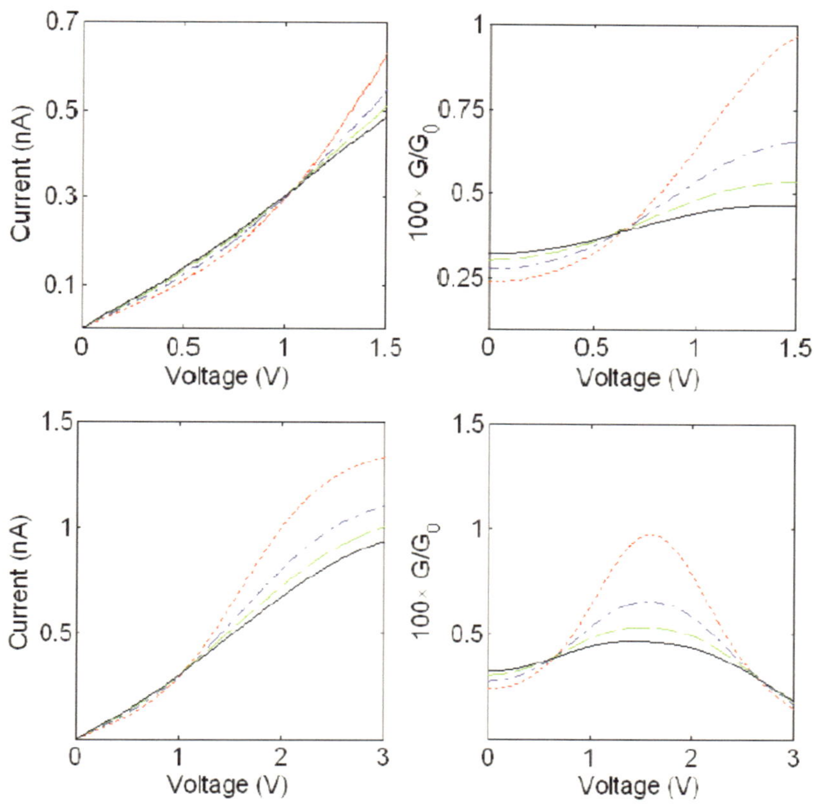

Fig. 4. Current (left panels) and conductance (right panels) versus voltage. The curves are plotted at $T_0 = 50K$, $T = 30K$, $E_1 = 40meV$, $\Delta_1 = \Delta_2 = 4meV$, $n = 1$, $\Delta V/V = 0.005$, assuming $J_0 = 9meV$ (solid line), $J_0 = 6meV$ (dashed line), $J_0 = 3meV$ (dash-dotted line), and $J_0 = 1.5meV$ (dotted line). $G_0 = 2e^2/h$ is the quantum conductance.

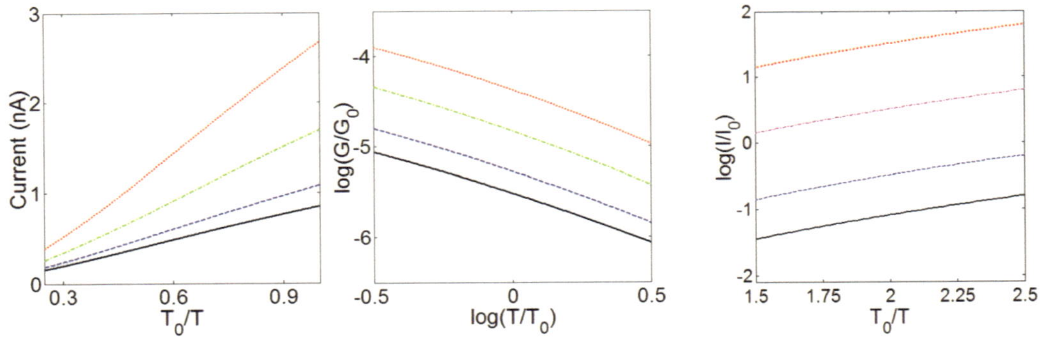

Fig. 5. Temperature dependence of current (left panel) and conductance (middle panel) at low voltage bias ($V = 0.3\,V$). Temperature dependence of the electron current at higher bias voltage ($V = 2\,V$) (right panel). The curves are plotted at $T_0 = 50K$, $J_0 = 9meV$ (solid line), $J_0 = 6meV$ (dashed line), $J_0 = 3meV$ (dash-dotted line), and $J_0 = 1.5meV$ (dotted line). The values of remaining parameters are the same in the figure 4. The current I_0 is computed using $J_0 = 3meV$, $T_0/T = 1.5$.

Now, we consider temperature dependencies of the electric current and conductance resulting from the intergrain electron tunneling via the intermediate localized state. These dependencies are shown in the Fig. 5. The curves in the figure are plotted at low bias voltage ($V = 0.3\,V$, $\Delta V/V = 0.005$) and $T_0 = 50K$, so $e\Delta V < kT$. This regime is chosen to compare the obtained temperature dependencies with those typical for the phonon assisted hopping transport discussed. We see that the temperature dependence of the tunnel current shown in the left panel of Fig. 5 crucially disagrees with the Mott's expression (??1). The tunnel current decreases as temperatute rises being proportional to $(T_0/T)^{\beta}$, and the exponent β takes on values close to unity.

Already it was mentioned that the drop in the conductivity upon heating a sample was observed in polymers and carbon nanotubes. However, such metallic-like behavior could originate from various dc transport mechanisms. Correspondingly, the specific features of temperature dependencies of the conductivity and/or current vary depending on the responsible conduction mechanism [39]. The particular temperature dependence of the electron tunneling current shown in Fig. 5 differs from those occurring due to other transport mechanisms. Such dependence was observed in the experiment on the electron transport in a single low-defect-content carbon nanotube rope, whose metallic-like conductivity was manifested within a wide temperature range ($T \sim 35 \div 300K$), as reported by Fisher et al [40]. The conductivity temperature dependence observed in this work could be approximated as $\sigma(T)/\sigma(300) \sim a + bT_0/T$ where A, b are dimensionless constants. The approximation includes the temperature independent term, which corresponds to the Drude conductivity. The second term is inversely proportional to the temperature in agreement with the results for the current shown in the Fig. 5 9left panel). It is also likely that a similar approximation may be adopted to describe experimental data obtained for chlorate-doped polyacetylene samples at the temperatures below $100K$ [41]. In both cases we may attribute the contribution proportional to $1/T$ to the resonance electron tunneling transport between metallic-like islands.

The conductance due to the electron intergrain tunneling reduces when the temperature increases, as shown in the middle panel of the Fig. 5. Irrespective of the electron-phonon coupling strength we may approximate the conductance by a power law $G \sim T^{\alpha}$ where α takes on values close to -1. This agrees with the results for the current. At higher bias voltage the temperature dependence of the current changes, as shown in the right panel of the Fig. 5. The curves shown in this panel could be approximated as $\log(I/I_0) \sim c + dT_0/T$, c, d being dimensionless constants. This resembles typical temperature dependencies of the tunneling current in quasi-one-dimensional metals which were predicted for conducting polymers being in a metal state (see e.g. Ref. [2]).

4. Effect of phonon induced electron states on the transport properties of conducting polymer fibers

Studies of dissipative effects in the intergrain electron transport in conducting polymers may not be restricted with the plain assumption of direct coupling of the bridge site to the phonon bath. Other scenarios can occur. In particular, analyzing electron transport in polymers, as well as in molecules, one must keep in mind that besides the bridge sites there always exist other nearby sites with close energies. In some cases the presence of such sites may strongly influence the effects of stochastic nuclear motions on the characteristics of electron transport. This may happen when the nearby sites somehow "screen" the bridge sites from direct interactions with the phonon bath. Here, we elucidate some effects which could appear in the electron transport in conducting polymer fibers in the case of such indirect coupling of the bridge state to the phonon bath.

We mimic the effects of the environment by assuming that the side chain is attached to the bridge, and this chain is affected by phonons. This model resembles those used to analyze electron transport through macromolecules [19]. The side chain is introduced to screen the resonance state making it more stable against the effect of phonons. We assume that electrons cannot hop along the side chain, so it may be reduced to a single site attached to the resonance site (the bridge).

Within the adopted model the retarded Green's function for the bridge acquires the form [19, 42]:

$$G^{-1}(E) = E - E_1 - \Delta_1 - \Delta_2 - w^2 P(E). \tag{26}$$

The first four terms in this expression represent the inversed Green's function for the resonance site coupled to the two grains, and the factor w is the hopping integral between the bridge and the attached side chain. The term $P(E)$ represents the effect of the phonons and has the form:

$$\begin{aligned} P(E) =& - i \int_0^{\infty} dt \exp\left[it(E - \tilde{E}_1 + \delta + i0^+)\right] \\ &\times \Big\{(1-f)\exp[-F(t)] + f\exp[-F(-t)]\Big\} \end{aligned} \tag{27}$$

with $\tilde{E}_1$ being the on-site energy for the side site, which is close to the bridge site energy E_1, $\exp[-F(t)]$ being a dynamic bath correlation function, and f taking on values 1 and 0 when the attached site is occupied and empty, respectively.

Characterizing the phonon bath with a continuous spectral density $J(\omega)$ given by Eq. (17) one may write out the following expressions for the functions $F(t)$ and δ:

$$F(t) = \int_0^\infty \frac{d\omega}{\omega^2} J(\omega) \left[1 - e^{-i\omega t} + \frac{2[1 - \cos(\omega t)]}{\exp(\hbar\omega/kT) - 1} \right], \tag{28}$$

$$\delta = \int_0^\infty \frac{d\omega}{\omega} J(\omega) = J_0. \tag{29}$$

Within the short time scale ($\omega_c t \ll 1$) the function $F(t)$ could be presented in the form [43]:

$$F(t) = \frac{J_0}{\hbar\omega_c} \left\{ \frac{1}{2} \ln[1 + (\omega_c t)^2] + i \arctan(\omega_c t) + K(t) \right\} \tag{30}$$

where

$$K(t) = (kT)^2 t^2 \zeta \left(2; \frac{kT}{\hbar\omega_c} + 1 \right). \tag{31}$$

Here, ζ (2; $kT/\hbar\omega_c$ + 1) is the Riemann ζ function. The asymptotic expression for $K(t)$ depends on the relation between two parameters, namely, the temperature T and the cut-off frequency ω_c of the phonon bath. Assuming $kT \gg \hbar\omega_c$

$$K(t) \approx \frac{kT}{\hbar\omega_c} (\omega_c t)^2. \tag{32}$$

In the opposite limit $\hbar\omega_c \gg kT$ we obtain:

$$K(t) \approx \frac{\pi^2}{6} \left(\frac{kTt}{\hbar} \right)^2 \tag{33}$$

Also, we may roughly estimate $K(t)$ within the intermediate range. Taking $kT \approx \hbar\omega_c$ we arrive at the approximation $K(t) \approx a^2(kTt/\hbar)^2$ where a^2 is a dimensionless constant of the order of unity. Correspondingly, within the short time scale we can omit the first term in the expression (30), and we get:

$$F(t) \approx \frac{J_0}{\hbar\omega_c} \{ i\omega_c t + K(t) \} \tag{34}$$

where $K(t)$ is given by either Eq. (32) or Eq. (33) depending on the relation between $\hbar\omega_c$ and kT.

Within the long time scale $\omega_c t \gg 1$, and provided that temperatures are not very low ($kT \gg \hbar\omega_c$), we may present the function $K(t)$ as:

$$K(t) = \frac{2kTt}{\hbar} \int_0^\infty \frac{dz}{z^2} (1 - \cos z) e^{-z/\omega_c t} \approx \frac{\pi kTt}{\hbar}. \tag{35}$$

Now, the term $K(t)$ is the greatest addend in the expression for $F(t)$, so the latter could be approximated as: $F(t) \approx \pi kTtJ_0/\omega_c\hbar^2$. The same approximation holds within the low temperature limit when $\omega_c \gg kT/\hbar \gg t^{-1}$.
Using the asymptotic expression (34), we may calculate the contribution to $P(E)$ coming from the short time scale ($\omega_c t \ll 1$). It has the form:

$$P_1(E) = -\frac{i}{2}\sqrt{\frac{\pi}{J_0 kT}}\exp\left[-\frac{(E-\tilde{E}_1)^2}{4J_0 kT}\right]\left\{1+\Phi\left[\frac{i(E-\tilde{E}_1)}{2\sqrt{J_0 kT}}\right]\right\} \tag{36}$$

where $\Phi(z)$ is the probability integral. When both $\hbar\omega_c$ and kT have the same order of magnitude the expression for $P(E)$ still holds the form (36). At $kT \ll \hbar\omega_c$, the temperature kT in the expression (36) is to be replaced by $\hbar\omega_c$. We remark that under the assumption $kT \gg \hbar\omega_c$ the function $P_1(E)$ does not depend on the cut-off frequency ω_c, whereas at $\hbar\omega_c \gg kT$ it does not depend on temperature. The long time ($\omega_c t \gg 1$) contribution to $P(E)$ could be similarly estimated as follows:

$$P_2(E) = \frac{1}{E-\tilde{E}_1+J_0+i\pi J_0 kT/\hbar\omega_c}. \tag{37}$$

Comparing these expressions (36) and (37) we see that the ratio of the peak values of $P_2(E)$ and $P_1(E)$ is of the order of $(\hbar^2\omega_c^2/J_0 kT)^{1/2}$. Therefore, the term $P_1(E)$ predominates over $P_2(E)$ when the temperatures are moderately high $\hbar\omega_c < kT$) and the electron-phonon interaction is not too weak $J_0/\hbar\omega_c \sim 1$. Usually, experiments on the electrical characterization of conducting polymer nanofibers are carried out at $T \sim 100 \div 300K$, so in further analysis we assume that $(\hbar^2\omega_c^2/J_0 kT)^{1/2} \ll 1$, and the term $P_2(E)$ could be omitted.
As shown in the Fig. 6 (left panel), the imaginary part of $P(E)$ exhibits a dip around $E = \tilde{E}_1$ and the width of the latter is determined by the product of the temperature kT (or $\hbar\omega_c$) and the constant J_0 characterizing the strength of the electron-phonon interaction. When either factor increases, the dip becomes broader and its magnitude reduces.
The presence of the term $w^2P(E)$ gives rise to very significant changes in the behavior of the Green's function given by the Eq. (26). Using the flat band approximation for the self-energy corrections and disregarding for a while all imaginary terms in the Eq. (26), we find that two extra poles of the Green's function emerge. Assuming $E_1 = \tilde{E}_1 = 0$ and $kT \gg \hbar\omega_c$, these poles are situated at:

$$E = \pm 2\sqrt{J_0 kT|\ln(2J_0 kT/w^2)|}. \tag{38}$$

The poles correspond to extra electron states which appear due to electrons coupling to the thermal phonons. These new states are revealed in the structure of the electron transmission $T(E)$ given by Eq. (8). The structure of $T(E)$ is shown in the right panel of the Fig. 6. Two peaks in the transmission are associated with the phonon-induced electronic states. Their positions and heights depend on the temperature and on the coupling strengths J_0, and w. The important feature in the electron transmission is the absence of the peak associated with the resonance state between the grains (the bridge site) itself. This happens due to the strong suppression of the latter by the effects of the environment. Technically, this peak is damped

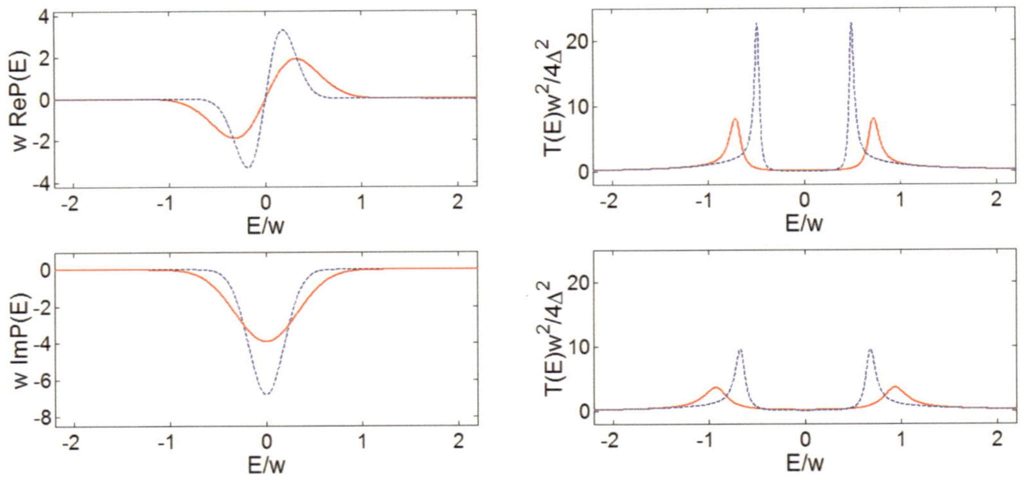

Fig. 6. Left panels: Energy dependence of the real (top panel) and imaginary (bottom panel) parts of $P(E)$. The curves are plotted at $J_0 = 20meV$, $kT \gg \hbar\omega_c$. Right panels: The renormalized electron transmission function vs energy. The constant J_0 equals $20meV$ (top panel) and $50meV$ (bottom panel). All curves are plotted assuming $w = 100meV$, $E_1 = \tilde{E}_1 = 0, T = 100K$ (dashed lines) and $T = 300K$ (solid lines).

for it is located at $E = 0$ where the imaginary part of $P(E)$ reaches its maximum in magnitude. To provide the damping of the original resonance the contribution from the environment (including the side chain attached to the bridge) to the Green's function (26) must exceed the terms $\Delta_{1,2}$ describing the effect of the grains. This occurs when the inequality

$$\Delta_{1,2} < w^2/\sqrt{J_0 kT} \tag{39}$$

is satisfied. When the coupling of the bridge to the attached side site is weak, the influence of the environment slackens and the original peak associated with the bridge at $E = E_1$ may emerge. At the same time the features originating from the phonon-induced states become small compared to this peak.

So, the effects of the phonons may lead to the damping of the original resonance state for the electron tunneling between the metallic islands in the polymer fiber. Instead, two phonon-nduced states appear to serve as intermediate states for the electron transport [44]. Environmental induced electron states were discussed in the theory of electron conductance in molecules Refs. [19, 45, 46]. For instance, it was shown that low biased current-voltage characteristics for molecular junctions with DNA linkers may be noticeably changed due to the occurrence of the phonon-induced electron states similar to these discussed in the present Section.

As we discussed before, one may treat a nanofiber as a set of parallel working channels, any single channel being a sequence of grains connected with the resonance polymeric chains. Accordingly, the voltage V applied across the whole fiber is distributed among sequental pairs of metallic-like islands included in a single channel. So, in practical nanofibers the voltage ΔV applied across two adjacent grains appears to be much smaller than V ($\Delta V/ V \sim 10^{-1}$ –10^{-3}). Experiments on the electrical characterization of the polymer fibers are usually carried out at moderately high temperatures ($T \div 300K$), so it seems likely that $kT > \hbar\omega_c$. Assuming that $w \sim 100meV$, and $J_0 \sim 20\div50meV$ we estimate the separation between the phonon-induced peaks in the electron transmission as $120\div170meV$ (see Fig. 6). This estimate is close to $e\Delta V$ when V takes on values up to 2÷3 volts. So, the phonon-induced peaks in the electron transmission determine the shape of the currentvoltage curves even at reasonably high values of the bias voltage applied across the fiber. The resulting current-voltage characteristics are shown in the Fig. 7. The $I - V$ curves exhibit a nonlinear shape even at room temperature despite the fact that the original state for the resonance tunneling is completely suppressed. This occurs because the intergrain transport is supported by new phonon-induced electron states.

It is worthwhile to discuss the temperature dependence of the peak value of the electron transmission which follows from the present results. Using the expression (26) for the Green's function and the expression (36) for $P(E)$ we may show that at low temperatures the transmission accepts small values, and exhibits rather weak temperature dependence. At higher temperatures ($T \sim 100K$) the transmission increases fast as the temperature rises and then it reduces as the temperature further increases. The peak in the transmission is associated with the most favorable conditions for the environment induced states to exist when all remaining parameters (such as J_0 and w) are fixed. At high temperatures the peaks associated with the environment induced states are washed out, as usual.

We may compare this result with the temperature dependence of the electron transmission function occurring when the bridge between two adjacent grains is directly coupled to the phonon bath. In this case the transmission peak value may be presented in the form determined by Eqs. (11), (24):

$$T(E) = 1 - \frac{\rho^4}{2(1 + \sqrt{1 + \rho^2})^2}. \tag{40}$$

The temperature dependencies are shown in the Fig. 8. Both curves are plotted at the same value of the electron-phonon coupling strength J_0. Comparing them we conclude that at higher temperatures the dependencies significantly differ. While the temperature rises, we observe a peak in the electron transmission assuming the indirect coupling of the bridge to the phonon bath, and we see the transmission to monotonically decrease when we consider the bridge directly coupled to the bath. Correspondingly, we may expect qualitative diversities in the temperature dependencies of the current, as well. These diversities originate from the difference in the effects of environment on the intergrain electron transport in the cases of direct and indirect coupling of the bridge site to the phonon bath. When the bridge is directly coupled to the bath, the stochastic motions in the environment only cause washing out of the peak in the electron transmission, and the higher is the temperature the less distinguishable is the peak. However, when the bridge is screened from the direct coupling with the phonons due to the presence of the nearby sites, the stochastic nuclear motions in the medium between the grains (especially those in the resonance chain) may take a very different part in the electron transport in conducting polymers at moderately low and room temperatures. Due to their influence, the original intermediate state for the resonance tunneling may be completely suppresed but new phonon-induced states may appear to support the electron transport between the metallic-like domains in conducting polymer nanofibers.

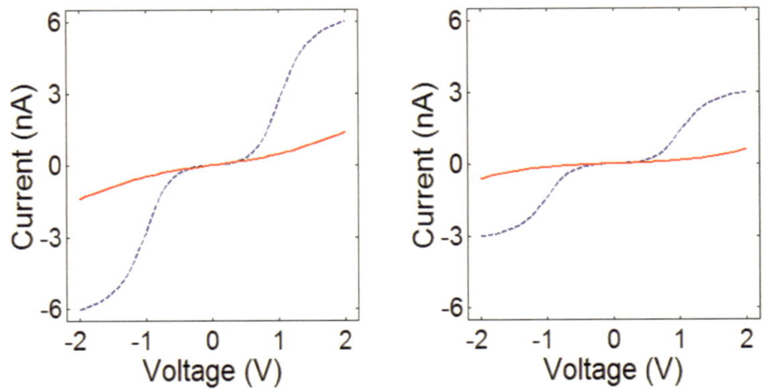

Fig. 7. The current-voltage characteristics (nA–V) plotted for $n = 10$, $\Delta_1 = \Delta_2 = 0.5meV$, $w = 100meV$, $J_0 = 20meV$ (left panel), and $J_0 = 50meV$ (right panel) at $T = 100K$ (dash lines) and $T = 300K$ (solid lines).

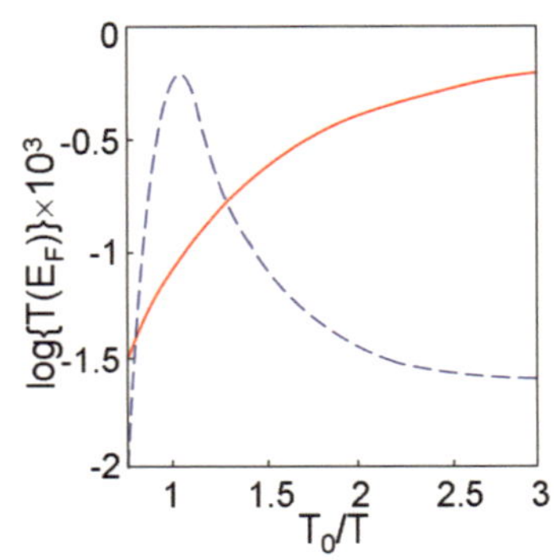

Fig. 8. Arrhenius plot of the peak value of the electron transmission function for $J_0 = 20meV$, $w = 100meV$, $\Delta_1 = \Delta_2 = 0.5meV$, and $T_0 = 100K$. The dashed line is plotted assuming the indirect coupling of the bridge to the phonon bath. The solid line is plotted assuming that the bridge is directly coupled to the phonons.

5. Conclusion

Studies of the electron transport in conducting polymers are not completed so far. Several mechanisms are known to control the charge transport in these highly disordered and inhomogeneous materials, and their relative significance could vary depending on both specific intrinsic characteristics of a particular material (such as crystallization rate and electronelectron and electron-phonon coupling strengths) and external factors such as temperature. Various conduction mechanisms give rise to various temperature dependencies of the electric current and conductance, which could be observed in polymer nanofibers/nanotubes being in conducting state. In the present work we aimed at finding out the character of temperature dependencies of both current and conductance provided by specific transport mechanism, namely, resonance tunneling of electrons.

Accordingly, we treated a conducting polymer as a kind of granular metal, and we assumed that the intergrain conduction occurred due to the electron tunneling between the metallic-like grains through the intermediate state. This scenario for the intergrain electron transport strongly resembles electron transport through molecules/quantum dots attached to the conducting leads. In the considered case the metallic-like islands work as the leads and the intermediate state in between acts as a single level quantum dot/molecular bridge. Basing on this similarity we did apply the well-known Landauer formula to compute the intergrain electron current. This bringed results which agreed with experiments on electrical characterization of doped polyaniline-polyethylene oxide nanofibers.

There are solid grounds to expect significant dissipative effects in the intergrain transport at moderately high temperatures. To take into account the effect of temperature we did represent the thermal environment stochastic nuclear motions as a phonon bath, and we introduced the coupling of the intermediate site to the thermal phonons. As shown in the previous studies of the electron transport through molecules, various dissipative effects may occur depending on characteristic features in the interaction of a propagating electron with the environment. Among these features we singled out the character of the electron coupling to the dissipative reservoir (phonon bath) as a very significant factor. It is likely that in practical conducting polymers both direct and indirect coupling of the intermediate state (the bridge) to the environment may occur.

We did analyze temperature dependencies of transport characteristics for both scenarios and we showed that these dependencies differ. Also, we showed that in general, resonance electron tunneling between the grains results in the temperature dependencies of transport characteristics, which differ from those obtained for other conduction mechanisms such as phonon-assisted hopping between localized states. Being observed in experiments on realistic polymer nanofibers, the predicted dependencies would give grounds to suggest the electron tunneling to predominate in the intergrain electron transport in these particular nanofibers. We believe the present studies to contribute to better understanding of electron transport mechanisms in conducting polymers.

6. Acknowledgments

Author thanks G. M. Zimbovsky for help with the manuscript. This work was supported by DoD grant W911NF-06-1-0519 and NSF-DMR-PREM 0353730.

7. References

[1] A. G. MacDiarmid, Rev. Mod. Phys. 73, 701 (2001).

[2] A. B. Kaiser, Rep. Prog. Phys. 64, 1 (2001).

[3] J. Joo, Z. Oblakowski, G. Du, J. P. Pouget, E. J. Oh, J. M. Wiesinger, Y. Min, A. G. MacDiarmid, and A. J. Epstein, Phys. Rev. B 49, 2977 (1994).

[4] J. P. Pouget, Z. Oblakowski, Y. Nogami, P. A. Albouy, M. Laridjani, E. J. Oh, Y. Min, A. G. MacDiarmid, J. Tsukamoto, T. Ishiguro, and A. J. Epstein, Synth. Met 65, (1994).

[5] N. F. Mott and E. A. Davis, *Electronic Processes in Non-Crystalline Materials* (Clarendon Press, Oxford, 1971).

[6] J. Voit, Rep. Progr. Phys. 58, 977 (1995).

[7] R. Mukhopadhyay, C. L. Kane, and T. C. Lubensky, Phys. Rev. B 64, 045120 (2001).

[8] A. Bachtold, M. de Jouge, K. Grove-Rasmussen, P. L. McEuen, M. Buitelaar, and C. Schonenberger, Phys. Rev. Lett. 87, 166801 (2001).

[9] A.N. Aleshin, H. J. Lee, K. Akagi, and Y. W. Park, Microelectronic Eng. 81, 420 (2005).

[10] A. N. Aleshin, H. J. Lee, Y. W. Park, and K. Akagi, Phys. Rev. Lett. 93, 196601 (2004).
[11] C. O. Yoon, M. Reghu, D. Moses, A. J. Heeger, and Y. Cao, Synth. Met. 63, 47 (1994).
[12] C. K. Subramanian, A. B. Kaiser, R. W. Gibberd, C-J. Liu, and B. Westling, Solid State Comm. 97, 235 (1996).
[13] M. S. Fuhrer, M. L. Cohen, A. Zettl, and V. Crespi, Solid State Comm. 109, 105 (1999).
[14] V. N. Prigodin and A. J. Epstein, Synth. Met. 125, 43 (2002).
[15] N. A. Zimbovskaya, A. T. Johnson, and N. J. Pinto, Phys. Rev. B 72, 024213 (2005).
[16] Yangxin Zhou, M. Freitag, J. Hone, C. Stali, A. T. Johnson Jr., N. J. Pinto, and A. G. MacDiarmid, Appl. Phys. Lett. 83, 3800 (2003).
[17] M. Galperin, A. Ratner, and A. Nitzan, J. Chem. Phys. 121, 11965 (2004).
[18] R. Egger and A. O. Gogolin, Phys. Rev. B 77, 1098 (2008).
[19] R. Gutierrez, S. Mandal, and G. Cuniberti, Phys. Rev. B 71, 235116 (2005).
[20] D. A. Ryndyk, P. D'Amico, G. Cuniberti, and K. Richter, Phys. Rev. B 78, 085409 (2008).
[21] R. G. Endres, D. L. Cox, and R. R. P. Singh, Rev. Mod. Phys. 76, 195 (2004).
[22] C. J. Bolton-Neaton, C. J. Lambert, V. I. Falko, V. M. Prigodin, and A. J. Epstein, Phys. Rev. B 60, 10569 (2000).
[23] D. A. Ryndyk and G. Cuniberti, Phys. Rev. B 76, 155430 (2007).
[24] E. Lortscher, H. B. Weber, and H. Riel, Phys. Rev. Lett. 98, 176807 (2007).
[25] R. Pelster, G. Nimtz, and B. Wessling, Phys. Rev. B 49, 12718 (1994).
[26] S. Datta, *Quantum Transport: Atom to Transistor* (Cambridge University Press, Cambridge, England, 2005).
[27] S. Skourtis and S. Mukamel, J. Chem. Phys. 197, 367 (1995).
[28] D. Segal, A. Nitzan, W. B. Davis, M. R. Wasielewski, and M. A. Ratner, J. Chem. Phys. 104, 3817 (2000).
[29] M. Buttiker, Phys. Rev. B 33, 3020 (1986).
[30] X. Q. Li and Y. Yan, J. Chem. Phys. 115, 4169 (2001); Appl. Phys. Lett. 81, 925 (2001).
[31] M. Kemp, V. Mujica, and M. A. Ratner, J. Chem. Phys. 101, 5172 (1994).
[32] J. L. D'Amato and H. M. Pastawski, Phys. Rev. B 41, 7411 (1990).
[33] N. J. Pinto, A. T. Johnson, A. G. MacDiarmid, C. H. Muller, N. Theofylaktos, D. C. Robinson, and F. A. Miranda, Appl. Phys. Lett. 83, 4244 (2003).
[34] N. A. Zimbovskaya, J. Chem. Phys. 123, 114708 (2005).
[35] G. D. Mahan, *Many-Particle Physics* (Plenum, New York, 2000).
[36] B. N. J. Persson and A. Baratoff, Phys. Rev. Lett. 59, 339 (1987).
[37] T. Mii, S. G. Tikhodeev, and H. Ueba, Phys. Rev. B 68, 205406 (2003).
[38] N. A. Zimbovskaya, J. Chem. Phys. 129, 114705 (2008).
[39] The matter is thoroughly discussed by Kaiser (see [2] and references therein).
[40] J. E. Fisher, H. Dai, A. Thess, R. Lee, N. M. Hanjani, D. L. Dehaas, and R. E. Smalley, Phys. Rev. B 55, R4921 (1997).
[41] Y. W. Park, E. S. Choi, and D. S. Suh, Synth. Met. 96, 81 (1998).
[42] U. Weiss in *Quantum Dissipative Systems*, Series in Modern Condensed matter Physics, Vol. 10 (World Scientific, Singapore, 1999).
[43] I. S. Gradshteyn and I. M. Ryzhik, *Tables of Integrands, Series and Products* (Academic, New York, 2000).
[44] N. A. Zimbovskaya, J. Chem. Phys. 126, 184901 (2007).
[45] F. L. Gervasio, P. Carloni, M. Parrinello, Phys. Rev. Lett. 89, 108102 (2002).
[46] R. G. Endres, D. L. Cox, and R. R. P. Singh, Rev. Mod. Phys. 76, 195 (2004).

18

Spectroscopy of Polyaniline Nanofibers

Gustavo M. Do Nascimento
Universidade Federal de Minas Gerais
Brazil

1. Introduction

1.1 Synthetic metals

The first materials, synthesized during the 60s, which have found values of conductivity above 1 Scm^{-1}, are called Krogman salts. These salts are square planar complexes of PtX_4 or IrX_4 (X = CN-or $C_2O_4^{-2}$), forming structures of chains, which conductivity reaches metallic values when exposed to oxidizing Br_2 vapors (Liepins & Ku, 1987). The oxidation of molecular chains causes a reduction of the distances between the atoms of the metals, allowing the filling of the conduction band formed by 5-D_{ZZ} orbitals of neighboring metal atoms in the chain. As a result, the conductivi3ty reaches values as higher as 10^{-7} to 10^{-2} Scm^{-1}, an increase of 10^5 times. Charge-transfer complex forms the second class of molecular substances that have shown high electrical conductivity. Tetrathiofulvalene (TTF) and Tetracyanoquinodimethane (TCNQ) and its derivatives are the most known complex transfers systems (Liepins & Ku, 1987).

The intrinsically conducting polymers (ICPs), more commonly known as "synthetic metals", forms the third class of molecular conductors. Initially, the study of conjugated polymers was hampered owing to their insolubility, infusibility and instability in the air. In the 70s, (Shirakawa & Ikeda, 1974; Shirakawa & Ikeda, 1971) synthesized more stable films of semiconducting poly(acetylene). However, just in 1977, (Chiang et al., 1977; Chiang et al. 1978; Shirakawa et al. 1977; Shirakawa, 2001; MacDiarmid, 2001) found that, when the poly(acetylene) is doped with acid (or base) of Lewis, it is possible to increase the conductivity by 13 orders of magnitude. Since this initial discovery, the development of the conducting polymer field has continued to grow up at an accelerated rate.

The concept of doping is unique and has central importance, because it is what differentiates the conducting polymers from all other types of polymers (Heeger, 2001; Nigrey et al. 1979; Han & Elsenbaumer, 1989). During the process of doping, an insulating or semiconducting organic polymer with low conductivity, typically ranging from 10^{-10} to 10^{-5} Scm^{-1}, is converted into a polymer which shows conductivity in a "metallic" regime (ca. 1-10^4 Scm^{-1}). The addition of non-stoichiometric chemical species in quantities commonly low ($\leq$10%), results in dramatic changes in electronic properties, electrical, magnetic, optical and the structural of the polymer. The term is used for conducting polymers in analogy to the process of doping of crystalline inorganic semiconductors. But, it is important to emphasize that the doping of inorganic semiconductors is of the order of ppm and the dopant causes practically no disturbance into the crystalline structure of the inorganic semiconductor. In the case of doping of the organic polymer, the dopant chemically reacts with the chain and causes disturbance in the crystalline structure of the polymer.

Doping is reversible, and the polymer can return to its original state without major changes in its structure. In the doped state, the presence of counter ions stabilizes the doped state. By adjusting the level of doping, it is possible to obtain different values of conductivity, ranging from the state or non-doped insulating state to the highly doped or metallic. The three major classes of conducting polymers (see Fig. 1) can be doped by p (oxidation) or n (reduction) through chemical and/or electrochemical process (Heeger, 2001; Nigery et al. 1979; Han & Elsenbaumer, 1989).

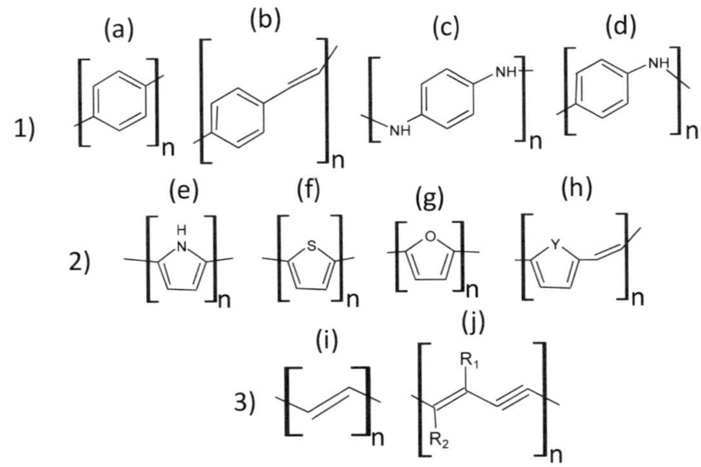

Fig. 1. Chemical representation of the most representative classes of conducting polymers. Group (1) is formed by polymers with benzene ring: (a) poly(*p*-phenylene), (b) poly(*p*-phenylene-vinylene), (c) poly(*p*-phenylenediamine), and (d) poly(aniline). Group (2) is constituted by polymers with cyclopentadiene ring containing a hetero atom: (e) poly(pyrrole), (f) poly(thiophene), (g) poly(furan), and (h) poly(heteroaromatic-vinylene) (where Y = NH, NR, S, O). Finally, the Group (3) is formed by acetylene units, being (i) poly(acetylene) and (j) poly(di-acetylene) (where R_1 and R_2 can be a large variety of organic groups (Song et al., 2004; Asefa et al., 1999; Peng et al., 2005)) the most important ones.

1.2 Polyaniline (PANI)

Polyaniline (PANI) is one of the most important conducting polymers owed to its easy preparation and doping process, environmental stability, and potential use as electrochromic device, as sensor and as corrosion protecting paint. These properties turned PANI attractive to use in solar cells, displays, lightweight battery electrodes, electromagnetic shielding devices, anticorrosion coatings and sensors. The recent research efforts are to deal with the control and the enhancement of the bulk properties of PANI, mainly by formation of organized PANI chains in blends, composites and nanofibers (MacDiarmid & Epstein, 1989a; MacDiarmid & Epstein, 1989b; MacDiarmid et al. 1987; MacDiarmid & Epstein, 1984). The fully reduced leucoemeraldine base form (LB, see Fig.2 for y = 1) and the fully oxidized pernigraniline base form (PB in Fig.2 for y = 0) are non-conducting forms of PANI. The half-oxidized emeraldine base (EB in Fig.2 for y = 0.5) is a semiconductor but after protonation it becomes a conducting emeraldine salt form of PANI (ES, see structure in Fig. 3) (MacDiarmid & Epstein, 1989a; MacDiarmid & Epstein, 1989b).

EB and ES can also assume two different types of crystalline arrangements depending on the synthetic route used (Pouget et al., 1991; Colomban et al., 1999).
Polyaniline can be synthesized by two main methods, by chemical or by electrochemical polymerization of aniline in acidic media. The chemical oxidation is commonly performed using ammonium persulfate in aqueous acidic media (hydrochloric acid, sulfuric, nitric or perchloric acid) containing aniline. This is the conventional synthetic route of PANI, but one of disadvantages of this route is the presence of excess of oxidant and salts formed during the synthesis, leading to a polymeric sample that is practically insoluble in majority of solvents, making its processing very difficult (Syed & Dineson, 1991).

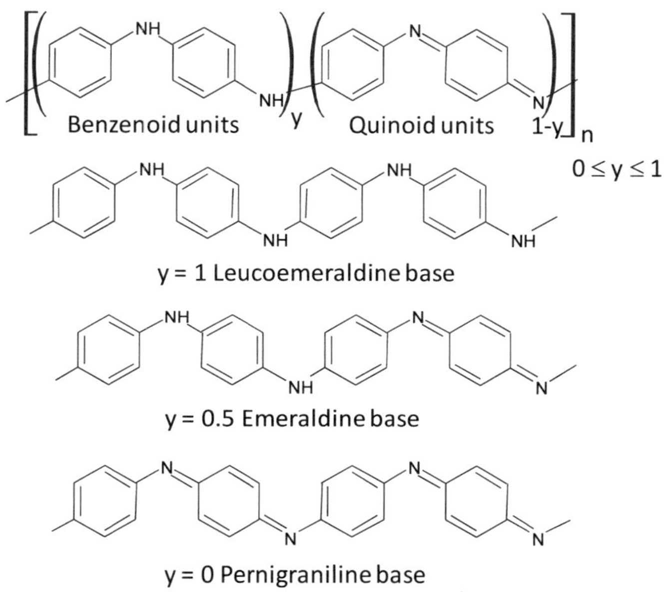

Fig. 2. Chemical representation of generalized PANI structure and its most common forms.

During the oxidative polymerization of aniline, the solution becomes progressively colored resulting in a solid dark green. The color of the solution is owing to the presence of soluble oligomers formed by coupling of radical cations of aniline. The intensity of the color depends on the environment and also the concentration of oxidant (Syed & Dineson, 1991). There are many variations of the chemical synthesis of PANI, however there is a certain consensus that there are four main parameters that affect the course of the reaction and the nature of final product, being: (1) nature of the synthetic medium, (2) concentration of the oxidant, (3) duration of the reaction, and (4) temperature of the synthetic medium (Syed & Dineson, 1991). The polymerization of aniline was observed as an autocatalytic process (Wei et al., 1989; Sasaki et al., 1986). Kinetic studies suggest that initial oxidation of aniline leads to the formation of dimeric species, such as *p*-amino-biphenyl-amine, N,N'-Biphenyl-hidrazine, and Benzidine (Mohilmer et al., 1962). (Bacon & Adams, 1968) and (Wawzonek & MacIntyre, 1967) showed that the oxidation of aniline and its derivatives in strongly acidic media favor the formation of benzidine, while in a slightly acid or neutral prevails *p*-amino-Biphenyl-amine, but in basic medium the formation of azo bonds, resulting from the head-head coupling is favored, (see Fig. 3). These dimeric species have lower oxidation potential than aniline and are oxidized immediately after its formation (the N,N'-biphenyl-hidrazine is converted to benzidine through rearrangement that occurs in acid medium (Geniés et al., 1990), resulting two types of charged quinoid-di-imine species. Afterwards, electrophilic attacks in these species, followed by deprotonation, are responsible for the growth of oligomers with subsequent formation of polymer chains of PANI (see Fig. 3).

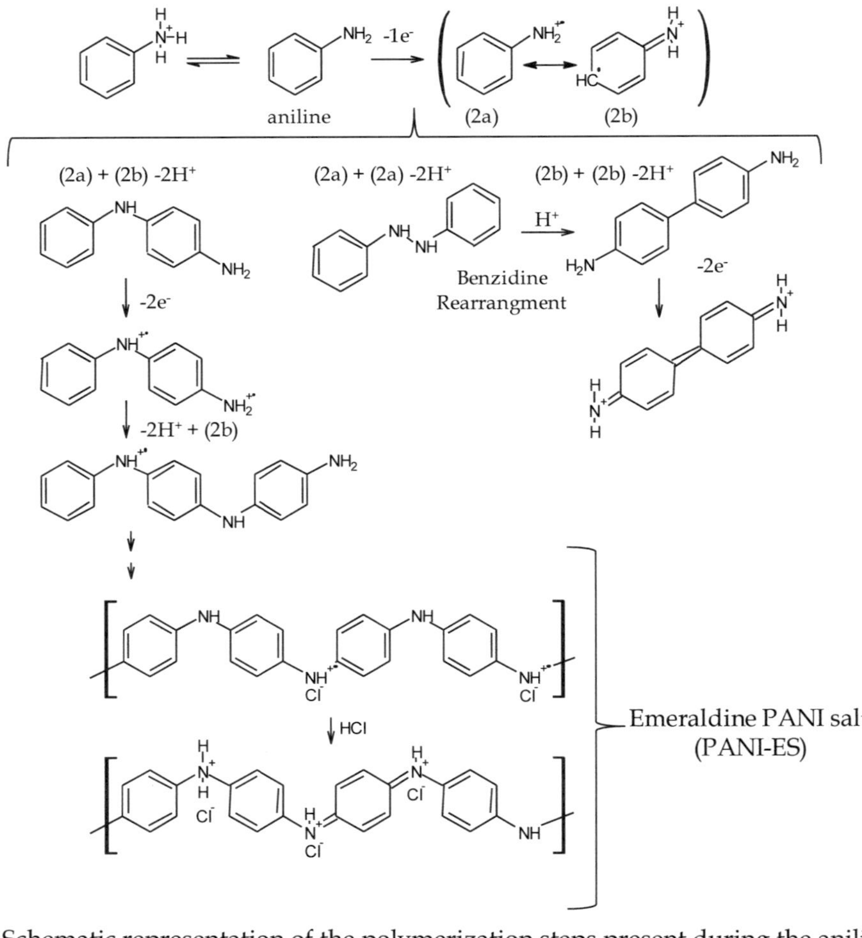

Fig. 3. Schematic representation of the polymerization steps present during the aniline polymerization. The structure of PANI is shown with radical cations (or polarons) and dications (bipolarons) segments.

1.3 Nanostructured polyaniline

The synthesis of nanostructured PANI, especially as nanofibers, can improve its electrical, thermal and mechanical stabilities. These materials can have an important impact for application in electronic devices and molecular sensors owing their extremely high surface area, synthetic versatility and low-cost. The conventional synthesis of polyaniline, based on the oxidative polymerization of aniline in the presence of a strong acid dopant, typically results in an irregular granular morphology that is accompanied by a very small percentage of nanoscale fibers (Huang & Kaner, 2004a; Huang & Kaner, 2004b). However, different approaches have been developed in order to produce PANI and many other polymers with nanostructured morphology. In this chapter will be analysed the synthetic routes that produce nanostructured PANI, mainly as nanofiber or nanotube morphology, without the use of rigid templates.

The nanostructured PANI has been prepared by different synthetic ways. Nevertheless, these approaches can be grouped into two general synthetic routes, as can be seen in the Fig. 4. Uniform nanofibers of pure metallic PANI (30-120 nm diameter, depending on the dopant) have also been prepared by polymerization at an aqueous-organic interface (Huang & Kaner, 2004a; Huang & Kaner, 2004b). In the first step (see item a) of the interfacial polymerization), the oxidant and monomers (aniline), dissolved in immiscible solvents, are put together without external agitation. Afterwards, some aniline monomers are oxidized in the interfacial region between the two solutions, being formed some oligomers (see item b) of the interfacial polymerization). It is hypothesized that migration of the product into the aqueous phase can suppress uncontrolled polymer growth by isolating the fibers from the excess of reagents. Afterwards, the initial chains grow up and more PANI chains are formed (see step c)). Interfacial polymerization can therefore be regarded as a non-template approach in which high local concentrations of both monomer and dopant anions at the liquid–liquid interface might be expected to promote the formation of monomer-anion (or oligomer-anion) aggregates. These aggregates can act as nucleation sites for polymerization, resulting in powders with fibrillar morphology. It has recently been demonstrated that the addition of certain surfactants to such an interfacial system grants further control over the diameter of the nanofibers. An important part that is frequently neglected or not deeply explained in details is the isolation of the nanostructured PANI from the solution. But, generally, the nanofibers are isolated by filtration in a nanoporous filters, being the isolated polymer washed with different solutions with the aim to clean it up. The solution can be also dialyzed and the cleaned solution containing the nanofibers is centrifugated in order to separate the nanofibers from the solution.

PANI nanofibers or nanotubes can be obtained by making use of large organic acids (see Fig. 4). These acids form micelles upon which aniline is polymerized and doped (see Fig. 4 steps (a), (b) and (c) of micellar polymerization). Fiber diameters are observed to be as low as 30-60 nm and are highly influenced by reagent ratios (Zhang et al., 2002; Qiu et al., 2001; Wei & Wan, 2002; Do Nascimento et al., 2006). Ionic liquids (ILs) have also been used as synthetic media for the preparation of nanostructured conducting polymers (Gao et al., 2004; Rodrigues et al., 2007a; Rodrigues et al., 2007b). Ionic liquids are organic salts with low lattice energies, which results in low melting points and many ILs are liquids at room temperature (Davis Jr et al., 2002). There is a large variety of ionic liquids and the most used ones are derived from imidazolium ring, pyridinium ring, quaternary ammonium and tertiary phosphonium cations. The usual differentiation between conventional molten salts and ionic liquids is based on the melting point. While most molten salts have melting points higher than 200°C, ionic liquids normally melt below 100°C (Wasserscheid & Keim, 2000). The most unusual characteristic of these systems is that, although they are liquids, they present features similar to solids, such as structural organization at intermediate distances (Dupont, 2004) and negligible vapor pressure (Earle, 2006). This structural organization can act as a template like system, and PANI nanofibers are obtained when the aniline is polymerized in these media.

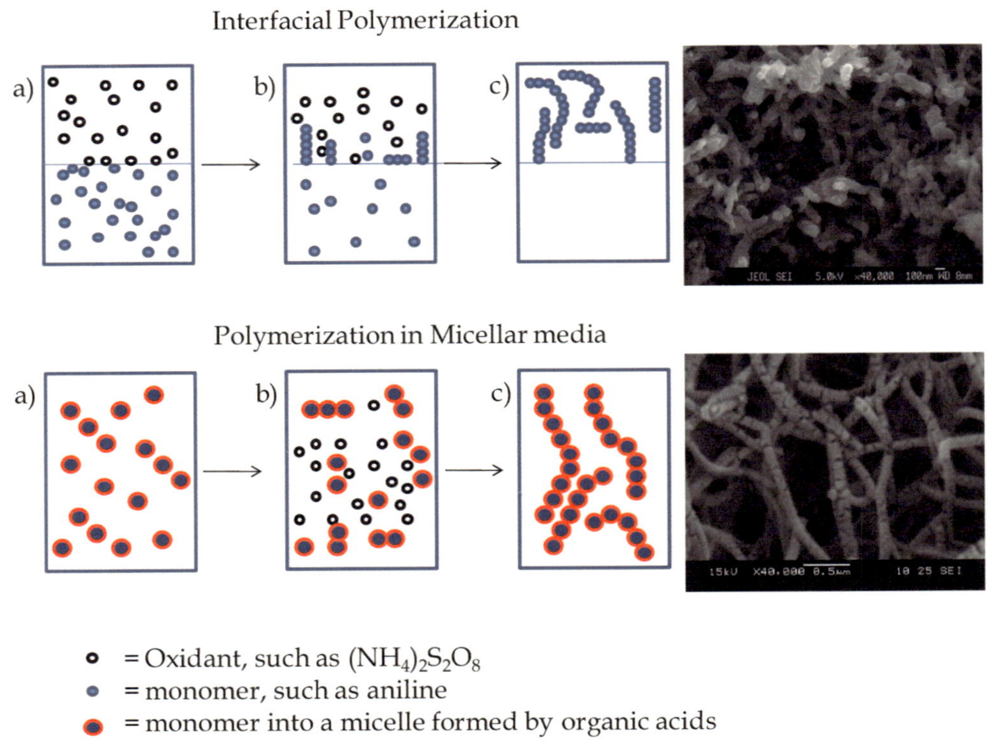

Fig. 4. Schematic representation of the two polymerization ways that are commonly employed in the preparation of PANI with nanofiber or nanotube morphology. -The interfacial polymerization can be drawn in three steps: a) The top layer is an aqueous solution of 1.0 M HCl acid and ammonium peroxydisulfate (others acids or oxidants can be used); the bottom layer is aniline dissolved in the organic solvent chloroform (others solvents immiscible in water can be used). b) Starting the polymerization and migration of oligomers from organic bottom layer to the aqueous top layer, and c) formation of PANI. The scanning electron microscopic (SEM) image was obtained from the PANI powder obtained from interfacial polymerization using HCl, $(NH_4)_2S_2O_8$, and chloroform. The nanofibers have ca. 30 nm of diameter. The SEM image was obtained from a powdered sample dispersed over a drop of silver glue, and recorded in a Field Emission Gun (SEM, JSM-6330), operated with a high-tension voltage of 5 kV. -The micellar polymerization can be visualized in three steps: a) solubilization of aniline in an aqueous solution containing organic acids that act as surfactants. b) addiction of oxidant and c) polymerization which depending on the concentration of aniline in solution, it is possible to form hollow nanofibers (as named nanotubes) or nanofibers. The SEM image was obtained from the PANI powder obtained from micellar polymerization using β-naphtalenesulfonic acid (β-NSA), $(NH_4)_2S_2O_8$, and molar ratio of β-NSA:aniline of 1:4. The nanofibers have ca. 93 nm of diameter. SEM image was obtained on JEOL low-vacuum SEM (JSM-5900LV operated with 15 kV), with samples covered with 16nm of sputtered gold film.

Thus, the template-free methods, such as interfacial, seeding and micellar can be employed as different "bottom-up" approaches to obtain pure PANI nanofibers. The possibility to prepare nanostructured PANI by self-assembly with reduced post-synthesis processing warrants further study and application of these materials, especially in the field of electronic nanomaterials. In this chapter this amazing new area of polyaniline nanofibers will be reviewed concerning the state-or-art results of spectroscopic characterization of their structural, electronic and vibrational features. Previous and new results of the spectroscopy of PANI nanofibers, obtained by our group, using Resonance Raman will be considered. Special attention will be given in the correlation of PANI nanofibers morphological stabity and their spectroscopic features. The main goal of this work is to contribute in the rationalization of some important results obtained in the open area of PANI nanofibers.

3. Vibrational spectroscopy of PANI

The resonance Raman spectroscopy (RR), unlike the absorption spectroscopy in the infrared region, uses as radiation source a laser with higher energy (usually light with wavelengths in the visible region of the electromagnetic spectrum) than the corresponding vibrational transitions. However, due to the scattering of the incident light (the Raman process), the vibrational frequencies (Raman bands) can be probed. Generally, the intensities of the Raman bands are linearly proportional to the intensity of the incident light (I_o, see Fig. 4), proportional to the fourth power of the wavelength of the scattered light (λ_s^4 or ν_s in wavenumber units, see Fig. 4), and proportional to the square of the polarizability tensor ($[\alpha]^2$) (Sala, 1995). The situation changes dramatically, when the laser line falls within the region of a permitted electronic transition. The Raman intensities associated with vibrational modes which are tightly coupled or associated with the excited electronic state can suffer a tremendous increase of about 10^5 powers; this is what characterizes the resonance Raman effect. (see Fig. 4). The mathematical and theoretical backgrounds used to the interpretation of the resonance Raman behavior can be found extensively in the literature (Clark & Dines, 1986; Batchelder, 1987; Batchelder, 1984; McHale, 1999). Generally, the tensor of polarizability is described as shown in the Fig.4. The equation is formed in the numerator part by transition dipole moment integrals between the electronic ground state (g, for the vibrational m or n states) and an excited electronic state (e, for any vibrational v states). The sum is done over all possible (e,v) states. In the denominator part is the difference or sum of the scattered and incident light, added by the dumping factor ($i\Gamma_{ev}$) that contents information about the lifetime of the transition states. The theoretical formalism developed by Albrecht et al. commonly employed (Clark & Dines, 1986; Batchelder, 1987; Batchelder, 1984; McHale, 1999). This enormous intensification makes, in principle, the Raman spectrum easy to be acquired. But, in a state of resonance, a lot of radiation is absorbed, leading to a local heating and frequently can be observed a decomposition of the conducting polymer. Despite of this problem, the RR spectroscopy has been largely used in the study of the different chromophoric units present in polyaniline and others conducting polymers, just by tuning an appropriate laser radiation to an electronic transition of the polymer.

In the pioneering work, (Sacricifti & Kusmany, 1987) showed that the intensity of the Raman spectra obtained for the electrochemically prepared PANI at 457.9 nm (E_{laser}= 2.73 eV) is increased when the PANI is reduced (see Fig. 6). The Raman spectrum of PANI in the fully reduced state (applied potential of -100 mV) was identified as being fully formed by benzenoid rings. In contrast, the intensity of the Raman spectra obtained for PANI at 632.8 nm (E_{laser}= 1.97 eV) increased when PANI was oxidized. In the oxidized form (potential

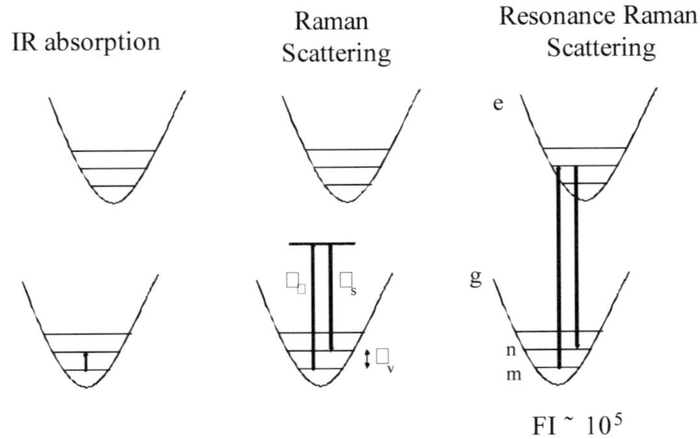

$$I_{Raman} \propto I_o v_s^4 \sum_{\rho\sigma} \left|[\alpha_{\rho\sigma}]_{gn,gm}\right|^2$$

$$[\alpha_{\rho\sigma}]_{gn,gm} = \frac{1}{hc}\sum_{ev}\left[\frac{\langle n|[\mu_\rho]_{ge}|v\rangle\langle v|[\mu_\sigma]_{eg}|m\rangle}{\nu_{ev,gm}-\nu_0+i\Gamma_{ev}} + \frac{\langle n|[\mu_\sigma]_{ge}|v\rangle\langle v|[\mu_\rho]_{eg}|m\rangle}{\nu_{ev,gn}+\nu_0+i\Gamma_{ev}}\right]$$

Fig. 5. Schematic representation of two electronic states (ground and excited) and their respective vibrational levels (the electronic and vibrational levels are not in the same scale). The arrows indicated the types of transitions that can be occurred among the different levels. It is important to say that in the case of Raman scattering, if the used laser line (λ_o, or as wavenumber, represent by v_o) has energy similar to one electronic transition of the molecule, the signal can be intensified by resonance process, know as resonance Raman Effect. In the Figure v_o and v_s (the scattered frequency is composed by: $v_{ev,gm}$ and $v_{ev,gn}$, the stokes and anti-stokes components, respectively) are the laser line and the scattered frequencies, respectively (for illustration purposes, just the Stokes, $v_s \langle v_o$, component is shown in the diagram). It was given the equations that describe the Raman Intensity and also the tensor of polarizability. The equation is formed in the numerator part by transition dipole moment integrals between the electronic ground state (g, for the vibrational m or n states) and an excited electronic state (e, for any vibrational v states). The sum is done over all possible (e,v) states. In the denominator part is the difference or sum of the scattered and incident light, added by the dumping factor ($i\Gamma_{ev}$) that contents information about the lifetime of the transition states.

applied +600 mV) three Raman bands (1160, 1490 and 1595 cm^{-1}) were identified as characteristics of the quinoid structure of PANI. Another important early work was carried out by (Furukawa et al., 1988; Furukawa et al. 1996). In this work the authors prepared PANI in different oxidation and protonation states and an extensive characterization was realized. Figure 6 presents the segments of PANI and its characteristic Raman bands at their corresponding exciting radiation.

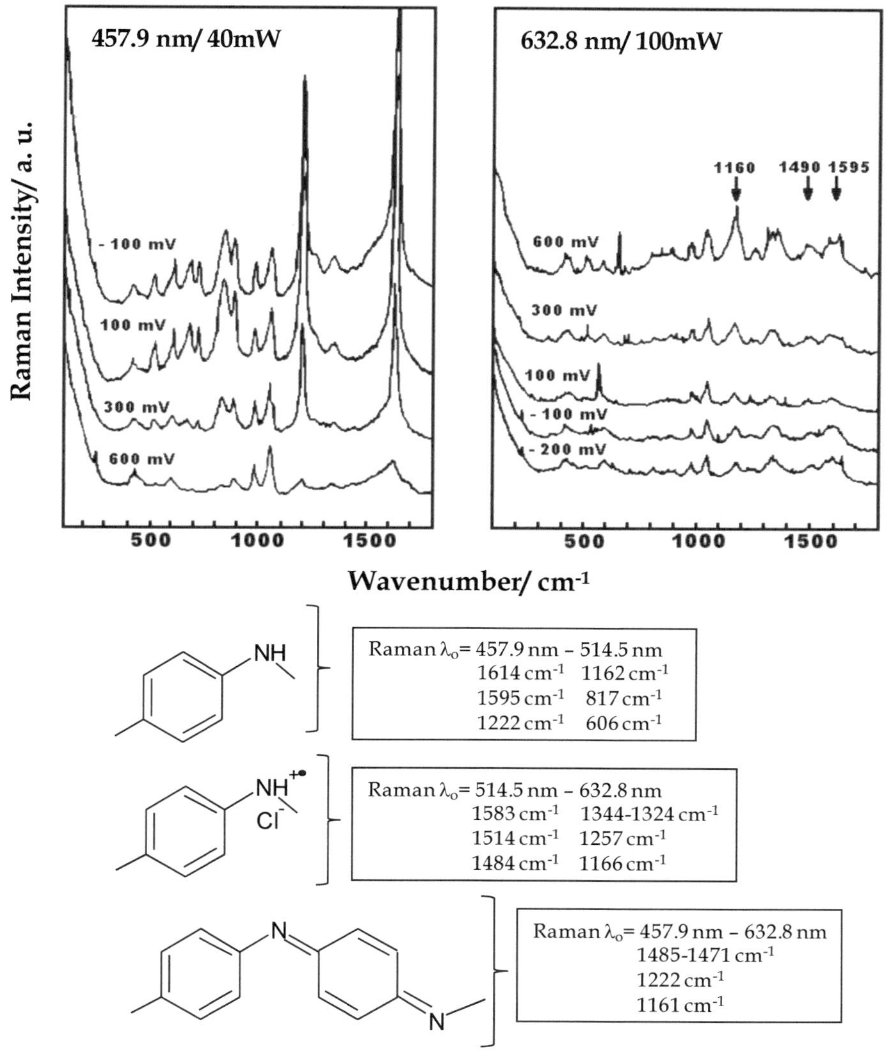

Fig. 6. Top: Raman spectra of PANI in different oxidation stages (the applied voltage is indicated in the figure) at indicated laser line (457.9 nm and 632.8 nm). Reproduction authorized by Elsevier (Sacricifti & Kusmany, 1987). Bellow: schematic representation of segments of PANI and its characteristic Raman bands at indicated laser lines (Furukawa et al., 1988).

(Quillard et al., 1994; Berrada et al. 1995) performed a normal coordinate analysis of PANI in its leucoemeraldine, emeraldine, and pernigraniline base forms. The leucoemeraldine base is characterized by the vibrational modes of the benzene ring in 1618 and 1181 cm^{-1}, which can be attributed to the νCC stretch and βCH angular deformation, respectively. The amine group is characterized by νCN stretch at 1220 cm^{-1}. For pernigraniline the βCH band value is at 1157 cm^{-1}, and the authors suggest that this band can be used as a qualitative measure of the degree of oxidation of the PANI chain. Another characteristic band of pernigraniline base is the stretch of C=N bond at 1480 cm^{-1}. Another way to determine the degree of oxidation of PANI, proposed by (Quillard et al., 1994; Berrada et al., 1995; Asturias et al., 1989), consists in determination of the intensities of the bands at about 1500 cm^{-1} for leucoesmeraldina (νCC) and the band around 1600 cm^{-1} for pernigranilina (νC=C) observed in the infrared spectra. The intensity ratio between these two bands (I(1600)/I(1500)) is a way to determine qualitatively the degree of oxidation in the chain of PANI.

(Louarn et al., 1996) performed a normal coordinate analysis of the emeraldine PANI salt, through the analysis of the infrared and Raman data of PANI with selective deuteration. The normal coordinate calculation confirms the assignments made by other authors and allowed to correlate the band at 1515 cm^{-1} to the angular deformation of NH group. The authors proposed the existence of bipolaronic segments (dications or protonated imines) to explain the Raman spectrum of emeraldine PANI salt obtained with laser line at 632.8 nm. The existence of these segments was also suggested by (Huang & MacDiarmid, 1993; McCall et al., 1990) through UV-VIS-NIR data and by other authors through spectroelectrochemical measurements monitored by *in situ* EPR (Geniés, 1987). The main issue posted by (Louarn et al., 1996) is that the doublet nature of the CN stretch (ca. 1320-1350 cm^{-1}) remains unclear. However, the authors suggest, like (Furukawa et al., 1988) that the origin of the doublet may be associated with the existence of two different conformations of PANI.

The Raman study of PANI doped with camphorsulfonic acid (CSA) and dissolved in *m*-cresol (Da Silva et al., 1999a; Da Silva et al., 1999b), named Secondary doping by (MacDiarmid & Epstein, 1995; MacDiarmid & Epstein, 1994; Xia et al., 1995), revealed a conversion of dications to radical cations. This behavior is justified by the increase of the polaronic band in ca. 1336 cm^{-1} and decrease of the intensities of the Raman bands at 1486 cm^{-1} and 1380 cm^{-1}, assigned to νC=N and νC=C modes of dications units, respectively using laser line at 632.8 nm. (M. Cochet et al., 2000) also investigated by resonance Raman spectroscopy, the secondary doping in PANI-CSA, trying to verify by normal modes calculations if the changes in the torsion angle between the benzene rings could explain the modifications in the Raman spectrum of PANI-CSA treated with *m*-cresol. But, the data could not be interpreted only in terms of flatness of the rings; the changes observed are also associated with changes in the electronic structure, leading to the appearance of new Raman bands and the modifications of others, due to, the high charge delocalization on the polymeric chains.

The major part of the Raman studies of PANI use laser lines in the visible region, (Sacricifti & Kusmany, 1987; Furukawa et al., 1988; Hugot-Le Goff & Bernard, 1993; Louarn et al., 1996; Colomban et al., 1999; Zhang et al., 2005; Trchová et al., 2006; Do Nascimento et al., 2002a; Do Nascimento et al., 2002b) but some Raman studies of PANI using near-infrared (NIR) laser line is also found (Engert *et al.*, 1994; Quillard *et al.*, 1995; Niaura et al., 2004). The most peculiar feature observed at 1064.0 nm is the presence of a sharp band around 1375 cm^{-1} in emeraldine base PANI spectrum, which was correlated by (Engert *et al.*, 1994) to polaronic segments localized at two benzene rings. On the other hand, (Quillard *et al.*, 1995) proposed that this band was not correlated with protonated segments but with over-oxidized segments such as those present in PANI as the pernigraniline base (PANI-PB). Some controversial aspects about the Raman spectra of poly(aniline) obtained at NIR radiation excitation were recently re-examined by (Do Nascimento & Temperini, 2008). The bands from 1324 to 1375 cm^{-1} were associated to νC–N of polarons with different conjugation lengths and with the presence of charged phenazine-like and/or oxazine-like rings in PANI-ES as chemically prepared. The formation of cross-linking structures is associated with the

ES form of PANI. The bands from 1450 to 1500 cm^{-1} in the PANI-EB and PANI-PB spectra were associated with the νC=N mode of the quinoid units having different conjugation lengths.

The thermal behavior of PANI has been also studied by vibrational spectroscopy. The Raman studies of PANI-CSA conducted by (Da Silva et al., 2000; Do Nascimento et al., 2002b) revealed that there is the appearance of intense bands at 574, 1393 and 1643 cm^{-1} in the Raman spectra at 632.8 nm during heating. These same bands were observed in the Raman spectra of a conducting polymer derived from PANI, the poly(diphenylamine) doped with HCSA (PDFA-CSA) during heating. It was observed that these bands are not present during the heating of polymers in vacuum. By comparing the results obtained from the thermal monitoring of PANI-CSA and PDFA-CSA, it was possible to assign these bands to the reaction of the polymer with oxygen, with formation of chromophores with oxazine-like rings. It was also demonstrated that the increase of laser power at 1064.0 nm causes deprotonation of PANI-ES and formation of cross-linking segments having phenazine and/or oxazine-like rings. The formation of cross-linking structures is associated with the ES form of PANI.

4. Results and discussion

4.1 PANI nanofibers obtained from micellar media

PANI has been largely prepared as nanofiber with the use of polymerization media containing organic acids that acts as surfactants (Zhang et al., 2002; Qiu et al., 2001; Wei & Wan, 2002; Do Nascimento et al., 2006; Zhang et al., 2009). A broad variety of organic acid has been employed in order to modulate the diameter of the resulting nanofibers. The FTIR spectra of PANI doped with various organic acids, containing sulfo-group (SO_3-H), show broad bands at about 3430 cm^{-1}, 1560 cm^{-1}, 1480 cm^{-1}, 1130 cm^{-1}, and 800 cm^{-1}, which are related to emeraldine PANI salt (Huang & Wan, 1999). The UV-vis spectra of all doped PANI samples show two polaronic absorptions around 400 and 800 nm. The position of polaronic bands shifts to a long wavelength when the size of organic dopant increases. For instance, the polaron absorption for the PANI doped with smaller dopant (α-NSA) is located at 800-900 nm. On the other hand, the polaron absorption for the doped PANI with larger dopant (β-NSA) is shifted to 1060-1118 nm.

The resonance Raman (RR) spectra (from 1000 to 1800 cm^{-1}) obtained for PANI nanofibers doped with NSA (β-naphtalenesulfonic acid), with different diameters, show similar spectral profile. This result indicates that the morphological differences in PANI-NSA nanofibers have small influence in the Raman spectra from 1000 to 1800 cm^{-1}. Comparing the RR spectra of PANI-NSA fibers to PANI-ES spectrum, bands at 1163 and 1330 cm^{-1} in PANI-NSA spectra can be associated with those at 1165 and 1317-1337 cm^{-1} in PANI-ES spectrum. These bands have been assigned to βC-H and νC-N of polaronic segments, respectively (Do Nascimento et al., 2006). Their relative intensities in PANI-NSA spectra increase as the molar ratio of β-NSA:aniline increases. Thus, the RR data of the PANI-NSA nanofibers show that the spectral changes observed among the as-prepared PANI-NSA samples are owing to differences in the protonation degrees. Similarly, (Wang et al., 2008) observed in that the bands at 1257 and 1337 cm^{-1}, assigned to C-N stretching modes of benzenoid and radical segments of doped PANI, respectively, increase their intensities with the increase of stearic acid used as soft template.

The Raman spectra of PANI nanofibers prepared in micellar media also show the presence of bands at ca. 578, 1400, and 1632 cm^{-1}. These bands were strictly correlated with the formation of cross-linking structures in PANI chains after heating in the presence of air (Do Nascimento et al., 2002b). Different studies show that the bands at ca. 578, 1400, and 1632 cm^{-1} are similar to those observed for dyes with phenoxazine ring. The presence of phenoxazine rings in PANI backbone was also observed by (Trchová et al., 2006 and Stejskal et al., 2006) in the study of formation of polyaniline nanotubes under different acidic media. The authors concluded that the presence of phenoxazine units is crucial for stacking and stabilization of the nanotube wall of PANI. Probably, the π-π stacking formed by phenoxazine rings, in the PANI backbone prepared in micellar media, is one of the driving forces for the formation of PANI chains with extended conformation and PANI particles with one-dimensional (needles and/or nanofibers) morphology. Thus, the presence of cross-linking structures in the PANI chains may be one of the characteristics of chains formed in micellar polymerization (Do Nascimento et al., 2008b).

On the comparison of the spectral behavior of PANI nanofibers/nanotubes prepared with NSA (β-naphtalenesulfonic acid) or with DBSA (dodecybenzenesulfonic acid) indicates that polymeric chains have a certain degree of extended conformation due to the presence of free-carrier absorption in the UV-VIS-NIR spectra. Hence, the presence of 609 cm^{-1} band in the PANI-NSA and PANI-DBSA Raman spectra indicates that these samples have a certain degree of extended conformation. The band at 609 cm^{-1} can be assigned to a vibrational mode related to benzene deformations or torsions. Probably, this mode is sensible to changes of the dihedral angle between neighbors benzene rings, or in other words, sensible to the conformation of the PANI chains (Do Nascimento et al., 2008b).

Electron microscopic images reveal the loss of the fibrous morphology of PANI after treatment of PANI-NSA samples with HCl solution in order to acquire higher doping state (Do Nascimento et al., 2006). However, further studies reveal that submitting the PANI-NSA to heating treatment at 200°C, occurs the formation of a high degree of cross-linking structures, verified by the appearance of characteristic RR bands at 578, 1398 and 1644 cm^{-1}, hence the fibrous morphology is retained after the doping process (Do Nascimento et al., 2008c). The structure of PANI-CSA (CSA is the 1*R*-(−)10-camphorsulfonic acid) nanofibers was also investigated by solid state ^{13}C and ^{15}N nuclear magnetic resonance (NMR) (Hopkins et al., 2008). The results (compared to the unstructured, granular form of PANI) revealed a slight variation in the structural features of the polymer that led to some differences in the chemical environments of the respective nuclei. The presence of two extra-sharp peaks at 96.5 and 179.8 ppm is a distinct feature found exclusively in the nanofiber spectra and the presence of a complete set of sharp NMR peaks is a consequence of the more ordered morphology in the PANI nanofibers.

PANI nanofibers synthesized in ionic liquids have been studied by Raman spectroscopy. (Wei et al., 2006) obtained nanofibers of PANI by electropolymerization of aniline in $BMIPF_6$ (1-butyl-3-methyl-imidazolium hexafluorophosphate). The Raman spectra show that the PANI is similar to the emeraldine salt form. However, the intensity of the quinoid ring stretching at 1578 cm^{-1} is higher than that of the benzenoid band at 1469 cm^{-1}, indicating the existence of a higher amount of quinoid structures. The authors (Wei et al., 2006) suggest that the PANI film synthesized in this ionic liquid media is formed by small amount of non-conducting forms such as emeraldine base and/or pernigraniline forms, which have more quinoid structures in the backbone. The bands at 1578, 1386, and 1340 cm^{-1} were assigned to emeraldine PANI base; and the bands at 1606, 1469, 1252, and 1177 cm^{-1} to emeraldine salt form.

4.2 PANI nanofibers obtained from interfacial polymerization

More recently, uniform nanofibers of pure metallic PANI (30-120 nm diameters, depending on the dopant) have also been prepared by polymerization at an aqueous organic interface (Huang & Kaner, 2004a; Huang & Kaner, 2004b). However, it was often observed that the quality of the products such as the shape and stability of their dispersions varies with different synthetic conditions, such as, stirring, temperature and pH, among others. (Colomban *et al.*, 1999; Cochet *et al.*, 2000a; Cochet et al., 2000b) have been demonstrated that the low energy Raman bands from 100 to 600 cm^{-1} in the PANI spectra are very sensible to the crystalline arrangement and also to the conformational changes of the PANI chains. The Raman spectra of PANI nanofibers prepared from interfacial polymerization were obtained at low wavenumbers. It was observed that the bands at 200 and 296 cm^{-1}, related to C_{ring}-N-C_{ring} deformation and lattice modes of polaron segments of PANI with type-I crystalline arrangement (Colomban *et al.*, 1999), practically disappears in the Raman spectra of PANI nanofibers. This effect is very pronounced for the nanofiber sample prepared using 5.0 $mol.L^{-1}$ HCl aqueous solution. The effects of HCl concentration over the bands at low wavenumbers (200 and 296 cm^{-1}) is strictly related to PANI samples prepared from interfacial polymerization. In addition, shifts of the bands from 393-427 cm^{-1} to 422-451 cm^{-1} are also seen for PANI nanofibers prepared with higher HCl concentration. (Cochet *et al.*, 2000a; Cochet et al., 2000b) demonstrated that mainly the bands at about 200-500 cm^{-1} are very sensible to the conformational changes. The shift to higher wavenumbers of the bands at about 400 cm^{-1} indicates the increase of the torsion angles of the C_{ring}-N-C_{ring} segments. Thus, the decrease in the intensities of the bands at 200 and 296 cm^{-1} and also the shift of the bands at about 400 cm^{-1} are due to the conformational changes in the PANI backbone. The FTIR spectra for PANI nanofibers display higher changes in the region from 2000 to 4000 cm^{-1} (Do Nascimento et al., 2008a). Mainly the bands related to NH_2^+ modes at 2480, 2830, and 2920 cm^{-1} increase in their intensities for PANI samples prepared with higher HCl concentration (higher than 1.0 $mol.L^{-1}$), consequence of the increase of protonated imine and amine nitrogens in the structure of PANI. The bands at about 3200 and 3450 cm^{-1} also change their relative intensities, being correlated to N-H stretching modes (the contribution of O-H stretching, due to the adsorbed water, is similar for all analyzed samples). The band at 3200 and 3450 cm^{-1} can be assigned to bonded N-H and free N-H stretching modes (Jana et al., 2003; Lunzy & Banka, 2000), showing that the contribution of the different types of hydrogen bonds are also changed by the increase of acidity of the synthetic media. The changes in the intensities associated with an increase in the torsion angles of C_{ring}-N-C_{ring} segments is owing to the formation of bipolarons (protonated, spinless units) in the PANI backbone higher than the PANI samples prepared by the conventional route (Geniés et al., 1990). The nanostructured surface of PANI permits major diffusion of the ions inside the polymeric matrix leading to a more effective protonation of the polymeric chain than the PANI prepared in the conventional way, leading to the reduction of crystallinity of PANI, and the decrease in the amount of nanofibers (Do Nascimento et al., 2008a).

5. Final remarks

The screening of the electronic and vibrational structure of the polyaniline nanofibers has been decisive in the studies related to the formation, interactions between the chains, properties and stabilities of the nanostructured polyaniline. Nowadays, two great approaches are used to acquire the PANI with nanostructured morphology without the use of rigid hosts: (i) polymerization of aniline in a micellar media and (ii) polymerization of aniline on the interface between two solvents. However, the morphology of PANI obtained without rigid hosts is more susceptible to the synthetic conditions (such as pH) and also post-synthesis procedures. Mainly, it is observed shifts in the vibrational frequencies of polyaniline and also variations in their intensities. The presence of bands owed to phenoxazine rings is observed in PANI backbone formed in micellar media. The presence of phenoxazine units is crucial for stacking and stabilization of the nanotube wall of PANI. Probably, The π-π stacking formed by phenoxazine rings, in the PANI backbone prepared in micellar media, is one of the driving forces for the formation of PANI chains with extended conformation and PANI particles with one-dimensional (needles and/or nanofibers) morphology. Thus, the presence of cross-linking structures in the PANI chains may be one of the characteristics of chains formed in micellar polymerization. The changes in the intensities of the vibrational spectra at low energies are associated with an increase in the torsion angles of C_{ring}-N-C_{ring} segments due to the formation of bipolarons (protonated, spinless units) in the PANI backbone higher than the PANI samples prepared by the conventional route. The nanostructured surface of PANI permits major diffusion of the ions inside the polymeric matrix leading to a more effective protonation of the polymeric chain than the PANI prepared in the conventional way, leading to the reduction of crystallinity of PANI, and the decrease in the amount of nanofibers.

6. References

Asefa, T.; MacLachan, M. J.; Coombs, N. & Ozin, G. A. (1999). Periodic mesoporous organosilicas with organic groups inside the channel walls. *Nature*, 402, 867-871.

Bacon, J. & Adams, R. N. (1968). Anodic oxidation of aromatic amines. 3. Substituted anilines in aqueous media. *J. Am. Chem. Soc.*, 90, 24, 6596-6599, 0002-7863.

Chiang, C. K.; Druy, M. A.; Gau, S. C.; Heeger, A. J.; Louis, E. J.; MacDiarmid, A. G.; Park, Y. W. & Shirakawa, H. (1978). Synthesis of highly conducting films of derivatives of polyacetylene (CH)x. *J. Am. Chem. Soc.* 100, 3, 1013-1015, 0002-7863.

Chiang, C. K.; Fincher Jr, C. R.; Park, Y. W.; Heeger, A. J.; Shirakawa, H.; Louis, E. J. & MacDiarmid, A. G. (1977). Electrical-conductivity in doped polyacetylene. *Phys. Rev. Lett.* 39, 17, 1098-1101, 31-9007.

Cochet, M.; Louarn, G.; Quillard, S.; Boyer, M. I.; Buisson, J. P. & Lefrant, S. (2000a). Theoretical and experimental vibrational study of polyaniline in base forms: non-planar analysis. Part I. *J. Raman Spectrosc.*, 31, 11, 1029–1039, 0377-0486.

Cochet, M.; Louarn, G.; Quillard, S.; Buisson, J. P. & Lefrant, S. (2000b). Theoretical and experimental vibrational study of emeraldine in salt form. Part II. *J. Raman Spectrosc.* 31, 11, 1041–1049, 0377-0486.

Colomban, Ph.; Folch, S. & Gruger, A. (1999). Vibrational study of short-range order and structure of polyaniline bases and salts. *Macromolecules*, 32, 9, 3080-3092, 0024-9297.

Davis Jr., J. H.; Gordon, C. M.; Hilgers, C. & Wasserscheld, P. (2002). Synthesis and Purification of Ionic Liquids, In: *Ionic Liquids in Synthesis*, Wassercheid, P. & Welton, T. (Eds.), 7-40, Wiley–VCH, 3-527-60070-1, New York.

Do Nascimento, G. M.; Constantino, V. R. L. & Temperini, M. L. A. (2002a). Spectroscopic characterization of a new type of conducting polymer-clay nanocomposite. *Macromolecules* 37, 20, 7535-7537, 0024-9297.

Do Nascimento, G. M.; Da Silva J. E. P.; De Torresi, S. I. C. & Temperini, M. L. A. (2002b). Comparison of secondary doping and thermal treatment in poly(diphenylamine) and polyaniline monitored by resonance Raman spectroscopy. *Macromolecules*, 35, 1, 121-125, 0024-9297.

Do Nascimento, G. M.; Constantino, V. R. L.; Landers, R. & Temperini, M. L. A. (2004). Aniline polymerization into montmorillonite clay: A spectroscopic investigation of the intercalated conducting polymer. *Macromolecules*, 37, 25, 9373-9385, 0024-9297.

Do Nascimento, G. M.; Silva, C. H. B. & Temperini, M. L. A. (2006). Electronic structure and doping behavior of PANI-NSA nanofibers investigated by resonance Raman spectroscopy. *Macromol. Rapid Commun.*, 27, 4, 255-259, 1022-1336.

Do Nascimento, G. M.; Kobata, P. Y. G. & Temperini, M. L. A. (2008a). Structural and vibrational characterization of polyaniline nanofibers prepared from interfacial polymerization. *J. Phys. Chem. B*, 112, 37, 11551-11557, 1520-6106.

Do Nascimento, G. M.; Silva, C. H. B.; Izumi, C. M. S. & Temperini, M. L. A. (2008b). The role of cross-linking structures to the formation of one-dimensional nano-organized polyaniline and their Raman fingerprint. *Spectrochim. Acta A*, 71, 3, 869-875, 1386-1425.

Do Nascimento, G. M.; Silva, C. H. B. & Temperini, M. L. A. (2008c). Spectroscopic characterization of the structural changes of polyaniline nanofibers after heating. *Polym. Degrad. Stab.*, 93, 1, 291-297, 0141-3910.

Do Nascimento, G. M. & Temperini, M. L. A. (2008). Studies on the resonance Raman spectra of polyaniline obtained with near-IR excitation. *J. Raman Spectrosc.*, 39, 7, 772-778, 0377-0486.

Dupont, J. (2004). On the solid, liquid and solution structural organization of imidazolium ionic liquids. *J. Braz. Chem. Soc.*, 15, 3, 341-350, 0103-5053.

Earle, M. J.; Esperança, J. M. S. S.; Gilea, M. A.; Lopes, J. N. C.; Rebelo, L. P. N.; Magee, J. W.; Seddon, K. R. & Wildegren, J. A. (2006). The distillation and volatility of ionic liquids. *Nature*, 439, 7078, 831-834, 0028-0836.

Engert, C.; Umapathy, S.; Kiefer, W. & Hamaguchi, H. (1994). Dynamic structure of charge-carrier in polyaniline by near-infrared excited resonance Raman spectroscopy. Chem. Phys. Lett. 218, 1-2, 87-92, 0009-2614.

Furukawa, Y.; Ueda, F.; Hyodo, Y.; Harada, I.; Nakajima, T. & Kawagoe, T. (1988). Vibrational-spectra and structure of polyaniline. *Macromolecules*, 21, 5, 1297-1305, 0024-9297.

Gao, H.; Jiang, T.; Han, B.; Wang, Y.; Du, J.; Liu, Z. & Zhang, J. (2004). Aqueous/ionic liquid interfacial polymerization for preparing polyaniline nanoparticles. *Polymer*, 45, 9, 3017-3019, 0032-3861.

Geniés, E. M.; Boyle, A.; Lapkowski, M. & Tsintavis, C. (1990). Polyaniline- A historical survey. *Synth. Met.*, 36, 2, 139-182, 0379-6779.

Han, C. C. & Elsenbaumer, R. L. (1989). Protonic acids- generally applicable dopants for conducting polymers. *Synth. Met.* 30, 1, 123-131, 0379-6779.

Heeger, A. J. (2001). Semiconducting and Metallic Polymers: The Fourth Generation of Polymeric Materials (Nobel Lecture) *Angew. Chem. Int. Ed.* 40, 14, 2591-2611, 1433-7851.

Hopkins, A. R.; Lipeles, R. A. & Hwang, S. J. (2008). Morphology characterization of polyaniline nano- and microstructures. *Synth. Met.* 158, 14, 594-601, 0379-6779.

Huang, J. & Kaner, R. B. (2004a). Nanofiber Formation in the chemical polymerization of aniline: a mechanistic study. *Angew. Chem. Int. Ed.*, 43, 43, 5817–5821, 1433-7851.

Huang, J. & Kaner, R. B. (2004b). A general chemical route to polyaniline nanofibers. *J. Am. Chem. Soc.*, 126, 3, 851–855, 0002-7863.

Huang, J. & Wan, M. (1999). Polyaniline doped with different sulfonic acids by in situ doping polymerization, *J. Polym. Sci.: Part A: Polym. Chem.*, 37, 1277–1284, 0887-624X.

Hugot-Le Goff, A. & Bernard, M. C. (1993). Protonation and oxidation processes in polyaniline thin-films studied by optical multichannel analysis and in-situ Raman-spectroscopy. *Synth. Met.*, 60, 2, 115-131, 0379-6779.

Jana, T.; Roy, S. & Nandi, A. K. (2003). High temperature d.c. conductivity of sulfonic acid doped thermoreversible polyaniline gels. *Synth. Met.* 132, 3, 257-264, 0379-6779.

Kusmany, H. & Sacriftci, N. S. (1987). In situ spectro-electrochemical studies of polyaniline. *Synth. Met.*, 18, 1-3, 353-358, 0379-6779.

Liepins, R. & Ku, C. C. (1987). *Electrical Properties of Polymers: chemical principles,* Hanser Publishers, 0029476313, New York.

Louarn, G.; Lapkowski, M.; Quillard, S.; Pron, A.; Buisson, J. P. & Lefrant, S. (1996). Vibrational properties of polyaniline - Isotope effects. *J. Phys. Chem.*, 100, 17, 6998-7006, 22-3654.

Lunzy, W. & Banka, E. (2000). Relations between the structure and electric conductivity of polyaniline protonated with camphorsulfonic acid. *Macromolecules* 33, 2, 425-429, 0024-9297.

MacDiarmid, A. G. (2001). "Synthetic Metals": a novel role for organic polymers (Nobel Lecture). *Angew. Chem. Int. Ed.* 40, 14, 2581-2590, 1433-7851.

MacDiarmid, A. G.; Chiang, J. C.; Richter A. F. & Sonosiri, N. L. D. (1987). *Conducting Polymers,* Alcácer, L. (Ed.), Reidel Publications, Dordrecht.

MacDiarmid, A. G. & Epstein, A. J. (1989a). Polyanilines- A novel class of conducting polymers. *Faraday Discuss. Chem. Soc.*, 88, 317-325, 1364-5498.

MacDiarmid, A. G. & Epstein, A. J. (1989b). The polyanilines: a novel class of conducting polymers In: *Conducting polymers, emerging technologies,* 27-35, Technical Insights, New Jersey.

MacDiarmid, A. G. & Epstein, A. J. (1994). Conducting polymers: Science and Technology, In: *Frontiers of Polymers and Advanced Materials,* Prasad, P. N. (Ed.), 251-261, Plenum Press, 978-0306447167, New York.

Mohilner, D. M.; Argersinger, W. J. & Adams, R. N. (1962). Investigation of kinetics and mechanism of anodic oxidation os aniline in aqueous sulfuric acid solution at a platinum electrode. *J. Am. Chem. Soc.*, 84, 19, 3618-3622, 0002-7863.

Niaura, G.; Mazeikiene, R. & Malinauskas, A. (2004). Structural changes in conducting form of polyaniline upon ring sulfonation as deduced by near infrared resonance Raman spectroscopy. Synth. Met. 145, 2-3, 105-112, 0379-6779.

Nigrey, P. J.; MacDiarmid, A. G. & Heeger, A. J. (1979). Electrochemistry of polyacetylene, (CH)X- Electrochemical doping of (CH)X films to the metallic state. *J. Chem. Soc.: Chem. Commun.* 14, 594-595, 0022-4936.

Peng, H.; Tang, J.; Pang, J.; Chen, D.; Yang, L.; Ashbaugh, H. S.; Brinker, C. J.; Yang Z. & Lu, Y. (2005). Polydiacetylene/Silica nanocomposites with tunable mesostructure and thermochromatism from diacetylenic assembling molecules. *J. Am. Chem. Soc.*, 127, 37, 12782-12783.

Qiu, H. J.; Wan, M. X.; Matthews, B. & Dai, L. M. (2001). Conducting polyaniline nanotubes by template-free polymerization. *Macromolecules*, 34, 4, 675-677, 0024-9297.

Quillard, S.; Berrada, K.; Louarn, G.; Lefrant, S.; Lapkowski, M. & Pron, A. In-situ Raman spectroscopic studies of the electrochemical behavior of polyaniline. (1995). *New J. Chem.*, 19, 4, 365-374, 1144-0546.

Rodrigues, F.; Do Nascimento, G. M. & Santos, P. S. (2007). Dissolution and doping of polyaniline emeraldine base in imidazolium ionic liquids investigated by spectroscopic techniques. *Macromol. Rapid Commun.*, 28, 5, 666-669, 1022-1336.

Sasaki, K.; Yaka, M.; Yano, J.; Kitani, A. & Kumai, A. (1986). Growth-mechanism in the electropolymerization of aniline and p-aminodiphenylamine. *J. Electroanal. Chem.*, 215, 1-2 401-407, 0022-0728.

Shirakawa, H. (2001). The discovery of polyacetylene film: the dawning of an era of conducting polymers (Nobel Lecture). *Angew. Chem. Int. Ed.* 40, 14, 2575-2580, 1433-7851.

Shirakawa, H. & Ikeda, S. (1971). Infrared spectra of poly(acetylene). *Polymer Journal* 2, 2, 231-244, 0032-3896.

Shirakawa, H. & Ikeda, S. (1974). Cyclotrimerization of acetylene by tris(acetylacetonato)titanium(III)-diethylaluminum chloride system. *J. Poylm. Sci. Chem.* 12, 5, 929-937, 0887-624X.

Shirakawa, H.; Louis, E. J.; MacDiarmid, A. G.; Chiang, C. K. & Heeger, A. J. (1977). Synthesis of electrically conducting organic polymers-halogen derivatives of polyacetylene, (CH)X. *J. Chem. Soc.: Chem. Commun.* 16, 578-580, 0022-4936.

Song, J.; Cisar J. S. & Bertozzi, C. R. (2004). Functional self-assembling bolaamphiphilic polydiacetylenes as colorimetric sensor scaffolds. *J. Am. Chem. Soc.*, 126, 8459-8465.

Stejskal, J.; Sapurina, I.; Trchova, M.; Konyushenko, E. M. & Holler, P. (2006). The genesis of polyaniline nanotubes. *Polymer*, 47, 16, 8253-8262, 0032-3861.

Syed, A. A. & Dineson M. K. (1991). Polyaniline- A novel polymeric material- review. *Talanta*, 38, 8, 815-837, 0039-9140.

Trchová, M.; Matejka, P.; Brodinova, J.; Kalendova, A.; Prokes. J. & Stejskal, J. (2006). Structural and conductivity changes during the pyrolysis of polyaniline base. *Polym. Degrad. Stab.*, 91, 1, 114-121, 0141-3910.

Trchová, M.; Syedenkova, I.; Konyushenko, E. N.; Stejskal, J.; Holler, P. & Ciric-Marjanovic, G. (2006). Evolution of polyaniline nanotubes: The oxidation of aniline in water. *J. Phys. Chem. B*, 110, 19, 9461-9468, 1520-6106.

Wang, X.; Liu, J.; Huang, X.; Men, L.; Guo, M. & Sun, D. (2008). Controlled Synthesis of Linear Polyaniline Tubes and Dendritic Polyaniline Fibers with Stearic Acid. *Polym. Bull.* 60, 1, 1–6, 0170-0839.

Wasserscheid, P. & Keim, W. (2000). Ionic liquids - New "solutions" for transition metal catalysis. *Angew. Chem. Int. Ed.*, 39, 21, 3773-3789, 1433-7851.

Wawzonek, S. & McIntyre, T. W. (1967). Electrolytic oxidation of aromatic amines. *J. Electrochem. Soc.*, 114, 10, 1025-1029, 0013-4651.

Wei, D.; Kvarnstrom, C.; Lindfors, T. & Ivaska, A. (2006). Polyaniline nanotubules obtained in room-temperature ionic liquids. Electrochem. Commun. 8, 1563–1566,

Wei, Y.; Sun, Y. & Tang, X. (1989). Auto-acceleration and kinetics of electrochemical polymerization of aniline. *J. Phys. Chem.*, 93, 12, 4878-4881, 0022-3654.

Wei, Z. X. & Wan, M. X. (2002). Hollow microspheres of polyaniline synthesized with an aniline emulsion template. *Adv. Mater.*, 2002, 14, 18, 1314-1317, 0935-9648.

Zhang, H.; Wang, J.; Wang, Z.; Zhang, F. & Wang, S. (2009). A Novel Strategy for the Synthesis of Sheet-Like Polyaniline. *Macromol. Rapid Commun.*, 30, 1577-1582, 1022-1336.

Zhang, J.; Liu, C. & Shi, G. (2005). Raman spectroscopic study on the structural changes of polyaniline during heating and cooling processes. *J. Appl. Polym. Sci.*, 96, 3, 732-739, 0021-8995.

Zhang, Z. M.; Wei, Z. X. & Wan. M. X. (2002). Nanostructures of polyaniline doped with inorganic acids. *Macromolecules,* 35, 15, 5937-5942, 0024-9297.

19

Fabrication of Ceramic Nanofibers Using Atrane Precursor

Bussarin Ksapabutr[1,2] and Manop Panapoy[1,2]
[1]*Department of Materials Science and Engineering, Faculty of Engineering and Industrial Technology, Silpakorn University, Nakhon Pathom 73000*
[2]*National Center of Excellence for Petroleum, Petrochemicals and Advanced Materials, Chulalongkorn University, Bangkok, 10330*
Thailand

1. Introduction

One-dimensional (1D) ceramic nanostructures, such as nanofibers, nanotubes, nanorods, nanowires, nanorings, nanobelts, and nanowhiskers, have been a very exciting and promising research topic in the last decades. This interest has been aroused by their unique properties and a wide variety of potential applications in future nanoelectronics and functional nanodevices, such as optoelectronics, photonics, cosmetics, health, bioengineering, gas and humidity sensors, mechanics, and catalysis (Xia et al., 2003; Wang, 2004; Carotta et al., 2007; Kumar & Ramaprabhu, 2007; Luo et al., 2007). A large number of techniques are available to prepare these materials, including vapour-phase method (Pan et al., 2001; Zhang et al., 2002), vapour-liquid-solid method (Morales & Lieber, 1998; Duan & Lieber, 2000), solution-liquid-solid method (Trentler et al., 1995; Markowitz et al., 2001), template-assisted method (Martin, 1994; Han et al., 1997; Zach et al., 2000; Barbic et al., 2002), solvothermal method (Heath & LeGoues, 1993; Wang & Li, 2002), arc discharge (Iijima, 1991), pulsed laser ablation (Sasaki et al., 2004), and precursor thermal decomposition (Wang et al., 2002a). Nevertheless, developing simpler and versatile approaches to the synthesis of one-dimensional ceramic nanostructures still remains a challenge. Among various approaches, electrostatic spinning (electrospinning) seems to be the simplest and most versatile technique capable of fabricating one-dimensional nanostructures including polymer, ceramic and composite nanofibers (Huang et al., 2003, Shao et al., 2003, Ksapabutr et al., 2005a ; Ksapabutr et al., 2005b; Li et al., 2006; Sigmund et al., 2006; Panapoy et al., 2008a; Panapoy et al., 2008b). Like a number of conventional spinning processes, electrospinning involves the ejection of a viscous solution (or melt) from an orifice and the subsequent drawing and solidification of the jet to form thin fibers. Unlike the conventional spinning processes, the exclusive driving force for the formation of ultrafine fiber in the electrospinning process is the electrostatic interaction. The fundamental principle of electrospinning is that a Taylor cone is formed by applying an electrical field to the viscous solution or melt hanging from a capillary tip, which causes jets of an electrically charged solution to be emitted when the applied electrostatic force exceeds the surface tension of the solution. A liquid jet shoots away from the capillary tip towards an oppositely charged electrode and solidifies with evaporation of solvent to form a mat on the grounded collector (Wang et al., 2002b; Berkland et al., 2004; Seong et al., 2005; Kalayci et al., 2005). Because of the small pore size and high surface to volume ratio inherent in electrospun nanofibers, these fibers show promise for interesting applications in fields of filtration, protective clothing, self-cleaning, drug delivery, tissue engineering scaffolds, wound dressings, biomimetic materials, composite reinforcement, electronic, optical and photonic devices, etc. (Remant et al., 2000; Guan et al., 2003; Li et al., 2003; Yuh et al., 2005; Zhang et al., 2007; Qi et al., 2008).

Zirconia (ZrO_2) is a well-known material with extreme refractoriness, high mechanical strength and fracture toughness, thermal insulation, wear and erosion resistance, chemical durability and alkali resistance. It has received intensive attention, due to its potential application in diverse fields, particularly in functional ceramics, electronic ceramics, constructional ceramics, biological ceramics, functional high-temperature coatings, heterogeneous catalysis processes, fuel cell technology, and so on (Cheung & Gates, 1997; Chadwick, 2000; Park et al., 2000; Feng et al., 2004). Its applications range from thin films to particles, platelets, whiskers and fibers. One-dimensional zirconia nanofibers have attracted considerable attention because of their surface dependant and size dependant properties and wide potential applications.

A variety of methods for the preparation of zirconia fibers have been reported, including relic process, in which organic polymer fibers are impregnated with zirconium salts, dried, and heated to burn out organic material and form zirconia fibers, spinning of the fiber precursor from organic polymer solution containing zirconium salt or fine zirconia particles with subsequent calcination process, and dry spinning of polyzirconoxane, or sol solutions prepared by sol-gel processing of zirconium acetate or alkoxide followed by calcination (Abe et al., 1998; Li et al., 2007). However, many efforts have been made to synthesize zirconia nanofibers by the electrospinning using zirconium salt or fine zirconia particles followed by calcinations process. For example, Shao et al. (2004) were the first to prepare zirconia nanofibers with diameters of 50–200 nm using the electrospun thin fibres of poly(vinyl alcohol)/zirconium oxychloride composites as precursor and through calcination treatment at 800°C. The viscous gel was prepared from zirconium oxychloride, poly(vinyl alcohol), and distilled water in the weight ratio of 1:2:20. Jing et al. (2005) prepared electrospun zirconia nanofibers having an average diameter less than 40 nm after calcination at 500°C using a viscous solution of zirconium oxychloride / poly(vinyl pyrrolidone) /ethyl alcohol /distilled water in the weight ratio of 1:1:3:3. Zhang and Edirisinghe (2006) also prepared zirconia nanofibers down to about 200 nm by electrospinning a mixture of a zirconia suspension and a poly(ethylene glycol)–poly(ethylene oxide) solution after calcinations at 600°-1200°C. An acetate-stabilized suspension containing 20 wt% of 5–10 nm size zirconia particles in water (pH = 3.5) was used. The polymer solution was prepared by dissolving poly(ethylene glycol):poly(ethylene oxide) (1:2 weight ratio) in a water–tetrahydrofuran (10:1 by volume). Two to 30 wt% of the zirconia suspension was added to the polymer solution.

The sol-gel process is a technique which is applicable for forming ceramic materials. In this process, liquid precursor materials are reacted to form a sol which then polymerizes into an inorganic polymeric gel. Advantages of this process over other techniques of ceramics are: compositional and microstructure control, wide variety in shape fabrication, and low processing temperatures. Furthermore, the metal alkoxide precursors used in the sol-gel process give homogeneity and high purity of metal oxide products although they are greatly moisture sensitive and more expensive. Unfortunately, the sol-gel technique traditionally applies metal alkoxide and organic solvent as raw materials and requires the careful addition of water to prevent uncontrolled hydrolysis of metal alkoxide. Chelating agents, such as acetylacetone, are sometimes added to control hydrolysis during the gelation process, and the reactions often are carried out in nitrogen atmosphere. The synthesis of novel precursor is necessary for the development of metal alkoxide chemistry and the investigation in sol-gel field (Charoenpinijkarn et al., 2001; Ksapabutr et al., 2004a; Ksapabutr et al., 2004b; Ksapabutr et al., 2005c). Generally, the atrane complexes are performed by means of trans-esterification reaction, using alkoxide derivatives in non-aqueous dried solvent under an inert atmosphere. The procedure is quite complicated. Furthermore, alkoxide starting materials used for the synthesis of atrane complexes are commercially expensive (Menge et al., 1991; Nugent et al., 1994). Anodic dissolution of the metal can provide an easy and straightforward way to scale up synthesis of alkoxides for many metals. Reactions between oxides or hydroxides and aminoalcohols can also be cost-effective routes to new metal alkoxides of some elements. This route is termed as the "oxide one-pot synthesis" (OOPS) process. The oxide one-pot synthesis process is a cheap, simple and straightforward process for preparing highly moisture stable and pure metal alkoxide precursors using inexpensive and readily available starting materials (Charoenpinijkarn et al., 2001; Opornsawad et al., 2001; Ksapabutr et al., 2004c; Panapoy & Ksapabutr, 2008c). Interestingly, metal derivatives of triethanolamine are comparatively more hydrolytic resistant than their alkoxide analogues. The chelating nature of the triethanolamine and coordinative saturation achieved by the central atoms in the final products appear to be the main factor for their hydrolytic stability to retard the hydrolysis and condensation reaction rates in order to obtain homogeneous gels rather than precipitates. Zirconatrane, which is one of atrane complexes, is an aminoalkoxide derivative of zirconium metal. Its structure possesses a remarkable stability property, leading to more controllable chemistry and minimizing special handling requirement. This advantage has made zirconatrane a promising candidate for use in ceramic and composite processing. Therefore, the use of another zirconia source that possesses higher stability towards air and moisture could overcome such obstacles. The present study is aimed at exploring the use of zirconatrane as the zirconia source for the fabrication of one-dimensional zirconia fibers via a combination of sol-gel technique and electrospinning method with subsequent calcination.

2. Synthesis of zirconatrane precursor

Zirconatrane used as a precursor for preparing zirconia nanofibers was synthesized directly from inexpensive and abundant raw materials, zirconium hydroxide and triethanolamine, via the oxide one-pot synthesis (Panapoy & Ksapabutr, 2008c). A mixture of 0.11 mol of zirconium hydroxide, 0.25 mol of triethanolamine, 0.15 mol of sodium hydroxide, and 20 mL of ethylene glycol was heated at 200°C to distill off ethylene glycol along with the removal of by-product water under nitrogen atmosphere. The reaction was completed within 5 h, and the resulting solution was slowly cooled down to room temperature. Dried methanol and acetonitrile used as precipitants were added into the solution and a white solid precipitated. After filtration, the white solid was washed with dried acetonitrile to remove unreacted ethylene glycol and triethanolamine and dried under vacuum at room temperature. The precursor obtained was identified using Fourier-transform infrared spectroscopy (FTIR), ^{1}H and ^{13}C nuclear magnetic resonance (NMR) spectroscopy, elemental

analysis (EA), mass spectrometry (MS). Fourier-transform infrared spectroscopic analysis was conducted using a Bruker Optik Vertex70 spectrometer with a spectral resolution of 4 cm^{-1} using transparent KBr pellets. ^{1}H and ^{13}C-nuclear magnetic resonance spectra were obtained on a Bruker AV300 spectrometer using deuterated dimethyl sulfoxide (DMSO-d_6) as solvent and reference for chemical shift measurements at room temperature. Elemental analysis was carried out on a CHNS/O analyzer (Perkin Elmer, PE-2400 series II). Matrix assisted laser desorption ionization time of flight mass spectrometry (MALDI-TOF MS) was performed on a ReflexIV instrument (Bruker Daltonics, Germany).
The reaction between a metal oxide or hydroxide and triethanolamine (or ethylene glycol) is a condensation reaction in which water is produced as a by-product of the reaction (Piboonchaisit et al., 1999; Opornsawad et al., 2001; Ksapabutr et al., 2004c). Therefore, the reaction was carried out at the boiling point of ethylene glycol under a nitrogen atmosphere in order to remove water and drive the reaction in the direction of the products. The reaction sequence in the synthesis of zirconatrane complex is illustrated in Fig. 1.

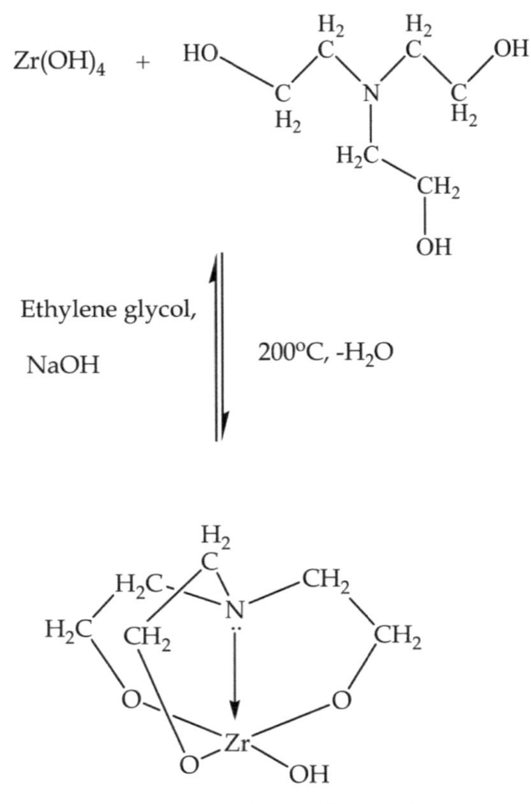

Fig. 1. Diagram of the reaction sequence in the synthesis of zirconatrane (adapted with permission from (Panapoy & Ksapabutr, 2008c))

The Fourier-transform infrared spectroscopic result showed the following peaks: 3000–3680 cm^{-1} (O-H), 2800–2976 cm^{-1} (C-H), 1240–1275 cm^{-1} (C-N), 1120 cm^{-1} (C-O) and 1000-1090 cm^{-1} (Zr-O-C). The ^{1}H-nuclear magnetic resonance spectra of product recorded in deuterated dimethyl sulfoxide solvent: the resonances at 2.3-2.7 and 3.0-3.5 ppm were assigned to N-

CH_2 and CH_2-O, respectively. The result of ^{13}C- nuclear magnetic resonance showed peaks at 57.6 and 60.1 ppm, corresponding to N-CH_2 and CH_2-O, respectively. Both elemental analysis and mass spectrometry were also carried out. The results are shown in Tables 1 and 2, respectively. For the elemental analysis, it was found that the obtained percentages of carbon, hydrogen and nitrogen are very close to those theoretically calculated (Table 1). The proposed structures and fragmentation patterns present in Table 2 clearly indicate the tetramer of zirconatrane (m/z 1019). According to the obtained fragmentation patterns, the expected products were successfully synthesized.

Element (%)	Experimental	Theoretical
C	28.49	28.32
H	5.47	5.11
N	5.25	5.50

Table 1. Percentages of carbon, hydrogen and nitrogen present in the synthesized zirconatrane

m/z	Intensity	Proposed structure
1019	63	$[Zr_4C_{24}H_{52}O_{16}N_4]+2H^+$
931	25	$[Zr_4C_{20}H_{44}O_{14}N_4]+2H^+$
887	20	$[Zr_4C_{18}H_{40}O_{13}N_4]+2H^+$
764	100	$[Zr_3C_{18}H_{39}O_{12}N_3]+H^+$

Table 2. Proposed structures and fragmentation patterns of zirconatrane (reproduced with permission from (Panapoy & Ksapabutr, 2008c))

3. Zirconia nanofibers

3.1 Fabrication of zirconia nanofibers

The configuration of the electrospinning setup used for the fabrication of zirconia nanofibers is schematically demonstrated in Fig. 2. The main equipment consists of a high-voltage power supply unit, a starting solution feeding unit, and a temperature control unit. The starting solution was pumped towards a nozzle by means of a syringe pump. When high positive voltage was applied to the starting solution, an electrostatic field was established between the metal capillary nozzle and the grounded collector. A small pendant drop is formed on the nozzle tip and the resulting electrostatic field is sufficient to overcome the surface tension of the starting solution and create a Taylor cone shape at the nozzle tip, whereupon a cone-jet of highly charged liquid particles is formed. During jet travel, the solvent evaporates and the remaining solid fibers are deposited on the grounded collector. Nanofibers were accumulated on the collector as long as the nozzle tip was continually supplied with the starting solution. Finally, an interconnected web of small filaments or a nonwoven fiber was created on the surface of the grounded collector.

Zirconia nanofibers were prepared according to an electrospinning in combination with sol-gel process and subsequently calcination (Panapoy & Ksapabutr, 2008c) with a slight modification. The spinning solutions were first prepared by mixing 12 mL of zirconatrane in crude form and 3 g of poly(vinyl pyrrolidone) with ethyl alcohol. The amount of ethyl alcohol was varied from 12 to 24 mL and the resulting mixture was continuously stirred for 30 min. The flow rate of the spinning solution was 1.0 mL/h, controlled by syringe pump. The stainless steel capillary tube (inner diameter: 0.394 mm, outer diameter: 0.711 mm) with a tilted angle of 15° at the end was used for the capillary nozzle. The spinning solution was electrospun under ambient air atmosphere at 300°C, as measured by a thermocouple placed on the heating plate. In all of the experiments, the applied voltage was maintained at 20 kV. A grounded metal screen covered by an aluminum foil was used as the counter electrode and was placed 15 cm from the tip of the capillary. As the jet accelerated towards the collector, the solvent evaporated, leaving only ultrathin fibers on the collector. Continuous fibers were deposited for 2 h and collected in the form of fibrous nonwoven mats. Subsequently, calcinations at 500°, 600° and 700°C for 2 h were also carried out for each sample after electrospinning.

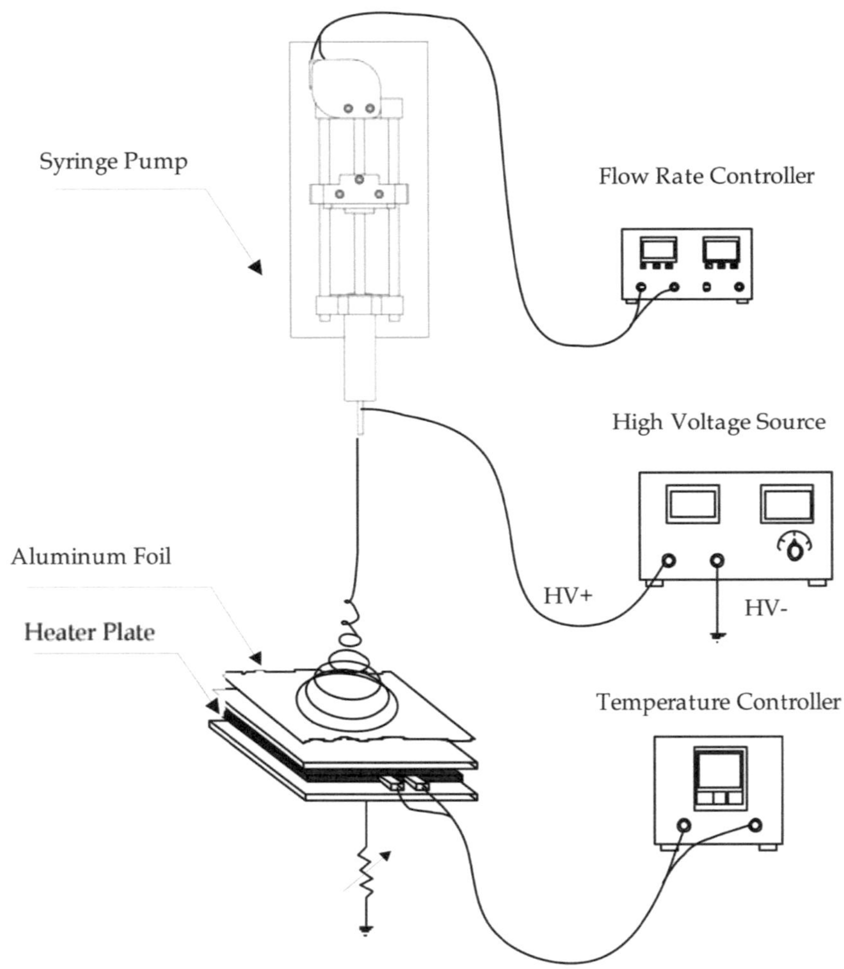

Fig. 2. Schematic illustrating the electrospinning setup used in this work

3.2 Properties of zirconia nanofibers

Morphology and size of the as-electrospun and calcined fibers were observed using a scanning electron microscopy (SEM, Model S3400N, Hitachi). Crystalline phase of the fibers was identified by an automated Rigaku D/Max 2000HV diffractometer with a CuKα radiation. Fourier-transform infrared analysis was measured on a Bruker Optik Vertex70 spectrophotometer with a resolution of 4 cm^{-1}. The solid samples were prepared by mixing 1% of samples with anhydrous potassium bromide, while the spinning solution was analyzed using a Zn-Se window cell. Characterization of the thermal behavior of the spinning solution was achieved by thermogravimetric analysis (Perkin Elmer, TGA7) with a ramp rate of 5°C/min in an air atmosphere. Fig. 3 shows thermogravimetric and derivative thermogravimetric (TG/DTG) curves of the spinning solution. The results revealed weight losses below 340°C, which corresponded to the evaporation of ethyl alcohol, ethylene glycol and some thermal decomposition of the poly(vinyl pyrrolidone) in the spinning solution. While the weight loss at approximately 550°C is probably due to decomposition of residual organic group such as 2-pyrrolidone group resulting from poly(vinyl pyrrolidone).

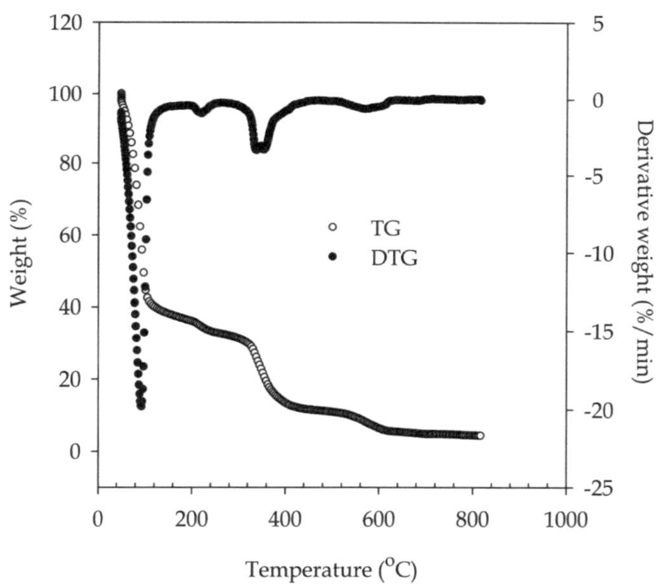

Fig. 3. Thermogravimetric and derivative thermogravimetric curves of the spinning solution

The sizes of as-spun and calcined fibers are compared in Table 3. The effect of ethyl alcohol content of the spinning solutions on morphology of the as-spun fibers is shown in Fig. 4. Obviously, ultrathin fibers with an average diameter of 359 nm were prepared from the spinning solution with an ethyl alcohol content of 24 mL. Increasing amount of ethyl alcohol led to a decrease in the fiber diameter. By viscosity measurement, it was found that the viscosity of the spinning solutions with the amount of ethyl alcohol of 12, 18 and 24 mL are 9.7, 7.4 and 4.3 mPa.s, respectively. The decrease in the fiber diameters with increasing ethyl alcohol content is likely a result of the decrease in the viscosity of the spinning solutions. A similar result was also reported by Jing et al. (2005) even though a different precursor, i.e., zirconium oxychloride, was used as the zirconia source.

Ethyl alcohol content in spinning solution (mL)	Average diameter of fiber			
	As-spun at 300°C (nm)	Calcination at 500°C (nm)	Calcination at 600°C (nm)	Calcination at 700°C (nm)
12	687 ± 124	234 ± 35	230 ± 25	222 ± 39
18	378 ± 131	195 ± 42	192 ± 19	189 ± 26
24	359 ± 110	124 ± 23	122 ± 15	121 ± 26

Table 3. Average diameter of fibers as a function of alcohol content and processing condition

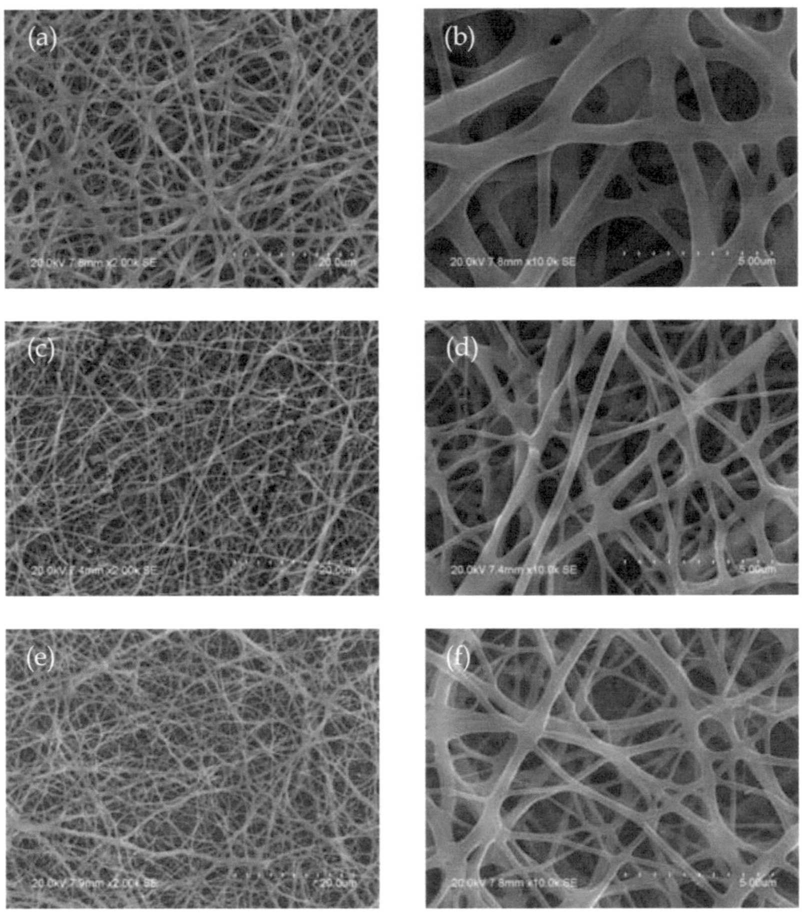

Fig. 4. Scanning electron microscopic images of as-spun zirconia fibers prepared using different amounts of ethyl alcohol (a,b) 12 mL; (c,d) 18 mL; (e,f) 24 mL

Considering the heat treatment process, it was found that the dimension and morphology of the calcined fibers at 500°, 600° and 700°C are different from those of the as-spun fibers at 300°C for all alcohol contents (see Figs. 5, 6 and 7). Moreover, the fiber shrinkage takes place by an increased calcination temperature. The shrinkage is mainly due to the removal of poly(vinyl pyrrolidone) from as-spun fibers during the calcination process. These results were supported by the thermogravimetric and Fourier-transform infrared analyses. Fig. 8 shows the Fourier-transform infrared spectra of the spinning solution and the prepared fibers. The band in the range of 950–2000 cm^{-1} is attributed to the bending and stretching vibrations of poly(vinyl pyrrolidone). Accompanying with the disappearance of these adsorptions, the vibration of Zr-O-Zr at 600 cm^{-1} gradually increased with calcination temperature. Furthermore, the scanning electron microscopic images (Fig. 5) reveal that nanothorns protrude from the nanofibers after heat treatment of as-spun fibers at 500°C, in particular as the amount of ethyl alcohol decreases. The formation of zirconia nuclei may be a driving force for the formation of thorn-like nanocrystals on the zirconia nanofibers upon heat treatment process. With increasing calcination temperature from to 500° to 700°C, it was noted that the thorn-like nanostructure on the nanofibers converted to flower-like architecture, probably due to the nucleation and growth of thorn-like nanocrystals.

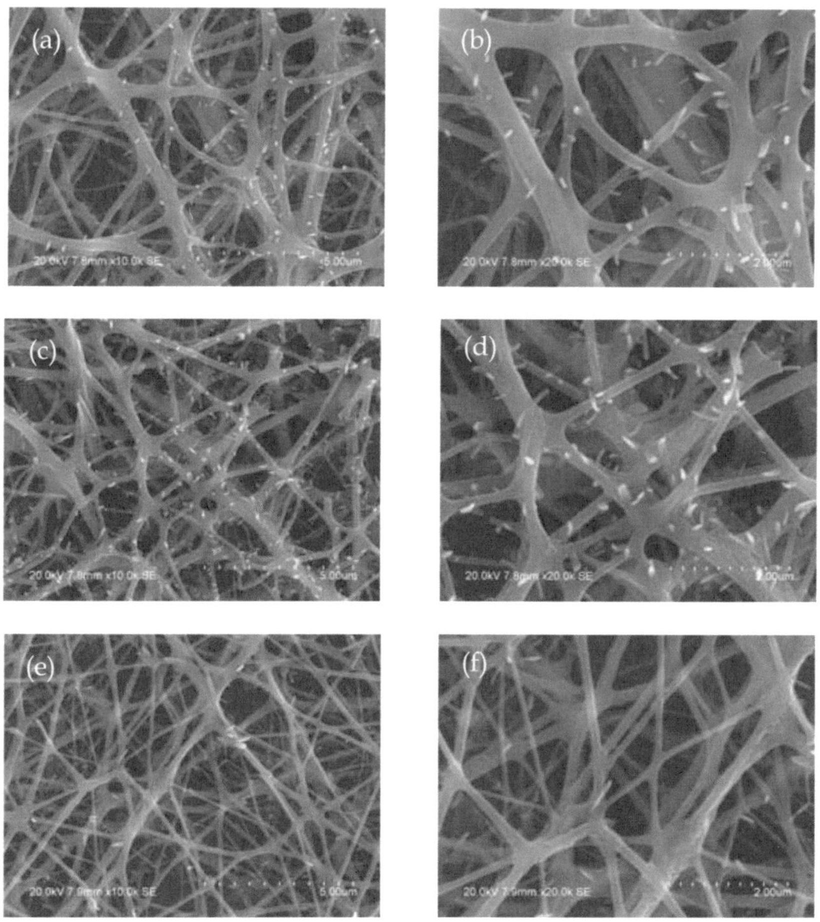

Fig. 5. Scanning electron microscopic images of zirconia fibers calcined at 500°C using different amounts of ethyl alcohol (a,b) 12 mL; (c,d) 18 mL; (e,f) 24 mL

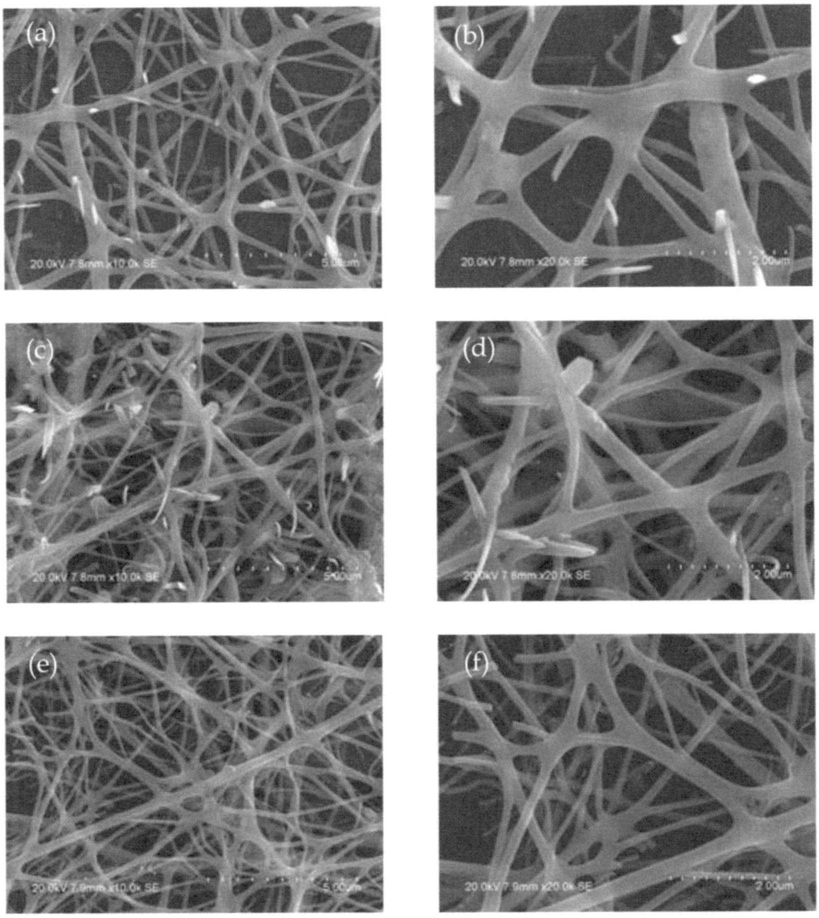

Fig. 6. Scanning electron microscopic images of zirconia fibers calcined at 600°C using different amounts of ethyl alcohol (a,b) 12 mL; (c,d) 18 mL; (e,f) 24 mL

This was supported by the X-ray diffraction analysis (see Fig. 9) that samples have a tendency to gradually crystallize upon further heat treatment. The crystallinity increased with increasing calcination temperature. Meanwhile, it can be seen that the average crystallite size of the calcined fibers at 500°C increased from 3.4 nm to 8.9 and 15.6 nm after heat treatment at 600° and 700°C, respectively. The major peaks and their intensities corresponding to the reflection lines of the tetragonal phase of zirconia are well matched with database in JCPDS (ICDD Card No. 17-923).

From knowledge of oxide one-pot synthesis process, zirconatrane can function as a zirconium alkoxide compound due to its higher stability and increased steric hindrance. Moreover, zirconatrane can be used successfully for the fabrication of one-dimensional zirconia nanofibers with controllable nanostructures. To have a clear understanding of the detailed mechanism of the thorn and flower-like structures, more in-depth studies are in progress. Moreover, the synthetic system providing an appropriate crystal growth environment for the formation of such a novel architecture will be studied and reported in detail later.

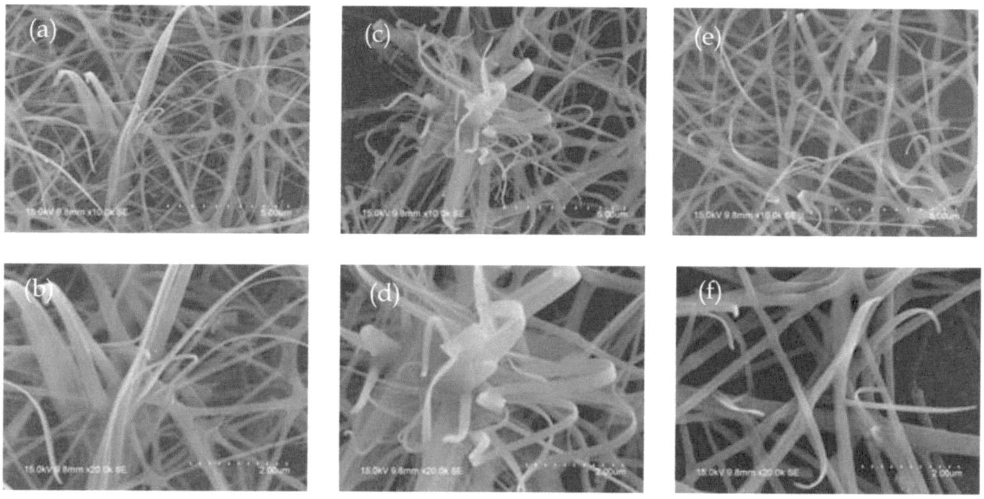

Fig. 7. Scanning electron microscopic images of zirconia fibers calcined at 700°C using different amounts of ethyl alcohol (a,b) 12 mL; (c,d) 18 mL; (e,f) 24 mL

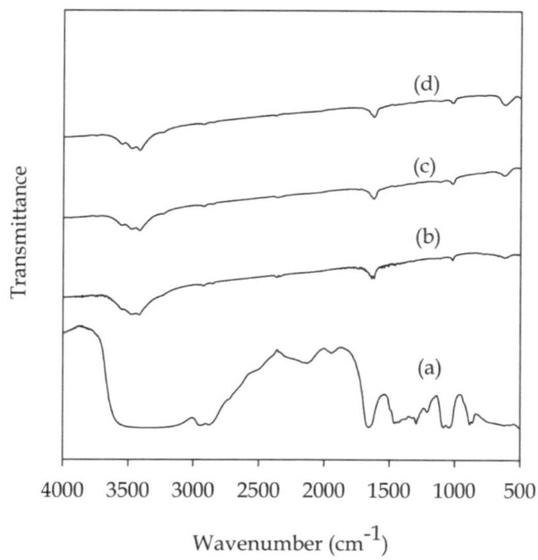

Fig. 8. Fourier-transform infrared spectra of (a) spinning solution, and zirconia fibers calcined at (b) 500°C; (c) 600°C; (d) 700°C using the ethyl alcohol content of 18 mL

4. Conclusion

Morphologically controlled synthesis of zirconia nanofibers prepared using an inexpensive and moisture-stable zirconatrane compound, was achieved by a combined sol-gel processing and electrospinning technique with subsequent heat treatment process. The diameters of the zirconia fibers decreased with an increase in the ethyl alcohol content of the spinning solution. The average diameter of theas-spun fibers (based on the amount of ethyl alcohol of 24 mL) decreased from 359 nm at a spinning temperature of 300°C to 121 nm at a

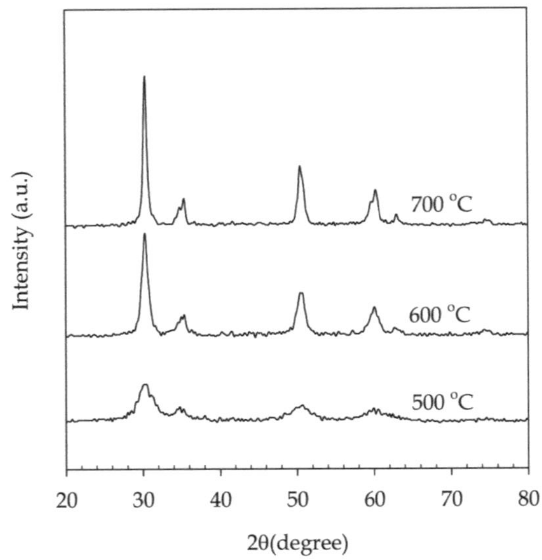

Fig. 9. X-ray diffraction patterns of zirconia fibers calcined at various calcination temperatures using the ethyl alcohol content of 18 mL

calcination temperature of 700°C. A morphological transition from thorn to flower-like architecture on zirconia nanofibers was observed upon increasing calcination temperature from 500° to 700°C. Thorn-like nanostructure on nanofibers prefers to form for lower heat-treated temperature and a smaller amount of ethyl alcohol, whereas flower-like nanostructure prefers to form for higher heat-treated temperature and a smaller amount of ethyl alcohol. The obvious crystal growth appears from 500° to 700°C, and the average size of the crystals in final fibers ranges between 3.4 and 15.6 nm. This present study provides a simple process to fabricate zirconia nanofibers with controllable architecture features, and could be applied to the fabrication of other metal oxide nanofibers.

5. Acknowledgements

The financial support of Silpakorn University Research and Development Institute, the Department of Materials Science and Engineering, the Faculty of Engineering and Industrial Technology, Silpakorn University and the National Center of Excellence for Petroleum, Petrochemicals and Advanced Material is gratefully acknowledged.

6. References

Abe, Y.; Kudo, T.; Tomioka, H.; Gunji, T.; Nagao, Y. & Misono, T. (2004). Preparation of Continuous Zirconia Fibres from Polyzirconoxane Synthesized by the Facile One-Pot Reaction. *Journal of Materials Science*, Vol. 33, No. 7, (November 2004) 1863-1870, ISSN 1573-4803.

Barbic, M.; Mock, J.J.; Smith, D.R. & Schultz, S. (2002). Single Crystal Silver Nanowires Prepared by the Metal Amplification Method. *Journal of Applied Physics*, Vol. 91, No. 11, (June 2002) 9341–9345, ISSN 0021-8979.

Berkland, C.; Pack, D.W. & Kim, K.K. (2004). Controlling Surface Nano-Structure Using Flow-Limited Field-Injection Electrostatic Spraying (FFESS) of Poly(D,L-lactide-co-glycolide). *Biomaterials*, Vol. 25, (November 2004) 5649-5658, ISSN 0142-9612.

Carotta, M.C.; Benetti, M.; Ferrari, E.; Giberti, A.; Malagù, C.; Nagliati, M.; Vendemiati, B. & Martinelli, G. (2007). Basic Interpretation of Thick Film Gas Sensors for Atmospheric Application. *Sensors and Actuators B: Chemical*, Vol. 126, No. 2, (October 2007) 672-677, ISSN 0925-4005.

Chadwick, A.V. (2000). Nanotechnology: Solid Progress in Ion Conduction. *Nature*, Vol. 408, No. 6815, (December 2000) 925-926, ISSN 0028-0836.

Charoenpinijkarn, W.; Suwankruhasn, M.; Ksapabutr, B.; Wongkasemjit, S. & Jamieson M.A. (2001). Sol-Gel Processing of Silatranes. *European Polymer Journal*, Vol. 37, No. 7, (July 2001) 1441-1448, ISSN 0014-3057.

Cheung, T.K. & Gates, B.C. (1997). Activation of Ethane in the Presence of Solid Acids: Sulfated Zirconia, Iron- and Manganese-Promoted Sulfated Zirconia, and Zeolites. *Journal of Catalysis*, Vol. 168, No. 2, (June 1997) 522-531, ISSN 0021-9517.

Duan, X.F. & Lieber,C.M. (2000). General Synthesis of Compound Semiconductor Nanowires. *Advanced Materials*, Vol. 12, No. 4, (February 2000) 298–302, ISSN 1521-4095.

Feng, X.; Bai, Y.J.; LÜ, B.; Zhao, Y.R.; Yang, J. & Chi, J.R. (2004). A Novel Reduction–Oxidation Synthetic Route to Cubic Zirconia Nanocrystallite, *Journal of Crystal Growth*, Vol. 262, No. 1-4, (February 2004) 420–423, ISSN 0022-0248.

Guan, H.; Shao, C.; Chen, B.; Gong, J. & Yang, X. (2003). A Novel Method for Making CuO Superfine Fibres via an Electrospinning Technique. *Inorganic Chemistry Communications*, Vol. 6, No. 11, (November 2003) 1409–1411, ISSN 1387-7003.

Han, W.; Fan, S.; Li, Q. & Hu, Y. (1997). Synthesis of Gallium Nitride Nanorods Through a Carbon Nanotube-Confined Reaction. *Science*, Vol. 277, No. 5330, (August 1997) 1287-1289, ISSN 0036-8075.

Heath, J.R. & LeGoues, F.K. (1993). A Liquid Solution Synthesis of Single Crystal Germanium Quantum Wires. *Chemical Physics Letters*, Vol. 208, No. 3-4, (June 1993) 263–268, ISSN 0009-2614.

Huang, Z.M.; Zhang, Y.Z.; Kotaki, M. & Ramakrishna, S. (2003). A Review on Polymer Nano-Fibres by Electrospinning and their Applications in Nanocomposites. *Composites Science and Technology*, Vol. 63, No. 15, (October 2002) 2223-2253, ISSN 0266-3538.

Iijima, S. (1991). Helical Microtubules of Graphitic Carbon. *Nature*, Vol. 354, No. 6348, (November 1991) 56-58, ISSN 0028-0836.

Jing, N.; Wang, M. & Kameoka, J. (2005). Fabrication of Ultrathin ZrO_2 Nanofibers by Electrospinning. *Journal of Photopolymer Science and Technology*, Vol. 18, No. 4, (August 2005) 503-506, ISSN 0914-9244.

Kalayci, V.E.; Patra, P.K.; Ugbolue, S.; Kim, Y.K. & Warner S.B. (2005). Charge Consequences in Electrospun Polyacrylonitrile (PAN) Nanofibers. *Polymer*, Vol. 46, No. 18, (August 2005) 7191-7200, ISSN 0032-3861.

Ksapabutr, B.; Gulari, E. & Wongkasemjit, S. (2004a). Sol–Gel Transition Study and Pyrolysis of Alumina-Based Gels Prepared from Alumatrane Precursor. *Colloids and Surfaces A: Physicochemical and Engineering Aspects*, Vol. 233, No. 1-3, (February 2004) 145–153, ISSN 0927-7757.

Ksapabutr, B.; Gulari, E. & Wongkasemjit, S. (2004b). Preparation of Zirconia Powders by Sol–Gel Route of Sodium Glycozirconate Complex. *Powder Technology*, Vol. 148, No. 1, (October 2004) 11 - 14, ISSN 0032-5910.

Ksapabutr, B.; Wongkasemjit, S. & Gulari, E. (2004c). One-Pot Synthesis and Characterization of Novel Sodium Tris(glycozirconate) and Cerium Glycolate Precursors and Their Pyrolysis. *Materials Chemistry and Physics*, Vol. 83, No. 1, (January 2004) 34–42, ISSN 0254-0584.

Ksapabutr, B.; Chalermkiti, T. & Panapoy, M. (2005a). Effect of Nozzle Shapes on the Formation of Taylor Cone and the Oscillation of Fibers During Electrospinning Process. *Chiang Mai University Journal*, Vol. 4, No. 1, (November 2005) 115-119, ISSN 1685-1994.

Ksapabutr, B.; Waikru, C. & Panapoy, M. (2005b). Effect of Target Shapes on Distribution of Polyacrylonitrile Nanofibers Prepared by Electrospinning Process. *Chiang Mai University Journal*, Vol. 4, No. 1, (November 2005) 109-113, ISSN 1685-1994.

Ksapabutr, B.; Wongkasemjit, S. & Gulari, E. (2005c). Rheology and Heat Treatment of Zirconia Based Gels Synthesized from Sodium Glycozirconate Precursor. *Materials Science Forum*. Vol. 480-481, (March 2005) 549-556, ISSN 0255-5476.

Kumar, M.K. & Ramaprabhu, S. (2007). Palladium Dispersed Multiwalled Carbon Nanotube Based Hydrogen Sensor for Fuel Cell Applications. *International Journal of Hydrogen Energy*, Vol. 32, No. 13, (September 2007) 2518-2526, ISSN 0360-3199.

Li, J.; Jiao, X. & Chen, D. (2007). Preparation of Zirconia Fibers via a Simple Aqueous Sol-Gel Method. *Journal of Dispersion Science and Technology*, Vol. 28, No. 4, 531–535, ISSN 0193-2691.

Li, D.; McCann, J.T. & Xiaw, Y. (2006). Electrospinning: A Simple and Versatile Technique for Producing Ceramic Nanofibers and Nanotubes. *Journal of the American Ceramic Society*, Vol. 89, No. 6, 1861–1869, ISSN 0002-7820.

Li, D. & Xia, Y. (2003). Fabrication of Titania Nanofibers by Electrospinning. *Nano Letters*, Vol. 3, No. 4, (March, 2003) 555–560, ISSN 1530-6984.

Luo, M.F.; Song, Y.P.; Wang, X.Y.; Xie, G.Q.; Pu, Z.Y.; Fang, P. & Xie, Y.L. (2007). Preparation and Characterization of Nanostructured $Ce_{0.9}Cu_{0.1}O_{2-\delta}$. Solid Solution with High Surface Area and Its Application for Low Temperature CO Oxidation. *Catalysis Communications*, Vol. 8, No. 5, (May 2007) 834–838, ISSN 1566-7367.

Markowitz, P.D.; Zach, M.P.; Gibbons, P.C.; Penner, R.M. & Buhro,W.E. (2001). Phase Separation in $Al_xGa_{1-x}As$ Nanowhiskers Grown by the Solution–Liquid–Solid Mechanism. *Journal of the American Ceramic Society*, Vol. 123, No. 19, (April 2001) 4502–4511, ISSN 0002-7820.

Martin, C.R. (1994). Nanomaterials: A Membrane-Based Synthetic Approach. *Science*, Vol. 266, No. 5193, (December 1994) 1961-1966, ISSN 0036-8075.

Menge, W.M.P.B. & Verkade, J.G. (1991). Monomeric and Dimeric Titanatrane. *Inorganic Chemistry*, Vol. 30, No. 24, (November 1991) 4628-4631, ISSN 0020-1669.

Morales, A.M. & Lieber, C.M. (1998). A Laser Ablation Method for the Synthesis of Crystalline Semiconductor Nanowires. *Science*, Vol. 279, No. 5348, (January 1998) 208–211, ISSN 0036-8075.

Nugent, W.A. & Harlow, R.L. (1994). Early Transition Metal Alkoxide Complexes Bearing Homochiral Trialkanolamine Ligands. *Journal of the American Chemical Society*, Vol. 116, No. 14, (July 1994) 6142-6148, ISSN 0002-7863.

Opornsawad, Y.; Ksapabutr, B.; Wongkasemjit, S. & Laine, R.M. (2001). Formation and Structure of Tris(alumatranyloxy-i-propyl)amine Directly from $Al(OH)_3$ and Triisopropanolamine. *European Polymer Journal*, Vol. 37, No. 9, (September 2001) 1877-1885, ISSN 0014-3057.

Pan, Z.W.; Dai, Z.R. & Wang, Z.L. (2001). Nanobelts of Semiconducting Oxides. *Science*, Vol. 291, No. 5510, (March 2001) 1947–1949, ISSN 0036-8075.

Panapoy, M.; Dankeaw, A. & Ksapabutr, B. (2008a). Electrical Conductivity of PAN-based Carbon Nanofibers Prepared by Electrospinning Method, *Thammasat International Journal of Science and Technology*, Vol. 13, (November 2008) 11-17, ISSN 0859-4074.

Panapoy, M.; Saengsil, N. & Ksapabutr, B. (2008b). Electrical Conductivity of Poly (3,4-ethylenedioxythiophene)-Poly(styrenesulfonate) Coatings on Polyacrylonitrile Nanofibers for Sensor Applications. *Advanced Materials Research*, Vol. 55-57, (August, 2008) 257-260, ISSN 1022-6680.

Panapoy, M. & Ksapabutr, B. (2008c). Fabrication of Zirconia Nanofibers Using Zirconatrane Synthesized by Oxide One-Pot Process as Precursor, *Advanced Materials Research*, Vol. 55-57, (August, 2008) 605-608, ISSN 1022-6680.

Park, S.; Vohs, J.M. & Gorte, R.J. (2000). Direct Oxidation of Hydrocarbons in a Solid-Oxide Fuel Cell. *Nature*, Vol. 404, No. 6775 (March 2000) 265-267, ISSN 0028-0836.

Piboonchaisit, P.; Wongkasemjit, S. & Laine, R. M. (1999). A Novel Route to Tris(silatranyloxy-i-propyl)amine Directly from Silica and Triisopropanolamine, Part.I. *ScienceAsia*, Vol. 25, No. 2, (June 1999) 113-119, ISSN: 1513-1874.

Qi, Q.; Zhang, T. & Wang, L. (2008). Improved and Excellent Humidity Sensitivities Based on KCl-Doped TiO_2 Electrospun Nanofibers, *Applied Physics Letters*, Vol. 93, (July 2008) 023105/1–023105/3, ISSN 0003-6951.

Remant, B K.C.; Kim, C.K.; Khil, M.S.; Kim, H.Y. & Kim, I.S. (2008). Synthesis of Hydroxyapatite Crystals Using Titanium Oxide Electrospun Nanofibers. *Materials Science and Engineering C*, Vol. 28, No. 1, (January 2008) 70–74, ISSN 0928-4931.

Sasaki, T.; Liang, C.; Nichols, W.T.; Shimizu, Y. & Koshizaki, N. (2004). Fabrication of Oxide Base Nanostructures Using Pulsed Laser Ablation in Aqueous Solutions. *Applied Physics A: Materials Science & Processing*, Vol. 79, No. 4-6, (July 2004) 1489–1492, ISSN 0947-8396.

Seong, O.H.; Son, W.K.; Youk J.H.; Lee, T.S. & Park, W.H. (2005). Ultrafine Porous Fibers Electrospun from Cellulose Triacetate. *Materials Letters*, Vol. 59, No. 24-25, (October 2005) 2998-3001, ISSN 0167-577X.

Shao, C.; Guan, H.; Liu, Y.; Gong, J.; Yu, N. & Yang, X. (2004). A Novel Method for Making ZrO_2 Nanofibres via an Electrospinning Technique. *Journal of Crystal Growth*, Vol. 267, No. 1-2, (June 2004) 380–384, ISSN 0022-0248.

Shao, C.; Kim, H.Y.; Gong, J.; Ding, B.; Lee, D.R. & Park, S.J. (2003). Fiber Mats of Poly (Vinyl Alcohol)/Silica Composite via Electrospinning. *Materials Letters*, Vol. 57, No. 9-10, (February 2003) 1579-1584, ISSN 0167-577X.

Sigmund, W.; Yuh, J.; Park, H.; Maneeratana, V.; Pyrgiotakis, G.; Daga, A.; Taylor, J. & Nino, J.C. (2006). Processing and Structure Relationships in Electrospinning of Ceramic Fiber Systems. *Journal of the American Ceramic Society*, Vol. 89, No. 2, 395–407, ISSN 0002-7820.

Trentler, T.J.; Hickman, K.M.; Geol, S.C.; Viano, A.M.; Gibbons, P.C. & Buhro,W.E. (1995). Solution-Liquid-Solid Growth of Crystalline III-V Semiconductors: An Analogy to

Vapor-Liquid-Solid Growth. *Science*, Vol. 270, No. 5243, (December 1995) 1791–1794, ISSN 0036-8075.

Wang, X. & Li, Y.D. (2002). Selected-Control Hydrothermal Synthesis of Alpha- and Beta-MnO_2 Single Crystal Nanowires. *Journal of the American Ceramic Society*, Vol. 124, No. 12, (March 2002) 2880–2881, ISSN 0002-7820.

Wang, Y.D.; Ma, C.L.; Sun, X.D.; Li, H.D. (2002a). Preparation of Nanocrystalline Metal Oxide Powders with the Surfactant - Mediated Method. *Inorganic Chemistry Communications*, Vol. 5, No. 10, (October 2002) 751-755, ISSN 1387-7003.

Wang, X.; Drew, C.; Lee, S.H., Senecal, K.J.; Kumar, J. & Samuelson, L.A. (2002b). Electrospinning Technology: A Novel Approach to Sensor Application. *Journal of Macromolecular Science, Part A: Pure and Applied Chemistry*, Vol. 39, No. 10, 1251-1258, ISSN 1060-1325.

Wang, Z.L. (2004). Functional Oxide Nanobelts: Materials, Properties and Potential Applications in Nanosystems and Biotechnology. *Annual Review of Physical Chemistry*, Vol. 55, (June 2004) 159–196, ISSN 0066-426X.

Xia, Y.; Yang, P.; Sun, Y.; Wu, Y.; Mayers, B.; Gates, B.; Yin, Y.; Kim, F. & Yan, H. (2003). One-Dimensional Nanostructures: Synthesis, Characterization, and Applications. *Advanced Materials*, Vol. 15, No. 5, (March 2003) 353–389, ISSN 1521-4095.

Yuh, J.; Nino, J.C. & Sigmund, W.M. (2005). Synthesis of Barium Titanate ($BaTiO_3$) Nanofibers via Electrospinning. *Materials Letters*, Vol. 59, No. 28, (December 2005) 3645 - 3647, ISSN 0167-577X.

Zach, M.P.; Ng, K.H. & Penner, R.M. (2000). Molybdenum Nanowires by Electrodeposition. *Science*, Vol. 290, No. 5499, (December 2000) 2120–2123, ISSN 0036-8075.

Zhang, Y.; Wang, N.; Gao, S.; He, R.; Miao, S.; Liu, J.; Zhu, J. & Zhang, X. (2002). A Simple Method to Synthesize Nanowires, *Chemistry of Materials*, Vol. 14, No. 8, (July 2002) 3564–3568, ISSN 0897-4756.

Zhang, Y.; Li, J.; Li, Q.; Zhu, L.; Liu, X.; Zhong, X.; Meng, J. & Cao, X. (2007). Preparation of In_2O_3 Ceramic Nanofibers by Electrospinning and Their Optical Properties. *Scripta Materialia*, Vol. 56, No. 5, (March 2007) 409–412, ISSN 1359-6462.

Zhang, H.B. & Edirisinghew, M.J. (2006). Electrospinning Zirconia Fiber from a Suspension. *Journal of the American Ceramic Society*, Vol. 89, No. 6, (April 2006) 1870–1875, ISSN 0002-7820.

20

Organic Fluorescent Nanofibers and Sub-micrometer Rods

Suzanne Fery-Forgues and Clara Fournier-Noël
Laboratoire des Interactions Moléculaires Réactivité Chimique et Photochimique, UMR CNRS 5623, Université Paul Sabatier 31062 Toulouse cedex 9, France

1. Introduction

Luminescent nanofibers are a particular class of nanofibers. Like the other members of this family, they are very small objects whose typical diameters range between 50 to 500 nm (it is noteworthy that beyond these values, the term of microfibers is more appropriate) and lengths can reach several hundred micrometers. Besides, they display special photophysical behavior. By definition, fluorescent nanofibers have the property to emit light after absorption of UV or visible radiations, and electroluminescent nanofibers emit light upon electrical excitation. Passive properties are also encountered. This is the case in particular for waveguiding, light being guided along the long axis when the fiber width is above a minimum critical value. It is thus easy to understand why these fibers are presently attracting increasing attention. One of the most promising applications is their use in nanotechnologies. This field has developed tremendously during the last few years and sub-micrometer organic materials that exhibit specific optical properties are now of high demand. Among them, luminescent nanofibers occupy a very special position, because their morphology makes then unique for use in miniaturized systems, especially photonic devices, where they can be incorporated individually and act as active component to generate or transmit light from one point to another (Schiek et al., 2008). Their use has been proposed as frequency doublers (Brewer et al., 2006) and nanolasers (Quochi et al., 2006). They are also prime candidates for applications in the fields of photovoltaic cells and organic light-emitting diodes (OLEDs) for color displays or white light illumination (Ner et al., 2008; Ner et al., 2009). Other original uses can be found. For example, some fluorescent nanofibers have been shown to act as chemosensors to detect traces of amine vapor (Che et al., 2008; Che & Zang, 2009), explosives (Long et al., 2009; Naddo et al., 2007; Naddo et al., 2008; Tao et al., 2007; Wang et al., 2001; Wang et al., 2002; Zang et al., 2009) and metal ions (Wang et al., 2001; Wang, Drew et al., 2002). As biosensors, nanofibers reveal the presence of pathogen bacteria (Memisevic et al., 2009) and proteases (Law et al., 2007), and could be used for diagnostic purpose. Generally, the sensing performance is improved by the extended effective surface area of the nanofibers. These structures could also play the role of antennae in light-harvesting nanomaterials (Channon et al., 2009; Channon et al., 2009;

Desvergne et al., 2006). More simply, nanofibers can also be stamped onto precious documents, such as banknotes, where they form a blue-light emitting mark, which appears more brilliant when seen from the side than from above, due to the particular waveguiding properties of the fibers (Schiek et al., 2008).

The advantage of using organic molecules instead of inorganic compounds to fabricate nanofibers is obvious. Organic compounds offer more variability and flexibility in molecular design than their inorganic homologues. A large number of chemical modifications can be introduced via synthesis to modulate or totally change the physical properties, and especially the photophysical properties that are of concern here. Besides, nanostructures based on organic molecules are able to provide high luminescence efficiency for quite a low material density, and easy and cheap processing since controlled self-assembled growth can be implemented.

It must be stressed that there is a special interest in crystallized organic materials. Actually, waveguiding is favored when the nanofibers present a high degree of crystallinity and preferential orientation of the molecules (Brewer & Rubahn, 2005). This is also the case for other photophysical processes, such as energy migration and energy transfer. These phenomena are encountered in materials that perform a light-harvesting function and, more simply, in materials where the color emission is tuned by energy transfer between two different types of molecules. Another consequence of crystallinity is that absorbance and fluorescence strongly depend on the polarization of the incident light with respect to the transition dipole of the molecules, and thus to the fiber axis. Finally, electrical conduction is also favored by a high crystal order (Henrichsen et al., 2007), which is of great interest for electroluminescent compounds.

2. Photophysical behavior and design of fluorescent nanofibers

The main problem in the design of fluorescent nanofibers is linked to the photophysical behavior of dyes in the solid state. Generally speaking, most dyes are strongly fluorescent when dissolved in their good solvents, but they lose their fluorescence upon aggregation. This is due to the intermolecular interactions that take place in the solid state and open deactivation pathways that compete with the fluorescence process (Birks, 1970; Silinsh, 1980). In particular, the frequently encountered plane-to-plane stacking of fluorophores (similar to *H*-aggregates) generates intermolecular coupling of electronic transition dipole moments and leads to virtually non-fluorescent solids. Particular excited species called excimers and excitons can also be formed between neighboring dye molecules, and often result in the loss of excitation energy in non-radiative processes. This behavior is not specific of fibers and is encountered in all other types of solids.

In these conditions, how can highly fluorescent nanofibers be obtained? All depends on the complexity degree that is sought after. The simplest approach is to get the fluorophores dispersed in a solid medium. As will be seen below, a good example is given by dye-doped polymers. In this case, the first requirement is to choose a highly fluorescent dye and to make sure that its fluorescence is not quenched by the surrounding medium. Many organic dyes can be dispersed in an organic medium and using them at low concentrations allows aggregation to be prevented. The dye molecule can also be individually inserted in a cage molecule before being incorporated in a solid matrix.

Another strategy is to take advantage of the fiber structure to drive the arrangement of the fluorophore molecules. This is the case for bionanofibers and some other fibers where the dye moiety is linked to an extensive alkyl moiety. Ideally, the dye molecules are arranged in the fibers as *J*-aggregates, forming a head-to-tail arrangement (Kobayashi, 1996). The arrangement of the dye molecules favors very interesting photophysical phenomena (actually quite close to those that quench fluorescence in the solids when they are not controlled). For instance, when the dye molecules are of the same chemical nature and placed in an identical environment, all of them have the same energy level. The excitation energy can then migrate from one molecule to the other. When two different types of molecule are used, the excitation energy is finally transferred to the molecule whose energy level is the lowest. If the molecule that collects the excitation energy is fluorescent, it readily emits light (Fig. 1). This phenomenon is known as a light-harvesting or antenna effect. In practice, pairs of fluorophores must be chosen so that the emission spectrum of the "donor" overlaps the absorption spectrum of the "acceptor". The process efficiency closely depends on the fluorophore separation distance and the orientation of their transition dipoles.

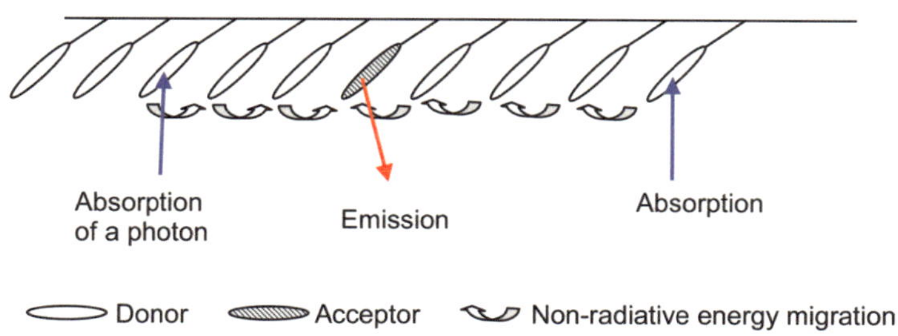

Fig. 1. Schematized energy transfer between donor and acceptor dyes: light-harvesting effect.

Maximum difficulties are probably encountered with crystallized dye materials. Actually, aggregation strongly affects fluorescence, but crystallization makes the things even worse. In a crystal, the molecules are closely packed, well aligned, and intermolecular interactions are very strong. The formation of excimers is favored and that of excitons, followed by the migration of the excitation energy to impurities traps or surface defects, also reduces drastically the emission efficiency. This is true for all organic solids, but takes a particular importance for submicron-sized solids. It has been shown recently that photophysical properties such as the photoluminescence quantum yield strongly differ according to the way the particles have been prepared (Katoh et al., 2009). The differences have been attributed to surface defects and impurities. It can be deduced that surface defects play a prominent role in nanofibers, whose surface area is wide by comparison with that of more compact structures. It must be noted that in the smallest nanocrystals, the number of surface molecules -in contact with the outer medium- is high enough not to be negligible with respect to the number of core molecules, and consequently, two populations of molecules can be distinguished from a spectroscopic viewpoint. But, this particular case will not be evoked here.

The best way to insure good fluorescence efficiency to crystallized nanofibers would be to choose a dye that is highly fluorescent in the solid state. Fortunately, although rare, this type of dyes exists. A small number of structures are already known and there is active research for creating new ones. These molecules generally display a particular molecular arrangement. For example, they form a brick-wall-like architecture (Cornil et al., 2000), are separated by bulky groups, or undergo cross dipole stacking (Xie et al., 2005). In fact, many requirements must be taken into account for the rational design of fluorescent crystallized nanofibers. Molecules must have the conjugated system of good fluorophores, spontaneously self-organize into nanofibers, and lead to a high degree of crystallinity with a special molecular arrangement, compatible with fluorescence. For the moment, very little is known about molecules that could satisfy all these specifications.

3. Overview of the different types of fluorescent nanofibers and their preparation methods

As for other nanomaterials, two main approaches are used to fabricate nanofibers. In the "top-down" approach, fibers are constructed from larger entities without atomic-level control. In the "bottom-up" approach, they are created through self-assembly processes of molecular components, according to the principles of molecular recognition. The chosen approach closely depends on the nature of the nanofibers.

3.1 Polymer-based nanofibers

The "top-down" approach is widely used for the fabrication of polymer-based nanofibers. Polymers can be intrinsically fluorescent due to an extended conjugated system (Chen et al., 2009; Kang et al., 2007; Long et al., 2009; Samitsu et al., 2009; Xin et al., 2009; Zhang et al., 2007; Zhao et al., 2007; Zhu et al., 2009). A good example is that of poly(phenylene vinylene) (PPV) derivatives, which emit from blue to yellow-green. Fluorescent polymers can sometimes be combined with nonfluorescent ones (Yoon & Kim, 2008). Polymers can also be dye-doped. In this case, the dye molecules can be either covalently fixed or simply dispersed in the bulk (Camposeo et al., 2007; DiBenedetto et al., 2008; Fantini & Costa, 2009; Lee et al., 2002; Liang et al., 2008; Lv et al., 2008; Tomczak et al., 2006; Tao et al., 2007; Yang et al., 2009; Wan et al., 2006; Wang et al., 2001; Wang, Drew & et al., 2002; Wang et al., 2007; Wang et al., 2008), incorporated in zeolite embedded in the polymer structure (Cucchi et al., 2007; Vohra et al., 2009), or even forming a distinct phase (Lee et al., 2006; Li et al., 2009).
Electrospinning is the most currently used fabrication method. It consists in exposing a polymer solution droplet to a high voltage, so that a fine liquid jet is formed, finally resulting in submicron or nanometer scale fibers. It has an advantage with its comparative low cost and relatively high production rate. However, other methods can be useful, for example the photopolymerization of an organogel (Kang et al., 2007) or the whisker method, according to which a nematic liquid crystal promotes the self-assembly and anisotropic crystallization of conducting polythiophene nanofibers (Samitsu et al., 2009). Polymer nanofibers were also fabricated via vapour deposition polymerization (VDP) using a porous aluminium oxide membrane as template. The treated template was then dipped in organic dye solution before additional VDP was carried out. The process results in encapsulating the dye molecules in the polymer and interesting energy transfer occurs when multiple organic dyes are used (Lee et al., 2006). Many polymer-based nanofibers display interesting optical properties. In particular, they open new perspectives in the field of low-cost fabrication technology of flexible nanoscale OLEDs.

3.2 Bionanofibers

Biological molecules that spontaneously assemble via non-specific hydrogen-bonding interactions offer an attractive starting point for the preparation of fluorescent nanofibers. They often allow a level of control over the distribution and orientation of the chromophores. They are relatively simple to produce and can be tailored to provide a wide range of photophysical functionality. Various preparation methods have been reported, pertaining either to the "top-down" approach or to the "bottom-up" approach.

For instance, nanofibers made of natural polymers such as DNA can be produced by electrospinning. This is the case for DNA molecules mixed with intercalating fluorescent dyes, the obtained fibers having a diameter of about 27 nm (Bellan et al, 2007). More elaborated systems, such as DNA complexed with a cationic surfactant, can be processed by the same method. This complex has unique material properties including thermal stability and solubility in organic solvents. It has been used as a matrix to allow homogeneous distribution of encapsulated multiple chromophores. Chromophores were chosen so that they act as pairs of energy donor and acceptor. In the present case, the optimized spatial organization provided by the DNA matrix favors efficient energy transfer and the simultaneous emission of various fluorophores results in white light, particularly interesting for use in OLEDs (Ner et al., 2008; Ner et al., 2009).

Another strategy is to synthesize labeled precursors that self-assemble spontaneously. For instance, a thymidylic acid derivative labeled with an anthracene dye arranged in a single-strand oligomer. The assembly with a complementary single-strand oligomer resulted in a helicoidal nanofiber of about 5 nm in diameter, presenting the spectral characteristics of *J*-aggregates (Iwaura et al., 2009).

Peptides are also very useful to drive the organization of chromophores. They are known to assemble into fibrillar "amyloid-like"architectures, which are generally about 10 nm in diameter and a few micrometers long. Various precursors containing a peptide component and an organic fluorophore were synthesized. They were allowed to self-assemble and gave fibrils that exhibited a β-sheet signature. In the cases where the fibrils were decorated with a single fluorophore, energy migration was shown to occur (Channon et al, 2008). When two types of well-selected fluorophores were used to label the peptide segments, the mixing of these precursors resulted in a fiber that bore both donor and acceptor groups. An efficient energy transfer process took place, so that the system was proposed as a route to light-harvesting nanomaterials (Channon et al., 2009; Channon et al., 2009). Finally, dye-labeled peptides were also incorporated into assemblies with nonfluorescent peptides able to bind with heparin, a biologically important macromolecule. When the system was put in the presence of fluorescein-tagged heparin, a selective transfer of energy was observed. The fluorescent peptide fibrils were then proposed as energy donors to study a range of interactions with biological molecules (Behanna et al., 2007).

3.3 Dendrimer-based nanofibers

Curiously, another class of macromolecules, specifically dendrimers, exhibits a facile one-dimensional self-assembling ability. Acetylene-linked stiff dendrimers were expected to assemble into high-order structures, due to intermolecular π–π stacking. They were composed of a pyrene core and carbazole/fluorine dendrons, with fluorinated groups at the periphery. The dendrimers were dissolved in a mixture of organic solvents. Since the "good" solvent evaporated faster than the "poor" solvent, the solubility of the dendrimer molecules was decreased and this induced aggregation. One of the compounds assembled

in nanofibers of heterogeneous size. Interestingly, the fluorescence of the fibers was blue-shifted with respect to that of the isolated dendrimer molecule (Zhao, Lee et al., 2008).
The formation of micrometer-long fluorescent nanofibers was also reported for polyphenylene dendrimers functionalized with one or multiple peryleneimide fluorophores. The fibers were prepared by simple evaporation of a solvent on various substrates such as silicon, glass or mica. The morphology of the fibers depends on the preparation method. When the fluorophore is embedded in the dendrimer core, the emission of the fiber is similar to that of the isolated chromophores, due to shielding effect. In contrast, when the fluorophore is attached to the surface of the dendrimer, it suffers interactions with other peripheral fluorophores (Liu et al., 2003).

3.4 Nanofibers made of low-molecular-weight molecules with appended fatty moieties

Let us now turn our attention on nanofibers made of low-molecular-weight molecules. The "top-down" approach used for organic macromolecules is not adapted. Besides, the fragile organic molecules are generally damaged by the microfabrication techniques, such as electron-beam/focused ion beam deposition, traditionally used for inorganic compounds. Thus, the best strategy to prepare quantitative amounts of nanofibers is probably the "bottom-up" approach, which turns to good account the self-association properties of the organic molecules.
From a general viewpoint, several organic systems have been reported to self-organize in solution into micro- or nanofibers (Shimizu et al., 2005). It must be noted that the fibers of concern here are potentially fluorescent, although this property is not always reported. The molecules generally contain a large aromatic moiety that is prone to stacking, and a lipophilic moiety that provides supplementary van der Waals attractions between adjacent molecules within the stacks, and between neighboring stacks to give nanostructures (Hoeben, 2005).
Some organic gelators spontaneously give fluorescent fibers in organic solvents. Among them are rod-shaped alkoxy-tetracene (**1**) and alkoxy-anthracene (**2**) (Del Guerzo et al., 2005; Desvergne et al., 2006; Shklyarevskiy et al., 2005; Reichwagen et al., 2005), and the cholesterol-appended derivatives of phenanthroline (**3**) (Sugiyasu et al., 2005) and salicylideneaniline (**4**) (Xue et al., 2007) (Fig. 2).

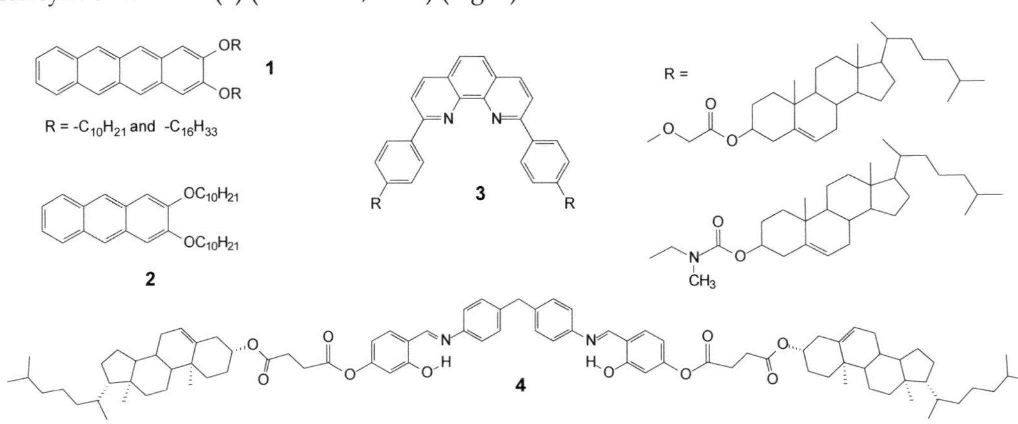

Fig. 2. Chemical structure of compounds **1-4**

Highly dipolar merocyanine dyes (**5**) (Würthner et al., 2003; Yao et al., 2004), *p*-phenylenevinylene (**6**) (Jeukens et al., 2005), perylene diimide trimers (**7**) (Yan et al., 2005), and a bi-oxadiazole derivative (**8**) (Qu et al., 2009), all associated to long alkyl chains, were also reported to give helical stacks, although the molecular arrangement is different for each of these compounds (Fig. 3).

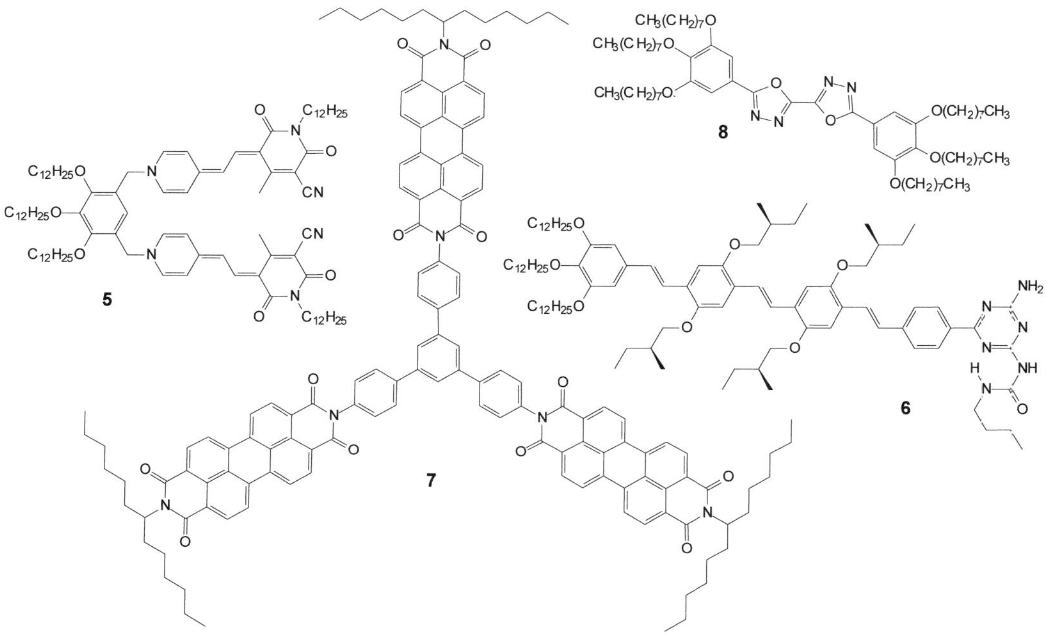

Fig. 3. Chemical structure of compounds **5-8**

For molecules like the carbazole-cornered tetracycle (**9**), π–π staking between molecules is extremely effective and micrometer-long nanofibers can be easily fabricated via surface casting (Balakrishnan et al., 2006; Naddo et al., 2007). For **10**, in contrast, the formation of fibers is driven by multiple hydrogen bonds between the six organic acid groups that are situated at both ends of the molecule. These rigid molecules favor high fluorescence efficiency in the solid state (Luo et al., 2009). It must also be noted that other compounds, such as hexa-peri-hexabenzocoronene linked to fatty chains (**11**), which are known to give columnar liquid crystals, can afford micrometer-long fibers (Pisula et al., 2004) (Fig. 4).

Amphiphilic compounds that self-organize into fluorescent nanofibers can be obtained by branching, for example, three poly(ether) segments to a pyrene core (**12**) (Kimura et al., 2009). In the T-shaped aromatic amphiphiles based on oligo(ethylene oxide) dendrons (**13**), the aromatic segments are stacked with dimeric association to maximize π–π interactions. Curiously, these molecules lead to a reversible phase transition from a fluid state to a gel with increasing temperature (Moon et al., 2007). The presence of an ionized group in the compound also bestows amphiphilic properties, and so water is allowed to be used as a solvent for thin film deposition. For instance, nanofibers were formed from aqueous solutions of a bola-amphiphile containing two carboxylic terminal groups and a central aromatic moiety (**14**) (Song et al., 2007). This is also the case for a cationic perylene diimide incorporating a hydrophobic ether tail (**15**) (Everett et al., 2006) (Fig. 5).

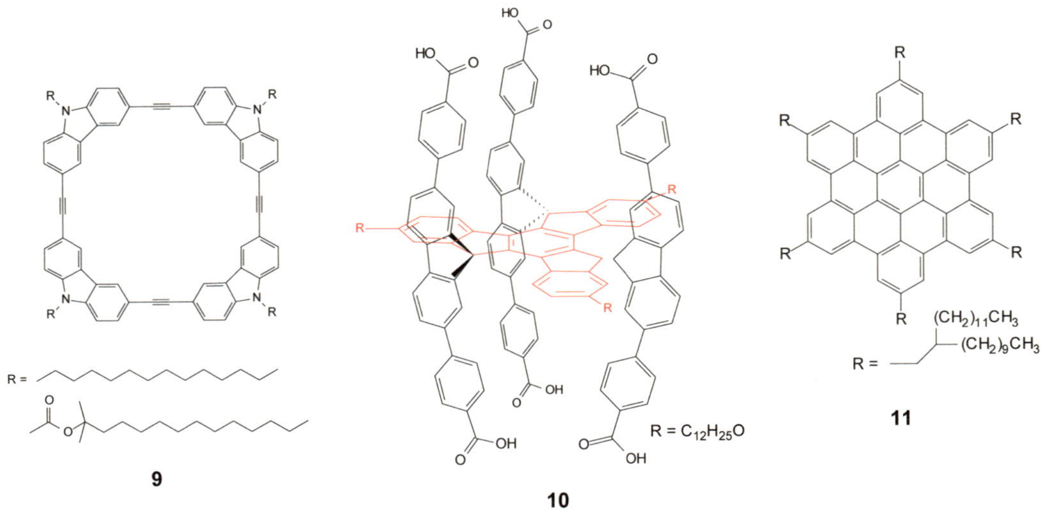

Fig. 4. Chemical structure of compounds **9** to **11**

These fibers could lead to interesting future applications. Some of them show chemical sensing (Naddo et al., 2007) and thermosensing ability (Moon et al., 2007), as well as light-harvesting properties (Desvergne et al., 2006). Their use has been proposed as semiconducting wires in organic electronics and photovoltaics (Everett et al., 2006; Luo et al., 2009). It is noteworthy that the fibers made of **12** were subsequently wrapped with an ultrathin silica wall, using a sol-gel process. The resulting fibers can then be considered as electronic cables that are composed of a conductive organic core and an insulating inorganic layer (Kimura et al., 2009).

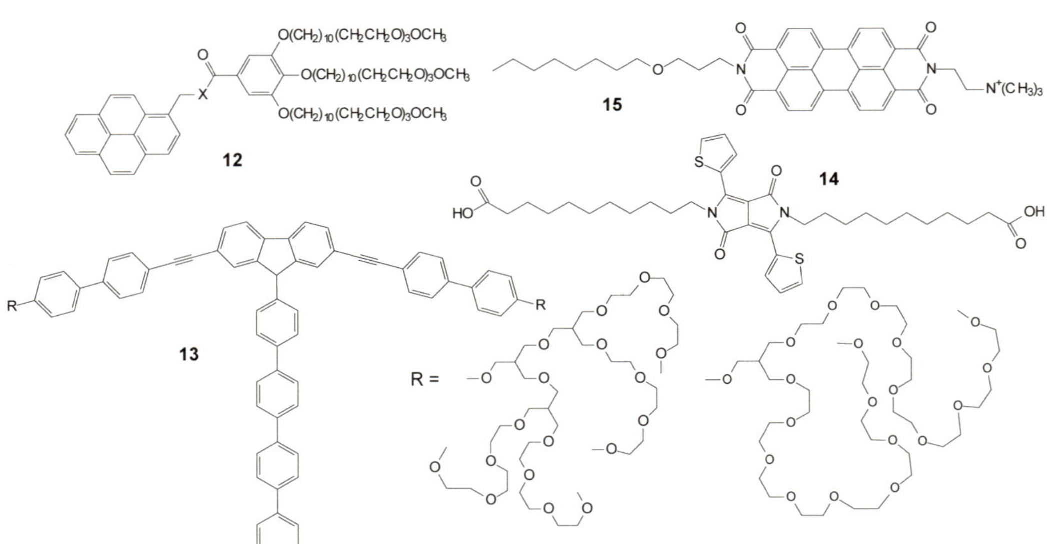

Fig. 5. Chemical structure of compounds **12** to **15**

3.5 Nanofibers made of low-molecular-weight molecules without appended fatty moieties

Most of the fibers made of molecules that contain a bulky fatty moiety are not crystallized. If the aim is to prepare crystallized materials, it is better to work with small molecules deprived of long alkyl chains. Many of them can give needle-like crystals that can be seen like nanofibers. But, the formation of these fibers is not easy and sometimes requires elaborated preparation methods. The fibers may be grown by either vapor or solution processes. The vapor growth process is usually carried out in a quartz tube by sublimating the compounds in a high temperature zone, and transporting the vapor with an inert gas till a low-temperature zone where it can crystallize again. This approach is widely used for oligomers and low-molecular-weight organic semiconductors, for example phthalocyanines (Tang et al., 2008). A variant is to sublimate the compound into a hot-wall tube loaded with a mineral substrate in a vacuum chamber, so that the molecules grow epitaxially on the substrate under conditions close to thermodynamic equilibrium (Yanagi et al., 2001 and refs cited). This method has been used independently by many teams (Balzer et al., 2003; Balzer & Rubahn, 2005; 1999; Hernandez-Sosa et al., 2009; Yanagi et al., 2001; Yanagi & Morikawa) to fabricate nanofibers of *p*-hexaphenyl (**16**) (Fig. 6). The structures obtained are morphologically well-defined and can be individually transferred by micromanipulation to be integrated in various devices (Kjelstrup-Hansen et al., 2006). They emit intense blue light upon electrical or UV-light stimulation. Light is polarized because of the crystallinity of the fibers. Interestingly, waveguiding and coherent random laser action have been observed in single and bundled nanofibers (Quochi et al., 2006; Yanagi & Morikawa, 1999; Yanagi et al., 2001). Functionalizing *p*-hexaphenyl would allow the morphology and the luminescence of the fibers to be modified. Unfortunately, the compound does not lend itself easily to chemical modifications, because of its low solubility. Functionalization has thus been achieved on quaterphenyl (**17**) (Brewer et al., 2006; Schiek et al, 2006; Schiek et al, 2007). Another strategy is to dope the nanofibers *p*-hexaphenyl with α-hexathiophene (**18**). Since the two materials are combined in a crystalline arrangement, their transition dipoles are highly aligned and the energy transfer efficiency is maximized. This results in a significant modification of the emission properties (Hernandez-Sosa et al., 2009). Fluorescent pyrene (**19**) and POPOP (**20**) nanowires with a diameter of about 70 nm were also obtained by classical vapor-deposition method, by introducing the organic fluorophore into a porous anodic aluminium oxide membrane and then dissolving the matrix (Lee et al., 2002). These examples show that single-crystalline nanofibers can indeed be obtained from small molecules. Their optical properties are extremely interesting for subsequent use in optoelectronic devices. It can also be noted that nanofibers of *p*-hexaphenyl have recently been coupled to bovine serum albumin, which could lead to a new type of protein sensors (Madsen et al., 2007).

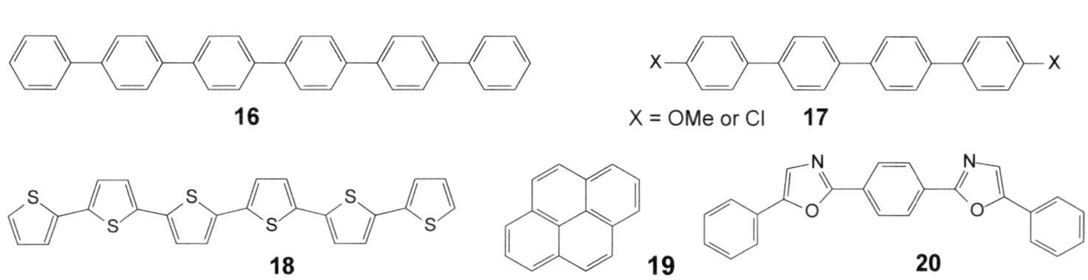

Fig. 6. Chemical structure of compounds **16** to **20**

However, the vapor-deposition method is only suitable for a small number of organic molecules, because many of them decompose upon heating. Therefore, a convenient alternative to generate nanofibers is to use a very simple and soft method based on solvent exchange. For instance, the perylene derivative **21** (Fig. 7) gives entangled fibers when a small volume of a concentrated solution of this compound is injected into a large volume of hexane (Che et al., 2008; Che & Zang, 2009). More often, the hydrophobic dye is dissolved in an organic solvent and water is used as a non-solvent. The dye very quickly precipitates and may crystallize. A very nice example of the use of the "reprecipitation" method has been given by Park's team for the cyanostilbene derivative **22**. Interestingly, these nanofibers are highly fluorescent while the isolated molecule itself is inherently nonfluorescent. This is a rare example of aggregation-induced emission enhancement (AIEE), which has been attributed here to the planarization and rigidification of the aromatic system in the solid state (An et al., 2004). Using analogues deprived of cyano and fluorinated groups, it was shown that a synergistic effect from these groups was necessary to induce the formation of fluorescent fibers. A strong π–π stacking interaction arises from the rigid rod-like aromatic segments, while additional intermolecular interactions are induced by CF_3 groups and the cyano group induces the head-to-tail type molecular arrangement.(An et al., 2009). Keeping these essential features, substituted molecules (**23-25**) were synthesized, all giving nanofibers that emit in different spectral range. It can be noted that the nanofibers actually display a hierarchical organization, since they consist of the assembly of nanofibril-like structural entities with particular orientation.

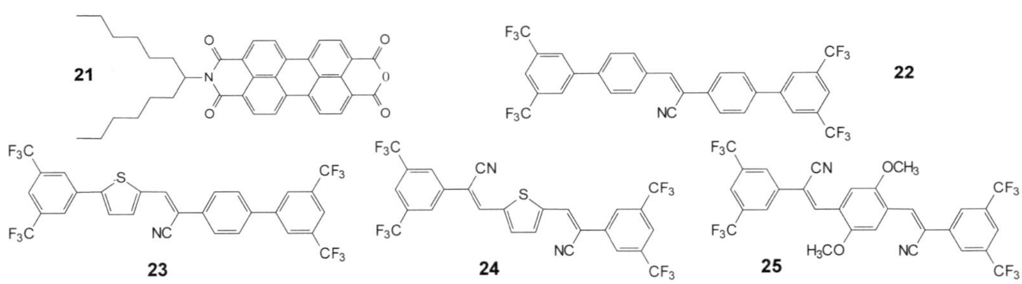

Fig. 7. Chemical structure of compounds **21** to **25**

The interest of the "reprecipitation" method is that many parameters such as dye concentration, temperature, aging time and the presence of additives, can be varied, thus inducing significant differences in the morphology of the particles obtained. This method has been used successfully in the literature to generate nanofibers from different molecules. For instance, the formation of nanowire crystals has been reported for 1,3-diphenyl-2-pyrazoline (**26**) (Fu et al., 2003) (Fig. 8). It is interesting to note that the morphology of the particles obtained strongly depends on the concentration of a surfactant, cetyltrimethylammonium bromide (CTAB), used as additive. It was proposed that the surfactant plays the role of both surface-active stabilizer and template, rod micelles directing the growth of cylindrical dye nanofibers. Similarly, sodiumdodecylsulfate (SDS) micelles induce the formation of pyrene (**19**) nanofibers (Zhang et al., 2005). In these cases, no 1D self-assembly is formed in the absence of micelles. In contrast, nanofibers of Sudan II (**27**) are formed in the presence of an amphiphilic triblock copolymer (F127) at concentrations below the critical micelle concentration. Curiously, the shape of the particles passes from nanofibers to microrods with increasing the copolymer concentration (Yu & Qin 2009). For a phenylenediamine derivative (**28**) (Li et al., 2004) and triphenylimidazole **29** (Zhao, Peng et al., 2008), the key factor is the aging time. Actually, nanoparticles convert spontaneously from spherical to rod-like aggregates, and then to nanofibers as time elapses. In some cases, many parameters must be optimized together, as for the polydiacetylene derivative (**30**) studied by the team of Nakanishi, a pioneer of the reprecipitation method (Onodera et al., 2001). This compound gives microcrystals after reprecipitation at 298 K in the absence of additive. Adding sodiumdodecylsulfate (SDS) and rising the temperature to 333 K induces the formation of nanowires, the diameter of which is about 50 nm. It must be added that fibers and rod-shaped microcrystals have also been obtained by this method from compounds such as fullerene (Masuhara et al., 2008; Tan et al., 2008), which are potentially fluorescent but whose optical properties have not been reported.

These few examples show that the formation of fluorescent nanofibers using the reprecipitation method has just begun to be investigated. The method has proved to be very efficient. It leads to the formation of homogeneously-sized nanofibers that are obtained as a suspension in an aqueous medium. They can be subsequently filtered and dried before being micromanipulated. It can just be noted that particular experimental conditions can be required to obtain such structures.

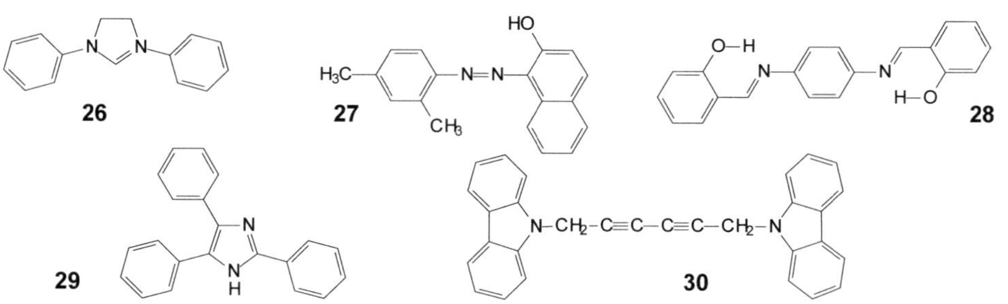

Fig. 8. Chemical structure of compounds **26** to **30**

It appears that the spontaneous formation of fluorescent nanofibers from small dyes without any fatty chains is rather uncommon. Some molecules, such as dye **26**, readily self-assemble into fluorescent nanofibers, whatever the preparation method used. Besides reprecipitation, different fabrication methods have been investigated, including drop casting, recrystallization and vacuum evaporation. All lead to nanofibers, although in different yield, diameter and aspect. However, this is far from being the case for the majority of molecules, for which appropriate fabrication methods must be implemented. And it is very difficult to predict that a molecule will give crystallized fibers or any other type of nano- or microparticle, and whether the fluorescence properties will be interesting or not. As an illustration, we report below a summary of the work that has been performed in our team using the reprecipitation method to prepare fibers from three different families of dyes.

4. Nanofibers in the coumarin series

The first study (Fery-Forgues et al., 2008; Mille et al, 2008) deals with four dyes of the coumarin series. This type of dyes is currently used for applications in fields as varied as the coloring of textile fibers, medical imaging, lasers, photonic devices and organic light-emitting diodes and their spectroscopic behavior in solution is well known. The selected compounds were made of the same chromophores with very little chemical modifications (Fig. 9). Since they are hydrophobic and very fluorescent in the solid state, the molecules appeared as good candidates to give fluorescent particles by the reprecipitation method.

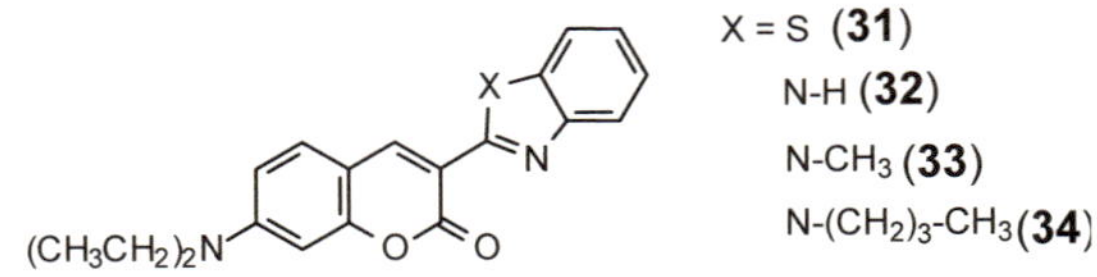

Fig. 9. Chemical structure of the coumarin derivatives **31-34**.

The reprecipitation process took between 4 and 40 min depending on the dye. When observed by fluorescence microscopy, all samples showed strongly emissive particles. It appeared that Coumarin 6 (**31**) lead to small rod-shaped microcrystals that were about 1-2 μm long. Coumarin 7 (**32**) gave significantly longer microcrystals (Fig. 10a). Coumarin 30 (**33**) gave straight individual nanofibers (Fig. 10b) and the butyl derivative **34** gave bundles of fibers.

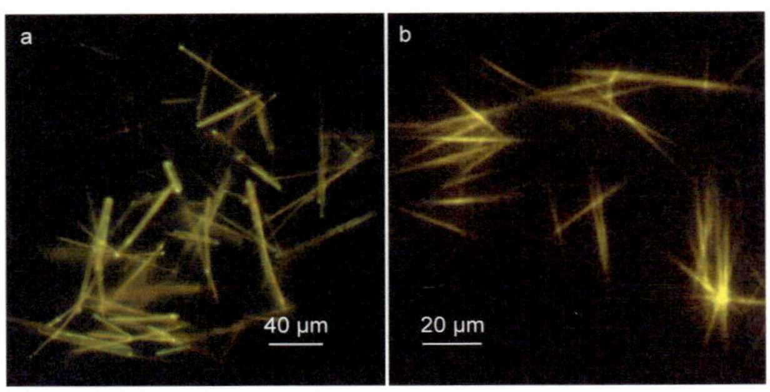

Fig. 10. Fluorescence microscopy images of compounds **32** (a) and **33** (b) after reprecipitation in water containing 2% ethanol. Dye concentration: 2 × 10^{-5} M.

The size of the particles was measured precisely by transmission (TEM) and scanning (SEM) electron microscopy. In particular, the straight fibers of Coumarin 30 were between 5 and 10 μm long, and less than 500 nm wide (Fig. 11a). For the butyl coumarin **34**, the fibers were about 1 μm long and 25 nm wide, but closely entangled (Fig. 11b). It was clear from the SEM images that all fibers were solid.

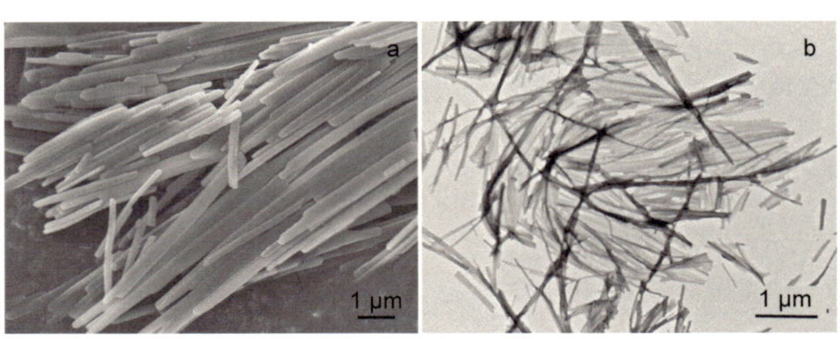

Fig. 11. a) Scanning electron microscopy image of the nanofibers formed in the aqueous suspensions of compound **33** after reprecipitation. b) Transmission electron microscopy image of the nanofibers formed by **34**.

Concerning the molecular arrangement, the crystal structures of Coumarin 6 and 7 has been published (Jasinski & Paight, 1995; Chinnakali et al. 1990). The molecules are almost planar and centrosymmetrically arranged in the crystal unit. The dipole moments are thus stabilized and this configuration is compatible with fluorescence. The crystal structure of the two other compounds was unknown, but for **33**, a clear powder diffraction spectrum ascertained the crystallinity of the nanofibers, which actually can be seen as a modification of the crystal. For **34**, close spectroscopic similarities suggested that the molecular arrangement was quite close from that encountered in **33**, and the anisotropic growth indicates a preferred direction of assembly and thus some degree of order.

The spectroscopic properties of the particles were investigated directly on the suspensions when reprecipitation was complete. The most striking feature was that the fluorescence lifetimes were shorter for the suspensions than for the dissolved dyes, and the quantum yields, all in the 10^{-2} range, were much lower. This can result from crystallization, but also from fluorescence quenching by water molecules, around or within the particles.

It was interesting to see through these examples that Coumarin 6 and 7 that tend to crystallize anisotropically, with a preferred growth axis, giving macroscopic rod-shape crystals, lead to microrods by the reprecipitation method. When tiny modifications were introduced in the chemical structure of Coumarin 7 by increasing the length of the alkyl chain, resulting in compounds **33** and **34**, the major crystallization trend was retained. Actually, the ultimate fate of these molecules appears to be mostly solid nanofibers. The presence of a small alkyl group that bring additional Van der Waals interactions between the molecules is thus enough to induce the formation of elongated crystals. Increasing the size of this alkyl chain by two carbon atoms favors the formation of fiber bundles. It must be underlined that the spontaneous formation of nanofibers was rather unexpected considering the chemical structure of the dye molecules.

5. Nanofibers in the 2-phenylbenzoxazole series

To avoid the problem of fluorescence quenching by water, attention was paid to another series of dyes, which were expected to be much less sensitive to hydrogen bonds. The derivatives of 2-phenyl-benzoxazole are rod-shaped molecules, with excellent fluorescence properties. They are used as textile whitening agents, electroluminescent materials in OLEDs, etc. They display very good chemical and photochemical stability and they are quite easy to synthesize. So, a series of compounds differently substituted on the phenyl group was prepared, as well as a naphthoxazole derivative (Fig. 12).

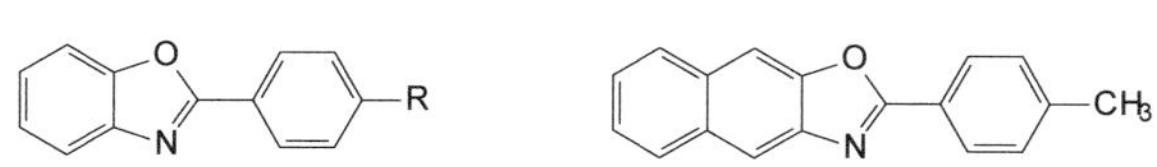

R = CH_3 (**35**), *n*-Bu (**36**), *t*-Bu (**37**), $-(CH_2)_7-CH_3$ (**38**)

39

Fig. 12. Chemical structure of the benzoxazole derivatives **35-39**

The particles obtained by reprecipitation were observed by fluorescence microscopy. It is very interesting to see that, among the compounds investigated, only the methyl derivative **35** gave long beautiful nanofibers. Compounds **36** to **38** gave platelets and shapeless microcrystals, agglomerated or not. The naphthoxazole derivative **39** gave microrods. If

comparing the particles obtained by the reprecipitation method with the crystals obtained by recrystallization in methanol, it was obvious that the reprecipitation method did not just reduce the size of the crystals, but also changed the shape. This is reminiscent of a former work done by our team on fluorescent microcrystals made of a nitrobenzoxadiazole (NBD) derivative. We showed that the shape of the microcrystals was strongly affected by the reprecipitation conditions, although the crystal packing mode seemed to be unchanged (Abyan et al., 2009). So, it not easy to predict which type of particles will be obtained by reprecipitation.
The X-ray analysis of a single crystal of the methyl derivative was obtained. In the unit cell the molecules are displayed by pairs, parallel to each others. They are arranged head-to-tail and the one is shifted with regard to the other one. A second pair of molecules is displayed perpendicularly to first one. This molecular arrangement may favor fluorescence. Actually, the spectroscopic study of the suspensions showed that particles are indeed very fluorescent. The emission spectra, in the ultra-violet range, were almost the same for the four benzoxazole derivatives. It was shifted to the visible for the naphthoxazole derivative, which emitted blue light as could be expected. The lifetimes were in the nanosecond range and the quantum yields were quite high. These properties indicate the interest of such molecules for subsequent use (D'Altério et al., to be published).

6. Nanofibers based on berberine ion pairs

The last series of dyes that was investigated is based on the berberine cation. This cation is an alkaloid, derived from several families of plants. It displays wide biological activity and has been used for long in traditional medicines. This aromatic heterocycle is also known for its interesting spectroscopic properties. In the bibliography, it appears that this cation has been associated to different inorganic anions, thus giving a crystallized, water-soluble salt. So, the idea was to associate this cation to organic anions such as palmitate, dodecylsulfate or laurate (Fig. 13), in order to promote aggregation in water, with the hope to keep crystallinity.

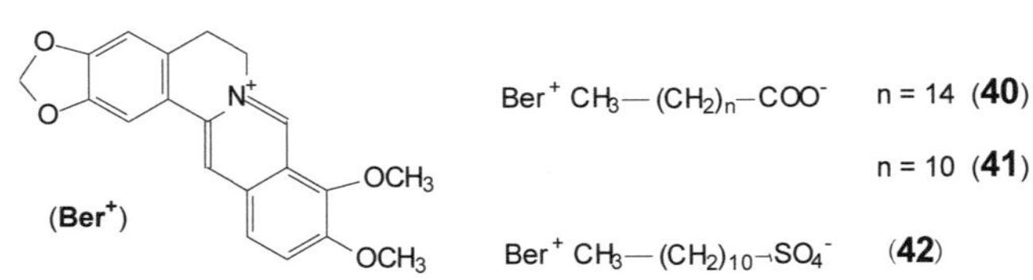

Fig. 13. Chemical structure of the berberine derivatives **40-42**

Reprecipitation of the three dyes was thus performed, the process being achieved in less than 8 min. The three suspensions contained microfibers that were visible with the fluorescence microscope, as well as nanofibers that were detected with TEM. Examples are given in Figure 14. Regarding the spectroscopic properties, it must be stressed that the berberine cation is very poorly fluorescent in water. We showed that this is due to a dynamic fluorescence quenching by water molecules. However, all the suspensions obtained by reprecipitation were very fluorescent. It was easy to show that fluorescence arises from the crystallized salt. This is a very nice example of aggregation induced emission enhancement.

The explanation must be found in the very unusual packing mode of these compounds. Actually, the berberine cations form a layer that is sandwiched between two layers of fatty chains. So, the fatty chains prevent fluorophore stacking. Meanwhile, the berberine cations are protected from the access of water molecules and this explains why the fibers are so strongly fluorescent (Chahine et al., to be published).

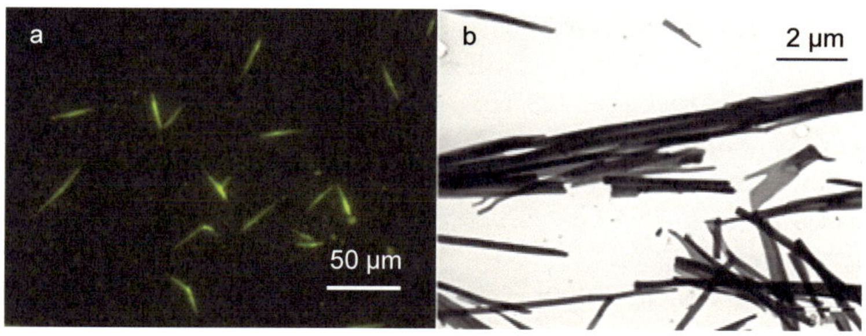

Fig. 14. Fluorescence (a) and transmission electron microscopy image (b) of the micro- and nanofibers formed in the aqueous suspensions of berberine laurate (**41**) after reprecipitation (dye concentration: 6.17×10^{-5} M in water containing 1.2% ethanol).

This fluorescent ion-pair concept is extremely interesting because of its large versatility and ease of implementation. Actually, a large combination of ion pairs can be investigated to modulate both the photophysical properties and the morphology of the particles obtained, in view of a desired application. There is no doubt that this strategy can be of high interest for the design of new molecular materials.

7. Conclusion

A cursory glance at the bibliography is enough to see that the interest for fluorescent nanofibers has very recently emerged. This field is still in its infancy, but rapidly expanding. Various preparation methods are available depending on the type of fibers that must be obtained. Progress can be expected in improving the physical properties of the fibers, which must be of homogeneous size and if possible deprived of surface defects. Concerning the photophysical properties, the limitations are now well understood, and there are numerous possibilities to cope with them. For example, encapsulation of dyes to produce fluorescent nanofibers is particularly promising and still poorly developed. Consequently, systems that display efficient light emission can be expected for a near future.

With slight improvements, luminescent nanofibers have potential for becoming important key elements for nanoscaled optoelectronic devices. These novel materials may also enable a number of new technologies. Their story is only beginning.

8. References

Abyan, M.; De Caro, D. & Fery-Forgues, S. (2009). Suspensions of organic microcrystals produced in the presence of polymers: Diversity of UV/vis absorption and fluorescence properties according to the preparation conditions. *Langmuir*, 25, 1651-1658.

An, B. K.; Lee, D. S.; Lee, J. S.; Park, Y. S.; Song, H. S. & Park, S. Y. (2004). Strongly fluorescent organogel system comprising fibrillar self-assembly of a trifluoromethyl-based cyanostilbene derivative. *J. Am. Chem. Soc.*, 126, 10232-10233.

An, B.-K.; Gihm, S. H.; Chung, J. W.; Park, C. R.; Kwon, S.-K. & Park, S. Y. (2009). Color-tuned highly fluorescent organic nanowires/nanofabrics: Easy massive fabrication and molecular structural origin. *J. Am. Chem. Soc.*, 131, 3950-3957.

Balakrishnan, K.; Datar, A.; Zhang, W.; Yang, X.; Naddo, T.; Huang, J.; Zuo, J.; Yen, M.; Moore, J. S.; Zang, L. (2006). Nanofibrils self-assembly of an arylene ethynylene macrocycle. *J. Am. Chem. Soc.*, 128, 6576-6577.

Balzer, F.; Bordo, V. G.; Simonsen, A. C. & Rubahn, H.-G. (2003). Optical waveguiding in individual nanometer-scale organic fibers. *Phys. Rev. B* , 67, 115408.

Balzer, F. & Rubahn, H.-G. (2005). Growth control and optics of organic nanoaggregates. *Adv. Funct. Mater.*, 1, 17-24.

Behanna, H. A.; Rajangam, K. & Stupp, S. I. (2007). Modulation of fluorescence through coassembly of molecules in organic nanostructures. *J. Am. Chem. Soc.*, 129, 321-327.

Bellan, L. M.; Strychalski, E. A. & Craighead H. G. (2007). Electrospun DNA nanofibers. *J. Vac. Sci. Technol. B*, 25, 2255-2257.

Birks, J. B. (1970). *Photophysics of Aromatic Molecules*, Wiley, London. ISBN-10: 0471074209.

Brewer, J.; Schiek, M.; Luetzen, A.; Al Shamery, K. & Rubahn, H.-G. (2006). Nanofiber frequency doublers. *Nano Lett.*, 6, 2656-2659.

Brewer, J.& Rubahn, H. G. (2005). Angular distribution of luminescence from quasi single crystalline organic nanofibers. *Phys. Stat. Sol. (c)*, 12, 4058-4061.

Camposeo, A.; Di Benedetto, F.; Stabile, R.; Cingolani, R. & Pisignano, D. (2007). Electrospun dye-doped polymer nanofibers emitting in the near infrared. *Appl. Phys. Lett.*, 90, 143115/1-143115/3.

Channon, K.J.; Devlin, G.L.; Magennis, S.W.; Finlatson, C.E.; Tickler, A.K.; Silva, C. & MacPhee, C.E. (2008). Modification of fluorophore photophysics through peptide-driven self-assembly. *J. Am. Chem. Soc.*, 130, 5487-5491.

Channon, K. J.; Devlin, G. L. & MacPhee, C. E. (2009). Efficient energy transfer within self-assembling peptide fibers: a route to light-harvesting nanomaterials. *J. Am. Chem. Soc.*, 131, 12520-12521.

Channon, K. J.; Devlin, G. L.; Magennis, S. W.; Finlayson, C. E.; Tickler, A. K.; Silva, C. & MacPhee, C. E. (2008). Modification of fluorophore photophysics through peptide-driven self-assembly. *J. Am. Chem. Soc.*, 130, 5487-5491.

Che, Y.; Yang, X.; Loser, S. & Zang, L. (2008). Expedient vapor probing of organic amines using fluorescent nanofibers fabricated from an n-type organic semiconductor. *Nano Lett.*, 8, 2219-2223.

Che, Y.& Zang, L. (2009). Enhanced fluorescence sensing of amine vapor based on ultrathin nanofibers. *Chem. Commun.*, 34, 5106-5108.

Chen, H.-C.; Wang, C.-T.; Liu, C.-L.; Liu, Y.-C. & Chen, W.-C. (2009). Full color light-emitting electrospun nanofibers prepared from PFO/MEH-PPV/PMMA ternary blends. *J. Polym. Sci. B*, 47, 463-470.

Chinnakali, K.; Sivakumar, K. & Natarajan, S. (1990). Structure of 3-(2-benzimidazolyl)-7-(diethylamino)coumarin. *Acta Cryst.*, C46, 405-407.

Cornil, J.; Beljonne, D.; Dos Santos, D. A.; Calbert, J. P.; Shuai, Z. & Brédas, J. L. (2000). A theoretical insight into the solid-state optical properties of luminescent materials: the supermolecular approach. *C. R. Acad. Sci. Paris, Series IV*, 403-408.

Cucchi, I.; Spano, F.; Giovanella, U.; Catellani, M.; Varesano, A.; Calzaferri, G. & Botta, C. (2007). Fluorescent electrospun nanofibers embedding dye-loaded zeolite crystals. *Small*, 3, 305-309.

Del Guerzo, A.; Olive, A. G. L.; Reichwagen, J.; Hopf, H. & Desvergne, J. P. (2005). Energy transfer in self-assembled [n]-acene fibers involving ≥ 100 donors per acceptor. *J. Am. Chem. Soc.*, 127, 17984-17985.

Desvergne, J.-P.; Olive, A. G. L.; Sangeetha, N. M.; Reichwagen, J.; Hopf, H. & Del Guerzo, A. (2006). Self-assembling and light-harvesting properties of fluorescent linear condensed aromatic gelators. *Pure Appl. Chem.*, 78, 2333-2339.

Di Benedetto, F.; Mele, E.; Camposeo, A.; Athanassiou, A.; Cingolani, R. & Pisignano, D. (2008). Photoswitchable organic nanofibers. *Adv. Mater.*, 20, 314-318.

Everett, T. A.; Twite, A. A.; Xie, A.; Battina, S. K.; Hua, D. H. & Higgins, D. A. (2006). Preparation and characterization of nanofibrous perylene-diimide-polyelectrolyte composite thin film. *Chem. Mater.*, 18, 5937-5943.

Fantini, D. & Costa, L. (2009). Dye, fluorophores and pigment coloration of nanofibers produced by electrospinning. *Polym. Adv. Technol.*, 20, 111-121.

Fery-Forgues, S.; El-Ayoubi, R. & Lamère, J.-F. (2008). *Fluorescent microcrystals obtained from coumarin 6 using the reprecipitation method. J. Fluoresc.*, 18, 619-624.

Fu, H.; Xiao, D.; Yao, J. & Yang G. (2003). Nanofibers of 1,3-diphenyl-2-pyrazoline induced by cetyltrimethylammonium bromide micelles. *Angew. Chem. Int. Ed.*, 42, 2883-2886.

Henrichsen, H. H.; Kjelstrup-Hansen, J.; Engstrøm, D.; Clausen, C. H.; Bøggild, P. & Rubahn, H.-G. (2007). Electrical conductivity of organic single-nanofiber devices with different contact materials. *Org. Electron.*, 8, 540-544.

Hernandez-Sosa, G.; Simbrunner, C. & Sitter, H. (2009). Growth and optical properties of a α-sexithiophene doped para-sexiphenyl nanofibers. *Appl. Phys. Lett.*, 95, 013306.

Hoeben, F. J. M.; Jonkheijm, P.; Meijer, E. W. & Schenning, A. P. H. J. (2005). About supramolecular assemblies of π-conjugated systems. *Chem. Rev.*, 105, 1491-1546.

Iwaura, R.; Ohnishi-Kameyama, M. & Iizawa, T. (2009). Construction of helical J-aggregates self-assembled from a thymidylic acid appended anthracene dye and DNA as a template. *Chem. Eur. J.*, 15, 3729-3735.

Jasinski, J. P. & Paight, E. S. (1995). 3-(2-benzothiazolyl)-7-(diethylamino)-coumarin. *Acta Cryst.*, C51, 531-533.

Jeukens, C. R. L. P. N.; Jonkheijm, P.; Wijnen, F. J. P.; Gielen, J. C.; Christianen, P. C. M.; Schenning, A. P. H. J.; Meijer, E. W. & Maan, J. C. (2005). Polarized emission of individual self-assembled oligo(*p*-phenylenevinylene)-based nanofibers on a solid support. *J. Am. Chem. Soc.*, 127, 8280-8281.

Kang, S. H.; Jung, B. M. & Chang, J. Y. (2007). Polymerization of an organogel formed by a hetero-bifunctional gelator in a monomeric solvent: preparation of nanofibers embedded in a polymer matrix. *Adv. Mater.*, 19, 2780-2784.

Katoh, R.; Suzuki, K.; Furube, A.; Kotani, M. & Tokumaru, K. (2009). Fluorescence quantum yield of aromatic hydrocarbon crystals. *J. Phys. Chem. C*, 113, 2961-2965.

Kimura, M.; Miki, N.; Suzuki, D.; Adachi, N.; Tatewaki, Y. & Shirai, H. (2009). Wrapping of self-organized fluorescent nanofibers with a silica wall. *Langmuir*, 25, 776-780.

Kjelstrup-Hansen, J.; Bøggild, P.; Hvam, J.; Majcher, A. & Rubahn, H.-G. (2006). Micromanipulation of organic nanofibers for blue emitting microstructures. *Phys. Stat. Sol. (a)*, 6, 1459-1463.

Kobayashi, T. (1996). *J-Aggregates*. World Scientific, Singapore. ISBN: 981-02-2737-X.

Law, B.; Weissleder, R. & Tung, C.-H. (2007). Protease-sensitive fluorescent nanofibers. *Bioconjugate Chem.*, 18, 1701-1704.

Lee, J. K.; Koh, W. K.; Chae, W. S. & Kim, Y. R. (2002). Novel synthesis of organic nanowires and their optical properties. *Chem. Commun.*, 138-139.

Lee, K. J.; Oh, J. H.; Kim, Y. & Jang, J. (2006). Fabrication of photoluminescent-dye embedded poly(methyl methacrylate) nanofibers and their fluorescence resonance energy transfer properties. *Adv. Mater.*, 18, 2216-2219.

Li, S.; He, L.; Xiong, F.; Li, Y.; Yang, G. (2004). Enhanced fluorescent emission of organic nanoparticles of an intramolecular proton transfer compound and spontaneous formation of one-dimensional nanostructures. *J. Phys. Chem. B*, 108, 10887-10892.

Li, F.; Zhao, Y.; Wang, S.; Han, D.; Jiang, L. & Song, Y. (2009). Thermochromic core-shell nanofibers fabricated by melt coaxial electrospinning. *J. Appl. Polym. Sci.*, 112, 269-274.

Liang, X. ; Li, Y. ; Peng, W. ; Bai, J. ; Zhang, C. & Yang, Q. (2008). Efficient method for fabrication of fluorescein derivative/PDAC composite nanofibers and characteristics of their photoluminescent properties. *Eur. Polym. J.*, 44, 3156-3162.

Liu, D.; De Feyter, S.; Cotlet, M.; Wiesler, U.-M. ; Weil, T. ; Herrmann, A.; Muellen, K. & De Schryver, F. C. (2003). Fluorescent self-assembled polyphenylene dendrimer nanofibers. *Macromolecules*, 36, 8489-8498.

Long, Y.; Chen, H.; Yang, Y.; Wang, H.; Yang, Y.; Li, N.; Li, K.; Pei, J. & Liu, F. (2009). Electrospun nanofibrous film doped with a conjugated polymer for DNT fluorescence sensor. *Macromolecules*, 42, 6501-6509.

Luo, J.; Lei, T.; Wang, L.; Ma, Y.; Cao, Y.; Wang, J. & Pei, J. (2009). Highly fluorescent rigid supramolecular polymeric nanowires constructed through multiple hydrogen bonds. *J. Am. Chem. Soc.*, 131, 2076-2077.

Lv, Y.-Y. ; Wu, J. ; Wan, L.-S. & Xu, Z.-K. (2008). Novel porphyrinated polyimide nanofibers by electrospinning. *J. Phys. Chem. C*, 112, 10609-10615.

Madsen, M.; Schiek, M.; Thomsen, P.; Andersen, N. L.; Luetzen & A.; Rubahn, H.-G. (2007). Organic nanofiber nanosensors. *Proc. SPIE*, 6769 (Nanosensing Materials, Devices and Systems III) 676908/1-676908/10.

Masuhara, A.; Tan, Z.; Kasai, H.; Nakanishi, H. & Oikawa, H. (2008). Fabrication and characterization of size- and shape-controlled C_{60} nano/microcrystals by solvent-induced reprecipitaiton process (SIRP). *Mater. Res. Soc. Symp. Proc.*, 1054, 1054-FF11-09.

Mille, M.; Lamère, J.-F.; Rodrigues, F. & Fery-Forgues S. (2008). Spontaneous formation of fluorescent nanofibers from self-assembly of low-molecular-weight coumarin derivatives in water.*Langmuir*, 24, 2671-2679.

Memisevic, J.; Riley, L. & Grant, S. (2009). Electrospun sol-gel fibers for fluorescence-based sensing. *Proc. SPIE, 7313* (Smart Biomedical and Physiological Sensor Technology VI), 73130I/1-73130I/8.

Moon, K.; Kim, H.-J.; Lee, E. & Lee, M. (2007). Self-assembly of T-shaped aromatic amphiphiles into stimulus-responsive nanofibers. *Angewandte Chem. Int. Ed.*, 46, 6807-6810.

Naddo, T.; Che, Y.; Zhang, W.; Balakrishnan, K.; Yang, X.; Yen, M.; Zhao, J.; Moore, J. S. & Zang, L. (2007). Detection of explosives with a fluorescent nanofibril film. *J. Am. Chem. Soc.*, 129, 6978-6979.

Naddo, T.; Yang, X.; Moore, J. S. & Zang, L. (2008). Highly responsive fluorescent sensing of explosives taggant with an organic nanofibril film. *Sensors Actuator B: Chem.*, B134, 287-291.

Ner, Y.; Grote, J. G.; Stuart, J. A. & Sotzing, G. A. (2008). Enhanced fluorescence in electrospun dye-doped DNA nanofibers. *Soft Matter*, 4, 1448-1453.

Ner, Y.; Grote, J. G.; Stuart, J. A. & Sotzing, G. A. (2009). White luminescence from multiple-dye-doped electrospun DNA nanofibers by fluorescence resonance energy transfer. *Angew. Chem. Int. Ed.*, 48, 5134-5138.

Pisula, W.; Kastler, M.; Wasserfallen, D.; Pakula, T. & Müllen, K. (2004). Exceptionally long-range self-assembly of hexa-peri-hexabenzocoronene with dove-tailed alkyl substituents. *J. Am. Chem. Soc.*, 126, 8074-8075.

Onodera, T.; Oshikiri, T.; Katagi, H.; Kasai, H.; Okada, S.; Oikawa, H.; Terauchi, M.; Tanaka, M. & Nakanishi, H. (2001). Nano-wire crystals of π-conjugated organic materials. *J. Crystal Growth*, 229, 586-590.

Qu, S.; Zhao, L.; Yu, Z.; Xiu, Z.; Zhao, C.; Zhang, P.; Long, B. & Li, M. (2009). Nanoparticles, helical fibers, and nanoribbons of an achiral twin-tapered bi-1,3,4-oxadiazole derivative with strong fluorescence. *Langmuir*, 25, 1713-1717.

Quochi, F.; Cordella F.; Mura, A.; Bongiovanni, G.; Balzer, F. & Rubahn, H.-G. (2006). Gain amplification and lasing properties of individual organic naofibers. *Appl. Phys. Lett.*, 88, 141106/1-041106/3.

Reichwagen, J.; Hopf, H.; Del Guerzo, A.; Belin, C.; Bouas-Laurent, H. & Desvergne, J. P. (2005). Synthesis of 2,3-substituted tetracenes and evaluation of their self-assembling properties in organic solvents. *Org. Lett.*, 7, 971-974.

Samitsu, S.; Takanishi, Y. & Yamamoto, J. (2009). Self-assembly and one-dimensional alignment of a conducting polymer nanofiber in a nematic liquid crystal. *Macromolecules*, 42, 4366-4368.

Schiek, M.; Luetzen, A.; Al-Shamery, K.; Balzer, F. & Rubahn, H.-G. (2006). Nanofibers from methoxy functionalized para-phenylene molecules. *Surf. Sci.*, 600, 4030-4033.

Schiek M.; Luetzen, A.; Al-Shamery, K.; Balzer, F. & Rubahn, H.-G. (2007). Organic nanofibers from chloride-functionalized *p*-quaterphenylenes. *Cryst. Growth Des.*, 7, 229-233.

Schiek, M.; Balzer, F.; Al-Shamery, K.; Brewer, J. R.; Lützen, A. & Rubahn, H.-G. (2008). Organic molecular nanotechnology. *Small*, 4, 176-181 and refs. cited.

Shimizu, T.; Masuda, M. & Minamikawa, H. (2005). Supramolecular nanotube architectures based on amphiphilic molecules. *Chem. Rev.*, 105, 1401-1443 and references therein.

Shklyarevskiy, I. O.; Jonkheijm, P.; Christianen, P. C. M.; Schenning, A. P. H. J.; Del Guerzo, A.; Desvergne, J. P.; Meijer, E. W. & Maan, J. C. (2005). Magnetic alignment of self-assembled anthracene organogel fibers. *Langmuir*, 21, 2108-2112.

Silinsh, E. A. (1980) *Organic Molecular Crystals*, Springer-Verlag, Berlin. ISBN: 3-540-10053-9.

Song, B.; Wei, H.; Wang, Z.; Zhang, X.; Smet, M. & Dehaen, W. (2007). Supramolecular nanofibers by self-organization of bola-amphiphiles through a combination of hydrogen bonding and π–π stacking interactions. *Adv. Mater.*, 19, 416-420.

Sugiyasu, K.; Fujita, N.; Shinkai, S. (2005). Fluorescent organogels as templates for sol-gel transcription toward creation of optical nanofibers. *J. Mater. Chem.*, 15, 2747-2754.

Tan, Z.; Masuhara, A.; Kasai, H.; Nakanishi, H.; Oikawa, H. (2008). Multibranched C_{60} micro/nanocrystals fabricated by reprecipitaiton method. Jpn. J. Appl. Phys. 47, 1426-1428.

Tang, Q.; Jiang, L.; Tong, Y.; Li, H.; Liu, Y.; Wang, Z.; Hu, W.; Liu, Y. & Zhu, D. (2008). Micrometer- and nanometer-sized organic single-crystalline transistors. *Adv. Mater.*, 20, 2947-2951.

Tao, S.; Li, G. & Yin, J. (2007). Fluorescent nanofibrous membranes for trace detection of TNT vapor. *J. Mater. Chem.*, 17, 2730-2736.

Tomczak, N.; Gu, S.; Han, M.; van Hurst, N. F. & Vancso, G. J. (2006). Single light emitters in electrospun polymer nanofibers: effect of local confinement on radiative decay. *Eur. Polym. J.*, 42, 2205-2210.

Vohra, V.; Devaux, A.; Dieu, L.-Q.; Scavia, G.; Catellani, M.; Calzaferri, G. & Botta, C. (2009). Energy transfer in fluorescent nanofibers embedding dye-loaded zeolite L crystals. *Adv. Mater.*, 21, 1146-1150.

Wan, L.-S.; Wu, J. & Xu, Z.-K. (2006). Porphyrinated nanofibers via copolymerization and electrospinning. *Macromolecular Rapid Commun.*, 27, 1533-1538.

Wang, S.; Yang, Q.; Du, J.; Bai, J. & Li, Y. (2007). Variety of photoluminescence intensity of fluorescent whitening agents introduced into polyacrylonitrile nanofibers. *J. Appl. Polym. Sci.*, 103, 2382-2386.

Wang, S.; Yang, Q.; Bai, J.; Du, J. & Li, Y. (2008). The influence of polystyrene and polyvinylpyrrolidone nanofiber onthe intensity of photoluminescence of fluorescent whitening agents. *J. Appl. Polym. Sci.*, 107, 1696-1700.

Wang, X.; Lee, S.-H.; Drew, C.; Senecal, K. J.; Kumar, J. & Samuelson, L. A. (2001). Electrospun nanofibrous membranes for optical sensing. *Polym. Mater. Sci. Eng.*, 85, 617-618.

Wang, X.; Drew, C.; Lee, S.-H.; Senecal, K. J.; Kumar, J. & Samuelson, L. A. (2002). Electrospun nanofibrous membranes for highly sensitive optical sensors. *Nano Lett.*, 2, 1273-1275.

Wang, X.; Lee, S.-H.; Ku, B.-C.; Samuelson, L. A. & Kumar, J. (2002). Synthesis and electrospinning of a novel fluorescent polymer PMMA-PM for quenching-based optical sensing. *J. Macromolecular Sci., Pure Appl. Chem.*, A39, 1241-1249.

Würthner, F.; Yao, S. & Beginn, U. (2003). Highly ordered merocyanine dye assemblies by supramolecular polymerization and hierarchical self-organization. *Angew. Chem. Int. Ed.*, 42, 3247-3250.

Xie, Z.; Yang, B.; Li, F.; Cheng, G.; Liu, L.; Yang, G.; Xu, H.; Ye, L.; Hanif, M.; Liu, S.; Ma, D. & Ma, Y. (2005). Cross dipole stacking in the crystal of distyrylbenzene derivative: the approach toward high solid-state luminescence efficiency. *J. Am. Chem. Soc.*, 127, 14152-14153.

Xin, Y.; Huang, Z.; Yang, P.; Jiang, Z.; Wang, C. & Shao, C. (2009). Controlling the deposition of light-emitting nanofibers/microfibers by the electrospinning of a poly(*p*-phenylene vinylene) polyelectrolyte precursor. *J. Appl. Polym. Sci.*, 114, 1864-1969.

Xue, P.; Lu, R.; Chen, G.; Zhang, Y.; Nomoto, H.; Takafuji, M. & Ihara, H. (2007). Functional organogel based on a salicylideneaniline derivative with enhanced fluorescnce emission and photochromism. *Chem. Eur. J.*, 13, 8231-8239.

Yan, P.; Chowdhury, A.; Holman, M. W. & Adams, D. M. (2005). Self-organized perylene diimide nanofibers. *J. Phys. Chem. B*, 109, 724-730.

Yanagi, H. & Morikawa, T. (1999). Self-waveguided blue light emission in *p*-sexiphenyl crystals epitaxially grown by mask-shadowing vapor deposition. *Appl. Phys. Lett.*, 75, 187-189.

Yanagi, H.; Ohara, T. & Morikawa, T. (2001). Self-waveguided gain-narrowing of blue light emission from epitaxially oriented *p*-sexiphenyl crystals. *Adv. Mater.*, 13, 1452-1455 and refs. cited.

Yang, Y.; Fan, X.; Long, Y.; Su, K.; Zou, D.; Li, N.; Zhou, J.; Li, K. & Liu, F. (2009). A simple fabrication of electrospun nanofiber sensing materials based on fluorophore-doped polymer. *J. Mater. Chem.*, 19, 7290-7295.

Yao, S.; Beginn, U.; Gress, T.; Lysetska, M. & Würthner, F. (2004). Supramolecular polymerization and gel formation of bis(merocyanine) dyes driven by dipolar aggregation. *J. Am. Chem. Soc.*, 126, 8336-8348.

Yoon, J. & Kim, J.-M. (2008). Fabrication of conjugated polymer supramolecules in electrospun micro/nanofibers. *Macromol. Chem. Phys.*, 209, 2194-2203.

Yu, H. & Qi, L. (2009). Polymer-assisted crystallization and optical properties of uniform microrods of organic dye Sudan II. *Langmuir*, 25, 6781-6786.

Zang, L.; Moore, J.; Naddo, T. & Zhang, W. (2009). Fluorescent organic nanofibrils as sensory materials for explosives detection. U.S. *Pat. Appl. Publ.*, US2009233374 A1 20090917.

Zhang, W.; Yan, E.; Huang, Z.; Wang, C.; Xin, Y.; Zhao, Q. & Tong, Y. (2007). Preparation and study of PPV/PVA nanofibers via electrospinning PPV precursor alcohol solution. *Eur. Polym. J.*, 43, 802-807.

Zhang X. J.; Zhang, X. H.; Shi, W. S.; Meng, X. M.; Lee, , C. & Lee, S. T. (2005). Morphology-controllable synthesis of pyrene nanostructures and its morphology dependence of optical properties. *J. Phys. Chem. B*, 109, 18777-18780.

Zhao, Q.; Xin, Y.; Huang, Z.; Liu, S.; Yang, C. & Li, Y. (2007). Using poly[2-methoxy-5-(2'-ethyl-hexyloxy)-1,4-phenylene vinylene] as shell to fabricate the highly fluorescent nanofibers by coaxial electrospinning. *Polymer*, 48, 4311-4315.

Zhao, Z.; Li, J.-H.; Chen, X.; Lu, P. & Yang, Y. (2008). Fluorescent conjugated dendrimers with fluorinated teminal groups: nanofiber formation and electroluminescence properties. *Org. Lett.*, 10, 3041-3044.

Zhao, Y. S.; Peng, A.; Fu, H.; Ma, Y. & Yao, J. (2008). Nanowire waveguides and ultraviolet lasers based on small organic molecules. *Adv. Mater.*, 20, 1661-1665.

Zhu, Z.; Zhang, L.; Smith, S.; Fong, H.; Sun, Y. & Gosztola, D. (2009). Fluorescence studies of electrospun MEH-PPV/PEO nanofibers. *Synth. Met.*, 159, 1454-1459.

21

Synthesis of Alumina Nanofibers and Composites

Mohamad Ridzuan Noordin[1] and Kong Yong Liew[1,2]
[1]*Faculty of Industrial Science and Technology, Universiti Malaysia Pahang, Kuantan,*
[2]*College of Chemistry and Materials Science,*
South Central University for Nationalities, Wuhan,
[1]*Malaysia*
[2]*China*

1. Introduction

Alumina, Al_2O_3, is a ceramic metal oxide of great importance as building material, refractory material, electrical and heat insulator, attributed to its high strength, corrosion resistance, chemical stability, low thermal conductivity, and good electrical insulation. One phase of alumina, the γ-phase is widely used as catalyst, catalyst support and adsorbent because its high porosity and surface area. It is conventionally prepared by heating the hydrated trihydroxides, gibbsite and bayerite, to temperature of > 1000°C to obtain the α-alumina, corundum, a material with great hardness and low surface area. The trihydroxides, gibbsite, bayerite, both are $Al(OH)_3 \cdot nH_2O$ and the monohydroxyl oxide, boehmite, $AlO(OH) \cdot aH_2O$, have the monoclinic crystal structure with similar lattice parameters (a=0.866 nm, b=0.506 nm, c=0.983 nm, β=94°34′ in boehmite, a=0.868 nm, b=0.507 nm, c=0.972 nm, β=94°34′ in gibbsite, and a=0.867 nm, b=0.506 nm, c=0.942 nm, β=90°26′ in bayerite [1, 2]. In the process of heat treatment, the trihydroxide undergoes a series of transformations. It loses the water of hydration, then dehydroxalate at < 300°C to form the monohydroxyl oxide boehmite, AlO(OH), which on further heating to increasingly higher temperatures, changes to the transition aluminas, including the η-, γ-, δ-, and θ-phases, which then transform finally to the α-form. These transition aluminas are crystalline solids with high porosity and surface areas as well as acidic and basic properties, which make them suitable as adsorbents, catalyst, catalyst support and fabricated into filtration membranes as well as used as fillers or components in polymer/ inorganic composite materials with enhanced mechanical properties. The conventional alumina as obtained are usually powder of particulates. These could be fabricated into alumina fibers by various method, through melt growth techniques including the internal crystallization method and extrusion, electrospraying [3] and electrospinning [4-8] have been developed for producing Al_2O_3 fibers.

The high surface areas of transition aluminas are due to the presence of pores. These pores are irregular in sizes and size distributions, and are mainly micropores of diameter less than 2 nm. Since the synthesis of mesoporous materials was first reported in the late 1980s , a number of methods for the synthesis of mesoporous alumina have been reported [8,9]. More recently, interest in nano-materials and nanotechnology has spurred a fury of investigations on the preparation and the applications of nano-materials including that of inorganic oxides such as aluminum oxide. Nanoparticles can be of 1, 2 or 3 dimensions. One dimensional nano-structured alumina may be described as fiber, wire, belt, rod, laths and tubes [9-17]. Nanofibers are one dimensional material with high aspect ratio of length to diameter of 1000:1 and diameter of 1 to 100 nm, which may be solid fibers or hollow nanotubes. The final properties of these nanofibers depend on the size and homogeneity of the fibrous structure, which shall enable tailoring Al_2O_3 nanofibers to advanced technological usage with the enhanced sintering ability, strength, catalytic activity, and absorption ability. However, for the present report, lower aspect ratio materials are also included.

2. Synthesis of alumina nanofibers

Various routes including hydrothermal or solvothermal, sol-gel techniques, electrospinning and extrusion, template techniques [19,20], chemical vapor deposition in flame[30], and mercury mediated technique [29] have been developed for producing Al_2O_3 nanofibers. Many of the nanfibers have been produced using hydrothermal treatment, with or without surfactant, [18-25] and sol-gel [17,27] techniques of various alumina precursors, subsequent calcinations dehydrate the solid from the sol-gel solution produced by the hydrothermal process. The solid, which consists of naonfibrous boehmite, may be calcined to obtain the α-alumina nanofibers. The gel has high viscosity and may be electrospun to obtain the nanofibers. The solid either from the hydrothermal or sol-gel processes or the electrospun fibers is dehydrated by heat treatment leading to transition aluminas and finally to the nanofibrous α-alumina. Other methods such as vapor–liquid–solid deposition to burn aluminum together with silica nanoparticles at high temperature between 1300 and 1700 °C, and by using other 1-D materials as hard template have also been reported [13-15, 19, 30].
Most of the earlier reports are on the generation of Al_2O_3 fibers of diameters in the submicrons to several tens of microns range. Only more recently did the synthesis of nanofibrial alumina appeared. However, information regarding the growth and particulate properties of Al_2O_3 nanofiber still remains very limited. For the hydrothermal and sol-gel processes, the aluminum precursors could be Al metal, the alkoxide or aluminum salts. The size and morphology of aluminum hydrate nanoparticles synthesized by these processes can be controlled by manipulating the pH value of the reaction mixture, the hydrothermal temperature and aging time, the solvent and the surfactant used. [2,3,20-27]

2.1 Hydrothermal and sol gel processes

The first step in the synthesis of nanofibrous alumina both these processes is the hydrolysis of the starting aluminum compound or elemental Al powder to the hydrated hydroxide of aluminum, boehmite, which are then aged subsequently. In the hydrothermal process, the aging could be done in an enclosed reactor under its own autogenic pressure, the colloidal solution obtained is either filtered and the solid further processed. In the sol gel process, the sol is transformed to a gel after the aging. The two processes are closely related and the reactions that take place are similar, these processes are discussed together. The boehmite formed in the first step has the formula, $AlOOH.nH_2O$. It may be amourphous particles initially which then transform into network structures linked by hydrogen bonds and forms nanosheets or sheets of crumpled nanostructures. The nanosheets could rolled up and be transformed into hollow tubular structures or nanofibers[17, 31], which may or may not be crystalline. The initial spherical particles of boehmite are sol solution with low viscosity. As it aged, the network structures and fibers are formed and the sol changed into gel with high viscosity. The solid in the gel is initially spherical nanoparticles [17], which on aging gradually transform into the crystalline boehmite. The solid on heating to higher temperature dehydrates and dehydroxylates forming the different phases of transition aluminas in stages at progressively higher temperature than 400°C, which finally transforms to the final α-alumina at temperature of 1000°C or higher. The reactions may be represented as follows [13-16, 18, 19, 21-27, 31]:

$$Al(OH)_3.\ nH_2O \rightarrow \text{amorphous } Al(OH)_4- \ +\ H_3O^+ + (n\text{-}1)\ H_2O \qquad (1)$$

$$Al(OH)_4\text{-} \leftrightarrow \gamma\text{-}AlOOH.nH_2O + (1\text{-}n)H_2O \qquad (2)$$

$$\gamma\text{-}AlOOH \cdot nH_2O \rightarrow \gamma\text{-}Al_2O_3 + 2(1 - n)H_2O \qquad (3)$$

$$\gamma\text{-}Al_2O_3 \rightarrow \rightarrow\rightarrow \alpha\text{-}\ Al_2O_3 \qquad (4)$$

The γ crystalline boehmite is formed under suitable synthesis condition and the morphology is retained in the various transition aluminas as well as in the final α-alumina as demonstrated by the TEM and Xrd characterizations, although the crystal structures are different. The boehmite formed nanosheets in the sol and gelling process and the curled up fibers or tubes were visible in the TEM images of the boehmite.

Alumina nanofibers can be synthesized with metallic aluminum particles [10,22, 43, 59], Al alkoxides [36; 46, 17] or its salts [27, 45, 47, 48, 49, 51] as the starting materials. Other starting materials that have been used include Al pentanedionate, [50], aluminum hydroxide, [52], alumina composite, [53] aluminum film, [42] and sodium aluminate. [54] However, whatever the starting materials used in the synthesis, the nanofibrous boehmite intermediate will have to be formed [22] for the formation of the nanofibers as mentioned above. The starting Al precursor is first hydrolyzed with water to form the hydroxide [17, 22, 32, 37, 38], aged to the viscous gel fibers of hydrated AlOOH, which can be calcined directly or electrospun or drawn from the viscous gel and converted to ceramic fibers after firing [16, 27, 31-37]. The metallic aluminum powder used for the process is prepared by exploding the metal wire by a high density current pulse normally produced by discharging a capacitor bank. The produced particles are spherical of about 80-120 nm [22, 43].

For the case in which metallic Al nanoparticles is used the first step of the reactions are depicted as follows:[17, 22]

$$2\ Al + 4\ H_2O \rightarrow 2\ AlO(OH) + 3\ H_2 \quad \text{without stirring} \qquad (5)$$

or

$$2\ Al + 6\ H_2O \rightarrow 2\ Al(OH)_3 + 3\ H_2 \quad \text{with stirring} \qquad (6)$$

reaction (5) leads to boehmite, while reaction (6) leads to formation of the trihydroxide, bayerite, which on dehydration via a series of transformation forms the boehmite:

$$Al(OH)_3 \rightarrow AlO(OH) + 2\ H_2O \qquad (7)$$

The boehmite will further dehydrate to convert to the transition aluminas which will further dehydrate to γ-Al_2O_3 up to 700°C without significant change in morphology. With the aluminum salts, hydrolysis occurs forming aluminum hydroxide, which then proceeds as in eq.7. Na aluminate reacts with acid to form the aluminum hydroxide and proceeds to be dehydrated similarly.

With alkoxides, $Al(OR)_3$, as the starting materials, the initial hydrolysis with water yield alcohol and the hydroxylated aluminum centers as follows: [17]

$$Al(OR)_3 + H_2O \rightarrow Al(OR)_2(OH) + ROH \quad (8)$$

$$Al(OR)_2(OH) + H_2O \rightarrow AlO(OH) + 2\ ROH \quad (9)$$

$$Al\text{-}OH + Al\text{-}OX\text{-}Al \rightarrow Al\text{-}O\text{-}Al + XOH\ (X=H \text{ or } R) \quad (10)$$

$$2\ AlO(OH) \rightarrow Al_2O_3 + H_2O \quad (11)$$

In the initial stage of the sol gel reaction, small three dimensional oligomers are formed with the -Al-OH on the outer surface. Linkange of –Al-OH groups occurs via polycondensation and results in oxypolymers. When sufficient interconnected Al-O-Al bonds are formed, they respond cooperatively to form the colloidal particles or a sol. These sol particles aggregate and increase in size with time forming three dimensional network and randomly distributed spherical nanoparticles as shown by TEM in the earlier stage of the process. With further aging, the nanosheets are shown to curl up forming the nanofibers. The morphology was seen to be preserved even after calcinations at 1200°C [17,21, 45, 55].

The size and morphology of the end product of the hydrolysis and subsequent dehydration of the intermediate boehmite from the hydrolysis depend on the conditions in which these processes are performed. The primary factors involved the nature of the starting material, whether a surfactant or templating agent is used, the amount of water used, the aging time in the hydrolysis, the pH value of the reaction mixture, and the hydrothermal temperature [17-19, 21-23, 45-47, 55-59]. Even a simple procedure of stirring the hydrolyzing reaction mixture of Al particles in water, affect the outcome of the resulting hydroxylate product. Without stirring, boehmite was produced, while with stirring, the trihydroxide, bayerite, was resulted.[22]. In the sol-gel synthesis using iospropoxide as the starting material [17], the reaction had to be carried out at the range of R_w=60-65, outside of this range no fibrous material was formed. Under acidic conditions, hydrolysis was enhanced and that the rate of polycondensation and aggregation to larger particles was increased too high for nanofibers formation. Additionally, the presence of organic solvent influences the reaction rate significantly. Of the solvent tested only isopropanol in the hydrolysis mixture led to nanofibers formation. Although the morphology is retained during the calcinations of the nanofibrous alumina, the physical property such as the pore size increased with calcinations temperature from the as prepared samples to 1000°C, however, the surface area decreased at temperature > 550°C. Electron microscopic observations of microstructure, together with studies of the evolution of X-ray and electron diffraction patterns, have been used to provide mechanistic information on the dehydration of boehmite to γ-Al_2O_3. Calcinations of boehmite at 400°C produces slow development of a fine, lamellar, porous microstructure, oriented parallel to $(001)_\gamma$, corresponding to the loss of oxygen atoms of the boehmite lattice. The mechanism proposed for the dehydration is controlled by diffusion in a direction determined by the hydrogen bond chains in the boehmite structure and involves counter migration of the Al cations and protons with formation of voids in a coherent cubic close-packed matrix. Final reorganization to give the spinel structure of γ-Al_2O_3 is suggested to involve gradual filling of the tetrahedral cation sites [3].

To control the hydrolysis and the polycondensation rates, much works have been reported for the synthesis of the nanofibers with the addition of various surfactants and other additives as templating agents to direct the growth of the fibers. Alumina nanofibers have been successful grown with the use of polyethylene oxide polyethylene glycol, hexamethyltetraamine, and organic solvents [14, 15, 17, 25, 41, 45, 48]. The surfactant directs the fibrils growth by forming rodlike micelles. Alumina nanofibers were successfully prepared by electrospinning technique using aluminum acetate and aluminum nitrate as alumina precursor or other different precursors with a combination of PVA and PEO as polymer precursor [48]. Recently, a one pot synthesis of high area alumina nanofibers were synthesized from aluminum isopropoxide and acetic acid in supercritical CO_2 (SCF) as the solvent has been described in which the water in the process was limited by the reaction between the alkoxide and the acetic acid [60]. Another report on the use of supercritical CO2 as a drying agent has also been described for the synthesis of unidirectional nanostructures such as nanotubes, nanofibers, and nanorods. The morphology of the final product depends on the nature of the employed surfactants [25].
Another method to ensure the formation of the nanofibrous alumina is by using the hard template route. Aluminum surfaces created by anodic oxidation electrically with nanosize pores could be used as the nanoreactors for the synthesis of the oxide fibers [14, 15, 19].

2.2 Other methods

2.2.1 Mercury mediated method

Alumina nanofibers were successfully synthesized in mercury media at room temperature. The procedure consisted of immersing high purity aluminum strips $HgCl_2$ solution for a few minutes, removed and exposed to air. After some time, the white alumina powder was formed on the surface of the aluminum strips. The as-grown aluminas were heated and calcined in air at 850°C for 2 h to. The product was initially amorphous with diameters of 5-15 nm and length of several μm but converted to γ-alumina after calcinations. During the process, the following reactions took place:

$$2\,Al + 3\,HgCl_2 \rightarrow 3\,Hg + 2AlCl_3 \tag{12}$$

$$4\,Al + 3\,O_2 + 2n\,H_2O = 2\,Al_2O_3\,nH_2O \tag{13}$$

The Al reacted with mercuric chloride forming Al amalgam by metathesis and got oxidized by air when removed from the solution. The dissolution of the Al into the Hg was a continuous process. The final product is the alumina nanofibers [28].

2.2.2 Flame aerosol method

Flame aerosol technology is used for large-scale manufacture of ceramic commodities such as pigmentary titania, fumed silica and alumina. The dynamics of formation and growth of these particles are reasonably well understood. The formation of alumina nanofibers in carbon-containing coflow diffusion flames with pure oxygen support, using trimethylaluminum vapor as the precursor has been reported recently. The alumina nanofibers formed were non-crystalline, typically 2–10 nm in diameter and 20–210 nm in length. These fibers were mixed with spherical nanoparticles along with carbon nanotubes formed at the same time. No alumina nanofiber formation was found in flames supported by air or in pure H_2 flames. The alumina nanofibers were formed from gas-phase aluminum-containing species in the flame. Gas-phase carbon-containing species such as CO or hydrocarbon were probably crucial to the formation of the alumina nanofibers. The nanofibers were formed in the region past the maximum temperature zone of the flame. Sufficiently high temperature may be due to the higher concentration of aluminum in the gas phase, while the presence carbon nanotubes or nanofibers may serve as templates or mediating agent for the alumina nanofibers formation [29, 64].

2.2.3 Vapor-Liquid-Solid process

Crystalline alumina nanowires were synthesized at elevated temperatures in a catalyst assisted process using iron as catalyst. Nanotrees formed by alumina nanowires were also found. The typical nanowires are crystalline of size around 50 nm in diameter and around 2 μm in length. The tree trunks of the nanotrees are around 100 nm in diameter and around 10 μm in length. The results are explained in terms of growth mechanism based on a vapor–liquid–solid (VLS) process. The process involved mixing commercially available aluminum powder, iron powder and silicon carbide powder in an appropriate ratio and then sprinkled on tungsten (W) boat. The boat was then placed at the center of the vacuum bell jar and evacuated down to about 5.0×10^{-2} Torr. The boat was gradually heated up by passing current through it, and kept at ~1700 °C for 1 h under flowing argon. The treated powders contained in the W boat were taken out and cooled down to room temperature in flowing argon [64]. Single-crystal α-Al_2O_3 fibers by vapor–liquid–solid deposition from aluminum and silica powder and Al_2O_3 nanowires were synthesized by heating a mixture of Al, SiO_2 and Fe_2O_3 catalyst. Al_2O_3 nanowires and nanotrees have also being grown on silicon carbide particles' surface in thermal evaporation process and using iron as catalyst [20].

2.3 Electrospinning

Electrospinning is a old technology for the production of polymer fibers. This process has experienced renewed interest for the synthesis of nanofibers of polymers, ceramics, and their composites. The process uses electrical force to produce fibers with nanometer-scale diameters from its solution. Nanofibers have a large specific surface area and a small pore are being used or finding uses in filtration, protective clothing, biomedical applications including wound dressings and drug delivery systems, structural elements in artificial organs, and in reinforced composites. Recently, this potentially commercially viable process has been much investigated for the production of ceramic including alumina nanofibers [4-7, 51, 53].

The process setup consists of a capillary tube, connected to a reservoir of the colloidal solution or melts, with a small orifice through which the fluid could be ejected and a collector. A high voltage is used to create an electrically charged jet of solution or melt out of the capillary solution fluid. This induces a charge on the surface of the droplet formed at the tip of the capillary and held by its surface tension. As the intensity of the electric field is increased, the hemispherical surface of the droplet at the tip of the capillary tube elongates to form a conical shape known as the Taylor cone. On further increasing the electric field, a critical value is attained with which the repulsive electrostatic force overcomes the surface tension and the charged jet of the fluid is ejected. The ejected colloidal solution jet undergoes an instabilization and elongation process, which causes the jet to become very long and thin. Meanwhile, the solvent evaporates, leaving behind a charged fiber. In the case of the melt the discharged jet solidifies when it travels in the air.

Many reports have appeared in the literature in the past few years on the fabrication of alumina nanofibers as the viscosity of the boehmite gel is well suited for the electrospinning technique [5,34, 48,50,51,53]. The gel solution for the spinning consisted of nanofibrous colloid prepared by the sol-gel or hydrothermal processes, however, the precursor mixture may also be used. A typical solution for electrospinning is prepared by mixing suitable concentration of the solution of an aluminum salt with a polymer solution such as polyvinyl alcohol or polyethylene oxide. The distance between the capillary tip and the collector electrode, the flow rate, the voltage were adjusted suitably. Polymer-alumina composite nanofibers were obtained which can then be processed by further drying or calcinations [48].

3. Alumina nanofiber composites

Alumina nanoparticles and nanofiber show interesting properties such as ability to be formed into structures that enhance the functions of osteoblast for bone replacement [65, 66], to form ultrathin alumina hollow fiber microfilm membrane [67] for separation processes, to form novel nanofilter for removal and retention of viral aerosols [68] and to serve as high performance turbidity filter [69]. The preparation of the final form of alumina fiber may involve the use of binders such as acid phosphate and silica colloid binder or polymers [70]. It the case where polymer was used it would be burnt off during high temperature treatment of the prepared.

Many composite materials comprising of alumina nanoparticles or nanofibers as minor or major component in the presence of polymer or inorganic substrate also show similar interesting properties. S. Sundarajan [71] for example explored the fabrication of nanocomposite membrane comprising of polymer and alumina nanoparticles for protection against chemical warfare stimulants and found alumina based materials to be prospective candidates. While traditionally alumina film has been used as protective film on metal substrate [72, 73], alumina presence thermoplastic and thermosetting polymeric materials are also gaining wide application as surface coatings. Landry [74] for example studied the preparation of alumina and zirconia acrylate nanocomposites for coating wood flooring. It was found that for both the alumina and zirconia nano-composites, the conversion of acrylate resin is faster and more important when silane is used as the coupling agent.

The addition of the nanoparticles and nanofibers is meant to enhance the mechanical and thermal properties compared to in absence of such constituents. It is generally agreed that the large surface to volume ratio of the nanoscale constituents plays a key role to the improvement.

3.1 Alumina in polymer substrate

The effect of alumina in polyaniline, diglycidyl ether of bisphenol A type epoxy resin, carbon fiber epoxy resin composite and PA1010 has been studied recently. Generally it has been found that the present of low loading of alumina nanoparticles and nanofibers tend to enhance the thermal and mechanical properties of the polymer matrix. In many instances the strength of the composites are below the strength of neat resin due to non-uniform particle size distribution and particle aggregation. Ash et.al (as noted in [75]) studied the mechanical behavior of alumina particulate/poly(methyl methacrylate) composites. They concluded that when a weak particle matrix interphase exists, the mode of yielding for glassy, amorphous polymers changes to cavitational to shear, which leads to a brittle-to-ductile transition.

Two challenges have been observed to be overcome to facilitate the enhancement of the properties of the polymer substrate. First is the need to disperse the nanoparticles and nanofibers uniformly throughout the polymer substrate and secondly to facilitate the interaction between the nanofibers and the molecules of the polymer substrate. At low loading the nanoparticles could be distributed uniformly across the substrate. However at high loading, there is the tendency for the fibers to clusters together and hence limit the enhancement of the mechanical properties and in fact lowered it. It has also been observed that while the tensile strength increases with reduction in particles size for micron-scale particles, the tensile strength decreased with reduced particle size for nano-scale particles [76].The changes were attributed not to the strength of bonding between the particulate with the matrix but more to the poor dispersion of particles.

Various preparative methods have been adopted to facilitate good dispersion. This include mechanical milling [77] and Mechanical milling followed by hot extrusion [78]. In both studies the alumina nanofibers was found in the as sintered product. M.I Flores-Zamora concluded that the presence of alumina based nanoparticles and nanofibers seemed to be responsible for the reinforcement effect.

Attempts to meet the second challenge involve functionalizing alumina particles such as in the on fiber and epoxy resin composites [79]. The functionalizing of the alumina surfaces is meant to enhance the miscibility of the alumina particles in the polymer substrate and also the facilitate bridging between alumina surfaces with the substrate. It was however observed that where the functionalised alumina, L-alumoxane is miscible with the resin, high loading results in a marked decrease in performance due to an increase in brittleness. This was proposed to be due to weak inter-phase bonding between resin and the alumina fiber.

3.2 Alumina in inorganic and carbon based substrate

The composites comprising of carbon and alumina is also of significant interest. Study on this category of composite material includes fabrication of macroscopic carbon nanofiber (CNF)/alumina composite by extrusion method for catalytic screening [80]. It was found that the synthesized composite possessed a mesoporous structure with a relatively high surface area (340 m^2/g) and a narrow particle size distribution, displaying a good thermal stability. A comparison of the surface acidity between the composite and commercial alumina demonstrated that the total number of acid sistes on composite was significantly increased along the distinct decrease in the number of strong acid sites, which may enhance the activity and anti coking property as a promising industrial catalyst support in petroleum industry.

Hirato [81] fabricate carbon nanofiber–dispersed alumina composites by pulsed electric-current pressure sintering and their mechanical and electrical properties. High bending and fracture strength were observed on the composites compared to that of monolithic alumina. The electrical resistivity of the composite material was also observed to reduce by 10^{17} order of magnitude. Xia [82] studied the fracture toughness of highly ordered carbon nanotube/alumina nanocomposites. The result of the study demonstrate that nanotube bridging/sliding and nanotube bridging necessary to induce nanoscale toughening, and suggest the feasibility of engineering residual stresses, nanotube structure and composite geometry to obtain high toughness nanocomposites.

3.3 Future development of alumina nanocomposite

There are wide applications of alumina nanofiber composites. These potential would continue to drive research interest in this field. In the case of applying alumina composites for application that is dependent on its strength, future work is like to further elucidate the factors that would first facilitate good dispersion of the fiber in the host matrix and second that would enhance the interphase interaction between alumina nanofiber and the matrix. The effect of additive that could facilitate dispersion or/and bonding between alumina fiber and the matrix may be of great interest. The application for catalysis may be of special interest especially the matrix between alumina and carbon nanofibers and similar substances. It may be argued that catalysts are in nearly all cases composite materials, however the interaction between purposely prepared nano size precursors may yield effect not seen when prepared through the traditional mode.

4. References

[1] H. Knoezinger, R. Ratnasamy, Catal. Rev. (1978), 17(1):31

[2] V. Saraswati, G.V.N. Rao, G.V. Rama Rao, Structural evolution in alumina gel, J. Mater. Sci. (1987), 22(7):2529;

[3] S.J. Wilson, The dehydration of Boehmite, γ-AlOOH, to γ--Al2O3, J. Solid State Chem. 30(2), (1979):247

[4] A. Jawaorek, A.T. Sobczyk, Electrospraying route to nanotechnology: an overview, J. Electrostatic (2008), 60:197

[5] A.M. Azad, Fabrication of transparent alumina nanofibers by electrospinning, Mater. Sci. Engineer. A. (2006), 435-436:468;

[6] T. Subbiah, G.S. Bhat, R.W. Tock, S. Parameswaran, S.S. Ramkumar, Electrospping of nanofibers, J. Appl. Polym. Sci., (2005), 96(2):557;

[7] D. Li, Y. Xia, Electrospinning of nanofibers: reinventing the wheel?, Advanced Mater. (2003), 16(14):1151

[8] K. Niesz, P. Yang, G.A. Somorjal, Chem. Comm., (2005), 1986,

[9] S.A. Bagshaw, T.J. Pinnavaia, Mesoporous Alumina Molecular Sieves, Angew. Chem. (Intern.Ed.English) (1996), 35 (10): 1102-1105.

[10] X. S. Peng, L. D. Zhang, G. W. Meng, X. F. Wang, Y. W. Wang, C. Z. Wang, G. S. Wu, Photoluminescence and Infrared Properties of α-Al2O3 Nanowires and Nanobelts, J. Phys. Chem. B, 2002, 106 (43)

[11] J.C. Hulteen, A general template-based metod for the preparation of nanometerials, J. Mater. Chem. (1997), 7(7):1075;

[12] L. Pu, X. Bao, J. Zou, D. Feng, Individual alumina nanotubes, Angew Chem. (2001), 40(8):1490;

[13] H. Hou, Y.Xie, Q. Yang, Q. Guo, C. Tan, Perapration and characterization of γ-AlOOH nanotubes and nanorods, Nanotechnol., (2005), 16(6):714;

[14] Y. Xia, Y. Yang, P. Sun, Y.Wu, B Mayers,., Gates, Y.Yin, , F.Kim, , H.Yan, One-dimensional nanostructures: Synthesis,characterization, and applications, Advanced Mater., 15(5),(2003):353

[15] J. G. Lu, P. Chang, Z. Fan, Quasi-one-dimensional metal oxide materials—Synthesis, properties and applications Mater. Sci. Engineer. R: Reports, 52(1-3), (2006): 49-91

[16] W. Cai, J. Yu, S. Mann, Template-free hydrothermal fabrication of hierarchically organized γ-AlOOH hollow microspheres, Microp. Mesop. Mater. 122(1-3), (2009):42

[17] G. L. Teoh, K. Y. Liew, W. A. K. Mahmood, Synthesis and characterization of sol–gel alumina nanofibers, J. Sol-Gel Sci. Technol., (2007) 44(3): 177

[18] M.G. Ma , J.F. Zhu, A facile solvothermal route to synthesis of γ-alumina with bundle-like and flower-like morphologies, Mater. Lett. 63 (2009): 881;

[19] Y. Zuo, Y. Zhao, X. Li, N. Li, X. Bai, S. Qiu, W. Yu Synthesis of alumina nanowires and nanorods by anodic oxidation method, Mater. Lett., 60(24),(2006):2937;

[20] Y.Z. Jin, Y.Q. Zhu, K. Brigatti, H.W. Kroto, D.R.M. Walton, Catalysed growth of novel aluminium oxide nanorods Appl. Phys. A: Mater. Sci. Proc., 77(1), (2003): 113.;

[21] H. C. Lee, H. J. Kim, S. H. Chung, K. H. Lee, H. C. Lee, J. S. Lee, Synthesis of Unidirectional Alumina Nanostructures without Added Organic *Solvents* J. Am. Chem. Soc., (2003), 125 (10):2882

[22] J. H. Park, M. K. Lee, C. K. Rhee, W. W. Kim, Control of hydrolytic reaction of aluminum particles for aluminum oxide nanofibers. Mater Sci.Eng. A, 375-377, (2004):1263

[23] W. Cai, H. Li, Y. Zhang, Influences of processing techniques of the H_2O_2-precipitated pseudoboehmite on the structural and textural properties of γ-Al_2O_3. Colloids Surf. A: Physicochem. Engin Asp, 295(1-3), (2007), 185

[24] V. Valcárcel, A. Pérez, M. Cyrklaff, F. Guitián, Novel Ribbon-Shaped α-Al_2O_3 Fibers, Advanced Mater. 10(16):1370

[25] H. Y. Zhu, X. P. Gao, D. Y. Song, Y. Q. Bai, S. P. Ringer, Z. Gao, Y. X. Xi, W. Martens, J. D. Riches, R. L. Frost, Growth of Boehmite Nanofibers by Assembling Nanoparticles with Surfactant Micelles. J. Phys. Chem. B, (2004), 108 (14): 4245

[26] W. Cai, H. Li, Y. Zhang, Azeotropic distillation-assisted preparation of macro-mesostructured γ-Al_2O_3 nanofibres of crumpled sheet-like morphology, Mater Chem. Phys., 96(1),(2006):136;

[27] J. Wang, Y. Wang, M. Qiao, S. Xie, K. Fan, A novel sol–gel synthetic route to alumina nanofibers via aluminum nitrate and hexamethylenetetramine, Mater. Lett., 61(28),(2007):5074;

[28] Q. Yang, Y. Deng, W. Hu, Synthesis of alumina nanofibers by a mercury-mediated method. Ceram. Intern 35(1), (2009):531

[29] B. Guo, H. Yim, Z.P. Luo Formation of alumina nanofibers in carbon-containing coflow laminar diffusion flames, J. Aerosol Sci., 40(4), (2009):379

[30] A. Jagminas, J. Kuzmarskytė, L. Malferrari, M. Cuffiani A new route of alumina template modification into dense-packed fibrilous material, Mater. Lett., 61(14-15), (2007):2896

[31] R. Ma, Y. Bando, T. Sasaki, Directly rolling nanosheets into nanotubes, J. Phys. Chem.B, 108(7), (2004):211

[32] J. Wang, Y. Wang, M. Qiao, S. Xie, K. Fan, A novel sol–gel synthetic route to alumina nanofibers via aluminum nitrate and hexamethylenetetramine, Mater. Lett., 61(28), (2007):5074

[33] X. Liu, Z. Wu, T. Peng, P. Cai, H. Lv, W. Lian Fabrication of alumina nanofibers by precipitation reaction combined with heterogeneous azeotropic distillation process, Mater. Res. Bul., 44(1), (2009):160-167

[34] V. Maneeratana, W. M. Sigmund Continuous hollow alumina gel fibers by direct electrospinning of an alkoxide-based precursor, Chem. Engineer. J., 137(1), (2008):137

[35] J. Chandradass, M. Balasubramanian, Extrusion of alumina fibre using sol-gel precursor J. Mater. Sci.,41(18), (2006):6026

[36] E. Elaloui, A. C. Pierre and G. M. Pajonk, Influence of the Sol-Gel Processing Method on the Structure and the Porous Texture of Nondoped Aluminas, J. Catal., 166(2), (1997):340

[37] J.Chandradass, D.S. Bae, D.S., M. Balasubramanian, Synthesis and characterization of sol-gel alumina fiber by seeding α-alumina through extended ball milling, Mater. Manuf. Proc. 23(8), (2008):786

[38] R. Venkatesh, P.K. Chakrabarty, B. Siladitya, M. Chatterjee, M. D. Ganguli, Preparation of alumina fibre mats by a sol-gel spinning technique, Ceramics Intern., 25(6), (1999):539-543

[39] G. Zhihong, N. Guilin, L. Yuan, C.Yu, Heat-induced transformation between nanospheres and nanofibers of boehmite, Proc. 1st IEEE-NEMS 2006, Article number 4135091, Pages 886-889

[40] R. Ma, Y. Bando, T. Sasaki, Directly rolling nanosheets into nanotubes, J. Phys. Chem.B, 108(7), (2004):2115

[41] Z. Zhu, H. Liu, H. Sun, D. Yang, PEG-directed hydrothermal synthesis of multilayered alumina microfibers with mesoporous structures, *Microp Mesop. Mater, 123(1-3), (2009): 39-44*

[42] Y. Kobayashi, T. Ishizaka, Y. Kurokawa, Preparation of alumina films by the sol gel method,. J. Mater Sci., (2005),40:263

[43] F. Tepper, L. Kaledin,Nanosize electropositive fibrous adsorbent, (2005) US patent:6838005;

[44] KMS. Khalil, Synthesis of Short Fibrous Boehmite Suitable for Thermally Stabilized Transition Aluminas Formation, J Catal. (1998),178(1):198

[45] J. Wang, Y. Wang, M. Qiao, S. Xie, K. Fan, A novel sol-gel route to alumina nanofibers via aluminum nitrate and hexamethylenetetramine, Mater. Lett. (2007), 61(28) :5074.

[46] T. Yogo, H. Iwahara, Synthesis of Synthesis of α-alumina from modified aluminum alkoxide precursor, J. Mater Sci., (1992), 27:1499;

[47] J.F. Hochepied, P. Nortier, Influence of precipitation conditions :pH and temperature on the morphology and porosity of boehmite particles, (2002), Powder Technol, 1 28(2-3), 268

[48] P.K. Panda, S. Ramakrisna, Eletrospinning of alumina nanofibers using different precursors, J. Mater Sci, (2007),42: 2189

[49] K. Song, K. Woo, Y. Kang, Preparation of alumina fibers from aluminum salts by the sol gel methods, Korean J. Chem. Eng. 16(1), (1999):75

[50] A.M. Azad, Fabrication of transparent alumina nanofibers by electrospinning, Meter. Sci. Eng. A, (2006), (435-473):468

[51] R.W. Tuttle, A. Chowdury, E.T. Bender, R.D. Ramsier, J.L. Rapp, M.P. Espe, Electrospun ceramic fibers: Composition, structure and the fate of precursors, Appl. Surf. Sci.,(2008), 254:4928;

[52] SC. Shen, W.K. Ng,Q.Chen, X.T. Zeng, R.B.H. Tan, Novel synthesis of lace-like nanoribbons of boehmite and γ-alumina by dry gel conversion method, Mater lett, (2007),61(21):4280 steam crystallization

[53] K.M. Sawicka, P.Gouma, Electrospun composite nanofibers for functional applications, J. Nanopart. Res.(2006), 8:769

[54] H.Y. Zhu, X.P. Gao, D.Y. Song, S.P. Ringer, Y.X. Xi, R.L. Frost, Manupilating the size and morphology of aluminum hydrous oxide nanoparticles by soft chemistry approaches, Microp. Mesop. Mater. (2005), 85(3):226

[55] H.Y. Zhu, J.D. Riches and J.C. Barry, γ-alumina nanofibers prepared from aluminum hydrate with poly(ethylene oxide) surfactant, Chem. Mater., 14, (2002): 2086

[56] X. Liu, Z. Wu, T. Peng, P. Cai, H. Lv, W. Lian Fabrication of alumina nanofibers by precipitation reaction combined with heterogeneous azeotropic distillation process, Mater. Res. Bul., 44(1), (2009):160

[57] H.Y. Zhu, X.P. Gao, D.Y. Song, Simon P. Ringer, Y.X. Xi, Ray L. Frost, Manipulating the size and morphology of aluminum hydrous oxide nanoparticles by soft-chemistry approaches, Microp. Mesop. Mater. 85(3), (2005):226

[58] W. Cai, H. Li, Y. Zhang, Influences of three post-precipitation techniques, Colloids Surf. A: Physicochem. Engin Asp, 295(1-3), (2007), 185,

[59] M. Thiruchitrambalam, V. R. Palkar, V. Gopinathan, Hydrolysis of aluminium metal and sol-gel processing of nano alumina, Mater Lett., 58(24), (2004):3063

[60] M. B. I. Chowdhury, R. Sui, R. A. Lucky, P. A. Charpentier, A One-Pot Procedure to Synthesize High Surface Area Alumina Nanofibers using Supercritical Carbon Dioxide, Langmuir (in press)

[61] H. C. Lee, H. J. Kim, C. H. Rhee, K. H. Lee, J. S. Lee, S. H. Chung, Synthesis of nanostructured γ-alumina with a cationic surfactant and controlled amounts of water, Microp. Mesop. Mater., 79(1-3), (2005): 61-68

[62] J. Zhang, F. Shi, J. Lin, S. Y. Wei, D. Chen, J. M. Gao, Z. H., X. X. Ding, C. Tang, Nanoparticles assembly of boehmite nanofibers without a surfactant, Mater. Res. Bul., 43(7), (2008): 1709

[63] S. E. Pratsinis, Flame aerosol synthesis of ceramic powders, Prog. Energy Comb. Sci., 24(3), (1998):197

[64] J. Zhou, S. Z. Deng, J. Chen, J. C. She, N. S. Xu, Synthesis of crystalline alumina nanowires and nanotrees, Chem. Phys. Lett.,365(5-6).(2002):505

[65] R.L. Price, K.M. Heberstroh, T.J. Webster, 2003, Enhanced functions of osteoblasts on nanostructured surfaces of carbon and alumina, Med. Biol. Eng. Comp., (2003):41, 372.

[66] T. J. Webster, E. L. Hellenmeyer, R. L. Price, Increased osteoblast function on theta + delta nanofiber alumina, Biomaterials 26, (2005): 953

[67] I. H. Choi, I. C. Kim, B. R. Min, K. H. Lee, Preparation and characterization of ultrathin alumina hollow fiber microfiltration membrane, Desalination, 193 (2006):256

[68] H. Li, C. Wu, F.Tepper, J. Lee, C. N. Lee, Removal and retention of viral aerosols by a novel alumina fiber filter, Aerosol Science 40, (2009): 65

[69] F. Tepper, L. Kaledin, C. Hartmann, High Performance Turbidity Filter, February 2005, 55-57

[70] J.A. Fernando, D.D.L. Chung, Thermochemical properties of alumina fiber membrane, Ceram. Intern., 31, (2005): 453

[71] S. Sundarajam, S. Ramakrishna, Fabrication of nanocomposite membranes from nanofibers and nanoparticles for protection against chemical warfare stimulants, J. mater. Sci., (2007),42: 8400

[72] C. Jing, X. Xu, J. Hou, Preparation of compact Al2O3 film on metal on metal for oxidation resistance by polyvinylpyrrolidone, J Sol-Gel Sci Technol (2007) 43: 321

[73] C. Jing, X. Xhao, Y. Zhang, Elimination of a zero-growth in thickness of Al2O3 protective film deposited by cycles of dip coating method, J Sol-Gel Sci Technol (2007) 42: 151

[74] V. Landry, B. Riedl, P. Blanchet, Alumina and zirconia acrylate nanocomposites coatings for wood flooring: Photocalorimetric characterization, Progr.Org. Coatings, 61, (2008):76

[75] E. T. Thostenson, C. Li, T. Chou, Nanocomposites in context, Composites Sci. Techn., 65 (2005): 491

[76] L. Sun, R. F. Gibson, F. Gordaninejad, J. Suhr, Energy absorption capability of nanocomposites: A Review, Composites Sci. techn. 69 (2009):2392

[77] A. Santos-Beltran, V. Gallegos-Orozco, I. Estrada-Guel, L. B. Gomez, F. E.Magana, M. M.Yoshida, R. M. Sanchez, J. Alloys Comp. 434-435,(2007):514

[78] M.I. Flores-Zamora, I. Estrada-Guel, J. Gonzalez-Hernandez, M. Miki-Yoshida, R. Martinez-Sanchez, Aluminium-graphite composites produced by mechanical milling and hot extrusion, J. Alloys Comp. 434-435, (2007):518

[79] N. Shahhid, R. G. Villate, A. R. Barron, Chemically functionalized alumina nanoparticles effect on carbon/epoxy composites, Composites Sci. Techn. 65 (2005):2250

[80] J.H. Zhou, C. Chen, R. Guo, X. C. Fang, X. G. Zhou, Synthesis and characterization of carbon nanofiber/alumina composites by extrusion casting, Carbon 47 (2009): 2077

[81] K. Hirota, Y. Takaura, M. Kato, Y. Miyamoto, J. Mater. Sci (2007), 42: 47920.
[82] Z. Xia, W.A. Curtin, B.W. Sheldon, Fracture toughness of highly ordered carbon nanotube/alumina nanocomposites, Transaction of ASME, 126,(2004):239

22

Core-Shell Nanofibers: Nano Channel and Capsule by Coaxial Electrospinning

Fengyu Li[1,2], Yong Zhao[1] and Yanlin Song[1]
[1]*Institute of Chemistry Chinese Academy of Sciences, Beijing,*
[2]*Bowling Green State University, Bowling Green, OH,*
[1]*China*
[2]*USA*

1. Introduction

In the last 30 years, nano science and technology have an unprecedented development. Nanofibers, one-dimensional (1D) nanostructures have been paid great attentions due to their unique properties and intriguing applications in many areas (Xia et al., 2003). In lots of synthetic and fabrication methods, electrospinning (a drawing process based on electrostatic interactions) provided the simplest approach to nanofibers with exceptionally long in length, uniform in shape, and diversified in composition (Li et al., 2004; Greiner et al., 2007). Following the fast and polymer common property, scientists invented complex electrospinning. Utilizing electrostatic interactions and through the elongation of viscoelastic jet device, electrospinning technique can copy the structure of jet device to nanoscale fibers from polymer solution or molten. Various nanofibers with different secondary nanostructure were fabricated. Especially, core-shell nanofibers pull more and more interesting and attention, because of the great potential applications and prospects for nanochannel, nanocapsule and small encapsulating devices. Coaxial electrospinning is an effective, fast and controlled technique to construct core-shell nanostructures into nanofibers (Li et al, 2004; Zhang et al., 2004 and Xia et al., 2006).

Here, based on complex electrospinning technique, we will describe nanofibers with nano-channel and core-shell nano-capsule fabrication, and how coaxial electrospinning is working to construct various nanostructures in the nanofibers. Furthermore, we will introduce some encapsulation of stimulating responsive materials into core-shell nanofibers, and the improvement of the responsibility and functional devices applications for the combination of core-shell nanofibers and stimulating responsive materials.

2. Coaxial electrospinning

Electrospinning, also known as electrostatic spinning, has been developed since 1990s as micro-nano fiber and particle fabrication technique (Reneker et al, 1996; Huang et al, 2003). It uses an electrical charge to draw a very fine (typically on the micro or nano scale) fibers from liquid. Figure 1 shows the common setup and working principle of electrospinning. A high voltage and a cuspidate spinneret provided by a point electrode creates a charge aggregation into the polymer solution or liquid at the spinneret of syringe. Under pulling of electrostatic force, polymer liquid is drawn and got a "Taylor cone" at the jut of spinneret, then is drawn out from spinneret as the "Liquid jet" part in Figure 1. Due to the electrostatic repulsion of the charges in the polymer liquid, the polymer liquid is drawn thinner and thinner, as the fiber spin part in Figure 1. In this process, the ultrathin fibers solidify or dry rapidly, and then the nanofibers are collected on the "Collector".

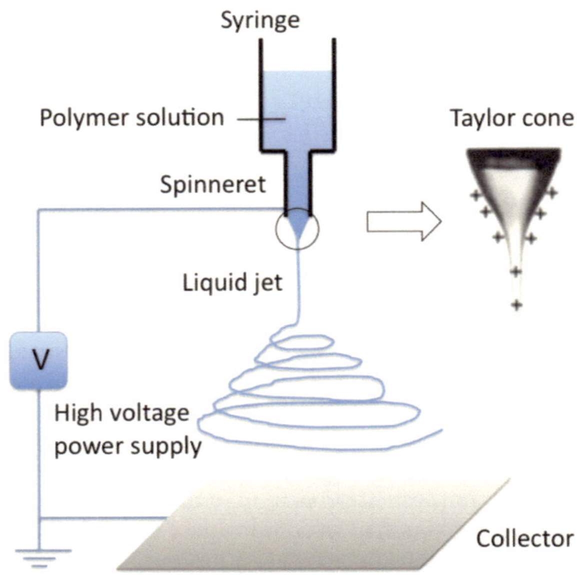

Fig. 1. The common setup and working principle of electrospinning.

Electrospinning shares characteristics of both electrospraying and conventional solution dry spinning of fibers. The spinning fibers process is non-invasive and does not require the use of coagulation invasion to produce solid threads from liquid. It makes the process particularly suitable for the production of fibres using macromolecule or polymer. In theory, electrospinning can be applied to all polymer to get nanofibers.

Nanofibers which be prepared by electrospinning, usually exhibit a solid interior and smooth surface. If an appropriate modification is employed to the electrospinning, it can prepare nanofibers with some specific secondary structure. Complex electrospinning was developed from basic electrospinning to fabricate nanofibers with secondary structures. Here we only discuss the core-shell nanofibers and the preparation method: "Coaxial electrospinning".

As shown in Figure 1, the process of fiber spinning after "Liquid jet", it is non-invasive, just thinner and thinner. It means, the "Taylor cone" plays an important role in the moulding of nanofibers. One can make a compound spinneret to modify "Taylor cone" to prepare nanofibers with complex secondary structures.

Figure 2 shows a basic setup for coaxial electrospinning and the fabrication process of common core-shell nanofibers. Based on the basic electrospinning setup, two syringes feed inter-separated and coaxial "Inner fluid" and "Outer fluid" to spinneret. Under high voltage, the electrospinning liquid is drawn out from spinneret and forms a "Compound Taylor cone" with a core-shell structure (Loscertales, et al. 2002). After "Coaxial jet", the core-shell structure will be built and kept in the fibers through spinning solid and be collected on the "Collector". In those processes, core-shell Taylor cone formation will decide the core-shell nanofibers forming and fabrication. In order to get a nice "Compound Taylor cone" with a core-shell structure, one has to make an utmost control and balance inject speed of inner and outer fluid, which keep the "Compound Taylor cone" in dynamic stabilization. A too high or too low speed of inner fluid is unfavourable. An appropriate injecting speed and rate of inner-outer fluid should be considered and investigated.

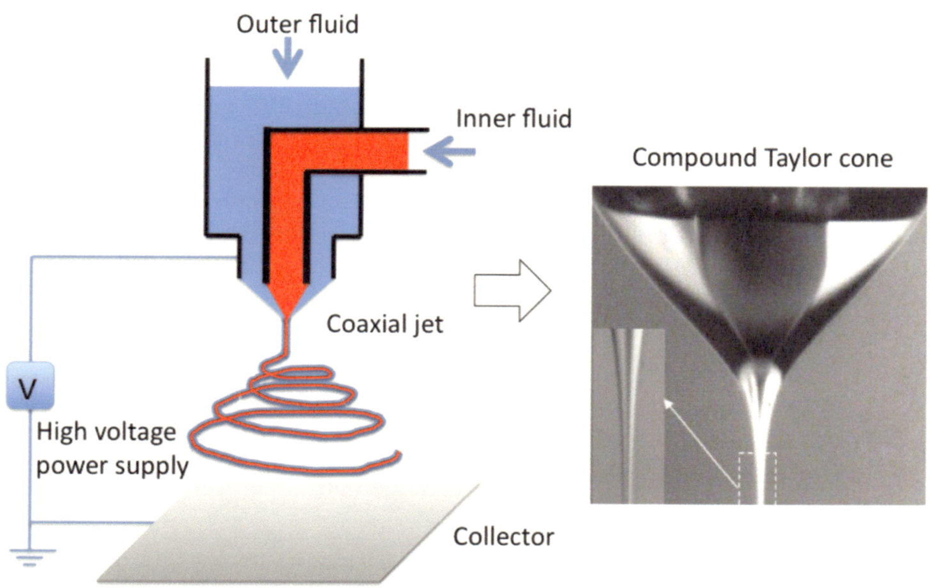

Fig. 2. The basic setup for coaxial electrospinning and fabrication process of common core-shell nanofibers.

Many scientists gave the abundant and excellent contribution on hollow and core-shell nanofibers by coaxial electrospinning after 2000s' respectively (Loscertales et al, 2002; Su et al, 2003; Li et al, 2004; Zhang, et al., 2004; Li et al, 2009). They fabricated hollow nanofibers with polyvinylpyrrolidone (PVP) and polycaprolactone (PCL) polymer system respectively. Figure 3 shows the hollow nanotube structure constructed by Prof. Xia's group. In order to prevent mixing between inner and outer fluid in Taylor cone, the mineral oil mostly was used as the inner fluid.

In the compound-jet electrospinning process, the outer polymer solution and inner paraffin oil were co-issued from respective nozzles, the outer liquid flowed through the gaps between the capillary and the formed liquid jacket enveloped the inner fluid. When the high voltage was applied, the conductive outer solution was charged and the compound liquid was stretched and whipped to a thin liquid thread in order to release the static electric repulsion by dispersing electric charge (Doshi et al, 1995). However, this repulse force could not act on inner liquids directly because of the insulation of paraffin oil, which means that the paraffin oil cannot be electrospun on its own. As a result, the inner fluids were subjected to the pressure transferred from the outer fluid and were compressed to thin liquid thread accompanied with outer liquid. At the same time, with the evaporation of solvent and gelation of polymer, the outer liquid shell solidified very quickly which suppressed the Reyleigh instability of the whipping jet. Consequently, solid tubes with independent channel were formed in which paraffin oil was enveloped.

After coaxial electrospinning, the fibers with a polymer shell and oily core were produced first. By removal of organics through calcinations, a TiO_2 hollow tube was obtained. In Figure 3, the walls of these tubes were made of a composite containing amorphous TiO_2 and PVP. TiO_2 (anatase) hollow fibers were obtained by calcining the composite nanotubes in air at 500 °C. SEM image of a uniaxially aligned array of anatase hollow fibers that were collected across the gap between a pair of electrodes. These fibers were fractured using a razor blade to expose their cross-sections. In the preparation of all these samples, the feeding rate for heavy mineral oil was 0.1 mL/h and the concentrations of $Ti(OiPr)_4$ and PVP were 0.3 and 0.03 g/mL, respectively. The voltage of electrospinning was 12 kV. Those fibers show an uniform, circular cross-section long and hollow nanostructure. They are well-separated from each other and can be conveniently transferred onto other substrates for a variety of applications. The circular cross-sections and well-controlled orientation of the hollow nanofibers should make them particularly useful as nanofluidic channels. The procedure can be extended to many other materials with potential applications in catalysis, sensing, encapsulation, and drug delivery.

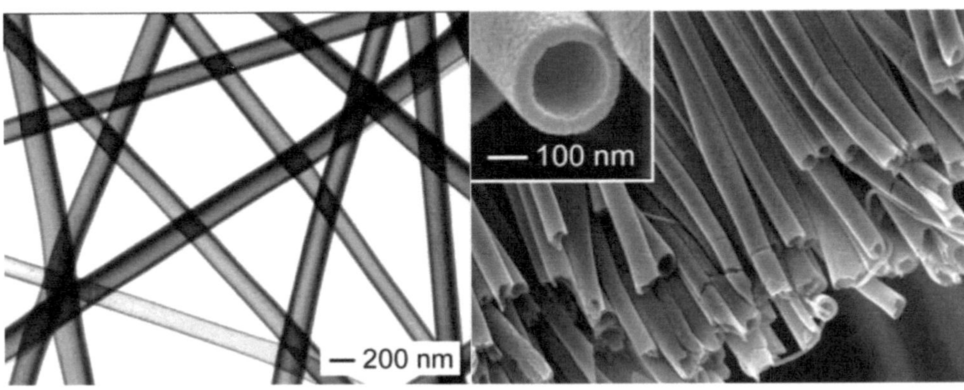

Fig. 3. TEM image (left) and SEM image (right) of TiO_2/PVP hollow fibers fabricated by coaxial electrospinning.

3. Multi-channel coaxial electrospinning and Nano-channel nanofibers

3.1 Multi-channel coaxial electrospinning

One-dimensional nanomaterials with hollow interiors have considerable applications in micro/nanofluidic devices, drug delivery, catalysis and others (Hu et al, 1999; Law et al, 2004). Coaxial electrospinning showed promising method for building ultralong nanotubes. However, most of the nanotubes are possess only one single inner channel. As the aspect and prospects of nano and functional materials research and development currently and future, the multifunctional, integrative and miniature devices research is urgently expected. Dr. Zhao and Prof. Jiang designed and carried out the strategies of building nanomaterials with more complex inner structures. Here we will describe a novel multi-fluidic compound-jet electrospinning technique in detail, which could fabricate hierachical multi-channel nanotubes in an effective way.

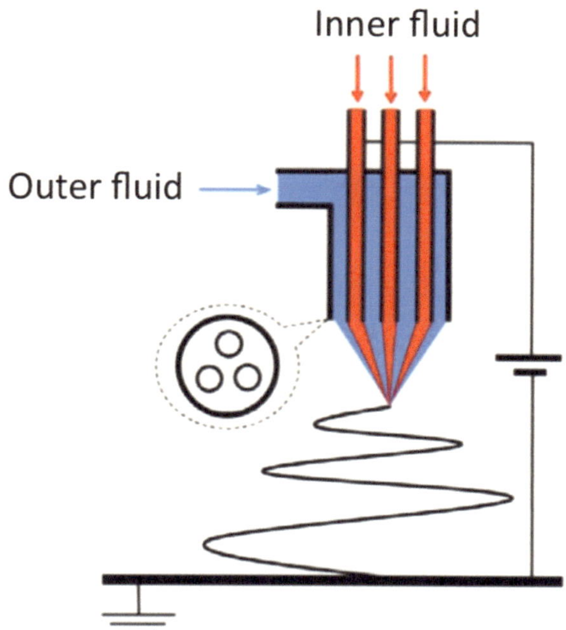

Fig. 4. Schematic illustration of multi-channel coaxial electrospinning system.

The experimental setup of multi-fluidic compound-jet electrospinnig is sketched in Figure 4. Several metallic capillaries with outer diameter (O.D.) of 0.4 mm and inner diameter (I.D.) of 0.2 mm were arranged at the several vertexes of an equilateral triangle. Then the bundle of capillaries was inserted into a plastic syringe (O.D. = 3.5 mm, I.D. = 2.0 mm) with gaps between individual inner capillaries and outer syringe. Two immiscible viscous liquids were fed separately to the three inner capillaries and outer syringe in an appropriate flow rate. A 20% Polyvinylpyrrolidone (PVP) ethanol solution, served as outer liquid, while a non-dissolution paraffin oil was chosen as inner liquid. Then a high voltage generator between three inner metallic capillaries and a metallic plate coated with a piece of aluminum foil acted as counter electrode provide the driving and controlling for the electrospinning. The immiscible inner and outer fluids (red for paraffin oil and blue for $Ti(OiPr)_4$ solution) were issued out separately from individual capillaries. With a suitable high voltage application, a whipping compound fluid jet is formed under the spinneret and then a fibrous membrane is collected on the aluminium foil.

A rational design of spinneret is utmost important for the successful fabrication of multi-channel fibers by coaxial electrospinning. The inject speed of inner and outer fluid must provide an elaborate control and balance. Further more, the gaps between each capillary and the capillary to the inner wall of outer syringe are equal, to ensure each inner capillary is isolated from other two capillaries and the outer nozzle. The outer fluid should surround the inner fluids effectively, and the inner liquids should flow out independently and don't mix with each other.

Because of surface tension, liquid column should prefer shrinking to cylinder to acquire the smallest surface area. But, the insection of inner channels of mulit-channel fibers is flabellate rather than circular, which represent the shape of inner liquids. Figure 5 shows the

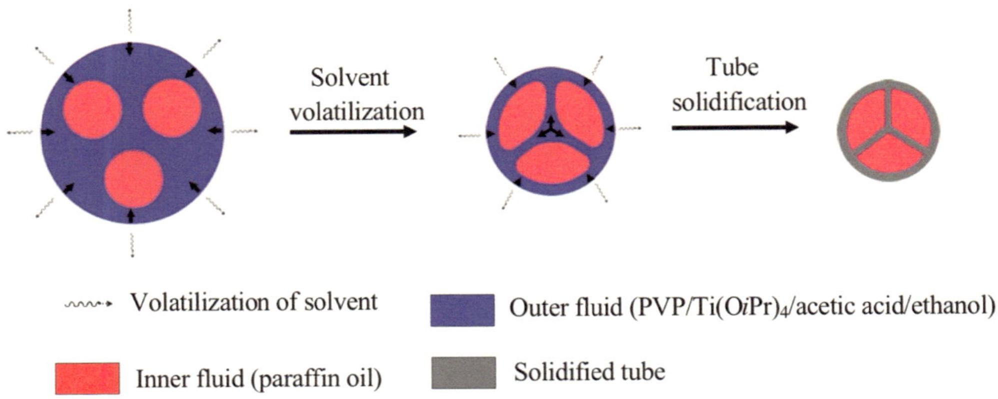

Fig. 5. Schematic illustration of multi-channel fiber shaping.

proposed formation process for multi-channel tube. (Here, a three-channel tube served as an example.) Two immiscible liquids with different properties are in the multi-fluidic compound electrospinning system. The outer liquid is a multi-component solution, in which the content of volatile solvent is in the majority, while the inner paraffin oil is a nonvolatile liquid. In the initial stage of electrospinning process that the compound fluid jet just leaves the spinneret, both outer and inner fluids are of cylinder shape according to the morphology of spinneret. With bending and whipping of the liquid jet in air, the volume of outer fluid shrinks remarkably by the lose of solvent. Then a dilemma emerge that either outer fluid or inner fluids must deform under the shrink pressure because the shrunk tube shell cannot hold the three cylinder inner fluids any more. Thermodynamic analysis indicates that the deformation of inner fluids is favorable for the stability of system. The interfacial tension of the outer and inner liquids is much smaller than outer liquid the surface tension of outer liquid (Adamson et al, 1997). The deformation of outer solution needs more energy than that of inner liquids. To lower the total free energy of the compound fluid, the outer fluid is drained to surface through the liquid node and border film (called Plateau border) between three inner fluids by capillary suction. When the Plateau border suction is counterbalanced by the disjoining pressure then reaches a mechanical equilibrium state, the border film between neighbouring inner fluid becomes flat (Exerowa et al, 1998; Höhler et al, 2005). With the drainage of outer fluid, the inner liquids evolve to three flabellate liquid columns. Plateau's law indicates that three angles between Plateau border are equal (120°) under thermodynamic equilibrium conditions, which agree well with the mutual angle of Y-shape inner ridge of three-channel nanotube. Generally speaking, to lower the free energy of the compound fluid system, the inner fluids transform to flabellate shape under the shrinkage pressure of outer liquid and form multi-channel nanotube with multi-pointed star shape inner ridge ultimately. The multi-fluidic compound-jet electrospinning technique breaks through the limit at two fluids system that could generate programmable multi-channel or multi-component 1D nanomaterials in an effective way.

3.2 Mulit-channel nanofibers

After an utmost control compound-jet electrospinning process and follow-up treatments (the inner channels of the tube correspond to the vacancy of the inner fluids after they were removed), Figure 6 and Figure 7 show the SEM images of the multi-channel fibers prepared by coaxial electrospinning. The fibers have uniform, flat and smooth surface. The side-view was checked to expose the cross sections of the multi-channel fibers. It can be clearly seen that most of tubes are of hollow structures with multi-cavum, circular and closed outer wall of the hollow tube. The diameter distribution of the tubes is relative uniform with average value of 2.3 μm (in a sub-nano size). The inner diameter of the channels is around 100-500 nm size. Decades nanometer walls made the compartment of the several cavums. The through cavums form the nano-channels in the polymer or inorganic fibers and give the huge surface area. The multi-furcate ridge embeds in the outer tube and exhibits an interesting mulit-pointed star "Y" or "X" shape, and the ridges partition the nanotubes into several flabellate parts. Like a scaffold, the multi-furcate ridges support the hollow structure, and make the hollow structure have higher intensity.

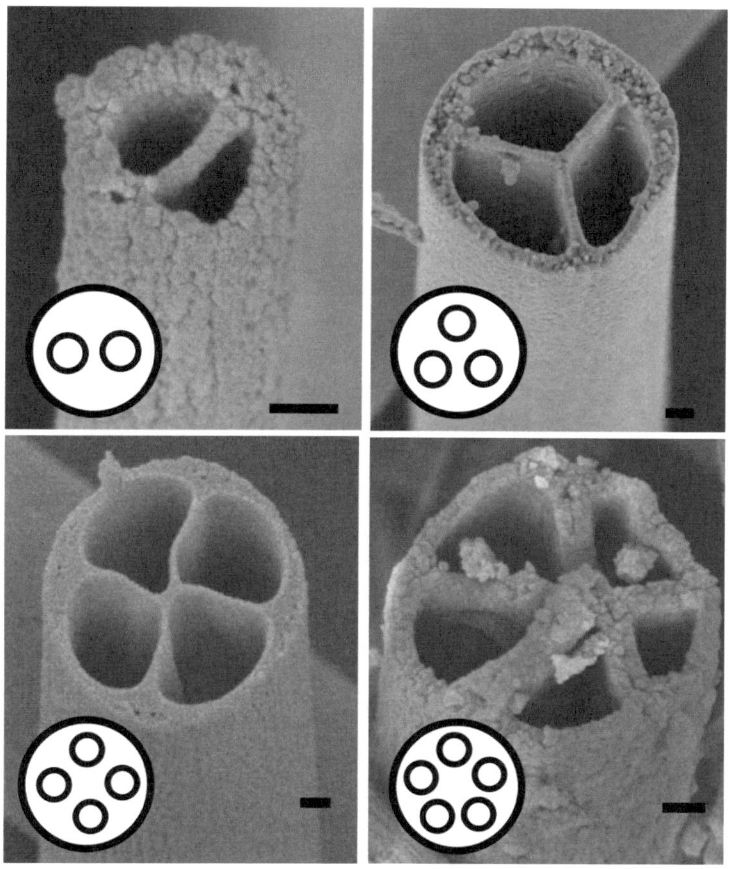

Fig. 6. The cross section SEM images of hollow fibers with two, three, four and five channels. The scale bars are 100 nm.

In Figure 6, nanotubes with two, three, four and five channels have been successfully fabricated by multi-channel coaxial electrospinning. All of the multi-channel nanofibers show good fidelity to the corresponding spinneret. With different inner fluid speed control of multi-fluidic compound-jet, one would fabricate the two or three channel nanofibers with different inner diameters. That reveals the efficient controllability to construct multi-channel tube with different shapes in polymer nanofibers. The multi-channel coaxial electrospinning has good diversity. It demonstrates that multi-channel coaxial electrospinning could fabricate all kinds of multi-channel nanofiber in theory.

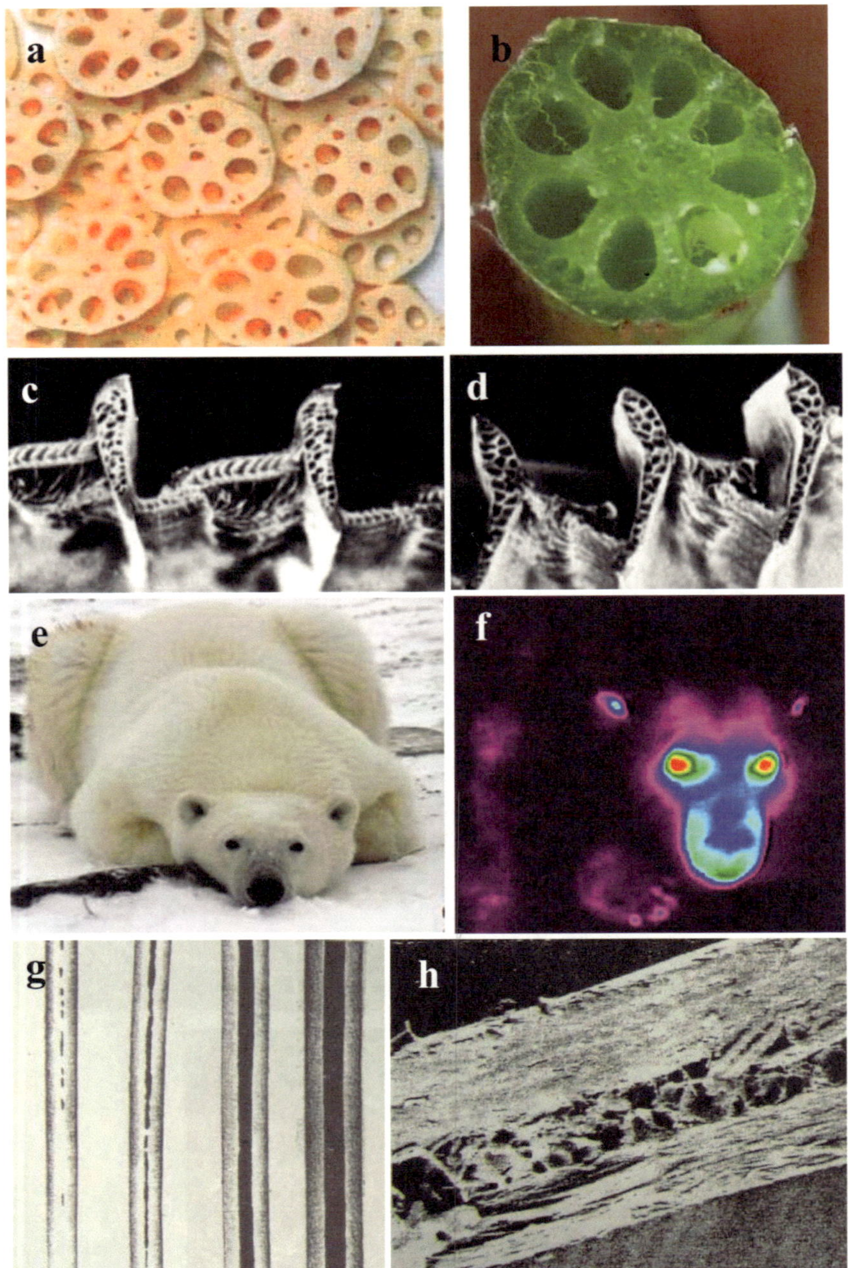

Fig. 7. Schematic of the multichannel structure in biology, which show great means of multichannel fibers in bionics. Materials delivery multichannel in lotus root a)-b), and multicavum structure for anti-cold in aves feather c)-d), and polar bear hair e)-h).

In application and biology, multichannel tubes or fibers have great importance and prospect. Figure 7 displays several typical examples of the multichannel structure in biology. There are multichannel in lotus root for the materials delivery. Multicavum structure in aves feather makes the feather ultra-light and high intensity. And multicavum structure in polar bear hair makes it have prefect temperature keeping and anti-cold property. (The infrared image of polar bear in Figure 7f shows that the heat energy losing only happens on the eyes, nose and ears, where are no or less hair covered parts.) High similarity with biological micro- or nano-structures and large areas, fast fabrication will give multi-channel fibers and multi-channel coaxial electrospinning wide prospects for research and application.

Compared with single channel, multi-channel structures may possess considerable advantages such as independent addressable channels, better mechanical stability, unique thermal properties and larger surface-to-volume area. Furthermore, by replacing inner fluids with other functional molecular, multicore-shell nanofibers can be created and different components can be integrated in nanodomain without interaction. Such nanofibers would have novel and improved properties that do not exist in each component. They might be promising candidates for a wide range of applications such as bionic super lightweight, thermo-insulated textiles, high efficiency catalysis, vessels for macro/nano fluidic devices in bioscience or lab-on-a-chip and multi-component drug delivery. This general method could be readily expanded to many other materials.

4. Melt coaxial electrospinning and Nano-encapsulation and capsule in nanofibers.

Multiplicity, controllability and applicability are the aspect and prospects of nano-science and nano-technology. At the same time for improving multiplicity and controllability, materials scientists pay great attention on the application of the new nano-technology or the new application of general nano-technology. Following diversification development of responsive materials and nano-technology, the combination of various functional materials with nano structures are drawing much attention for the great prospect in smart materials and devices, which always can generate new materials with prominent functions (Gil et al, 2004; Lu et al, 2007). The core-shell nanomaterials give small capsule to encapsulate the responsive or functional materials. And the coaxial electrospinning technique is an easy and fast process to build kinds of core-shell nanofibers.

In the fore part of this chapter, we have discussed coaxial electrospinning can perform good control, and fabricate core-shell and hollow nanofibers fast and simply. In the coaxial electrospinning process of fabricating polymer nanofibers, the inner fluid and outer fluid should be delaminated and without mutual mixing, for example water and dichloromethane (DCM), to keep a clear interface between core and shell of the nanofibers. However, for good conductivity and solubility, most good solvents for electrospinning are amphiphilic, for example dimethylformamide (DMF), dimethyl sulfoxide (DMSO), tetrahydrofuran (THF) and ethanol. But as a fast process of electrospinning, if the outer fluid polymer solution was dry and the inner fluid was solidify before the mixture, the nice core-shell structure still can be bullied in the polymer nanofibers.

4.1 Melt coaxial electrospinning

In 2006, based on coaxial electrospinning, Prof. Xia's group made the first try and invented melt coaxial electrospinning to fabricated phase change materials encapsulation core-shell nanofibers (McCann et al., 2006). They appended a heating system on the conventional coaxial electrospinning setup to provide a thermal atmosphere for the fluidic inner fluid. The melt coaxial electrospinning experimental setup is shown in Figure 8. The heating tape with a temperature controller device on the inner fluid syringe was used to keep the inner fluid molten and fluid. Two syringe pumps were used to perform the utmost control of the inner fluid (melt hydrocarbon phase change materials) and outer fluid (PVP/Ti(OiPr)$_4$ solution) respectively. The two fluids met at the metallic needle spinneret, which was built to coaxial cannula. Electrospinning relied on the use of a high-voltage electric field to draw a viscous droplet into an elongated jet. Under high-voltage service, the liquid was pulled out from spinneret to fibers thinner and thinner. In this process, the inner melt hydrocarbon phase change materials fluid froze rapidly, and the outer fluid polymer solution dried and solidified fast.

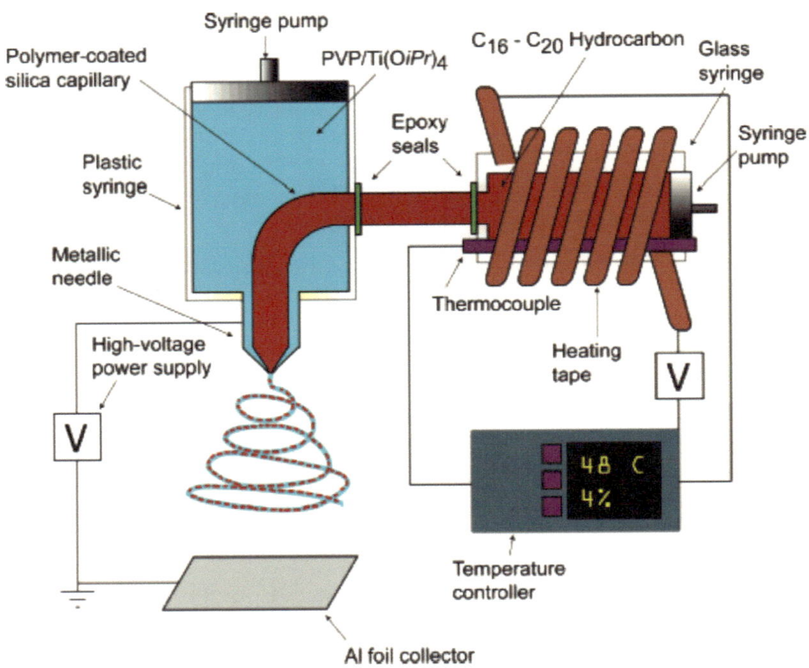

Fig. 8. Schematic of the melt coaxial electrospinning setup used for fabricating TiO_2-PVP nanofibers loaded with hydrocarbon PCMs.

2009, Dr. Li and Prof. Song made an improvement for the melt coaxial electrospinning (Li et al, 2009). The melt coaxial electrospinning experimental setup is shown in Figure 9. In addition to the high voltage generator supply and the syringe pump controller, a whole thermal atmosphere heating system was build into the conventional coaxial electrospinning setup. Two injectors with different diameter and needles constructed the outer and inner dopes loading setup. In practice, the whole thermal atmosphere of loaded system was proved more propitiously to prevent the inner dope's freeze by jamming of the needle. (For low phase change temperature materials, an infrared lamp will be easier to supply a whole thermal atmosphere for the loaded system.) Keeping the inner inject materials (the phase change materials) fluid before it is spun out from spinneret should be treated with an utmost care and control. It is the key factor to keep the propitious encapsulation of phase change materials into the inner of the fibers and get the Phase Change Materials (PCMs)-Polymer core-shell fibers with a high yield filling.

A dynamic instability resulted in whipping and stretching which was responsible for the attenuation of the jet into long fibers with ultrathin diameters. Electrospinning was remarkably simple and versatile and capable of producing nano- and microscale fibers in large quantities. Polymer solutions were predominantly used in this process, though composites, sol-gels, and surfactant-based solutions had also been included to fabricate nanofibers with a broad range of compositions, morphologies, and properties.

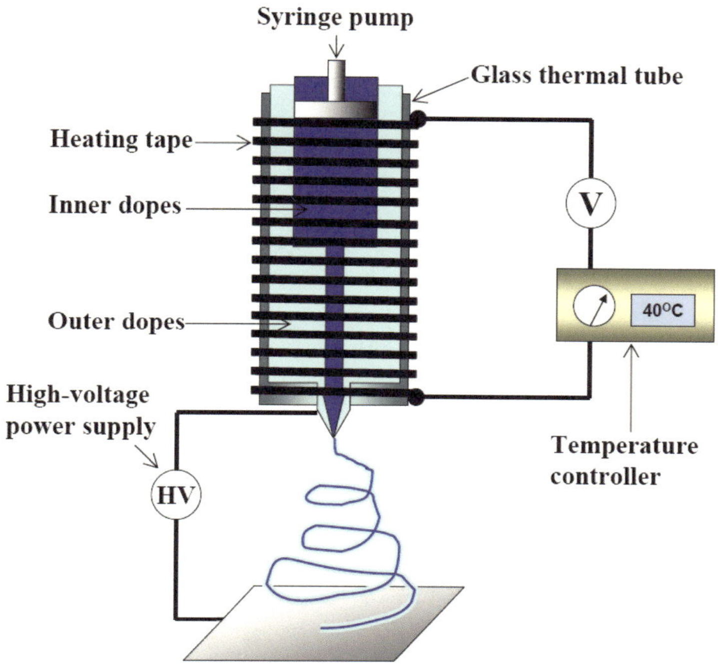

Fig. 9. The melt coaxial electrospining setup with a whole thermal atmosphere for the loaded system used for fabricating polymer nanofibers loaded with phase change materials.

4.2 Nano-encapsulation and capsule in nanofibers.

4.2.1 Phase change materials encapsulation in nanofibers.

Scanning electron microscopy (SEM) images and transmission electron microscopy (TEM) images can clearly recal the secondary nanostructures of the fibers electrospun with a melt coaxial spinneret. Figure 11 shows the Octadecane@TiO_2-PVP nanofibers electrospun with a melt coaxial spinneret by Prof. Xia's setup (Figure 8). The sheath consisted of a TiO_2-PVP composite while the core was octadecane. The core material was heated up to melt temperature for injection. In Figure 10 (A) and (C) show the SEM images of the as-prepared fibers with different hydrocarbon materials loading, and (B) and (D) show the corresponding TEM images of the fibers after they have been soaked in hexane for 24 h to remove the hydrocarbon core. Those fibers were 100-200 nm in average diameter. The TEM images indicate that the octadecane formed spherical droplets or elongated, compartmentalized domains along the long axis of the fiber.

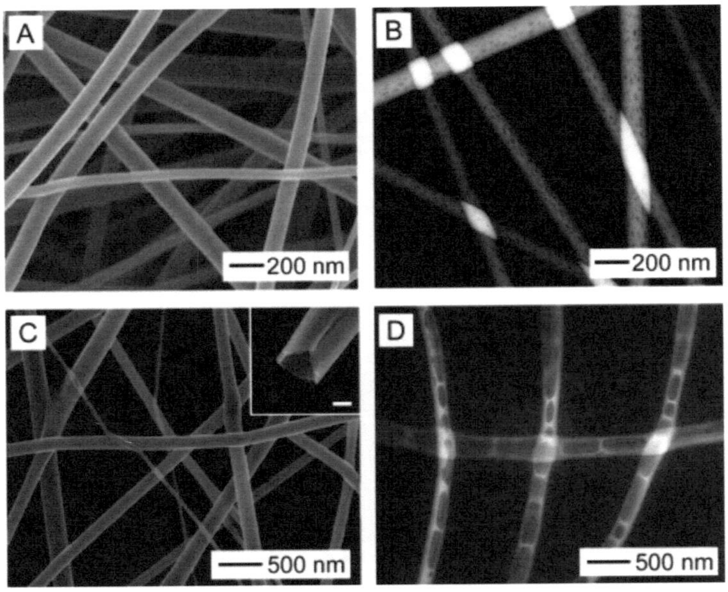

Fig. 10. Octadecane@TiO_2-PVP nanofibers electrospun with a melt coaxial spinneret. SEM images and TEM images of the nanofibers with 7% (A)-(B) and 45% (C)-(D) octadecane loading.

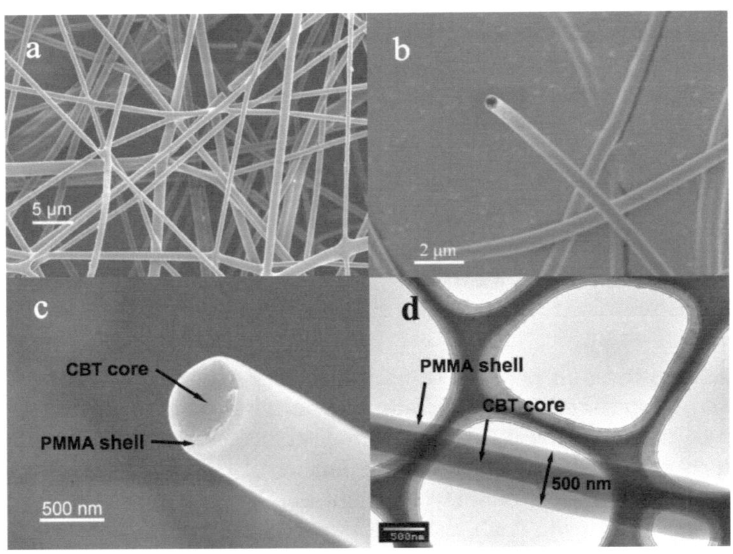

Fig. 11. Tetradecanol@PMMA nanofibers electrospun with a melt coaxial spinneret. SEM images of nanofibers in wide area (a) and side-view of core-shell nano- fiber lateral sections (b)-(c), and core-shell structure in TEM images (d).

Figure 11 is the Tetradecanol@PMMA nanofibers electrospun with a melt coaxial spinneret by Prof. Song's setup (Figure 9), which shows the actual loading state of core-shell nanofibers. (No core-removing) The sheath consisted of optical transmission polymer ploy(methyl methacrylate) (PMMA) while the core was 1-tetradecanol and phase transformation developer composite (CBT). Figure 11a indicates the SEM image of the fibers with smooth surface and 500 nm to 2 μm average diameter. In the side-view SEM image of Figure 11b and 11c, core-shell structure of nanofiber lateral sections was displayed. TEM image in Figure 11d reveals the clear interface between 1-tetradecanol core and PMMA polymer shell, and 200 nm inner diameter and 500 nm outer diameter. It is the actual core-shell nano-fibers. Further more, it indicates that the core-material was encapsulated independently and phase separated from polymer matrix shell wall. That will be the most important fact to keep the thermo-responsive, energy-storage and management properties of the phase-transformation.

4.2.2 Stimulation chromic materials encapsulation in nanofibers.

As a kind of classic responsive material, phase change materials (PCMs) were attracted much attention for the phase transformation absolutely reversibility and good energy storage and management property. (Muligan, et al., 1996; Zalba, et al, 2003) However, the fluidity of the phase change materials after melting made the PCMs hard to fix and stabilize, which limited the practical application. It is necessary to stabilize PCMs in a solid matrix. The melt coaxial electrospining just gives a suitable and ideal process and method to encapsulate and stabilize the PCMs into the nanomatrix of core-shell polymer fibers. The core-shell structure gives the free space out of the polymer matrix for their phase change. As illustrated in Figure 12, the PCMs were encapsulated and stabilized in the centre cavum, where it could perform the melt and crystallization independently and no interruption. And with the nano-encapsulation, the fluidity of the melt PCMs will be utmost limited by the strong capillarity of the nanotubes. While huge surface of nanofibers provide huge heat area when the temperature change, which should make a more sensitive thermo-responsive property of the PCMs. It provides the new insight into the preparation of temperature sensors, calefactive materials with energy absorption, retention, and release.

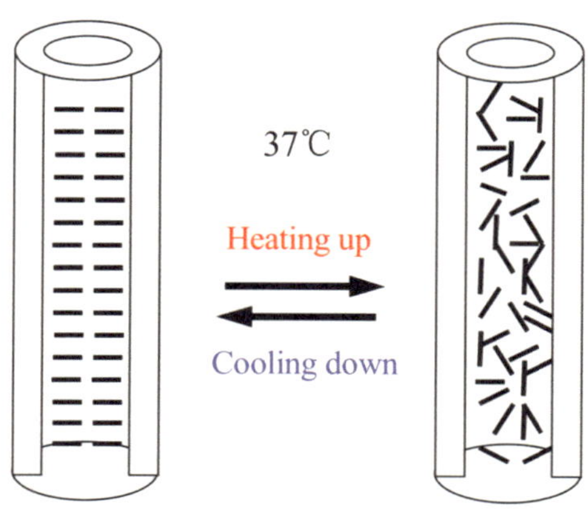

Fig. 12. Reversible phase transformation process in core-shell nanofiber of phase change materials.

Figure 13 displays the experiment for the capability of these PCM Octadecane@TiO_2-PVP nanofibers to stabilize temperature. A borosilicate glass vial was covered with a different insulation jacket, filled with quantitative 60 °C water and then allowed to cool in a 4 °C environment. The water temperature in the vial was measured using a thermal couple every

30 s and recorded until it reached 20 °C. Curve A was no insulation jacket on the glass vial; curve B and C was half and whole insulated with a 2 mm thick layer of octadecane@TiO_2-PVP fibers sandwiched between Al foils on the vial; and curve D was the vial covered with an 8 mm thick jacket of fiberglass fibers. Supercooling of vial A was not observed in the temperature history curve for the PCM nanofibers. And the fiberglass fibers were as effective as the PCM nanofibers in insulating the vial with a 4 times thicker cover. The PCM fibers cover had a temperature stabilization time (close to the melting point of octadecane) for 5 min. It gave an evident energy release in the cooling process, which indicated a practical applicable energy storage and management character.

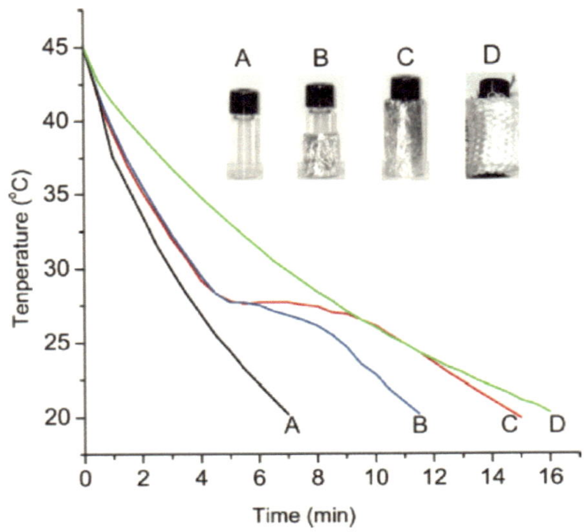

Fig. 13. Demonstration of thermal insulation capability of octadecane@TiO2-PVP nanofibers, where 1 cm^3 of water at 60 °C was allowed to cool in a 4 °C environment in glass vials covered with different insulation jackets.

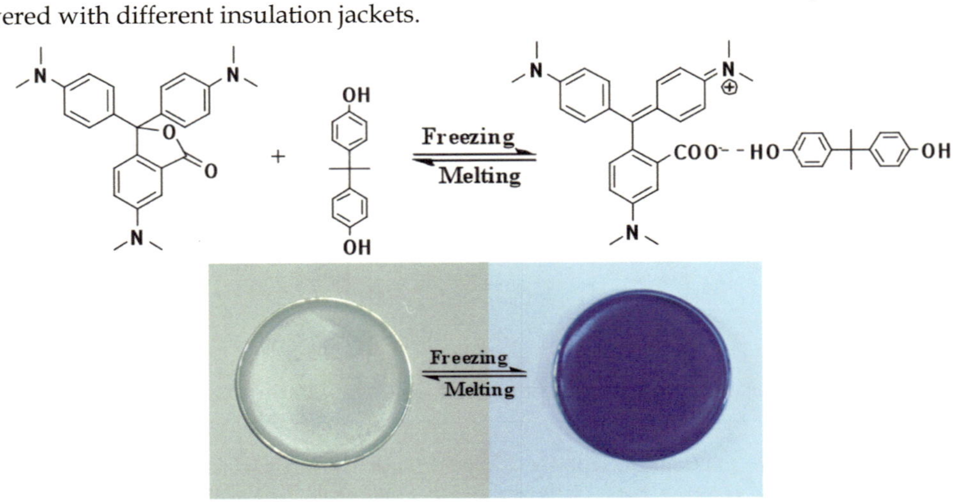

Fig. 14. Schematic of phase change thermochromic material CVL, bisphenol Ae and 1-tetradecanol (CBT) thermochromism in the phase-change process.

To realize more application, Dr. Li and Prof. Song introduced the thermochrom into the PCMs core-shell nanofibers. They used the phase change thermochromic materials (PCTMs) to make an improvement to PCMs. Displaying in Figure 14, PCTM system of crystal violet lactone (CVL) as dye and bisphenol A as developer mix in fatty alcohol or fatty acid was chosen (Burkinshaw, et al., 1998; Hirata, et al., 2006; MacLaren, et al., 2005). As a traditional PCTM, it has a simple component and stable thermochromic property, and the thermochromic temperature can be adjusted by changing fatty alcohol or fatty acid filling. (Here, 1-Tetradecanol 37~ 39 °C PCM was chosen for the good prospect in intelligent senors and devices of body temperature materials.) And CVL, bisphenol A and 1-tetradecanol (the mixture system was abbreviated to CBT) were chosen as the inner loading material for the melt coaxial electrospinng.

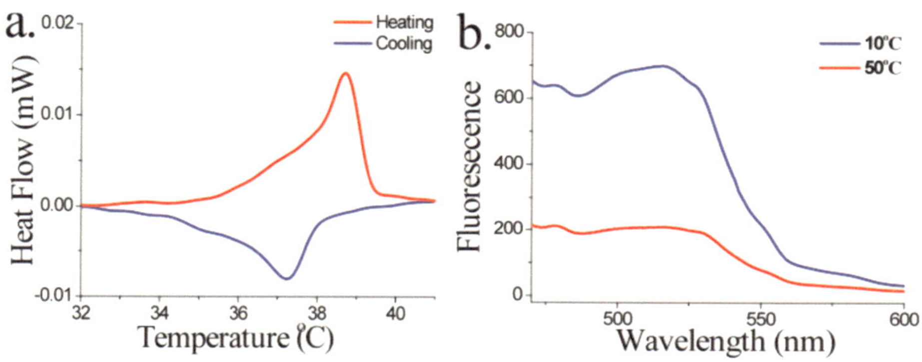

Fig. 15. a) DSC measurements cycle curve and b) Fluorescence spectra of the CBT-PMMA nanofibers at 10 °C and at 50 °C.

After successful encapsulation by melt coaxial electrospinng, (Figure 11) these CBT-core and PMMA-shell nanofiber non-woven materials show some excellent properties. Figure 15 indicates the thermo-responsive property. Differential scanning calorimetry (DSC) experiment gave the DSC cycle curve of CBT-PMMA fibers, which revealed an obvious absorption and release process in the DSC measurement cycle. (Figure 15a) The acuate peak and vale of DSC cycle curve indicates that the CBT-PMMA core-shell nanofibers have a more sensitive phase-transformation behaviour than a bulk CBT mixture. (McCann et al., 2006; Li et al, 2009) Figure 15b shows the fluorescence spectra below and beyond the phase change temperature. At the freezing state of CBT at 10 °C, the fibers have strong fluorescent emission at 503 nm, which is the characteristic fluorescence of CVL in CBT system. With temperature increased to 50 °C, the emission intensity of the fibers has a great decrease. When temperature is decreased to 10 °C, the emission intensity was reverted again. The fluorescent change of the fibers in the heating-cooling process showed the obvious thermochromism.

PMMA polymer was used as the out shell material to encapsulate the thermochromic material, due to its good optical transparent property, which always be used for organic glass. Figure 16 displays the fluorescent signal and image under fluorescence microscope. With UV light exciting, the CBT-PMMA nanofibers showed good green emission and got clear fluorescence image. That makes the encapsulation of fluorescence thermo-chromic materials in core-shell nanofibers have an insight into the thermo-responsive senor.

Fig. 16. (a) Optical and (b) fluorescent images of CBT–PMMA nanofibers of fluorescence microscope.

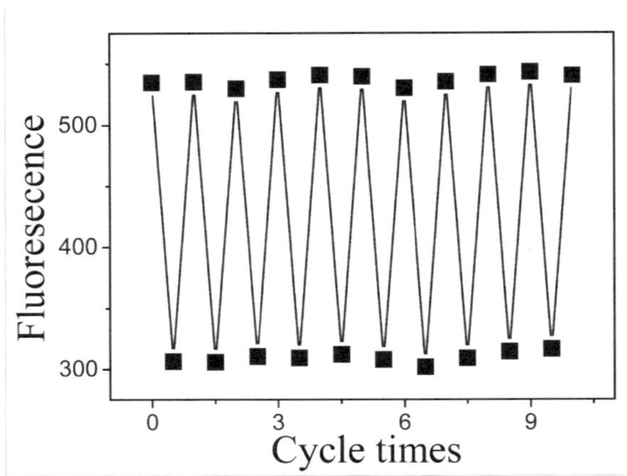

Fig. 17. The thermochromic cycle reversibility experiment of CBT-PMMA nanofibers' film. In each cycle, the fluoresecent emission at λ = 503 nm of the samples was monitored at 10 °C and 50 °C, respectively.

The thermochromic reversibility cycle experiments of the CBT-PMMA fibers film were investigated to check the responsive stability. The fluidity after melting is the main limiting factor for the PCTMs practical application. The encapsulation of PCTMs in micro/nano matrix to stabilize the PCTMs could solve the problem. In Figure 17, ten heating-cooling cycles between 10-50 °C were performed, and the fluorescent maximal emission at 503 nm of the CBT-PMMA fibers film was monitored. There was not any essential loss in fluorescent characteristics during the repeated thermochromism processes. It proved that the CBT-PMMA core-shell nanofibers showed good fluorescence thermochromic reversibility. The encapsulation of CBT in PMMA nanofibers realizes the device and practical application of PCTMs CBT. It has new insight into the preparation of temperature sensors with good fluorescence signal, and body temperature calefactive materials with intelligent thermal energy absorbing, retaining and releasing.

Melt coaxial electrospinning is one good example to fabricate functional core-shell nanofiber materials. By introducing different responsive or functional materials as the core and choosing adaptable polymer, we could accomplish the novel functionality and function modification. Thus we could perform versatile modification and control to realize multi-function and diversification, for the multi-channel coaxial electrospinning example. As the extension and development of coaxial electrospinning, melt coaxial electrospinning shows good application performance and controllability. It indicates the generality of electrospinning for one-dimensional nanomaterials fabrication.

4.2.3 Future of core-shell nanofibers: Multi-encapsulation and Multi-responsive materials.

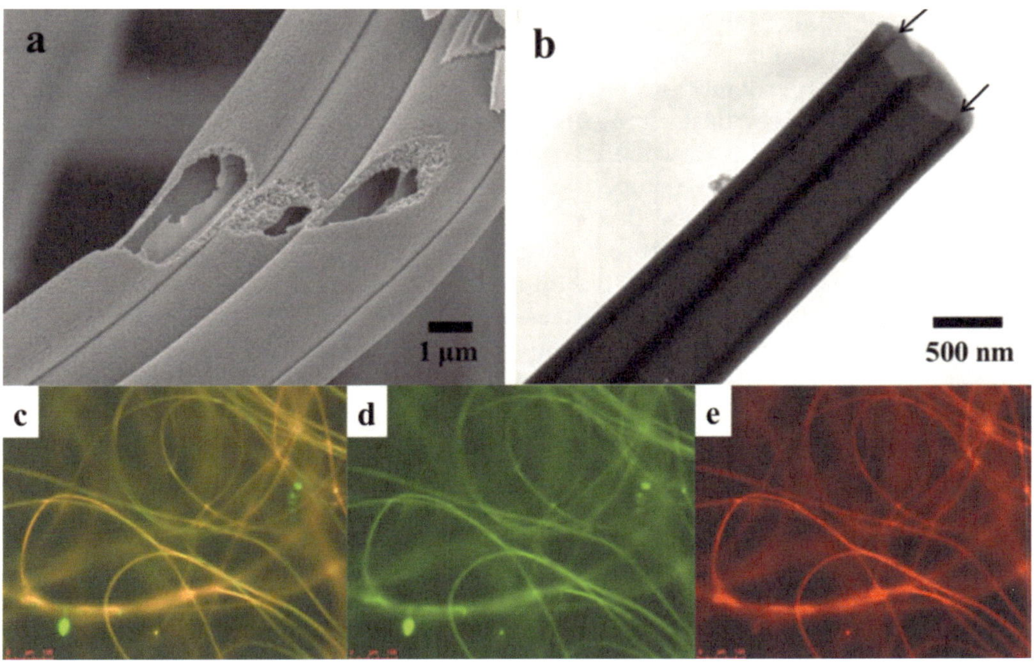

Fig. 18. SEM (a) and TEM (b) image of multichannel nanofibers, which could be loaded with different fluorescence materials, and colourful fluorescence images (c)-(e) of multi-encapsulation core-shell nanofibers.

Diversification and integration is the pilot of the science research on philosophy. And for the aspects and prospects of the new materials developing currently and future, the multifunctional, integrative and miniature devices researches are greatly and urgently expected. At the same time for core-shell nanofibers, diversification and multifunction will be the main aspects in the future. Coaxial electrospinning provides the flexible and facilitate method to construct and fabricate diversiform nano- or micro- encapsulation materials and core-shell nano- or micro- devices. Figure 18 shows some primary research of the multi-encapsulation and multi-responsive materials in the core-shell nanofibers by coaxial electrospinning. In the cavums centre of the fibers, rhodamine B and fluorescein isothiocyanate (FITC) et al. dyes were used as the core loading materials to make the multi-channel stain. Then, colourful fluorescence images of multi-loading integrative core-shell nanofibers were obtained as shown in Figure 18 (c)-(e). We could believe that: with more versatile responsive materials loading or encapsulation, one can obtain the more multifunction nanofibers materials by coaxial electrospinning.

5. Conclusion

Compared with self-assembly of molecular building blocks or template printing et al. methods, coaxial electrospinning can be used to prepare various organic or inorganic tubular nanostructures fast and facilely. With better controlling, coaxial electrospinning can realize diversification and encapsulation of nanofibers with tubular or core-shell second nanostructures. Multichannel nanotubes have ultra-large specific surface area, isolation nanostructure and continuous nanotube. In core-shell nanofibers, as core, varied responsive materials were independently encapsulated into polymer-shell. The materials were fixed and protected, but the responsive properties were kept. With nano-space encapsulating and ultra-large specific surface area, the responsive core-shell nanofibers materials are more sensitive on stimuli-responsive properties. These core-shell nanofibers or nanotubes have great applications in catalysis, fluidics, ptrification, separation, gas storage, energy conversion and storage, drug release, sensing, and environmental protection. Creating and accurate controlling 1D nanomaterials with multicompartmental inner structures is still a great challenge. It is believed that the core-shell nanofibers will give a wide space to scientists to show more creativity at the nano-channel and nano-encapsulation domain.

6. Acknowledgments

The authors thank the Natural Science Foundation of China (NSFC), the Ministry of Science and Technology (MOST) of China, for continuous financial support.
The authors are grateful to Prof. Lei Jiang for great help on technology and knowledge. They are also grateful to the Chinese Academy of Sciences.

7. References

Adamson, A. W., Gast, A. P. (1997). *Physical Chemistry of Surface*, Sixth Edition, 1997, John Wiley & Sons, Inc.

Burkinshaw, S. M.; Griffiths, J.; Towns, A. D. (1998). Reversibly thermochromic systems based on pH-sensitive functional dyes. *J. Mater. Chem.*, Vol. 8, 1998, pp. 2677–2683.

Doshi, J.; Reneker, D. H. (1995). Electrospinning process and applications of electrospun fibers, *J. Electrostat.*, Vol. 35, 1995, pp. 151–160.

Exerowa, D., Kruglyakov, P. M. (1998). *Foam and foam films: Theory, experiment, application.* 1998, Elsevier.

Gil, E. S.; Hudson, S. M. (2004). Stimuli-reponsive polymers and their bioconjugates, *Prog. Polym. Sci.*, Vol. 29, 2004, pp. 1173–1222.

Greiner, A.; Wendorff, J. H. (2007). Electrospinning: A fascinating method for the Preparation of Ultrathin Fibers, *Angew. Chem. Int. Ed.*, Vol. 46, 2007, pp 5670-5703.

Hirata, S.; Watanabe, T. (2006). Reversible thermoresponsive recording of fluorescent images (TRF), *Adv. Mater.*, Vol. 18, 2006, pp. 2725–2729.

Höhler, R., Cohen-Addad, S. (2005). Rheology of liquid foam, *J. Phys.: Condens. Matter.* Vol. 17, 2005, pp. 1041-1069.

Hu, J. T., Odom, T. W., Lieber, C. M. (1999). Chemistry and physics in one dimension: synthesis and properties of nanowires and nanotubes, *Acc. Chem. Res.*, Vol. 32, 1999, pp. 435-445.

Huang, Z.-M.; Zhang, Y.-Z.; Kotaki, M.; Ramakrishna, S. (2003). A review on polymer nanofibers by electrospinning and their applications in nanocomposites, *Compos. Sci. Technol.*, Vol. 63, 2003, pp. 2223–2253.

Law, M., Goldberger, J., Yang, P. D. (2004). Semiconductor nanowires and nanotubes, *Ann. Rev. Mater. Res.*, Vol. 34, 2004, pp. 83-122.

Li, D.; Xia, Y. (2004). Electrospinning of nanofibers: Reinventing the wheel? *Adv. Mater.*, Vol. 16, 2004, pp 1151-1170.

Li, D.; Xia, Y. (2004). Direct fabrication of composite and ceramic hollow nanofibers by electrospinning, *Nano Letters,* Vol. 4, 2004, pp 933-938.

Li, F.; Zhao, Y.; Wang, S.; Han, D.; Jiang, L.; Song, Y. (2009). Thermochromic core-shell nanofibers fabricated by melt coaxial electrospinning, *J. Appl. Polymer Sci.*, Vol. 112, 2009, pp 269-274.

Loscertales, I. G., Barrero, A., Guerrero, I., Cortijo, R., Marquez, M., Ganan-Calvo, A. M. (2002). Micro/nano encapsulation via electrified coaxial liquid jets, *Science*, Vol. 295, 2002, pp.1695-1698.

Lu, Y.; Liu, J. (2007). Smart nanomaterials inspired by biology: Dynamic assembly of error-free nanomaterials in response to multiple chemical and biological stimuli, *Acc. Chem. Res.*, Vol. 40, 2007, pp. 315–323.

MacLaren, D. C.; White, M. A. (2005). Design rules for reversible thermochromic mixtures, *J. Mater. Sci.*, Vol. 40, 2005, pp. 669-676.

McCann, J. T.; Marquez, M.; Xia, Y. (2006). Melt coaxial electrospinning: A versatile method for the encapsulation of solid materials and fabrication of phase change nanofibers, *Nano Letters,* Vol. 6, 2006, pp. 2868–2872.

Mulligan, J. C.; Colvin, D. P.; Bryant, Y. G. (1996). Microencapsulated phase-change material suspensions for heat transfer in spacecraft thermal systems, *J. Spacecraft Rockets*, Vol. 33, 1996, pp. 278-284.

Reneker, D. H.; Chun, I. (1996). Nanometre diameter fibres of polymer, produced by electrospinning, *Nanotechnology*, Vol. 7, 1996, pp. 216–223.

Sun, Z. C., Zussman, E., Yarin, A. L., Wendorff, J. H., Greiner, A. (2003). Compound core-shell polymer nanofibers by co-electrospinning, *Adv. Mater.*, Vol. 15, 2003, 1929-1932.

Zalba, B.; Marín, J. M.; Cabeza, L. F.; Mehling, H. (2003). Review on thermal energy storage with phase change: materials, heat transfer analysis and applications, *Appl. Thermal. Eng.*, Vol. 23, 2003, pp. 251–283.

Zhang, Y.; Huang, Z-M.; Xu, X.; Lim, C. T.; Ramakrishna, S. (2004). Preparation of core-shell structured PCL-r-Gelatin bi-Component nanofibers by coaxial electrospinning, *Chem. Mater.*, Vol. 16, 2004, pp 3406-3409.

Zhao, Y.; Cao, X.; Jiang, L. (2007). Bio-mimic multichannel microtubes by a facile method, *J. Am. Chem. Soc.*, Vol. 129, 2007, pp 764-765.